The Palgrave Handbook of Arctic Policy and Politics

Ken S. Coates • Carin Holroyd
Editors

The Palgrave Handbook of Arctic Policy and Politics

palgrave
macmillan

Editors
Ken S. Coates
Johnson-Shoyama Graduate School of
Public Policy
University of Saskatchewan
Saskatoon, SK, Canada

Carin Holroyd
Department of Political Studies
University of Saskatchewan
Saskatoon, SK, Canada

ISBN 978-3-030-20559-1   ISBN 978-3-030-20557-7   (eBook)
https://doi.org/10.1007/978-3-030-20557-7

This Palgrave Macmillan imprint is published by the registered company Springer Nature Switzerland AG.
The registered company address is: Gewerbestrasse 11, 6330 Cham, Switzerland

CONTENTS

NOTES ON CONTRIBUTORS

Nigel Bankes is Professor of Law and Chair of Natural Resources Law at the University of Calgary, Canada, and an adjunct professor of law with the KG Jebsen Centre for the Law of the Sea, at University of Tromsi - Norway's Arctic University, Norway.

Lawson W. Brigham is Distinguished Fellow and Faculty at the International Arctic Research Center at University of Alaska Fairbanks, USA, and a fellow of the USA Coast Guard Academy Center for Arctic Study and Policy, USA.

Else Grete Broderstad is Professor of Indigenous Studies at UiT (the University of Tromsø), the Arctic University of Norway, Norway, and one of Norway's foremost scholars of Indigenous affairs. She is a long-standing Saami activist with many appointments and commissions on Indigenous affairs by the Saami Parliament and the Government of Norway. She has been heavily involved with international collaborative research initiatives, including the TUNDRA project.

Ken S. Coates is Canada Research Chair in Regional Innovation, Johnson-Shoyama Graduate School of Public Policy and Munk Senior Fellow with the Macdonald-Laurier Institute, Canada. He has published extensively on northern and Indigenous issues and is active in circumpolar comparative studies. His most recent book is the co-authored *From Treaty People to Treaty Nation: A Roadmap for All Canadians*. He is the co-Chair of the Thematic Network on Circumpolar Innovation for the University of the Arctic.

Elena Dybtsyna is Associate Professor of Business Administration at Business School and the High North Center for Business and Governance at Nord University, Norway, where she is also Program Director for MSc Degree Program in Energy Management: International Governance and Business. She has a PhD from Bodø Graduate School of Business. Her research interests are in accounting, management, governance, and internationalization of education and research.

Sveinung Eikeland is Vice Rector at UiT (the University of Tromsø), the Arctic University of Norway, Norway. He is a sociologist and an alumnus of the University of Trondheim (today part of the Norwegian University of Science and Technology). He has previously worked as a senior researcher at the Norwegian Institute for Urban and Regional Research, and as a director of research at the Northern Research Institute (Norut). He was elected as rector at Finnmark University College in 2011, and following the merger with UiT in 2013, he has served as vice rector at UiT. His research interests are primarily regional development, regional politics, and Sami politics.

Aileen A. Espiritu is a senior researcher at The Barents Institute at the University of Tromsi - Norway's Arctic University. She has ongoing research on sustainable development in the Arctic regions, notably its urban areas; mining and questions of trust; comparative region-building in the Arctic and the Barents Region, and Asia; identity politics in Indigenous and non-Indigenous Northern communities. She also continues with research on the impact of industrialization and post-industrialization on mono-industry towns in the High North; and the politics of community sustainability in the Russian North and Siberia in comparative perspective. She is co-creator and co-editor of the international refereed journal Barents Studies, with partners from Finland and Russia.

Heather Exner-Pirot is Management Editor of the Arctic Yearbook and was formerly the Strategist for Outreach and Indigenous Engagement at the University of Saskatchewan, Canada. She is a consultant with the National Coalition of Chiefs. She is a regular contributor to Radio Canada's Eye on the Arctic website, a Board member for both The Arctic Institute (TAI) and the Saskatchewan First Nations Economic Development Network, Editorial Board member for the *Canadian Foreign Policy Journal*, and Chair of the Canadian Northern Studies Trust. She has held previous positions at the International Centre for Northern Governance and Development and the University of the Arctic. She completed her doctoral degree in political science at the University of Calgary in 2011, focusing on Arctic regionalization and human security.

Gail Fondahl is Professor of Geography at the University of Northern British Columbia. Her research focuses especially on Indigenous territorial rights, and more broadly on sustainability issues, in the Russian North. Fondahl co-edited the second *Arctic Human Development Report* (2015, with J.N. Larsen) and *Northern Sustainabilities: Understanding and Addressing Change in the Circumpolar World* (2017, with G.N. Wilson). She served as President of the International Arctic Social Sciences Association from 2011 to 2014, and represented Canada on the International Arctic Science Committee's Social and Human Working Group from 2011 to 2018.

Geir Gotaas is a senior adviser to the university rector and university director at UiT (the University of Tromsø), the Arctic University of Norway, Norway. He holds a PhD in Zoo-Physiology from UiT. He has formerly served as both secretary (six years) and vice-chair (three years) for the council of the University of the Arctic, and as secretary for an Expert Committee appointed by the Norwegian government to provide input to the development of Norway's Arctic/High North policy.

Adam Grydehøj is Director of the Island Dynamics research organization, Editor of *Island Studies Journal*, and Research Associate at the University of Prince Edward Island, Canada. He is a human geographer who considers the intersection of culture, politics, and economy in island communities. His recent research has focused on the Arctic, East Asia, and issues of indigeneity.

Heather M. Hall is Assistant Professor of Economic Development and Innovation in the School of Environment, Enterprise and Development at the University of Waterloo, Canada. She holds a PhD in Geography from Queen's University, an MA in Planning from the University of Waterloo, and a BA in Geography from Laurentian University. Her research focuses on innovation and entrepreneurship in rural and northern contexts; regional economic development planning, policy and practice; community readiness & community impacts related to natural resource development; and planning in slow-growth and declining communities. Hall is the co-author of *Planning Canadian Regions* and she is the vice-lead of the University of the Arctic's thematic network on the Commercialization of Science and Technology for the North.

Stephen Haycox is Distinguished Professor Emeritus at the University of Alaska Anchorage, USA, where he taught American cultural history, Alaska history, and Alaska Native history for 40 years. His publications include a new narrative history of Alaska and analyses of the politics of environmental protection and economic development. He continues to publish and to teach part-time; he also writes a biweekly op/ed column for the state's largest circulation news outlet.

Timothy Heleniak is a senior research fellow at Nordregio, Sweden. He has researched and written extensively on population trends and regional development in the Arctic, Russia, and the Nordic countries. He previously worked at the US Bureau of the Census, the World Bank, UNICEF, and George Washington University. He holds an MBA and PhD from the University of Maryland (USA).

Carin Holroyd is Associate Professor in the Department of Political Studies at the University of Saskatchewan, Canada. She has published widely on political economy in East Asia, including *Government, International Trade and Laissez Faire Capitalism, Green Japan: Environmental Technologies, Innovation Policy and the Pursuit of Green Growth* and the co-authored books *Japan and the Internet Revolution, Innovation Nation: Science and Technology in 21st Century Japan* and *Digital Media in East Asia: National Innovation and the Creation of a Region*. She was involved with the establishment of the University of Northern British Columbia where she negotiated agreements with several Arctic partners, and has traveled extensively in the Canadian and circumpolar north.

Anne Husebekk is Rector at UiT (the University of Tromsø), the Arctic University of Norway, Norway, where she was Professor in Immunology. She focuses on the university as a main contributor to developing the High North.

Valur Ingimundarson is Professor of Contemporary History at the University of Iceland, Iceland. He holds a PhD in History from Columbia University in New York. He has authored, co-authored, and edited several books and has published extensively on topics such as Arctic geopolitics, governance, and national policies; Icelandic foreign, security, and Arctic policies; Iceland's defense and security relations with the USA and NATO; the politics of justice and memory in Europe; US-German relations and US-European relations; and post-conflict politics in the former Yugoslavia.

Aytalina Ivanova is a research docent at the North Eastern Federal University, Yakutsk, Russia. She specializes in legal anthropology, Arctic Indigenous peoples law, and environmental governance. Her publications have focused on the relations between Indigenous and rural livelihoods, processes of Arctic industrialization, and the governance of the Arctic. Ivanova has worked in a project on Indigenous territorial governance at the Arctic University of Norway (UiT) (2016–2018), and is leading a Russian-Finnish research project on the legal anthropology of Arctic Youth (2018–2020).

Noor Johnson is a research scientist with the National Snow and Ice Data Center at the University of Colorado Boulder, USA, where she works on projects related to food sovereignty, community-based monitoring, and Indigenous data management. She is a cultural anthropologist whose research focuses on the politics and practices of environmental knowledge in the Arctic. Her work has examined movements for Indigenous sovereignty through activism and bureaucratic governance. From 2015 to 2016, she was an inaugural Fulbright Arctic Initiative Scholar researching offshore development and renewable energy. In addition to her academic research, she has served as a science policy advisor and consultant to non-profit and Indigenous organizations, including the National Geographic Society, the Smithsonian Institution, and the Inuit Circumpolar Council.

Peter Kikkert is the Irving Shipbuilding Chair in Arctic Policy and Assistant Professor in the Mulroney Institute of Government at St. Francis Xavier University, Canada. His research focuses on historic and contemporary sovereignty, security, and safety issues in the Arctic and Antarctic.

Stefan Kirchner is Associate Professor of Arctic Law at the Arctic Centre of the University of Lapland, Finland. Formerly a practising lawyer in private and governance practice, his research interests include the law of the sea, human rights, environmental law, areas beyond national jurisdiction, and international economic law.

Pirjo Kleemola-Juntunen is a senior researcher in the Northern Institute for Minority and Environmental Law/Artic Centre at the University of Lapland, Finland. She has a doctoral degree in international law and her work focuses on international law in particular the law of the sea and international environmental law. She is a lawyer by training. She has published several peer-reviewed writings related to the law of the sea, navigational rights, and marine pollution areas.

Timo Koivurova is a research professor and Director of the Arctic Ventre at the University of Lapland, Finland, who has specialized in various aspects of international law applicable in the Arctic and Antarctic region. In 2002, Koivurova's doctoral dissertation "Environmental impact assessment in the Arctic: a Study of International Legal Norms" was published by Ashgate. Increasingly, his research work addresses the interplay between different levels of environmental law, legal status of Indigenous peoples, law of the sea in the Arctic waters, integrated maritime policy in the EU, the role of law in mitigating/adapting to climate change, the function and role of the Arctic Council in view of its future challenges, and the possibilities for an Arctic treaty. He has been involved as an expert in several international processes globally and in the Arctic region and has published on the above-mentioned topics extensively.

P. Whitney Lackenbauer is the Canada Research Chair (Tier 1) in the Study of the Canadian North and a professor in the School for the Study of Canada at Trent University, Peterborough, Ontario, Canada. His research focuses on Arctic policy, sovereignty, security, and governance issues; modern Canadian and circumpolar history; military history and contemporary defense policy; and Indigenous-state relations in Canada.

Joan Nymand Larsen is Professor of Economics and Arctic Studies at the University of Akureyri, Iceland; Senior Scientist and Research Director at the Stefansson Arctic Institute, Akureyri, Iceland; and Adjunct Professor in the Department of Economics and Business at the University of Greenland, Greenland. Her research focus is on Arctic economies and resource development; Arctic human development and monitoring; and climate change impacts and adaptation. Among field research is her lead role in *Arctic Youth and Sustainable Futures*; her work on adaptation and mitigation to thawing permafrost along the Arctic coast in *Nunataryuk* (EU H2020); *Resource Extraction and Sustainable Arctic Communities* (REXSAC); and other sustainability projects. Notable publications include Arctic Human Development Report (2014); The New Arctic (2015); Arctic Social Indicators (2014); IPCC—*Polar Regions* in Impacts, Adaptation, and Vulnerability (2014); Arctic Marine Resource Governance and Development (2018). Among board and committee work are the Scientific Steering Committee of Future Earth Coasts; Scientific Advisory Group to Abisko Research Station; President of IASSA, 2008–2011; and IASC Social and Human Working Group.

Frode Mellemvik is Director of the High North Center for Business and Governance and Professor at Nord University Business School, Bodø, Norway. Mellemvik holds PhD, and has written many books and articles associated with the fields of accounting, management, government, education, research, industrial development, international cooperation, and on topics related to the High North (Arctic). He has worked at academic institutions in Europe and the USA, and is Honorary Doctor and Honorary Professor at several universities. He is honored by the Russian Foreign Ministry with the highest medal a foreign citizen can achieve for Contribution to International Cooperation.

Andrei Mineev is a researcher at the High North Center for Business and Governance at Nord University Business School, Bodø, Norway. He serves as coordinator for the international project Business Index North that aims at increased awareness of business opportunities and development challenges in the Arctic. Mineev has a Master's degree from Baltic State Technical University (Russia, Saint Petersburg), and a Master's and PhD from Nord University Business School. His areas of research are management control and High North regional business development.

Maria Madalena das Neves is Associate Professor and Academic Director of the LLM Programme in Law of the Sea at the KG Jebsen Centre for the Law of the Sea, Faculty of Law, University of Tromsi - Norway's Arctic University, Norway.

Dwight Newman is Professor of Law and Canada Research Chair in Indigenous Rights in Constitutional and International Law at the University of Saskatchewan, Canada. He has published a hundred articles or chapters and a dozen books mainly on constitutional law, international law, and Indigenous rights topics. His writings on collective rights and on Indigenous consultation have been cited extensively in judicial decisions. He has been a recent visitor at Cambridge, Montréal, Princeton, Oxford, and Western Australia. He is a member of the College of the Royal Society of Canada and was designated a Queen's Counsel in 2018.

Heather N. Nicol is Professor of Geography and School of the Environment, and Acting Director School of the Study of Canada, Trent University. She has a long-standing research interest in the political geography of the circumpolar world, focusing on Canada-US relations in the Arctic. She is interested in cross-border relations and the exercise and assertion of sovereignty in northern regions. Nicol held the Visiting Fulbright Chair at the University of Washington. She is the co-author, with Victor Konrad, of the 2016 book *Beyond Walls: Re-inventing the Canada-United States Borderlands*.

Tony Penikett is a Senior Associate with the Morris J. Wong Centre for DIalogue in Vancouver and has been a mediator with Tony Penikett Negotiations Ltd. since 2001. He was Deputy Minister of Negotiations and Labour, Government of British Columbia (1997–2001), senior policy advisor, Cabinet Planning Unit, Saskatchewan Government (1995–1997), and a member of the Yukon Legislative Assembly (1978–1995) and premier, Yukon Territory, (1985–1992). He is the author of *Reconciliation: First Nation Treaty Making* (2006) and *Hunting the Northern Character* (2017).

Andrey N. Petrov is the President of the International Arctic Social Sciences Association (IASSA) and Director of the ARCTICenter and Associate Professor of Geography at the University of Northern Iowa, USA. He is also serving as Vice-Chair of the International Arctic Science Committee Social and Human Working Group. His expertise is in economic geography and regional development in the Arctic, Arctic sustainability science, indicators of well-being in Arctic communities, and impacts of resource development, benefit sharing, and knowledge economy in remote regions. He leads US flagship science cooperation initiatives in the Arctic Social Sciences: Arctic-FROST, and Arctic-COAST. Petrov is the lead author of the recent books *Arctic Sustainability Research: Past, Present and Future* (2017) and *Arctic Sustainability: A Synthesis of Knowledge* (forthcoming in 2019), and has been a lead contributor to the Arctic Human Development Report and Arctic Social Indicators.

Rebecca Pincus is Assistant Professor in the Strategic and Operational Research Department of the Center for Naval Warfare Studies at the U.S. Naval War College, USA. From 2014 to 2018, she worked at the Center for Arctic Study and Policy (CASP) at the US Coast Guard Academy. Her research addresses security dimensions of climate change in the maritime context, with a focus on the Arctic Ocean.

Kenneth Ruud is Pro Rector specializing in Research at UiT (the University of Tromsø), the Arctic University of Norway, Norway. He holds a PhD from the University of Oslo in Theoretical Chemistry, and came to UiT as an associate professor in 2001 and since 2003 full professor.

Eeva Turunen is a junior research fellow and cartographer at Nordregio, Sweden. She has an MSc degree in Environmental and Natural Resource Economics from the Faculty of Forestry and Agriculture at the University of Helsinki (Finland). She has been working with Russian Arctic socio-economic data analyses in the Nunataryuk project.

Warwick F. Vincent is Professor of Biology at Laval University in Quebec City, Canada, where he holds the Canada Research Chair in Aquatic Ecosystem Studies. He is former Director/Scientific Director (2008–2016) of the Centre for Northern Studies (CEN), an interuniversity center of excellence in northern research and training. He holds PhD in Ecology from the University of California Davis, and his research is on aquatic ecosystems and climate change issues in the Canadian North. Vincent is a founding member of Sentinel North, ArcticNet and the joint international laboratory Takuvik (Québec-France), and he has published several books and many scientific articles on polar ecosystems. He is a recipient of the Polar Medal (Governor General of Canada, 2016) and the Martin Bergmann Medal (Royal Canadian Geographical Society, 2016) for leadership in Arctic science and collaboration, and is a fellow of the Royal Society of Canada (Academy of Science) and the Royal Canadian Geographical Society.

Ron R. Wallace is a fellow of the Canadian Global Affairs Institute, Calgary, Alberta, Canada. He has extensive research, consulting, and regulatory experience in the Canadian and Siberian Arctic regions. He has worked with the World Bank (Washington), the Asian Development Bank (Manila), and the European Bank for Reconstruction and Development (London) on international resource management projects. He is a retired Permanent Member of the National Energy Board and resides in Calgary, Alberta.

Shinan Wang is a cartographer/GIS analyst at Nordregio. She is specialized in GIS and cartography, data management and spatial analysis, and has been working on data analyses and mapping activities in the Arctic, Nordic, Baltic regions, and Europe. She holds an MBA from KTH Royal Institute of Technology (Sweden).

David A. Welch is University Research Chair and Professor of Political Science at the University of Waterloo, and teaches at the Balsillie School of International Affairs. His 2005 book *Painful Choices: A Theory of Foreign Policy Change* is the inaugural winner of the International Studies Association ISSS Book Award for the best book published in 2005 or 2006, and his 1993 book *Justice and the Genesis of War* is the winner of the 1994 Edgar S. Furniss Award for an Outstanding Contribution to National Security Studies. He is co-author of *Understanding Global Conflict and Cooperation*, 10th ed., with Joseph S. Nye, Jr. He received his PhD from Harvard University in 1990.

Gary N. Wilson is a professor in the Department of Political Science and the Acting Chair of the Department of First Nations Studies at the University of Northern British Columbia. His research interests are in the area of comparative Indigenous and regional governance in the circumpolar north and resource development in the Canadian provincial north. He also serves as the President of the Association of Canadian Universities for Northern Studies.

Matthew S. Wiseman is a postdoctoral fellow in the Social Sciences and Humanities Research Council of Canada (SSHRC) with the Bill Graham Centre for Contemporary International History and the Department of History, at the University of Toronto, Canada. His research explores the social and environmental consequences of military activity in the circumpolar north. He has published about the health implications of military-funded science in the post-1945 period, and recently completed a forthcoming book that examines the militarization of Arctic research in Canada during the Cold War.

Barry Scott Zellen is the Class of 1965 Arctic Scholar at the United States Coast Guard Academy (USCGA)'s Center for Arctic Study and Policy (CASP). He has been selected as the Spring 2020 Fulbright Scholar at the Polar Law Program at the University of Akureyri, and was appointed a research affiliate of the Anchorage-based Institute of the North (IoN) in 2012, the

Center for Australia, New Zealand and Pacific Studies (CANZPS) at the Georgetown University School of Foreign Service (SFS) in 2014, and the Department of Geography at the University of Connecticut in 2018. He got his PhD in International Relations (*eximia cum laude approbatur*) from the University of Lapland (2015), his MA in Political Science (with distinction in International Relations) from the University of California, Berkeley (1985), and his BA in Government (magna cum laude) from Harvard University (1984). His postdoctoral research comparing indigenous borderlands in the Arctic and Tropics was funded by the Kone Foundation from January 2016 through December 2017. Zellen lived in Canada's Western Arctic and Yukon from 1990–1999, where he managed several native media organizations funded by the Northern Native Broadcast Access Program (NNBAP), including Northern Native Broadcasting, Yukon (NNBY) (1998–1999), the Native Communications Society of the NWT (NCS-NWT) (1995–1998), and *Tusaayaksat*, the newspaper of the Inuvialuit Communications Society (ICS) (1990–1993). In 1994, he was a research associate at the NYU Center for War, Peace and the News Media.

List of Figures

LIST OF TABLES

CHAPTER 1

Introduction: Circumpolar Dimensions of the Governance of the Arctic

Ken S. Coates and Carin Holroyd

Over 20 years ago, noted Arctic scholars Gail Oscherenko and Oran Young published a book with the compelling title *The Hot Arctic*. If the Arctic was hot at that time—and in comparative terms it was—it is a raging inferno now. Consider just a small subset of the issues currently at play: unchecked climate change, the largely unmoderated introduction of transformational technologies, the near collapse of traditional languages and severe cultural erosion among some Indigenous peoples, the redevelopment of Arctic spaces into playgrounds for wealthy outsiders, the rapid outmigration of northern residents, including Indigenous peoples, continued economic marginalization, the decline in harvestable wildlife, tragic levels of Indigenous suicide, local violence, HIV AIDS, and many other social, cultural and environmental challenges. There are offsetting and more positive developments, to be sure, including the rise of Indigenous internationalism, the continued success of the Arctic Council, the rapid growth in Indigenous economic development, greater stability among the Arctic non-Indigenous settlers, the continued growth of the Far Northern research and development capacity, more supportive southern interests in the region, international concern about northern ecological vulnerabilities, global interest in Arctic ecological sustainability, community engagement with renewable energy systems and the sustained rise of regional political voices. These are complex, promising and troubling times.

K. S. Coates
Johnson-Shoyama Graduate School of Public Policy, University of Saskatchewan, Saskatoon, SK, Canada

C. Holroyd (✉)
Department of Political Studies, University of Saskatchewan, Saskatoon, SK, Canada

© The Author(s) 2020
K. S. Coates, C. Holroyd (eds.), *The Palgrave Handbook of Arctic Policy and Politics*, https://doi.org/10.1007/978-3-030-20557-7_1

The old debates, about whether or not the Northwest Passage through the Arctic archipelago was an international passageway, Russian military aspirations in the Far North, the need to protect Indigenous rights in the region and the sweeping challenges of northern social and cultural change and the like, seem comparatively minor in comparison to the current set of issues. Most of the former challenges remain on the table. Russian saber-rattling remains an issue and adds to Arctic uncertainty. The legal status of the Northwest Passage remains in limbo, although the debates have somewhat passed from the diplomatic halls to the academic conference rooms as the opening of the Northeast Passage to commercial shipping has sidelined concerns about the route across the top of North America. Regional activists and national governments continue to wrestle with social changes in Indigenous communities, particularly in North America, and poverty and political marginalization remain in evidence across much of the region. A complex struggle has emerged between Indigenous peoples and environmentalists, with the latter promoting the fight against climate change in the region and the former irritated at times from the NGOs penchant for speaking for Indigenous peoples. The struggle for Indigenous rights and the recognition of the authority of the traditional owners of the Far North has seen some improvements, particularly with modern treaties and political restructuring in the Canadian North and Alaska, but this progress is incremental and often more symbolic than effective.

The Arctic faces numerous policy challenges, both in terms of the identification of problems and potential solutions and with the logistical, administrative and financial challenges of implementing real and systematic change. There are productive developments, to be sure, including the autonomy movement in Greenland, the strengthening of Saami political engagement in Scandinavia, the outspokenness of regional governments in the Far North, the growing collective understanding of the scope and extent of environmental threats and the development of a culture of open debate and discussion across the Circumpolar World that is perhaps without parallel. There are points of serious conflict and uncertainty, particularly across the Russian North, at the social and cultural level in Indigenous communities, and in the unresolved questions about the impact and nature of resource development in the region. Many of the prominent issues are international, if not global, in scope while the locale for debate remains largely locked in the structures and processes of the nation-states. Indigenous peoples created space and opportunity for Circumpolar collaboration, an approach subsequently followed by academics, national governments and international organizations.

The world is paying attention, as it long has done so. Southern fascination with the Arctic pre-dates European exploration in the region and is now demonstrated in a continued interest in Arctic literature and Indigenous artistic expression, the continued growth in adventure tourism, the popularity of Arctic cruise ships and a fascinating transformation of corporate relationships with Indigenous communities and companies. Ongoing efforts by non-Arctic states to join the Arctic Council is one example, as is the expansion of scientific research in the region. As the now-iconic symbol of global environmental vulnerability—the misrepresentation of the final stages of the life of a polar bear has irritated northerners as much as it has inspired southern environmentalists—the Far North has re-entered the global consciousness, with at least a superficial sense of urgency. But the accommodations are, at best, partial. The European assault on the Arctic fur trade had devastating impacts at the community level. Clearly, the European community was more comfortable with their view of the world and the environment than in having a sustained and respectful conversation with Indigenous peoples about their lifeways and the relationships with the land and natural world, although conversations have improved in recent years. Interest in Arctic perspectives is, at best, partial and episodic, leaving the region vulnerable to new forms of colonization, including by some environmental groups who have been less than respectful in their relations with Indigenous peoples.

For this collection, we assembled an impressive group of northern scholars, drawn from across the region and from around the world, and challenged them to reflect on the policy issues and political forces shaping the Far North. They came to the task from a variety of thematic and disciplinary perspectives, some writing from the North and others reflecting the insights gleaned from a career-long engagement with the region. The essays presented in this volume are all thoughtful reflections on the past, present and future challenges facing the region, reflecting the diversity of thought and perspective that continue to animate debate about the Arctic. We did not have the space to tackle all of the topics that could have been addressed. The collection is, of necessity, representative rather than absolutely complete. There are many other themes that could have been covered and that, we trust, will be addressed by scholars and policy developers in the years to come. Collectively, this book reveals the complexity and interdependency of Arctic issues, Arctic peoples and Northern and Southern regions. The world is more alert to Arctic realities than in the past, but a great deal remains in terms of policy development, regional investment, Indigenous empowerment, political compromise and environmental action if the North is to be put on the track toward regional autonomy, self-sufficiency and ecological sustainability.

We trust that the ideas raised and perspectives presented in these contributions illustrate the range of ideas and fundamental challenges facing the Far North. We hope that they challenge academics, Indigenous politicians, government officials, business leaders and the general public to look anew at the contemporary Arctic. In so many ways, the Far North has emerged as the front lines of the global future, the place where local people, national governments, non-governmental organizations and international agencies will first have to determine how to actually address the greatest challenges of our age.

The policy and politics track record to date has been less than stellar. New difficulties are emerging faster than effective solutions can be found for long-standing challenges. Robust, focused and urgent policy debates are therefore crucial to the future of the Arctic and, further, to all of humanity. The path forward is far from easy. Nor is it clear that there are obvious strategies that will produce the solutions needed in the Arctic and for the world as a whole. Furthermore, it is far from clear that what is in the best interests of the Far North will automatically serve southern and global priorities. The Arctic may be politically hot, but it is also increasingly messy in ecological, economic, social and cultural terms.

But an effort must be made. And as this collection, produced by a subset of scholars who devote their careers to understanding the Arctic, makes clear, there is an impressive international group of analysts determined to figure out the best way to work with and for the Arctic. The goal is to address, as a top priority, the needs of the region and to ensure that the Southern and global actors understand their collective responsibility to reverse and correct the patterns and policies of the past. More than anything, the chapters collected here make it clear that there are policy and political options, many of them urgent, most of them expensive, and all requiring a collaborative approach with the peoples of the Arctic. If the new North is messy and complex as well as hot, it is supported by creative, optimistic people, the Indigenous communities foremost among them, who understand that change is inevitable, that locally and regionally controlled policy is preferred and that coordinated and collective action is urgently required. If this collection moves the northern-centered agenda forward and generates public policy debate about a new and regionally controlled future for the Arctic, the primary objective of this collective effort will have been realized.

The chapters in this collection are divided into seven sections:

INDIGENOUS PEOPLES AND ARCTIC SOCIAL DYNAMICS

Over the past 40 years, Indigenous peoples have emerged as primary players in the political evolution of the Far North. They are redefining Arctic politics and reformatting the northern policy agenda. They are facing, with non-Indigenous residents, intense pressures to adapt to new economic and political realities. Indigenous peoples also seek to expand their international connections and determine how best to handle dramatic shifts in state-corporate relations.

ECONOMIC DEVELOPMENT

Twenty years ago, the Arctic was slated to experience a massive development boom. That has happened unevenly, with rapid growth in parts of Russia and Alaska, nation-leading development in Nunavut and continued strong economic activity across the Scandinavian North. But the new Arctic is markedly different from the past. An innovation economy is taking hold in major Arctic centers, Arctic tourism is growing and companies are recasting their relationships with Indigenous and non-Indigenous northern communities and governments. Major economic disruptions, as through technological change and the transformation of the world of work, are being led by greater collaborations between business and government, expanded Indigenous engagement in the market economy, and the rapid growth of northern post-secondary systems.

POLICIES OF ARCTIC NATIONS

The tendency to see the Arctic as a single region has obscured the nature of national political and policy differences between the Circumpolar nations. The variation across the North is significant, ranging from the complex and often confusing realities of Russia to the Indigenous-led developments in Greenland and Nunavut. As collaboration increases across the region, and as nations learn from each other and work cooperatively on Circumpolar solution, the diversity of the region has produced an impressive laboratory for policy and political innovation and is gradually developing a tapestry of program innovation, government and private investments, and societal responses that will strengthen Arctic policy development in future years.

THE ARCTIC AND INTERNATIONAL RELATIONS

The emergence of the global Arctic has transformed the region in recent decades. Half a century ago, Arctic issues did not play out on the global stage. Now, with a combination of climate change, expansive northern resource development, Indigenous political mobilization and the growing interest of non-Arctic nations in creating a role for themselves in Circumpolar affairs have thrust regional concerns onto the international agenda. Major issues loom from Arctic militarization and the Indigenous assertiveness to the search for the means of ameliorating northern climate change. The way forward will not be easy, but the policy and program debates will occur in an atmosphere of engagement with Indigenous peoples, intra-regional cooperation and Arctic leadership of key national debates.

ARCTIC LEGAL AND INSTITUTIONAL SYSTEMS

While most of the policy debates within the Arctic occur within national political environments, the Far North operates within a unique body of international law and Circumpolar institutions led by the Arctic Council. The effort to control the Arctic Ocean, to regulate the Arctic, to connect Indigenous rights to new international legal conventions and to expand the role of the Arctic Council will build on existing political and legal structures and will enhance the Arctic-specific governance system for one of the world's most compelling regions.

ARCTIC SECURITY

Since the end of the Cold War, national governments and international agencies struggle to understand the nature of Arctic militarization. While the demilitarization of the Far North greeted the thawing of USSR-USA tensions, Russia under President Vladimir Putin has re-empowered the country's Arctic military. Small regional tensions—Russia and Canada, Russia and the USA, Russia and Norway—have not yet escalated into a reprise of the Cold War, but all Circumpolar nations have to determine how to define and defend their strategic interests in the Arctic. It is not the same Arctic, of course. Global influences and climate change have international Arctic strategic considerations and added to the complexity of policy development in this fast-moving field.

REFLECTIONS ON FUTURE OF THE ARCTIC

The Arctic has an intriguing past, a complex present and an uncertain future. It is clear that the decades ahead will be turbulent and, in political and policy terms, unpredictable. Multiple forces—the return of Russia, technological change, East Asian interests in the Arctic, economic transitions, Indigenous political re-emergence, ecological change and the like—are reshaping the North, each bringing a new form of unease and opportunity to the region. The path dependence of much national political action and policy development suggests that the Arctic could be poorly prepared for the extensive and interwoven transformations that lie ahead. As nations and international organizations wrestle with emerging and expanding challenges, they will draw on the lessons from the past and from other nations and will draw, one hopes, from the rich cultures and spirit of the Arctic, preserving its special character and responding to the aspirations of northern peoples who continue their decades-long effort to reclaim control of the Far North.

PART I

Indigenous Peoples and Arctic Social Dynamics

CHAPTER 2

Indigenous Peoples of the Arctic: Re-taking Control of the Far North

Ken S. Coates and Else Grete Broderstad

In one of the most remarkable and peaceable political transitions in recent decades, Indigenous peoples across much of the Circumpolar area have been gradually reasserting their presence in their traditional homelands. Fifty years ago, most Indigenous peoples had been pushed to the political, economic and social margins within various nation-states, relegated by the dominance of the resource economy and smothered by the intrusions of the activist and southern-based welfare state. By the beginning of the twenty-first century, Indigenous peoples had established a substantial international presence, captured a great deal of media attention and secured (outside of Russia) a significant measure of self-government and the beginnings of influence over northern policy. Indeed, in much of the region, Indigenous peoples have power and authority that belies their small numbers while still lacking the financial and political resources to resume control over traditional lands (Coates 2004).

First Peoples of the Circumpolar World

Indigenous peoples have occupied the Circumpolar World for thousands of years through processes that are comprehensively described in Indigenous creation stories and oral history and that have slowly been documented through western science, particularly via the disciplines of archeology, anthropology,

K. S. Coates (✉)
Johnson-Shoyama Graduate School of Public Policy, University of Saskatchewan, Saskatoon, SK, Canada

E. G. Broderstad
University of Tromsø, Tromsø, Norway
e-mail: else.g.broderstad@uit.no

© The Author(s) 2020
K. S. Coates, C. Holroyd (eds.), *The Palgrave Handbook of Arctic Policy and Politics*, https://doi.org/10.1007/978-3-030-20557-7_2

history and genetics. In the process of inhabiting the Arctic, many Indigenous peoples—for example, the Nenets, Khanty, Evenk, Chukchi, Aleut, Yupik, Inuit and Saami—created impressively well-adapted and resilient socio-economic orders (Nuttall 2012). They developed technologies of housing, clothing, transporting and harvesting. They were protected from external interventions, in large measure, by the extreme cold, absence of agricultural potential (in most parts of the Arctic) and the inability of outsiders to identify commercial opportunities in what southerners typically saw as snow-covered, forbidding and unappealing lands. These historic societies were small, mobile and vulnerable to extreme weather events or changes in food supplies (Damas 1984).

Separated from each other for generations due to limited trade and rare cultural encounters (except from Fennoscandia where there was contact between the Saami and the Norse population) (Zachrisson 2008), Indigenous peoples and newcomers struggled to make sense of each other. Outsiders were entranced by their first glimpses of Indigenous societies that lived in such seemingly inhospitable lands. They described the "Eskimos" (Inuit), "Laplanders" (Saami) and others in terms that were at once flattering and dismissive (Beach 1988; Huhndorf 2000). They admired the ability of Indigenous peoples to live where few others could survive but criticized the mobility of most peoples. For the Indigenous peoples themselves, rooted on lands they had occupied for thousands of years, they were far from overwhelmed by their initial encounters with newcomers. Those from the south or from distant lands typically struggled with the vast expanses, the limited food supplies and the extreme cold of winter or the ever-bright but often bug-dominated summers. The outsiders adapted slowly to the region and significant parts of their material culture, particularly clothing, were ill-suited for the Far North (Nuttall 2005).

While there are similar experiences across the Circumpolar North, there are also huge variations in colonial histories and nation building processes. The overseas colonization that Indigenous peoples in the Americas were exposed to were different from the internal colonization of Fennoscandia. There were significant differences in contact. The subjugation of the Saami through colonization, taxation, Christian missions and political and economic integration happened over centuries; these relationships can be traced back as far as the Norse era (Hansen and Olsen 2014).

There was, in the generations before economic and social integration in the nineteenth and twentieth century, no single Indigenous North. The Inuit peoples of North America shared a fair bit in common, with major adaptations connected to local harvestable resources, which ranged from abundant fish supplies, domesticated reindeer, massive caribou herds and substantial quantities of marine mammals (whales, seals, walrus and others). Local climate, the dependability (or lack thereof) of food supplies, pressure from outside people and intrusions from external missionaries, traders and harvesters shaped both the patterns of relationships and the nature of Indigenous societies.

Move forward to the twenty-first century. In many parts of the North, Indigenous languages are in sharp decline, although Inuktitut and Saami are comparatively strong. Cultural loss has been significant, spurred on by the substantial outmigration of people from their traditional territories and the cumulative effects of dozens of government interventions and attempts to change, if not destroy, Indigenous cultures. The spread of the resource economy, particularly after World War II, brought thousands of newcomers into Indigenous homelands, such as Yukon, Northwest Territories and Alaska, often overwhelming the local population (Coates 1991a, b). Rendered as minorities in their traditional lands, Indigenous peoples often worked hard to maintain traditional lifeways in the face of rapid modernization and a demographic wave. Add to this wide-ranging effects of militarization, technological change and cultural intrusion mediated by radio, television and print media brought about dramatic changes.

In many parts of the Far North, although less so in Arctic Scandinavia, Indigenous peoples demonstrate many signs of social threat. Health conditions generally lag well behind southern and urban populations, often dramatically so. Rates of Indigenous suicide, alcoholism and drug abuse, teenage pregnancy, HIV/AIDs and other social pathologies are distressingly high. Across the Russian North, much of rural Alaska and northern Canada housing is typically poor and over-crowded, with major deficiencies in educational and health services. While Scandinavia has national quality facilities and infrastructure, including such contemporary basics as the Internet and electricity, many of the Indigenous villages across the Arctic struggle to secure safe drinking water and social safety for their residents (Howkins 2015; Einarsson 2014).

The situation is not totally bleak with Fennoscandinavia leading the way and with Greenland showing gains in Indigenous lifeways and standard of living (Lehtola 2002). While many of the primary demographic and social markers among the Indigenous peoples are unfavorable, there have been significant improvements in some quarters. Some Indigenous communities are sending more young people through high school and into post-secondary education and universities. Cultural and language revitalization programs have had positive impacts as have the revival of traditional sports and artistic, musical and harvesting activities. It would be wrong to portray Arctic Indigenous peoples as being in perpetual crisis, just as it would be inappropriate to overlook the substantial challenges and socio-cultural threats being addressed in the North.

The Long Darkness

Understanding the re-empowerment of Arctic Indigenous peoples requires an appreciation for the long and painful descent under the influence of external social, economic, cultural and political processes. When European explorers, traders, military personnel, government officials and others entered into the region, the vanguards of a much broader occupation that lay in the offing, they came with ignorance and even animosity toward the Indigenous peoples. The

Arctic peoples did not share the religious and political views of the newcomers; the outsiders looked on the Indigenous residents with a combination of awe, disdain and sorrow. They saw it to be within their power, particularly driven by their religious leaders and sense of spiritual superiority, to impose their culture and their will on the Indigenous peoples. The newcomers displayed a callous disregard for Indigenous spirituality and world views and imposed economic systems, government control and new forms of education on the Arctic residents. Even as they marveled at the ability of Indigenous peoples to live off the land, the outsiders dismissed, if they did not outright mock, harvesting activities as being non-economic, anti-capitalistic and of little substantive utility (Damas 1984; Forsyth 1994; Minde 2003).

These processes resulted, over time, in the extensive marginalization of Indigenous peoples across the Arctic. The outsiders exploited commercially viable resources, typically leaving the region when ore, fish stocks, whale herds and others declined in value. Economic booms and busts swept over many parts of the North from aggressive harvesting of fur-bearing animals and whales to gold stampedes and hard-rock mining. As newcomers swept into the region, Indigenous peoples were pushed to the margins of the new economy, often seeing their traditional economy disrupted by the invasion and activities of the outsiders. The apparatus and colonial values of the expansionist nation-state governments typically defined the Indigenous peoples as being legally and politically inferior, subject to domination by state officials. The now-dominant societies held contradictory assumptions about the Indigenous peoples, at once fascinated by their ability to thrive in the Far North and convinced of their "backwardness" and cultural irrelevance.

A "cant of conquest" suffused the outsiders' entrance into the Arctic region. But this was not a conquest agenda backed by armies and navies, although they were available, but rather through commercial engagement, imported diseases that ravaged Indigenous populations, spiritual colonialism in the form of Christian intrusions and the physical occupation of traditional territories, largely through the targeted development of natural resources. In many parts of the North, the economic imperialism started with the fur trade, an international enterprise that required a continuation of Indigenous lifeways in order to flourish and that provided a measure of material opportunities. Other harvesting ventures, including the domestication of reindeer in Scandinavia and an aggressive whaling industry that worked its way systematically through Arctic regions, followed (Heikkilä 2006; Caulfield 1997).

The development of extractive industries, particularly mining, brought new realities in the North. Newcomers came in droves, initially in the dozens and hundreds and, later, with major discoveries, in the thousands and tens of thousands. The arrival of more than 40,000 prospectors, gold miners and camp followers in the Far Northwest during the Klondike Gold Rush in North America changed international images about the wealth and potential prosperity of the Far North. But in the short term, the gold rush overwhelmed the Indigenous peoples in the mining regions and pushed them to the margins in their own homelands.

The commercialization and industrialization of the Arctic took place in the late nineteenth and early twentieth century. Mines opened in the more southerly areas, typically close to tidewater and with ready access to markets. Even Svalbard, which became the most northerly inhabited part of the world, hosted mining activity starting in 1899, with several countries opening coal operations on the island. In Kiruna in Northern Sweden, iron mining started around 1900. Hard-rock gold, silver and other mines opened across the Canadian North and Alaska in the 1920s and 1930s. Regional supply centers opened up and, in a small number of places, the private sector and national governments built port facilities, roads and railways that "opened" up significant portions of resource-rich territories.

Governments moved slowly into the Far North, content initially to defend national boundaries, however, roughly drawn, and later to assert sovereignty in the Far North. Concerns about Indigenous peoples lagged well behind preoccupation with protecting and asserting national interests. The vast Arctic expanses of the Canadian North, which attracted little commercial attention before the 1950s, were left under the supervision of the North West Mounted Policy (later the Royal Canadian Mounted Police), who handled administrative responsibility by way of occasional and long patrols across the region. In the Union of Soviet Socialist Republics, Stalin built large and infamous *gulags* (prison camps) that held thousands of political and cultural prisoners and that provided a tragic foundation for the development of Siberia. Only during World War II, when Germany invaded northern Norway and the United States rushed to the defense of Alaska against Japanese aggression by building the Northwest Staging Route, the Alaska Highway and the CANOL oil pipeline and refinery project, did the militarization of the North begin in earnest. The onset of the Cold War saw a massive expansion in military investments in Alaska and Russia, producing infrastructure that underpinned a substantial expansion of population and economic activity. The North had, by the mid-1950s, entered the strategic and military mainstream, with substantial investments in Scandinavia and more minor commitments (including radar systems) in Northern Canada.

Before the post-war era, governments generally left Indigenous peoples alone, allowing traditional economic activity, including hunting, trapping, fishing and, in Scandinavia and Russia, reindeer husbandry, to continue. Impressed with the Scandinavian achievements, Canada even launched several bold efforts to bring commercial reindeer herding to the Far North in an attempt to bring more economic activity to the region (Össbo and Lantto 2011). There were some efforts at Indigenous education through partnerships with Christian churches in North America, including the boarding schools in the Canadian North, state schools in Scandinavia and Communist-inspired "red tent" mobile schooling in Russia. (Continuing Indigenous mobility limited the impact of colonial educational and cultural institutions, ensuring that Indigenous languages and cultural activities generally flourished into the post-war era.) Small bureaucracies were created in some jurisdictions, but national governments (except from the Nordic countries) had few officials on the ground and their minor efforts had limited impact.

The post-war era saw dramatic transitions in Indigenous lifeways after World War II. The combination of industrial expansion, particularly in the form of mineral development and oil and gas exploration and exploitation, rapid militarization and growing concerns about national sovereignty, resulted in a much greater non-Indigenous presence across the Far North. The imperatives of northern development combined with a surge in social awareness and concern about entrenched racism in the western democracies to unleash a flurry of social welfare programs in the 1950s and 1960s. New government priorities focused on "improving" Indigenous conditions, providing mainstream education and replacing mobile lifestyles with sedentary existences dominated policy-making in the Arctic (Hamilton 1994). New towns opened primarily to service mines, oil and gas projects or government officials with a variety of southern-style amenities and services. Major national infrastructure investments resulted in the electrification of even remote communities (with the USSR/Russia lagging well behind the western nations), an incomplete network of highways and airfields, and improved communications systems.

The transition from traditional lifeways—trapping and hunting in Northern North America and Russia, marine mammal harvesting in coastal zones, reindeer husbandry and small-scale farming and fishing in Scandinavia and Russia—to regular wage employment in the government and natural resource economy proved difficult for many Indigenous peoples. Education and training levels for Indigenous peoples, save for the Scandinavian North, lagged well behind national norms. Industrial firms hired comparatively few Indigenous workers, meaning that the economic benefits from resource developments flowed primarily to the growing number of permanent non-Indigenous residents in the Arctic and a steady stream of southern-based "fly in, fly out" workers. In most instances, external firms and national governments secured more of the benefits from resource activity than did Indigenous and local residents and authorities.

As general standards of living spiked upward through the 1950s and beyond in the Arctic states, Indigenous communities found themselves dislocated and transformed. In their settlements, unemployment was widespread, as was Indigenous poverty. Poorly built government settlements settled into despair. Language loss accelerated, as did the decline in traditional harvesting activities and cultural practices. Some of the government intrusions—state and religious education, highly regulated social welfare and sedentary communities being the best examples—proved to be extremely disruptive to Indigenous peoples and cultures. Significant problems emerged, with alcohol and drug abuse, encounters with the criminal justice system and even teenage suicide becoming distressingly common, particularly in Arctic North America. There were promising elements led by the general prosperity and cultural strength of the Saami people across Scandinavia (Eidheim 1997) and marked by successful community-owned companies in Alaska and the Canadian North, a growing number of Indigenous northerners in post-secondary institutions and dozens of high-profile Indigenous businesses that provided local jobs and supported regional economic development.

Government actions and economic marginalization of Indigenous peoples were matched by the social discrimination of the now-dominant newcomer societies. Negative stereotypes about the Indigenous peoples abounded, particularly after World War II, and the "problematization" of Indigenous ways became more common. Media coverage and a fair amount of the academic research focused on Indigenous challenges, building the general sense that Indigenous peoples were being left behind in the economic and social development of Arctic regions (Wilson Rowe 2013). These attitudes spilled over into hiring, housing practices and even government policies, which focused on social ills and issues of poverty and had difficulty matching the aspirations and expectations of Indigenous peoples in the Arctic with the general portrayal of communities in disarray.

By the 1960s and early 1970s, Indigenous peoples started to push back, capitalizing on the greater acceptance in western democracies of social and cultural protests and the growing frustration with the effects of development and non-Indigenous protests. Approaches varied widely. Greenlanders criticized Denmark's stable, financially significant but culturally insensitive colonial administration. The Saami, led by a Saami-environmental alliance protests against the Alta Dam, took their frustrations to the streets of Oslo and the grounds of the Norwegian parliament (Minde 1985; Somby 2000). In Alaska, the Eskimo and American Indians recoiled at plans to use an atomic bomb to make a port on the Bering Strait and rallied in response to the development of North Slope oil and a pipeline from Point Barrow to Valdez (O'Neill 1994). Canadian groups, part of a nation-wide effort to combine political protests with legal challenges, took the Government of Canada to court on numerous occasions, particularly on matters related to resource development, and launched demands for the negotiation of northern treaties across the entire northern reaches of the country (Gallagher 2012). In sharp contrast, Indigenous peoples in Russia, collectively described as the Small Peoples of the North, struggled to get government attention and coped with the rapid industrialization and resource development of their homelands, marked by some of the most severe environmental damage across the Arctic (Slezkine 1994; Xanthaki 2004).

Governments adapted to the new Indigenous activism as, more slowly, did corporations in North America, which realized that they had to engage more collaboratively with Indigenous communities if they wanted their projects to proceed promptly. National authorities expanded their financial contributions, invested more substantially in community infrastructure, and offered expanded educational programs and post-secondary educational opportunities, new cultural and language preservation initiatives, and a variety of economic and social development options. Governments established substantial bureaucracies, endeavoring to hire Indigenous peoples to staff the offices and supervise the programs. Money and programs did not bring about immediate and dramatic changes, although the Saami in Scandinavia and the Inuit in Greenland built constructively upon a stronger and broader level of public support than experienced by Indigenous peoples in other parts of the Circumpolar World (Lantto and Mörkenstam 2008; Loukacheva 2007).

The efforts by national authorities seemed sincere, albeit often shaped by entrenched colonial attitudes and prevailing non-Indigenous attitudes about Indigenous peoples and cultures. Various initiatives, although often well-funded and broad in aspiration, rarely extended to empowering Indigenous peoples to make their own decisions and operate their own programs. These new initiatives, designed to address the social, economic and cultural challenges facing the Indigenous communities, introduced the heavy hand of governments into communities that previously had been left largely alone (particularly in the Canadian north). Funding for Indigenous governments in Canada was typically subject to government oversight, and the proliferation of programs and strategies placed additional administrative and governance burdens on Indigenous communities, creating a deeper dependency on national governments and capturing a great deal of Indigenous time and effort.

RE-EMPOWERING THE INDIGENOUS PEOPLE OF THE ARCTIC

By the last decades of the twentieth century, if not before, it became clear that Indigenous peoples and communities were not comfortable with a government-centric approach to Indigenous affairs. Canadian authorities opened negotiations on modern treaties in the early 1970s and, over the following 30 years, signed agreements with Indigenous peoples across much of the territorial North, northern Quebec and Labrador. These sweeping agreements provided substantial cash settlements, resource revenue arrangements, Indigenous roles in environmental management and the approval of natural resource projects and, most importantly, substantial scope for Indigenous self-government. Negotiating the treaties took years and drained Indigenous financial and political resources; implementing the agreements provided to be a formidable challenge on its own, requiring extensive and ongoing engagement with both the government and the court. In Canada, Indigenous people also won a series of major court victories, particularly the 2004 Supreme Court decisions in Haida and Taku, that gave them sharply increased authority, under "duty to consult and accommodate" provisions, over natural resource development on their traditional territories (Alcantara 2007).

The Indigenous peoples in the State of Alaska, operating under the 1971 Alaska Native Claims Settlement Act, had wrestled for several decades with the difficult challenge of setting up community corporations. By the early twenty-first century, several of the most northerly Indigenous governments in Alaska had developed successful corporations and were actively involved in business, local governance and regional development (Hirschfield 1991; Anders and Anders 1986). The Inupiat Eskimo in Utqiagvik became quite wealthy, although far from free from social and cultural challenges. The Saami had secured substantial recognition of their political and community aspirations, particularly through Saami parliaments in Norway, Sweden and Finland. These were consultative, as opposed to legislative agencies, but they raised the profile of Indigenous rights across Scandinavia (Henriksen 2008; Josefsen 2004).

The situation was far from uniform across the Arctic. In Greenland, the Kingdom of Denmark gradually relaxed its hold over regional affairs, putting the Greenlandic Inuit on a path that will likely lead to independence. Russia proceeded in a different direction. Government restrictions on Indigenous rights and sharp criticism of RAIPON (Russian Association of the Indigenous Peoples of the North), the national Indigenous organization for the Arctic, limited the ability of the political arm to represent Indigenous issues and press for continued reform. The Inuit of the Canadian North pressed for the division of the Northwest Territories and the establishment of the new territory of Nunavut in 1999 (the Inuvialuit in the Western Arctic remained in the recon-figured Northwest Territories). The Inuit exercised effective control of the territory, but continued to struggle with economic marginalization and a wide range of social and cultural challenges. Across the rest of the Canadian Far North, modern treaties, self-government initiatives and strong Indigenous-territorial partnerships gave Indigenous peoples and communities a much greater role in controlling their lives. In Alaska, the economic prosperity of several of the Arctic communities provided a foundation for autonomy and permitted substantial investments in community priorities. In contrast to half century earlier, Alaskan Indigenous peoples played a much more substantial role in state affairs.

Over 200 years, Indigenous peoples had lost control of their traditional ter-ritories. While the impact of state intervention and the degree of the disruption of Indigenous lifeways varied significantly across the Arctic, the general pattern was much the same: political and legal marginalization, social and economic challenges and substantial crises in language and culture. Over the past 50 years and working largely independently of each other, Arctic Indigenous peoples fought for recognition, authority and autonomy. While the achievements fell short of aspirations, the reality is that Indigenous re-empowerment had expanded significantly in a relatively short period of time.

INDIGENOUS INTERNATIONALISM

While Indigenous communities worked within national and regional boundaries, they found inspiration and guidance in the work of other Indigenous organiza-tions. The Saami Council (earlier the Nordic Saami Council), established as early as 1956, may be the oldest international indigenous organization in a circumpo-lar context. There was some Arctic involvement in various Fourth World organi-zations, like the World Council of Indigenous Peoples, which the Saami played a significant part in establishing (Crossen 2014). Indigenous peoples in the Far North did not engage a great deal with southern organizations within their host countries. They did, however, engage with other Arctic peoples, starting with an Arctic Indigenous conference held in Copenhagen in 1973 and continuing with the Inuit Circumpolar Conference (established in 1977) and the Arctic Athapaskan Council, developing a level of intra-regional and international engagement that surpassed most regions in the world (Gerhardt 2011; Wilson 2007).

Indigenous leadership, matched with the growing cooperation among Arctic nations, pressed for a formal mechanism for intra-regional collaboration. Acting on the terms of the Ottawa Declaration, the Arctic countries created the Arctic Council in 1996, with the extraordinary element of including Indigenous peoples as full members, or Permanent Participants, of the Council. (International pressures tied to the potential resource wealth of the Far North and growing concern about Arctic climate change resulted in non-Arctic nations pressing, in many cases successfully, for Observer status on the Arctic Council.) While the Council has limited responsibilities and is primarily advisory, it has had a significant impact in pressing Indigenous interests, developing environmental strategies and encouraging Circumpolar collaboration. More significantly, the symbolic importance of a major international organization that has permanent, Circumpolar-wide Indigenous involvement cannot be under-estimated (English 2013; Koivurova and Heinämäki 2006).

Indigenous collaboration across the Arctic is not restricted to political arrangements. There are extensive cultural and artistic engagements in the region, including in a variety of festivals and cultural events. The Arctic Winter Games both draws northerners together from across much of the North and has had the added element of hosting and promoting Indigenous games and contests. The level of engagement extends to academic cooperation, including the truly international University of the Arctic, various exchange programs for faculty and staff, numerous conferences and collaborative projects and a smaller number of business outreach initiatives. There is growing Indigenous interest in cultural exchanges, cooperation on tourism promotion and extensive consultations on the growing environmental challenges facing the Far North. Arctic Indigenous peoples share a strong interest in sustaining traditional lifeways, including long-standing harvesting practices, particularly in the face of concerted external criticism of Indigenous harvesting like seal harvesting. There are no other places in the world where Indigenous peoples representing dozens of ethnic and cultural groups, from seven different countries, have moved well beyond identifying the common cause and are working collaboratively on a number of different political, economic and cultural files.

The Indigenous populations in the Arctic are small and widely scattered. The region lacks local economies of scale, with the limited exception of the Saami in Scandinavia. Deficiencies in numbers have been made up, significantly, through international Indigenous cooperation, but their collective authority is limited. By drawing international attention to the issues and challenges of the North, by taking a strong, collaborative and consistent stand on major issues like climate change and Indigenous harvesting, Arctic Indigenous peoples have developed an out-sized international media and political presence, have secured attention from national governments and have created Circumpolar institutions that have had real, if less than dramatic, policy impacts. Maintaining regional, national and international engagement has placed major demands on Indigenous communities and their leaders, but some four decades of collective and sustained effort has given the Indigenous peoples of the Arctic a substantial international profile.

WEALTH CREATION, TRADITIONAL ECONOMIES AND NEW BUSINESS MODELS

The search for Indigenous influence over the future of the Arctic has taken a commercial twist in recent years. This shift can be traced to three main sources: growing corporate concern with social responsibility, empowerment of Indigenous communities through court decisions and political agreements and Indigenous frustration with the long-term and disheartening effects of the welfare state. The relationship between the welfare state and the Saami is the exception, where the Saami have been more strongly integrated as individuals and have benefited from the rights and services provided by an advanced welfare state in such fields as education or healthcare. Many Indigenous communities, particularly in Greenland, the Canadian North and Alaska, have focused their attention on building wealth alongside sustained support for supporting traditional economies.

The process was slowed, for several decades, by government intervention both in terms of the control of Indigenous communities and an expansion of national financial support measures. The government involvement combined with discrimination to limit Indigenous access to jobs in mainstream economies. The effects of "geographic luck" put a small number of Indigenous communities near major resource projects but left most Indigenous peoples hundreds of miles from the closest market-driven economic activity. From the 1960s to the early twenty-first century, Indigenous communities devoted most of their collective effort to fight against national governments for recognition of their legal and political rights and for protection of their land and resources. This left, understandably, little human capital to invest in business and general economic development.

The situation shifted in the twenty-first century, with some efforts before that time, as Indigenous peoples began to engage more aggressively in the market economy. This happened on an individual level, with more Indigenous people securing an advanced education and starting businesses and with community-wide efforts to collaborate with resource companies, infrastructure firms and governments. Major court victories and treaty settlements, particularly in Alaska and the Canadian North, provided Indigenous authorities with pools of investment capital and allowed them to expand their role in regional economic activity. Resource revenue sharing arrangements imbedded in Canadian treaties provided a revenue stream when resource projects proceeded. Impact and benefit/collaboration agreements with resource firms offered communities cash payments, job guarantees and preferential business opportunities to ensure their engagement with the projects.

The transition to greater market engagement was far from smooth. Most Indigenous communities lacked the skilled labor to capitalize in full on commercial opportunities. Resource development was episodic and unreliable; northern resource projects often follow uneven trajectories tied to the high costs of northern production and market volatility. Most Indigenous investments were in support services and infrastructure systems, which provided greater flexibility and higher levels of employment for local residents. But some North American examples—North Slope oil and the Red Dog Mine in Alaska, diamond mines in the Northwest Territories, oil sands production in northern Alberta, the Baffinland project in Nunavut and the Voisey Bay mine in Labrador—produced sizeable and sustained returns to the regional Indigenous populations. But Indigenous business development generally expanded slowly, substantially in some areas, and altered long-standing assumptions about Indigenous aversion to commercial engagement. At times, however, northern Indigenous groups protested against resource and business development and emphasized environmental protection over commercial return. This has, in particular, been the case in Fennoscandinavia (Nygaard 2016).

Indigenous community-wide engagement in the often-controversial development of natural resources in their homelands marks many Indigenous areas across the Canadian North, Alaska and Russia. This is also evident in Greenland, where Inuit groups and the government of Greenland have been encouraging community-centered economic development. In Scandinavia, Saami economic engagement has been more individual than collective, although Saami politicians have been insisting on greater collective benefit from resource developments. In the Scandinavian context, Indigenous lands and waters often serve as a venue for the clash between the traditional use of renewable resources and large-scale economic development, as emphasized by the 2016 report by the UN Special Rapporteur on the rights of Indigenous peoples. The conflicts are the same as three to four decades ago, and the Saami continue to face obstacles in their efforts of safeguarding land rights. But the dynamics have changed due to changed political and legal frameworks. Across the Canadian North, the emergence of Indigenous economic development corporations has been one of the most significant transitions in the region.

Where there have been land claims settlements and modern treaties, the financial contributions have been allocated to community-owned commercial entities, which have a mandate for local reinvestment, employment and community support. The most successful entities manage substantial sums, in some instances over $1 billion in total assets. Even the smaller development corporations assume control of key local businesses, including hotels and stores, invest in regional infrastructure (such as airlines and local energy systems) and support training and business development programs. Individually, the development corporations play major roles in their communities; collectively, these Indigenous-controlled firms have sizeable pools of investment capital and stand to figure prominently in the economic future of the Far North.

Co-production of Policy: Political Partnerships and Collaboration

Perhaps the most significant long-term change in Indigenous affairs in the Far North rests with regional engagement in policy-making. For generations, in a classic Arctic manifestation of colonial governance, policy was developed in southern capitals and imposed, typically with little or no consultation, on the Indigenous peoples in the North. For generations, the governance of Arctic, especially in the American Arctic, regions was marked by neglect or disinterest. After World War II, welfare-statism took hold, perceived by governments and southern peoples as acts of generosity and as a commitment to bringing the Indigenous peoples of the North into the national mainstream. Indigenous peoples pushed back against the intrusions of the state, but with a decline in traditional economies and the dislocations associated with major resource development, population growth and major infrastructure projects, along with the expansion of southern and international cultures, brought substantial and external-imposed changes into the North. The battle lines were many including stopping or changing large-scale development, fighting for the recognition of Indigenous rights, coping with linguistic and cultural change and managing the aggressiveness and cultural weight of non-Indigenous peoples in the North. At the same time, in many areas resources were limited, including an over-stretched leadership cadre, few local-controlled funds and limited government support. Paternalism proved resilient and national governments resisted efforts to accommodate Indigenous aspirations.

Over time, the combination of Indigenous self-determination, national and international political developments, growing non-Indigenous support for Indigenous aspirations and new legal, political and treaty arrangements, as well as international law, resulted in dramatically different approaches to Arctic governance and management. Through processes that varied by country and region, Indigenous peoples became more actively involved in issues of national and Indigenous governance, implying policy influence and interchange, as well incorporation of indigenous perspectives into mainstream political arrangements (Broderstad 2014). There were many aspects to this change, ranging from treaty negotiations, self-government agreements and self-governance arrangements to Indigenous educational and training programs, natural resource management and socio-cultural programming.

The central development in Indigenous governance has been the emergence of the co-production of policy and political partnerships across the Arctic. From the Arctic Council, where Indigenous representations work closely with national government representations, through national and regional government collaborations with Indigenous authorities, co-production of policy has emerged as a major force for governance transformation. The concept is relatively straight-forward. With the processes varying between countries, Indigenous organizations and institutions work with national and regional governments to identify areas for policy development, agree on general program and policy parameters, develop collaborative policy-making and implementation processes, and monitor the policy activity.

The most successful early examples came largely in wildlife and land management, but governments increasingly realized the folly of proceeding without full and sustained indigenous engagement. In almost all areas of Arctic public policy, from health care and education to economic development and climate change initiatives, Indigenous peoples are engaged, to a greater or lesser extent, on policy design and development. The arrangements are less pronounced in Russia, where national authority remains extremely strong, but are central to policy efforts in Greenland and the Canadian North. Across Scandinavia, the Saami Parliaments in Finland, Norway and Sweden stand out as prominent political institutions of Saami-state interaction and enhanced institutionalization. None of the Saami parliaments have any legislative power. But it is fair to say that the actual influence of the Norwegian Saami Parliament in relation to national political institutions is more comprehensive compared to the two Nordic siblings. In Norway, consultations have become the main mechanism in the governance of Saami affairs. As emphasized by the Norwegian Saami Parliament, the right to self-determination is more than the right to be consulted. Simultaneously, the right to be consulted is a central element of implementing self-determination on areas affecting both Saami and others. In Alaska, community engagement is notable in areas related to economic development, energy infrastructure and education.

Across the Arctic, in one of the most important transitions in recent history, Indigenous peoples and communities have become increasingly involved in the co-development of public policy and the general management of government programming. Colonialism dies hard, as history has shown, and there are profound policy-making challenges that remain. But the Far North has produced, largely at the insistence of Indigenous peoples, a "laboratory" for the understanding of Indigenous re-empowerment. That the process remains far from complete and quite uneven across the Circumpolar World, there are promising signs that Indigenous engagement in policy production and, indeed, the co-production of Arctic policy at the regional, national and international levels can produce substantial improvements in Indigenous conditions.

Future Prospects

The re-empowerment of Indigenous peoples in the Arctic comes at a time of continuing and rapid change. In the coming decades, Arctic Indigenous peoples are going to have to respond to a continuation of current issues and challenges, particularly relating to resource development, economic marginalization and cultural loss. They will have to cope with the disruptions associated with widespread climate change and major shifts in work, commerce and society related to the introduction of new technologies. In many ways, particularly related to climate change, Arctic Indigenous peoples have become global symbols of twenty-first-century vulnerability. Indigenous communities continue to suffer through outmigration, particularly of young people, and the painful dislocations of language loss and culture change. Arctic Indigenous political leaders, arguably among the most accomplished Indigenous leadership in the world, have fought for generations to secure a greater role in decision-making, governance, land rights and economic development. The re-empowerment of Indigenous communities stands as the most important transition in Arctic affairs, but it remains a work in progress. The search for true and sustainable equality and equity continues, with the primary focus on legal equality taking precedence over the downstream emphasis on equity.

References

Alcantara, Christopher. 2007. To Treaty or Not to Treaty? Aboriginal Peoples and Comprehensive Land Claims Negotiations in Canada. *Publius: The Journal of Federalism* 38 (2): 343–369.

Anders, Gary C., and Kathleen K. Anders. 1986. Incompatible Goals in Unconventional Organization: The Politics of Alaska Native Corporations. *Organization Studies* 7 (3): 213–233.

Beach, Hugh. 1988. *The Saami of Lapland*. London: Minority Rights Group.

Broderstad, Else Grete. 2014. Implementing Indigenous Self-Determination: The Case of the Sámi in Norway. In *e-International Relations*.

Caulfield, Richard A. 1997. *Greenlanders, Whales, and Whaling: Sustainability and Self-Determination in the Arctic*. London: Dartmouth College Press.

Coates, Ken S. 1991a. *Best Left as Indians: Native-White Relations in the Yukon Territory, 1840–1973*. Montreal: McGill-Queen's Press-MQUP.

Coates, Peter A. 1991b. *The Trans-Alaska Pipeline Controversy: Technology, Conservation, and the Frontier*. London: Lehigh University Press.

Coates, Ken. 2004. *A Global History of Indigenous Peoples: Struggle and Survival*. London: Palgrave Macmillan.

Crossen, Jonathan. 2014. *Decolonization, Indigenous Internationalism, and the World Council of Indigenous Peoples*. Doctoral Thesis of Philosophy in History, University of Waterloo.

Damas, David. 1984. *Arctic. Handbook of North American Indians*. Vol. 5. Washington, DC: Smithsonian Institution.

Eidheim, Harald. 1997. Ethno-Political Development Among the Sami After World War II: The Invention of Selfhood. *Sami Culture in a New Era: The Norwegian Sami Experience*, pp. 29–61.

Einarsson, Níels. 2014. *Arctic Human Development Report (AHDR)*. Dordrecht: Springer.

English, John. 2013. *Ice and Water: Politics Peoples and the Arctic Council*. Toronto: Penguin.

Forsyth, James. 1994. *A History of the Peoples of Siberia: Russia's North Asian Colony 1581–1990*. Cambridge: Cambridge University Press.

Gallagher, Bill. 2012. *Resource Rulers: Fortune and Folly on Canada's Road to Resources*. Bill Gallagher.

Gerhardt, Hannes. 2011. The Inuit and Sovereignty: The Case of the Inuit Circumpolar Conference and Greenland. *Politik* 14(1).

Hamilton, John David. 1994. *Arctic Revolution: Social Change in the Northwest Territories, 1935–1994*. Toronto: Dundurn.

Hansen, Lars, and Bjørnar Olsen. 2014. *Hunters in Transition. An Outline of Early Sámi History*. Leiden, Boston: Brill.

Heikkilä, Lydia. 2006. *Reindeer Talk: Sámi Reindeer Herding and Nature Management*. Rovaniemi: University of Lapland.

Henriksen, John B. 2008. The Continuous Process of Recognition and Implementation of the Sami People's Right to Self-Determination. *Cambridge Review of International Affairs* 21 (1): 27–40.

Hirschfield, Martha. 1991. The Alaska Native Claims Settlement Act: Tribal Sovereignty and the Corporate Form. *Yale LJ* 101: 1331.

Howkins, Adrian. 2015. *The Polar Regions: An Environmental History*. Hoboken, NJ: Wiley.

Huhndorf, Shari. 2000. Nanook and His Contemporaries: Imagining Eskimos in American Culture, 1897–1922. *Critical Inquiry* 27 (1): 122–148.

Josefsen, Eva. 2004. *The Saami and the National Parliaments*. Resource Centre for the Rights of Indigenous Peoples.

Koivurova, Timo, and Leena Heinämäki. 2006. The Participation of Indigenous Peoples in International Norm-Making in the Arctic. *Polar Record* 42 (2): 101–109.

Lantto, Patrik, and Ulf Mörkenstam. 2008. Sami Rights and Sami Challenges: The Modernization Process and the Swedish Sami Movement, 1886–2006. *Scandinavian Journal of History* 33 (1): 26–51.

Lehtola, Veli-Pekka. 2002. *The Sami People: Traditions in Transitions*. Translated by Linna Weber Müller-Wille. Aanaar-Inari: Kustannus-Puntsi Publisher.

Loukacheva, Natalia. 2007. *The Arctic Promise: Legal and Political Autonomy of Greenland and Nunavut*. Toronto: University of Toronto Press.

Minde, Henry. 1985. *The Sami Movement, the Norwegian Labour Party and Sami Rights*.

———. 2003. The Challenge of Indigenism: The Struggle for Sami Land Rights and Self-Government in Norway 1960–1990. *Indigenous Peoples: Resource Management and Global Rights*: 75–104.

Nuttall, Mark. 2005. *Protecting the Arctic: Indigenous Peoples and Cultural Survival*. London: Routledge.

———. 2012. *Encyclopedia of the Arctic*. London: Routledge.

Nygaard, Vigdis. 2016. Do Indigenous Interests Have a Say in Planning of New Mining Projects? Experiences from Finnmark, Norway. *The Extractive Industries and Society* 3 (1): 17–24.

O'Neill, Dan. 1994. H-Bombs and Eskimos: The Story of Project Chariot. *The Pacific Northwest Quarterly* 85 (1): 25–34.

Össbo, Åsa, and Patrik Lantto. 2011. Colonial Tutelage and Industrial Colonialism: Reindeer Husbandry and Early 20th-Century Hydroelectric Development in Sweden. *Scandinavian Journal of History* 36 (3): 324–348.

Slezkine, Yuri. 1994. *Arctic Mirrors: Russia and the Small Peoples of the North*. Ithaca: Cornell University Press.

Somby, Ande. 2000. The Alta Case in Norway. A Story About How Another Hydro-Electric Dam Project Was Forced Through Norway. *Dams, Indigenous People and Vulnerable Ethnic Minorities*.

Wilson, Gary N. 2007. Inuit Diplomacy in the Circumpolar North. *Canadian Foreign Policy Journal* 13 (3): 65–80.

Wilson Rowe, Elana. 2013. A Dangerous Space? Unpacking State and Media Discourses on the Arctic. *Polar Geography* 36 (3): 232–244.

Xanthaki, Alexandra. 2004. Indigenous Rights in the Russian Federation: The Case of Numerically Small Peoples of the Russian North, Siberia, and Far East. *Human Rights Quarterly* 26: 74.

Zachrisson, Inger. 2008. The Sámi and Their Interaction with the Nordic Peoples. In *The Viking World*, ed. Stefan Brink and Neil Price. London: Routledge.

CHAPTER 3

Indigenous Internationalism in the Arctic

Gary N. Wilson

INTRODUCTION

One of the most important developments in Arctic politics over the last several decades has been the emergence of organizations representing Indigenous peoples in various circumpolar and international forums. Transnational Indigenous organizations such as the Inuit Circumpolar Council and the Saami Council have taken a leadership role in promoting the concerns and issues that resonate with Indigenous peoples in the circumpolar region. In doing so, they have added an important and necessary voice to discussions on a range of topics including resource development, environmental sustainability and protection and human security. The ability of Indigenous organizations to continue their role as the "conscience of the Arctic" is challenged by the complexity of the policy environment in which they operate, as well as by broader global changes over which they have very little control. Indigenous peoples in the Arctic, however, have demonstrated an innate ability to adapt to a harsh and constantly changing environment. This experience will be invaluable as they respond to future challenges in the Arctic.

Drawing on the existing literature on Indigenous diplomacy and the engagement of Indigenous peoples and organizations with various actors both inside and outside the circumpolar region, this chapter explores the scope and impact of Indigenous internationalism in the Arctic. It begins with a general overview of the common characteristics and circumstances of Indigenous peoples in the circumpolar region and a discussion of the values that underpin their approach to international diplomacy. This section will also provide an overview of the organizations that have emerged to promote Indigenous internationalism. Section two

G. N. Wilson (✉)
University of Northern British Columbia, Prince George, BC, Canada
e-mail: Gary.wilson@unbc.ca

© The Author(s) 2020   27
K. S. Coates, C. Holroyd (eds.), *The Palgrave Handbook of Arctic Policy and Politics*, https://doi.org/10.1007/978-3-030-20557-7_3

examines the historical development of Indigenous internationalism in more depth, looking at relations between Indigenous peoples and outsiders during the pre-contact and colonial periods. A particular focus of this section is the changing domestic and international context that has mobilized Indigenous peoples to engage in international diplomacy. The last part of the chapter looks at the main challenges and issues faced by Indigenous organizations as they seek to consolidate and strengthen their position in relation to a highly complex and changing international context.

INDIGENOUS PEOPLES OF THE ARCTIC

The Arctic is home to numerous Indigenous peoples who have lived in their traditional territories for thousands of years, as supported by both oral histories and archeological evidence. It goes without saying that these Indigenous peoples are distinct in many different ways. The purpose of this section, therefore, is not to provide an exhaustive overview of the Indigenous peoples of the Arctic; rather it is to outline some of their common characteristics and conditions, as well as discuss the particular groups who have been active on the international stage.

Generally speaking, the Arctic is comprised of small, sparsely populated and remote communities. There are, of course, exceptions to this rule, but the majority of communities in the Arctic have populations of less than 5000 people. Often, these communities are not connected by roads or railways, and transportation in and out is limited to air in the winter and air and water in the summer. These geographical circumstances contribute to the isolation of Arctic communities; although it is important to note that, in recent years, this isolation has been partially diminished by advances in communication technology.

In many cases, the majority (or at least a higher proportion compared to communities in the south) of the population in Arctic communities is Indigenous. This is a result of a number of factors including distance and remoteness. In some cases, Indigenous peoples in the Arctic have had a shorter experience with colonization. As such, there has been a relatively smaller influx of non-Indigenous settlers. In Canada, for example, the colonization of most parts of the Arctic occurred much later than it did in the south. That is not to say Indigenous peoples in the Arctic have suffered any less than Indigenous peoples elsewhere. Indeed, in certain respects, their experience with colonialization has been much more intense. As the Inuit activist and diplomat, Sheila Watt-Cloutier (2015: viii) has observed, "In a sense, Inuit of my generation have lived in both the ice age and the space age. The modern world arrived slowly in some places in the world, and quickly in others. But in the Arctic, it appeared in a single generation."

Indigenous peoples around the world have strong connections to the land and the marine environment. In the Arctic, traditional activities such as hunting, fishing and gathering are not only important parts of the local economy and the sustenance of individuals and families, they carry a cultural and spiritual significance that cannot be quantified. These traditional activities are a critical link to the time before colonization and the establishment of settled communities. Maintaining and strengthening these traditions is absolutely essential to the survival of Indigenous peoples in the Arctic.

The diversity of Arctic Indigenous peoples supports a richness of cultural traditions and governance models that is unparalleled in many other parts of the world. At the same time, this diversity has also created political barriers between different peoples and communities that prevent Indigenous peoples from confronting non-Indigenous governments and other outside interests with a unified voice. That said, one of the most interesting developments over the last several decades is the emergence of pan-Arctic Indigenous organizations that link distinct Indigenous communities and their representatives across multiple states. In the case of the Inuit and Saami, for example, these developments support the idea that these groups constitute multi-state nations.

Drawing on an earlier work by Bennett and Rowley (2004) on Inuit values, Abele and Rodon (2007: 48) identified a number of general attitudes and practices that characterize Inuit diplomacy. In many respects, these attitudes and practices can be used to describe the approaches of other Indigenous peoples in the Arctic in terms of their relationship with external actors. Arctic Indigenous peoples exhibit collective persistence and an ability to overcome barriers while maintaining focus on long-term goals. The struggles that they have endured over the last several decades and the progress they have made at both the domestic and international levels are a testament to their persistence. They are politically realistic, pragmatic and adaptable to changing circumstances. Such adaptability has served them well over many thousands of years as they have adjusted to a changing and harsh environment. But it is also an important set of characteristics that have allowed them to make progress in political negotiations with non-Indigenous governments and other organizations. As will be outlined in more detail in the next section, these values have helped Arctic Indigenous peoples adapt to the political, economic and social changes brought by colonization, and to build a complex and sophisticated set of governance bodies that promote their domestic and international interests.

The following section will focus on the six Indigenous Permanent Participants in the Arctic Council. These are the Inuit Circumpolar Council; the Saami[1] Council; the Russian Association of Indigenous Peoples of the North; the Aleut International Association; the Arctic Athabaskan Council and the Gwich'in Council International. Collectively, these six transnational organizations represent the majority of Indigenous peoples in the Arctic, and, as the next section will illustrate, they are also important players in the international and regional forums that oversee Arctic governance.

---

[1] The term Saami is spelt differently depending on the country and context. Other variations include: Sami and Sámi.

The *Inuit Circumpolar Council* (ICC) was founded in 1977 and represents approximately 160,000 Inuit in four different countries across the Arctic: Denmark (Kalaallit Nunaat—Greenland); Canada (Nunatsiavut, Nunavik, Nunavut and the Inuvialuit Settlement Region); the United States (Alaska) and Russia (Chukotka). The ICC's Executive Council is comprised of representatives from each of these countries, but each country has its own ICC organization. The stated goals of the ICC are to strengthen unity among Inuit of the circumpolar region; promote Inuit rights and interests on an international level; develop and encourage long-term policies that safeguard the Arctic environment; and seek full and active partnership in the political, economic and social development of circumpolar regions (Inuit Circumpolar Council 2018). In addition to being active in various international forums, mainly through the ICC, Inuit have made considerable progress in terms of self-government and self-determination at the national level (Abele and Rodon 2007; Wilson 2007). For example, Kalaallit Nunaat is a self-ruling region within Denmark and its government has political jurisdiction over most domestic matters. Nunavut has the status of a territory within the Canadian federation and has devolved authority over a number of different policy areas.

The *Saami Council* or Sámiráđđi was founded in 1956 and, like the ICC, represents Saami in four different Arctic countries: Norway, Sweden, Finland and Russia. The Council is comprised of representatives from Saami communities in all four countries and gathers twice a year. Its main tasks are to promote "Saami rights and interests in the four countries where the Saami are living" and "to consolidate the feeling of affinity among the Saami people, to attain recognition for the Saami as a nation and to maintain the cultural, political, economic and social rights of the Saami in the legislation of the four states (Norway, Sweden, Russia and Finland) and in agreements between states and Saami representative organizations" (Saami Council 2018). Like the Inuit, the Saami have also developed self-governing institutions that allow them to influence politics and policies at the domestic level. An example of this is the Sámi Parliament (Sámediggi), a representative body for people of Sámi heritage in Norway (Falch et al. 2016). Similar parliaments have also been established in Sweden and Finland.

The *Russian Association of Indigenous Peoples of the North* (RAIPON) was founded in 1990 as the First Congress of Indigenous Peoples of the North of the USSR. It represents over 270,000 Indigenous people from 41 different Indigenous groups throughout northern Russia, Siberia and the Russian Far East. RAIPON works with the national government and legislature (State Duma) "to protect indigenous peoples' human rights, defend their legal interests, assist in solving environmental, social, economic, cultural and educational issues, and to promote their right to self governance" (Arctic Council 2018b). The Congress of RAIPON meets every four years, but in between those meetings, the President and the Presidium (with representatives from different regions in Russia) run the affairs of the Association (Russian Association of Indigenous Peoples of the North 2018). Indigenous peoples in Russia have had much less success than their Inuit and Saami counterparts in establishing institutions of self-governance. In large part, this is because of the legacy of Soviet rule and the challenges facing Russia since the collapse of the Soviet Union. It is also important to note that certain Indigenous groups such as the Saami of northwestern Russia and the Yupik (Inuit) of Chukotka in the Russian Far East also participate in other transnational Indigenous organizations (Saami Council and Inuit Circumpolar Council, respectively).

The *Aleut International Association* (AIA) was established in 1998 and is a transnational organization comprised of representatives from Aleut communities on the Aleutian, Pribilof and Commander Islands of the United States (Alaska) and Russia (Kamchatka). The AIA was originally established by the Aleutian/Pribilof Islands Association in Alaska and the Association of the Indigenous Peoples of the North of the Aleut District of the Kamchatka Region of the Russian Federation (Arctic Council 2018c). The Association is governed by a Board of Directors comprised of equal numbers (four) of representatives from the United States and Russia and a President. The AIA was established "to address environmental and cultural concerns of the extended Aleut family whose wellbeing has been connected to the rich resources of the Bering Sea for millennia" (Ibid.).

The *Arctic Athabaskan Council* (AAC) was founded in 2000 and represents Indigenous peoples of Athabaskan descent in the Arctic and sub-Arctic regions of the United States (Alaska) and Canada (Yukon Territory and Northwest Territories). The traditional territories of the Athabaskan peoples, however, extend over 3 million square kilometers and into the Canadian provinces of British Columbia, Alberta, Saskatchewan and Manitoba (Arctic Athabaskan Council 2018). In between AAC meetings, the Council is directed by the AAC Secretariat, comprised of an International Chairperson and Vice-Chair, and an Executive Director. The stated goal of the AAC is "to defend the rights and further the interests internationally of American and Canadian Athabaskan members First Nation governments in the eight-nation Arctic Council and other international fora" (Ibid.).

The *Gwich'in Council International* (GCI) was established in 1999 to represent Gwich'in peoples in the United States (Alaska) and Canada (Yukon Territory and Northwest Territories). Its Board of Directors is comprised of four members from Alaska and two each from the Yukon Territory and the Northwest Territories. The Chair of the Council rotates between the Gwich'in Tribal Council and the Vuntut Gwich'in First Nation in Canada. The Vice-Chair position is held by a representative from Alaska (Gwich'in Council International 2018a). The GCI is primarily involved in two working groups at the Arctic Council: the Sustainable Development Working Group, which focuses on the human elements of the Arctic; and the Conservation of Arctic Flora and Fauna, which focuses on biodiversity (Gwich'in Council International 2018b).

Historical Overview of Indigenous Internationalism
in the Arctic

The origins of Indigenous internationalism in the Arctic extend back to the period when this region was colonized and forcefully incorporated into settler states, often without the knowledge or approval of the Indigenous peoples who had lived there for millennia. Like many other Indigenous peoples throughout the world, the Indigenous peoples of the Arctic were profoundly affected by European colonization, first in the form of trade and incorporation into western economic systems and later in the form of settlement and assimilation.

At first, Europeans who traveled to this region were dependent on Indigenous peoples for survival; indeed, as the famed Franklin expedition demonstrated, those who failed to adapt to the Arctic environment by following the ways and practices of the Indigenous inhabitants of the region were doomed to failure. For the most part, initial contact involved the establishment of trading relationships. While these relationships were often mutually beneficial, over time, they drew Indigenous peoples into the structures of European colonial domination, and eventually created dependencies between Indigenous peoples and the institutions of the various settler states.

Although the particular circumstances and timing of this transition from independence to dependency differed across the Arctic, the means through which it occurred are similar. In many parts of the Arctic, the intensification of renewable and non-renewable resource exploitation in Indigenous territories eventually led to the establishment of permanent settlements, first in the form of trading posts and later settled communities. While the trading posts primarily created economic dependencies, the settled communities established a permanent European presence in the Arctic. Moreover, forced resettlement and the abandonment of traditional, nomadic lifestyles intensified the process of colonialization and assimilation. These processes were assisted by other institutions of colonial domination including religious and educational organizations. A particularly egregious example were the residential or boarding schools in which children were forced to abandon their languages and cultures and often suffered horrific abuse that would have a lasting intergenerational impact on Indigenous communities (Truth and Reconciliation Commission of Canada 2015).

This short summary of the impacts of European colonization provides a backdrop to conditions that faced Indigenous peoples in the Arctic at the start of the post-war period. It was during this period that Indigenous peoples started to mobilize politically in an effort to regain political and economic control. In addition to pursuing self-determination at the domestic level, Indigenous peoples also started to engage on an international level. It is important to note that Indigenous mobilization in the post-war period was part of a broader series of changes that transformed the political order from the local to global levels. Decolonization or the independence of former colonies occurred in many parts of the world in the 1950s and 1960s. This was paralleled in many western democracies by civil rights movements that sought to address inequality and racism in established democracies. Indigenous peoples throughout the world, including the Arctic, were both inspired by and contributed to these changes. As Coates and Holroyd (2014: 6) have noted, "Indigenous peoples, in turn, been influenced by global development in Aboriginal rights. The American Indian Movement (AIM), in the 1960s, politicized and radicalized indigenous demands for self-determination, and attracted adherents in Canada." In the circumpolar north, specifically, "the creation of the Nordic Sami Council in 1956 was the first tangible political result of the pan-Sami movement" (Henriksen 2008: 29) and the first of a number transnational organizations that would emerge to promote the interests of Arctic Indigenous peoples.

The 1970s was a time of great change and great beginnings for the Indigenous peoples of the Arctic, both internationally and domestically. On an international level, the Arctic Peoples Conference, held in Copenhagen in 1973, was the genesis of modern Indigenous internationalism. The conference provided an opportunity for Indigenous peoples from across the Arctic to discuss issues of mutual concern and build networks that would later develop into a series of international organizations (Jull 1999). As noted earlier, the Saami of Fennoscandia formed the Saami Council in 1956. In 1977, Inuit established the Inuit Circumpolar Conference, the forerunner of the current Inuit Circumpolar Council. Together, these two organizations represented Indigenous peoples in seven of the eight Arctic states.[2] They would be followed in the 1990s and 2000s by the establishment of other Indigenous organizations that today comprise the Permanent Participants to the Arctic Council.

On a domestic level, changes were occurring that would further contribute to the mobilization of Indigenous peoples in the Arctic (Loukacheva 2009). Resource development projects that threatened the traditional territories of Indigenous peoples throughout the Arctic region served as a catalyst for demands for autonomy. The response to these demands played out differently depending on the region in question but, in general, they created a series of organizations and institutional structures which provided a foundation for greater self-determination. For example, in Canada, the proposed Mackenzie Valley pipeline in the western Arctic and the James Bay hydroelectric project in northern Québec were instrumental in mobilizing Inuit and other Indigenous peoples to successfully demand land claims agreements or modern treaties (Sabin 1995; Rodon and Grey 2009). In Norway, Saami opposition to the development of a hydroelectric power plant in Alta was the first step toward the establishment of the Samediggi (Sami Parliament) in 1989 and the Finnmark Act in 2005, which would entrench Saami rights and territorial autonomy (Falch et al. 2016). One of the most important developments at this time was the establishment of Home Rule in Greenland in 1979. Home Rule allowed Greenlanders (the vast majority of

---

[2] This does not include Iceland, whose Indigenous population are the descendants of Norse settlers who originally came to the uninhabited island in the ninth century.

whom are Inuit) greater autonomy from Denmark over domestic affairs and set the stage for a more enhanced form of autonomy through Self-Rule in 2007 (Nuttall 2008). Collectively, these developments created a momentum that not only inspired Indigenous peoples across the Arctic but in other parts of the world too.

The backdrop for many of these political developments involved legal changes at both the domestic and international levels. In countries such as Canada, a series of judicial rulings in favor of Indigenous rights emboldened and strengthened the legal case for land claims and self-government (Havemann 2004). On an international level, the creation of the International Covenant on Civil and Political Rights (ICCPR) by the United Nations in 1966 expanded the international legal architecture established by the Universal Declaration of Human Rights in 1948. As Henriksen (2008: 37) has observed in the case of the Saami, "the United Nations Human Rights Committee (UNHRC), which is mandated to monitor the implementation of the [ICCPR], has on several occasions addressed the rights of the Sami people with a reference to the right to self-determination under article 1 of the Covenant." More recently, one of the most important legal developments was the establishment of the International Labour Organization's (ILO) Indigenous and Tribal Peoples Convention (number 169) in 1989. Despite only being ratified by a handful of countries (including only Norway among the Arctic countries), this convention was an important step in promoting the recognition of Indigenous rights at the international level (International Labour Organization 2018). ILO 169 is seen as the forerunner to the United Nations Declaration on the Rights of Indigenous Peoples (UNDRIP), which was signed in 2007.

In terms of geopolitical changes, the most significant for the Arctic and the Indigenous peoples living there was the collapse of the Soviet Union in 1991. Initially, the Soviet collapse and the end of the Cold War ushered in a new era of peace in a region that had been heavily militarized and controlled for most of the post-war period. While this change was certainly anticipated by Soviet leader Mikhail Gorbachev's famous Murmansk speech in 1987, it was the fall of the Soviet Union that provided the impetus for greater collaboration among Arctic states and peoples (Axworthy 2012). For example, the establishment of the Arctic Council in 1996 was a direct result of this geopolitical development.

More importantly, for the purposes of this chapter, was the impact that the Soviet collapse had on Indigenous peoples in the Arctic. As noted earlier, a number of Indigenous peoples in the Arctic live in multiple countries, including Russia. For much of the 20th century, however, the broader geopolitical struggle between the Soviet Union and the West separated Indigenous peoples living in the Soviet Union from their brethren in North America and northern Europe (Henriksen 2008). The end of the Cold War allowed them to reconnect after decades of separation. It also allowed representatives from Indigenous groups in Russia to become active in international Indigenous organizations such as the Saami Council, the Inuit Circumpolar Council and the Aleut

International Association. Within Russia itself, Indigenous peoples took advantage of the political thaw generated by Gorbachev's reforms and, in particular, his policy of *glasnost* (openness) (Gray 2004). As noted earlier, they established the Russian Association of Indigenous Peoples of the North (RAIPON) in 1990 to promote the interests of Indigenous peoples at the regional and national levels.

Following the years of communist authoritarian rule, the transition to democracy and a market economy was extremely difficult, not least for Indigenous peoples living in remote and northern regions of the new Russian Federation (Thompson 2009). Indigenous peoples outside Russia were aware of the predicament faced by their Russian brethren and organizations such as the Inuit Circumpolar Council, in collaboration with the Canadian government, made efforts to send humanitarian aid (Wilson 2007). The transition (both before and after the collapse of the Soviet Union) also encouraged a similar Indigenous mobilization in Russia that had started earlier in other parts of the Arctic. The economic and political challenges confronting Indigenous peoples in Russia, however, have meant that they have not enjoyed the same level of progress on issues such as autonomy and self-determination (Wilson and Kormos 2015).

In addition to breaking down barriers to interaction between Indigenous peoples, the collapse of the Soviet Union and the establishment of a democratic Russian Federation under the leadership of Boris Yeltsin created opportunities for greater collaboration between the countries of the circumpolar north. This collaboration started just prior to the Soviet collapse with the creation of the Arctic Environmental Protection Strategy (AEPS) in 1991 (Bloom 1999). The main objective of AEPS was to provide an overarching mechanism for protecting and monitoring the Arctic environment. To achieve this goal, the signatories to the strategy created four working groups: Arctic Monitoring and Assessment Program (AMAP); Conservation of Flora and Fauna (CAFF); Emergency Prevention, Preparedness and Response (EPPR); and Protection of the Arctic Marine Environment (PAME) (Bloom 1999). AEPS would lead to the establishment of the Arctic Council in 1996, the purpose of which was to promote "cooperation, coordination and interaction among the Arctic States, Arctic indigenous communities and other Arctic inhabitants on common Arctic issues, in particular on issues of sustainable development and environmental protection in the Arctic" (Arctic Council 2018a).

One of the most innovative and progressive aspects of the AEPS and the Arctic Council is the involvement of transnational organizations representing Indigenous peoples in both the establishment of these bodies and their operations. The six Indigenous organizations mentioned earlier in this article are Permanent Participants to the Arctic Council. Although they do not have a vote within the Council (this is reserved for the eight member states[3]), "they have the right to participate in all meetings and activities of the Council, and their representatives sit alongside Ministers and [Senior Arctic Officials]. Like states, they also have the right to present proposals for cooperative activities" (Bloom 1999: 716). This structure not only lends a significant Indigenous voice to the proceedings of the Council, it also facilitates interactions and dialogue between different Indigenous organizations and between Indigenous organizations and the governments of the member states, the latter of which "has historically proven difficult for domestic reasons" (Bloom 1999: 717).

This section has provided a general overview of the historical development of Indigenous internationalism in the Arctic in the post-war period. In many respects, the involvement of Indigenous peoples and the organizations that represent them in Arctic and international affairs is part of a broader trend which challenges the monopoly that nation-states have over international relations. This new multilevel governance environment is inhabited not only by traditional actors such as state governments, but increasingly by new actors including transnational and non-governmental organizations, supra-national governments and bodies and regional and local governments. Indigenous peoples have been active at all levels and in doing so have made a strong contribution to Arctic governance and international leadership.

ISSUES AND CHALLENGES OF INDIGENOUS INTERNATIONALISM IN THE ARCTIC

Although Indigenous peoples and organizations have played an instrumental role in shaping the political development of the Arctic, there are several issues and challenges that affect their involvement in Arctic politics and governance. Observers of Indigenous governance in the circumpolar north have noted the complex matrix of organizations and governments that represent the interests of Indigenous peoples such as the Saami and the Inuit (Henriksen 2008; Wilson 2017). For the most part, the multilevel governance structure in each Indigenous group is well-organized. For example, in the case of the circumpolar Inuit, governments and organizations at the sub-national level are often connected, both structurally and in terms of representation, to national and international organizations. These connections allow ideas to flow from the local to the global and in a way that is consistent with the values of consensus and collaboration that lie at the heart of Inuit governance. Despite this collegiality, however, disagreements between actors at different levels have arisen; indeed, this is to be expected as Inuit governments at the regional level become more autonomous and develop different priorities and goals compared to their counterparts at the international level (Wilson and Smith 2011). Such differences of opinion are normal in any kind of multilevel governance system and

---

[3] The eight member states of the Arctic Council are Canada, the United States, the Russian Federation, Finland, Sweden, Norway, Denmark and Iceland.

thus far they have not threatened the overall cohesion of the Inuit governance system. Nevertheless, as this governance system continues to evolve, the Inuit will need to maintain the lines of communication between different organizations and governments in order to resolve disputes as they arise. In this regard, the Inuit experience is insightful to other Indigenous peoples in the Arctic.

A second issue concerns the place of Indigenous peoples in a rapidly changing and expanding Arctic governance context. A number of non-Arctic states and entities are keenly interested in opportunities in the Arctic and have gained or have applied for observer status within the Arctic Council. The expansion of Arctic governance and development to include these non-Arctic states is concerning not only for the member states but also for the Permanent Participants "who are anxious because of outsiders' lack of understanding regarding their culture and traditions...This disquiet is further strengthened by an uncertainty surrounding their privileged position within the [Arctic Council] and whether it might be retained if powers such as the [European Union] and China were to gain a greater presence within [Arctic Council] proceedings" (Graczyk and Koivurova 2013). Such pressures will likely only increase as climate change makes the Arctic more accessible for resource development and maritime transportation. Indigenous organizations will have to remain vigilant about the intentions and agendas of non-Arctic observer states and their relationship with the member states of the Arctic Council.

With the expansion of Indigenous governance in response to calls for greater self-determination and autonomy, both at the domestic and international levels, it is not surprising that Indigenous organizations in the Arctic face challenges in the area of capacity and, in particular, human capacity. There are many examples of strong Indigenous leaders who have cut their political teeth at the local and regional levels and then become formidable advocates for Indigenous interests in international organizations such as the Arctic Council and the United Nations. In a previous work, I have referred to these leaders as rooted cosmopolitans; individuals who are grounded in their local cultures and identities and, at the same time, who are also perfectly comfortable interacting with high-level state officials on the international stage (Wilson 2007; see also Tarrow 2005). It is important to recognize, however, the enormous toll that all of this activity places on individuals, their families and their communities. Sheila Watt-Cloutier (2015), among others, has spoken extensively about these pressures. The population of Indigenous peoples in the Arctic is relatively small and, therefore, the pool of people from which to draw is also small. Increasingly, educated and experienced Indigenous leaders wear many hats and are being pulled in multiple directions by the private sector, non-governmental organizations and a myriad of government agencies, Indigenous and non-Indigenous. As such, maintaining effective governance and representation at all levels, but in particular at the international level, will be a constant challenge in the years to come.

Lastly, in most cases, Indigenous peoples in the Arctic have benefitted from a changing attitude in government that, over time, has become more open to the idea of Indigenous autonomy and representation. While there is much work that needs to be done in the area of reconciliation and self-determination, and significant barriers to the realization of Indigenous self-determination remain, the expansion of Indigenous governance at the domestic level and the enhanced representation of Indigenous organizations in international forums such as the Arctic Council provide evidence of changing political and societal attitudes toward Indigenous peoples in many Arctic states. The fact that seven out of the eight Arctic states are open and established democracies with well-developed systems for protecting rights and freedoms has helped promote the interests of Indigenous peoples. The only Arctic state with a recent history of authoritarian rule is the Russian Federation. In comparison to Indigenous peoples in other Arctic states, Indigenous peoples in Russia have struggled to assert political control over their territories. The Soviet Union provided some limited autonomy for Indigenous groups, but that autonomy was constrained by powerful forces within the state apparatus. Even in the post-Soviet period, Indigenous peoples have made little or no progress with regards to self-government or representation within the political system (Wilson and Kormos 2015). As noted earlier, one positive development was the emergence of RAIPON as a representative organization for Indigenous peoples across the Russian north and in international forums such as the Arctic Council. Even RAIPON, however, has experienced pressure from the Russian state to curb its activities. As recently as 2012, the organization was temporarily suspended by the Russian government because of its associations with international organizations (Survival International 2012). Although it appears that other transnational Indigenous organizations with Russian members did not suffer the same fate, this suspension harkens back to the days of the Cold War, when Indigenous peoples from the Soviet Union were prevented from interacting with their western counterparts.

Conclusion

This chapter has provided a general overview of the ways in which Indigenous peoples in the Arctic have engaged internationally over the last several decades. This engagement demonstrates some of the redeeming values of Arctic Indigenous peoples, including resilience, endurance and pragmatism in the face of powerful global forces and organizations. Indigenous internationalism in the Arctic is built upon a foundation of local and regional governance systems. Indeed, many of the leaders who have represented Indigenous peoples on an international stage gained their initial political experience working for local and regional organizations.

Indigenous transnational organizations such as the Inuit Circumpolar Council and the Saami Council have played an important role in Arctic governance. These and other Permanent Participants in the Arctic Council represent the interests of Indigenous peoples in multiple states. They also lend an Indigenous voice to deliberations and policy discussions about important matters such as the environment, human security and emergency preparedness. The participation of Indigenous transnational organizations in the Arctic Council has encouraged the development of stronger ties between Indigenous peoples and between Indigenous organizations and state actors, which could lead to closer relations and other benefits at the domestic level in the future.

While Indigenous peoples have made a great deal of progress in terms of projecting their influence in the Arctic, they also face a number of issues which will challenge their ability to shape the political agenda in the Arctic in years to come. These include managing competing agendas and initiatives at the domestic and international levels, continuing to find ways to effectively use their limited capacity and being able to project their voices among an increasingly crowded political environment. Despite these challenges, Indigenous peoples and the organizations that represent them will continue to monitor and protect their Arctic homeland as they have done for many thousands of years.

References

Abele, Frances, and Thierry Rodon. 2007. Inuit Diplomacy in the Global Era: The Strengths of Multilateral Internationalism. *Canadian Foreign Policy* 13 (3): 45–63.

Arctic Athabaskan Council. 2018. https://arcticathabaskancouncil.com/wp/. Accessed September 16, 2018.

Arctic Council. 2018a. https://arctic-council.org/index.php/en/about-us. Accessed October 1, 2018.

———. 2018b. *Russian Association of Indigenous Peoples of the North*. https://arctic-council.org/index.php/en/about-us/permanent-participants/raipon. Accessed September 14, 2018.

———. 2018c. *Aleut International Association*. https://arctic-council.org/index.php/en/about-us/permanent-participants/aia. Accessed September 16, 2018.

Axworthy, Thomas S. 2012. *Changing the Arctic Paradigm from Cold War to Cooperation: How Canada's Indigenous Leaders Shaped the Arctic Council*. Paper Prepared for the Fifth Polar Law Symposium, Arctic Centre, Rovaniemi, Finland, September 6–8, 2012.

Bennett, John, and Susan Rowley, eds. 2004. *Uqalurait: An Oral History of Nunavut*. Montréal and Kingston: McGill-Queen's University Press.

Bloom, Evan T. 1999. Current Developments: Establishment of the Arctic Council. *The American Journal of International Law*. 93 (3): 712–722.

Coates, Ken, and Carin Holroyd. 2014. Indigenous Internationalism and the Emerging Impact of UNDRIP in Aboriginal Affairs in Canada. In *The Internationalization of Indigenous Rights: UNDRIP in the Canadian Context – Special Report*. Waterloo, Ontario: Centre for International Governance Innovation.

Falch, Torvald, Per Selle, and Kristin Strømsnes. 2016. The Sámi: 25 Years of Indigenous Authority in Norway. *Ethnopolitics* 15 (1): 125–143.

Gray, Patty A. 2004. *The Predicament of Chukotka's Indigenous Movement: Post-Soviet Activism in the Russian Far North*. Cambridge: Cambridge University Press.

Graczyk, Piotr, and Timo Koivurova. 2013. A New Era in the Arctic Council's External Relations? Broader Consequences of the Nuuk Observer Rules for Arctic Governance. *Polar Record* 50 (254): 225–236.

Gwich'in Council International. 2018a. https://gwichincouncil.com/. Accessed September 16, 2018.

————. 2018b. https://gwichincouncil.com/sites/default/files/2017-06-23%20 GCI%20Background%20Presentation_0.pdf. Accessed September 16, 2018.

Havemann, Paul (ed). 2004. *Indigenous Peoples' Rights in Australia, Canada and New Zealand*. Melbourne: Oxford University Press.

Henriksen, John B. 2008. The Continuous Process of Recognition and Implementation of the Sami People's Right to Self-Determination. *Cambridge Review of International Affairs* 21 (1): 27–40.

International Labour Organization. 2018. *C169 – Indigenous and Tribal Peoples Convention, 1989 (no. 169)*. https://www.ilo.org/dyn/normlex/ en/f?p=NORMLEXPUB:12100:0::NO::P12100_ILO_CODE:C169. Accessed September 20, 2018.

Inuit Circumpolar Council. 2018. http://www.inuitcircumpolar.com/. Accessed September 14, 2108.

Jull, Peter. 1999. Indigenous Internationalism: What Should We Do Next? *Indigenous Affairs* 1 (January–March): 12–17.

Loukacheva, Natalia. 2009. Arctic Indigenous Peoples' Internationalism: In Search of a Legal Justification. *Polar Record* 45 (232): 51–58.

Nuttall, Mark. 2008. Self-Rule in Greenland: Towards the World's First Independent Inuit State. *Indigenous Affairs* 3 (4): 64–70.

Rodon, Thierry, and Minnie Grey. 2009. The Long and Winding Road to Self Government: The Nunavik and Nunatsiavut Experiences. In *Northern Exposure: Peoples, Powers and Prospects in Canada's North*, ed. Frances Abele, Thomas J. Courchene, F. Leslie Seidle, and France St-Hilaire. Montréal: Institute for Research on Public Policy.

Russian Association of Indigenous Peoples of the North. 2018. http://www.raipon. info/index.php#/. Accessed September 14, 2018.

Saami Council. 2018. http://www.saamicouncil.net/en/. Accessed September 14, 2018.

Sabin, Paul. 1995. Voices from the Hydrocarbon Frontier: Canada's Mackenzie Valley Pipeline Inquiry (1974–1977). *Environmental History Review* 19 (1): 17–48.

Survival International. 2012. *Russian Indigenous Peoples' Organization Ordered to Close*. https://www.survivalinternational.org/news/8845. Accessed October 3, 2018.

Tarrow, Sidney. 2005. *The New Transnational Activism*. New York: Cambridge University Press.

Thompson, Niobe. 2009. *Settlers on the Edge: Identity and Modernization on Russia's Arctic Frontier*. Vancouver: UBC Press.

Truth and Reconciliation Commission of Canada. 2015. http://nctr.ca/reports.php. (Date accessed: July 8, 2019)

Watt-Cloutier, Sheila. 2015. *The Right to Be Cold. One Woman's Story of Protecting Her Culture, the Arctic and the Whole Planet*. Toronto: Allen Lane.

Wilson, Gary N. 2007. Inuit Diplomacy in the Circumpolar North. *Canadian Foreign Policy* 13 (3): 65–80.

————. 2017. Nunavik and the Multiple Dimensions of Inuit Governance. *American Review of Canadian Studies* 47: 148–161.

Wilson, Gary N., and Jeffrey J. Kormos. 2015. At the Margins: Political Change and Indigenous Self-Determination in Post-Soviet Chukotka. In *Arctic Yearbook 2015: Arctic Governance and Governing*, ed. Lassi Heininen, Heather Exner Pirot, and Joël Plouffe, 158–173. Akureyri, Iceland: Northern Research Forum.

Wilson, Gary N., and Heather Smith. 2011. The Inuit Circumpolar Council in an Era of Global and Local Change. *International Journal* 64 (4, Autumn): 909–921.

CHAPTER 4

Demographic Changes in the Arctic

Timothy Heleniak, Eeva Turunen, and Shinan Wang

INTRODUCTION

The settlements and regions of the Arctic share locations at high latitudes, cold and harsh climatic conditions, sparse settlement patterns, and long distances to larger urban settlements and centers of economic activity. While all Arctic regions are in more advanced countries, there are considerable differences in population size, growth rates, and settlements patterns as well as in fertility, epidemiological, and migration patterns. An important distinction is demographic differences between Arctic native or indigenous populations and non-native populations. The chapter begins by explaining the demographic transition theory and the associated transitions. The next two sections analyze demographic changes across the Arctic at the regional and settlement levels, respectively. The final section concludes with a discussion of population projections for the Arctic.

Understanding the size, composition, spatial distribution, and growth rates of the population of the Arctic regions is a necessary input to population policy. The Arctic regions and states have well-developed statistical systems so that information about the current demographic situation is available to policymakers at national, regional, and local levels. The next step in the formulation of

Research for this article is part of a project titled *Polar Peoples: Past, Present, and Future* supported by a grant from the US National Science Foundation, Arctic Social Sciences Program (award number PLR-1418272), and Nunataryuk, a project under the EU Horizon 2020 program (EU grant agreement No. 773421).

T. Heleniak (✉) • E. Turunen • S. Wang
Nordic Centre for Spatial Development, Stockholm, Sweden
e-mail: timothy.heleniak@nordregio.org; eeva.turenen@nordregio.org;
shinan.wang@nordregio.org

41

K. S. Coates, C. Holroyd (eds.), *The Palgrave Handbook of Arctic Policy and Politics*, https://doi.org/10.1007/978-3-030-20557-7_4

population policy is to assess what the future size, composition, and spatial distribution of the Arctic populations would be if there was no intervention. The projections of the Arctic populations described in the final section are a useful tool for policymakers to assess whether the expected demographic situation in their region differs from the desired situation. As shown, several Arctic states already have population policies aimed at trying to influence population size and distribution.

DEMOGRAPHIC TRANSITION

The framework used for analyzing population change in the Arctic is the demographic transition theory (Weeks 2008). It is derived from modernization theory to describe the transition that societies undergo as they modernize from traditional to more modern societies. Demographically, this is the transition from high birth and death rates, where there is little control over fertility and mortality to where there is increased control over fertility through modern contraception and death rates from communicable and infectious diseases have been reduced. Declines in mortality usually precede declines in fertility leading to a period of rapid population growth. This chapter describes recent demographic trends in the Arctic and expected trends in the future.

The demographic transition is accompanied by several other transitions which occur simultaneously. Usually, the first to occur is the *health and epidemiological mortality transition,* the shift from deaths to infants, children, and mothers, where communicable diseases are the most common, to degenerative and lifestyle causes of death being dominant and deaths occurring at older ages. This is typically followed by the *fertility transition,* the shift from high and uncontrolled to low and more controlled fertility. These two transitions lead to an *age transition* with an older average age and relatively more people in the older ages than in the younger age groups. This is often followed by a *migration transition* due to overpopulation in rural areas leading to out-migration to urban areas. This leads to an *urban transition* where increasing shares of the populations reside in urban areas and the center of economic activity is focused in urban areas. Another is the *family and household transition* of smaller families and postponement of marriage brought about by other related transitions.

Arctic Indigenous Populations and the Demographic Transition

Arctic indigenous populations are typically behind those of non-indigenous populations in the demographic transition and have demographic indicators which set them apart. This is due in part to Arctic indigenous peoples often engaging in traditional economic activities, a major factor driving the demographic transition. Arctic indigenous populations tend to have higher birth and death rates, larger families, younger age structures, and reside more in rural areas.

When analyzing population change in the Arctic, an important distinction is between Arctic indigenous populations and others. The Arctic states differ in how they classify people and whether they have a concept of indigenous peoples. The United States classifies people based on race. In Alaska, about 19 percent of the population identify themselves as Alaskan Native or American Indian. Canada classifies people based on ethnic origin, which includes three groups of aboriginal peoples—Inuit, Métis, and First Nations. In the Yukon territory, about 24 percent of the population belong to one of these groups as does about half of the population in the Northwest Territories. Nunavut, which separated from the Northwest Territories in 1999, is a predominantly Inuit region where 86 percent identify themselves as Inuit. Greenland categorizes people based on place of birth. This distinction can be thought to be native Greenlanders or non-Greenlanders, or Inuit or non-Inuit. Most recently, 87 percent of the population are Inuit (Statistics Greenland 2018). The Faroe Islands and Iceland were uninhabited until the 800s and have no indigenous populations. Norway, Sweden, and Finland are considered together because the indigenous peoples in the northern regions are the same, the Sami. All ceased recording ethnicity in the censuses after World War II. The current total number of Sami is estimated at between 80,000 and 110,000, including 60,000 in Norway, 36,000 in Sweden, and 10,000 in Finland (Hassler et al. 2008).

The Soviet Union created the concept of *natsionalnost'* (ethnicity) to divide people into different groups, which is still used in post-Soviet Russia (Hirsch 2005). Of these, 26 groups with populations less than 50,000 were designated as *malo-chislenny narod severa* (Small Numbered Peoples of the North), a number which has since grown to 37. Thus, in the Arctic and Siberia, there are both Small Numbered Peoples of the North and larger groups such as Yakuts, Komi, and Karelians. Several of the northern or Arctic regions are designated homelands of these groups. In the Nenets okrug 28 percent of the population are indigenous, in Yamal-Nenets 9 percent, in the Khanty-Mansi okrug 2 percent, in the Taymyr okrug 25 percent, in the Evenki okrug 23 percent, in Sakha 54 percent, in the Koryak okrug 41 percent, and in the Chukotka okrug 35 percent.

DEMOGRAPHIC CHANGE AT THE REGIONAL LEVEL IN THE ARCTIC

There are different definitions of the Arctic in the natural and social sciences.[1] In this chapter, a broad definition of the Arctic from the ArcticStat database is used, which allows comparison of population change across the 33 regions highlighted in Table 4.1.[2]

---

[1] The definition in the Arctic Human Development Report (AHDR) is a common one used when analyzing social and economic issues in the Arctic (Larsen and Fondahl 2015).

[2] The rationale according to the ArcticStat website (http://www.arcticstat.org/) is that the territory of ArcticStat is as inclusive as possible. It covers all the populations living in an Arctic region as well as the populations having characteristics that are similar to those of Arctic populations or living in a similar environment.

Table 4.1   Total population and population change, 1990–present

	Total population		Population change, 1990–2018 (percent)		
	1990	*2018*	*Total*	*Natural increase*	*Net migration*
World	5,240,735,117	7,632,819,325	45.6	45.6	0.0
United States	252,529,950	322,179,605	27.6	16.6	10.9
Alaska	553,171	737,080	33.2	37.0	−3.8
Canada	27,692,680	36,624,199	32.3		
Canadian Arctic					
Yukon	27,797	35,874	29.1		
Northwest Territories	40,845	44,597	9.2		
Nunavut	27,498	37,996	38.2	41.4	−6.1
Greenland	55,558	55,877	0.6	24.8	−23.1
Iceland	253,785	348,450	37.3	27.2	10.0
Faroe Islands	47,773	50,498	5.7	17.4	−11.6
Norway	4,233,116	5,295,619	25.1	10.7	14.4
Norwegian Arctic					
Nordland	239,532	243,335	1.6	4.1	−2.5
Troms	146,594	166,499	13.6	11.9	1.6
Finnmark	74,148	76,167	2.7	12.2	−9.5
Svalbard	3544	2637			
Sweden	8,527,036	10,120,242	18.7	4.8	13.9
Swedish Arctic					
Västerbotten	250,134	268,465	7.3	2.7	4.7
Norrbotten	262,838	251,295	−4.4	−1.5	−2.9
Finland	4,974,383	5,513,130	10.8	5.2	5.5
Finnish Arctic					
Lappi	199,973	179,223	−10.4	2.1	−12.5
Norra Österbotten	348,292	411,856	18.3	18.6	−0.3
Kainuu	92,458	73,959	−20.0	−2,9	−17.2
Russia	147,665,081	144,204,000	−2.3	−8,7	6.5
Russian Arctic					
Karelian Republic	791,719	627,000	−20.8	−12.5	−8.3
Komi Republic	1,248,891	850,000	−31.9	−1.9	−30.1
Arkhangel'sk Oblast	1,575,502	1,166,000	−26.0	−8.5	−17.5
Nenets Autonomous Okrug	51,993	44,000	−15.4	8.8	−24.2
Murmansk Oblast	1,191,458	757,000	−36.5	−2.3	−34.1
Khanty-Mansi Aut. Okrug	1,267,030	1,646,000	29.9	23.1	6.8
Yamal-Nenets Aut. Okrug	489,161	536,000	9.6	24.8	−15.2
Taymyr Autonomous Okrug	51,867	34,432	−33.6	7.8	−41.4
Evenki Autonomous Okrug	24,005	16,253	−32.3	6.5	−38.8
Sakha Republic (Yakutia)	1,111,480	963,000	−13.4	15.5	−28.8
Chukotka Autonomous Okrug	162,135	50,000	−69.2	4.3	−73.5
Kamchatka oblast	476,911	315,000	−33.9	0.5	−34.4
Koryak Autonomous Okrug	37,622	18,759	−50.1	−1.8	−48.3
Magadan Oblast	390,276	146,000	−62.6	−0.3	−62.3

Sources and notes: National and regional statistical offices

Population Change in the Arctic from Natural Increase and Net Migration

Population change for any country or region consists of two components: natural increase—the difference between the number of births and deaths; and net migration—the difference between people migrating to a region and those leaving. The Arctic regions are demographically advanced meaning birth rates do not exceed death rates by much, and in some regions, the number of deaths exceeds the number of births. Because of their often-small population sizes and narrow economic structures, migration plays a larger role, both positive and negative in population change of Arctic regions than in larger, more diversified regions. The age structure of a region also plays a significant role in population change. A region with a relatively younger population will have more people in the childbearing years and thus will grow faster. A region with more people in the older ages will have a higher death rate and grow more slowly.

The population of the world grew by 46 percent since 1990, obviously all due to natural increase. The population of the United States has grown by 28 percent since 1990, faster than most other developed countries. This was due to a combination of higher natural increase than most other developed countries and continued high rates of immigration. The population of Alaska grew by one-third because of higher natural increase and moderate out-migration. Like other Arctic regions, migration to and from Alaska is quite volatile based on relative economic conditions in Alaska and elsewhere in the United States. There has been net out-migration from Alaska since 2012, and in 2017, the population of the state declined for the first time in three decades.

The population of the three Canadian northern territories continued to grow. Since 2001, the population of the Northwest territories grew by 9 percent, mostly from natural increase as net migration was close to zero. The population of the Yukon increased by 29 percent between 1991 and 2016, through roughly equal contributions of natural increase and net migration. Nunavut grew the most of the three by 41 percent since 2000, all due to natural increase as there was some out-migration. The high population growth in Nunavut was due to higher fertility and the younger age structure of the predominantly Inuit population.

There has been considerable population growth of 38 percent in Iceland since 1990. Over that period, three-quarters of the growth was from natural increase and one-quarter from net immigration. However, in recent years, the contribution of migration to population increase has been much greater. Fertility has declined to its lowest level in the country's history. Like other Arctic regions, migration in Iceland fluctuates considerably, some of the reasons being unique to Iceland. Prior to the banking crisis, from 2005 to 2008, there was considerable net migration into the country followed by a 4-year period from 2009 to 2012 of out-migration. In 2013, net migration into Iceland became positive again and has increased each year since. In 2017, there was the highest ever recorded migration into the country of 14,929 persons, equivalent to more than 4 percent of the population.

The population of Greenland has remained remarkably constant over time varying by less than 1000 plus or minus of 56,000. This is because any excess of births over deaths is balanced by roughly the same amount of net out-migration. Because it consists of a largely indigenous population, Greenland has a younger age structure and higher fertility and grew by 25 percent between 1990 and 2018 from natural increase.

The Faroe Islands had moderate population growth of 5.7 percent since 1990 and a similar pattern to Greenland of natural increase, nearly offset by net out-migration. Net migration in the Faroe Islands fluctuates considerably being quite negative in the early 1990s following a banking crisis and then vacillating between periods of positive and negative net migration.

Since 1990, the population of Norway has increased by over one million, over 25 percent. Over the entire period since 1990, natural increase accounted for 40 percent of this increase and net migration 60 percent. However, in the past decade, net migration into Norway has been the main contributor to population growth. The populations of the three Arctic regions increased by much less, Nordland by 1.6 percent, Troms by 13.6 percent, and Finnmark by 2.7 percent. Nordland has had low natural increase offset by roughly the same amount of net out-migration. Troms has grown through having natural increase at about the national level and moderate net in-migration. Finnmark has also had a natural increase about the national level, but this was offset by significant out-migration keeping population increase low. The population of the Norwegian settlements in Svalbard has declined over this period from 3544 in 1990 to 2583 in 2017, much of this attributable to out-migration.

The population of Sweden has grown significantly since 1990 by 19 percent, with one-quarter of the growth from natural increase and three-quarters from net immigration. Like other Nordic countries, much of the more recent population increase has been driven by historically high levels of immigration. The population of Västerbotten grew by much less than the national rate by 7.3 percent since 1990, with roughly the same proportions of natural increase and net migration. The population of Norrbotten has declined since 1990 by 4.4 percent through a combination of negative natural increase and net out-migration.

Since 1990, the population of Finland has increased by 11 percent, growing from just under 5 million to 5.5 million in 2018. Over the period since 1990, the contributions of natural increase and net immigration have been roughly equal. However, over the past decade, the contribution of net migration to population increase has been much higher. Of the three Arctic regions of Finland, only Pohjoil-Pohjanmaa (North Ostrobothnia) grew over this period, increasing by 18 percent, all due to natural increase. Kainuu declined by 19 percent and Lappi by 10 percent. In both, nearly all of the decline was attributable to out-migration.

The breakup of the Soviet Union, the transition to a market economy, and the liberalization of society resulted in significant demographic upheaval in Russia and the Russian north. In Russia, life expectancy plunged as a result of the psychosocial stress of the transition, which impacted males much more than females and which impacted the northern regions more than the rest of the country. Fertility fell to extremely low levels. There was a major redistribution of the population across the post-Soviet space. Since 1990, the population of Russia has declined by 2 percent because of a 9 percent decline from natural decrease partially compensated for by net immigration into the country.

The population of the Russian north adjusted to the new economic conditions by declining by 20 percent through a combination of slight natural increase and large-scale out-migration of nearly one-quarter of the population. There was decline in the population size and settlement structure as a number of settlements across the Russian Arctic were either closed or abandoned when they became depopulated. Out-migration and population decline were greater in regions further east and in some smaller regions, particularly in some of the ethnic homelands of Arctic indigenous peoples. In the Far East, the population of Kamchatka declined by one-third, the Koryak okrug in the northern portion of the Kamchatka peninsula declined by half, the Magadan oblast by nearly two-thirds, and the Chukotka okrug by nearly 70 percent. In all of these regions, out-migration of sizeable portions of the population was the driving factor behind the steep population declines. In the post-Soviet period, two different sets of northern regions have emerged, both economically and demographically. The Khanty-Mansi and Yamal-Nenets okrugs, the oil and gas regions in west Siberia, make up the bulk of the economic output of the Russian north and are also the only two regions which are growing in population size.

Health and Mortality Transition in the Arctic

While all of the Arctic countries and regions have made the transition to low death rates, there are considerable differences in current levels and patterns of mortality. First group consists of Arctic regions with high levels of life expectancy of around 80 for men and 84 for women, including Iceland, the Faroe Islands, Arctic Norway and Sweden (Fig. 4.1). In these regions, life expectancy is among the highest in the world, male-female differences are small, and for those Arctic regions which are a part of a larger country, differences from the national levels are minimal. In these regions, there are either no indigenous populations or mortality differences between indigenous populations and others are minimal.

The second group consists of Alaska, Yukon, and the Northwest Territories where levels of life expectancy are slightly lower than in the Nordic Arctic regions. The third group consists of the predominantly Inuit regions of Nunavut and Greenland with life expectancies of 69 for men and 74 for women.

The fourth group consists of the Russian northern regions where life expectancy lags behind the national averages in a country where mortality is much higher than countries at comparable income levels. Life expectancy for males in Russia dropped by 6 years from an already-low level of 63.7 years in 1990 to 57.5 in 1994 before beginning a slow recovery to 66.5 years currently, possibly the highest level ever recorded in Russian history, but still 15 years lower than the Nordic average. Russian women have much higher levels of life expectancy, and the female advantage in life expectancy is among the highest in the world. One factor keeping

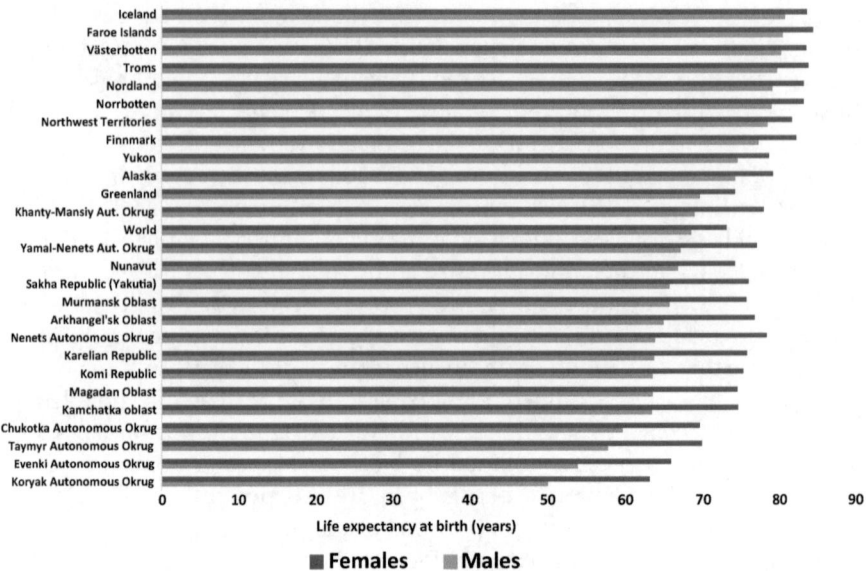

Fig. 4.1   Life expectancy at birth in selected Arctic regions, 2015–2017. Sources and notes: Data are from national and regional statistical offices of Arctic regions

overall life expectancy low in the northern regions of Russia is that women outlive men by 10 years or more in most. Apart from the more prosperous northern regions of Khanty-Mansi and Yamal-Nenets, all northern regions have levels of life expectancy below the national average. Russian northern regions with large indigenous populations have pitifully low levels of life expectancy. In Chukotka, life expectancy for men remains below 60, 6 years less than the national level. When data were last available for all the autonomous okrugs, in 2008, life expectancy for males in the Taymyr okrug was 58, in Evenki it was 54, and in the Koryak okrug, it was just 50. This level was up from a low of 46 in 2004.

Fertility Transition in the Arctic

Another transition is the fertility transition to lower birth rates. Most of the Arctic countries and regions have long made the shift to lower fertility rates. For some, rates are at levels demographers describe as very low fertility because of delayed marriage and childbearing and increasing childlessness. The total fertility rate is the hypothetical number of children a woman would have if she were to pass through her childbearing years at the current age-specific fertility rates. Allowing for some mortality, a rate of 2.1 children per woman is the replacement level, the level at which a population would just replace itself over the long term. Deviations from this level have a large influence on population growth, positive or negative.

The fertility rates in nearly all Arctic countries and regions have been declining over the past few decades and most have fertility rates at or below the replacement level. Arctic regions with high shares of indigenous populations tend to have higher fertility rates, including Nunavut, Greenland, the Nenets

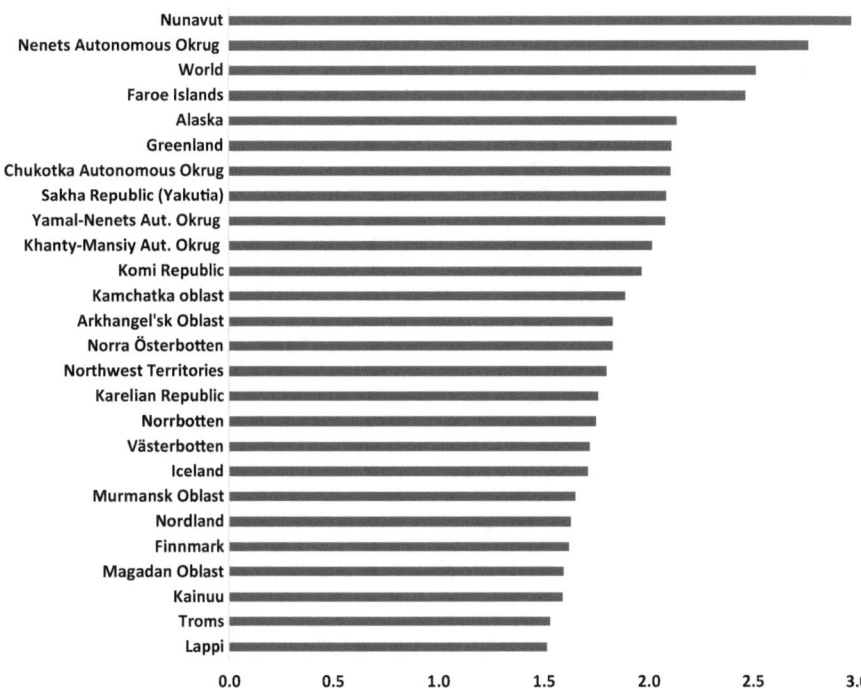

Fig. 4.2 Total fertility rate in selected Arctic regions, 2017. Sources and notes: Data are from national and regional statistical offices of Arctic regions

okrug, and Chukotka (Fig. 4.2). The Arctic regions of Norway, Lappi, and several regions of Russia have extremely low fertility rates. Part of the 'demographic crisis' in Russia, including the north, during the 1990s was a steep decline in childbearing when the fertility rate declined from 1.89 in 1990 to 1.16 in 1999 before increasing to 1.79 in 2016.

DEMOGRAPHIC CHANGE AT THE SETTLEMENT LEVEL IN THE ARCTIC

The size, composition, and spatial distribution of the settlements within Arctic regions are determined by a variety of factors. These include the political status of the region, development history, climatic conditions, transport connections, and natural resources. A broad overview of the physical geography, economic situation, and population distribution is presented before examining population change at the settlement level. This is part of the important parallel transitions of the migration and urban transitions when the structure of economies change, economic activity becomes focused in urban areas, and people move from rural to larger urban settlements.

Figure 4.3 provides an overview of population change during the period 2000–2017 for settlements with 500 inhabitants or more. They are shown as circles with an area corresponding to their total population. The colors indicate

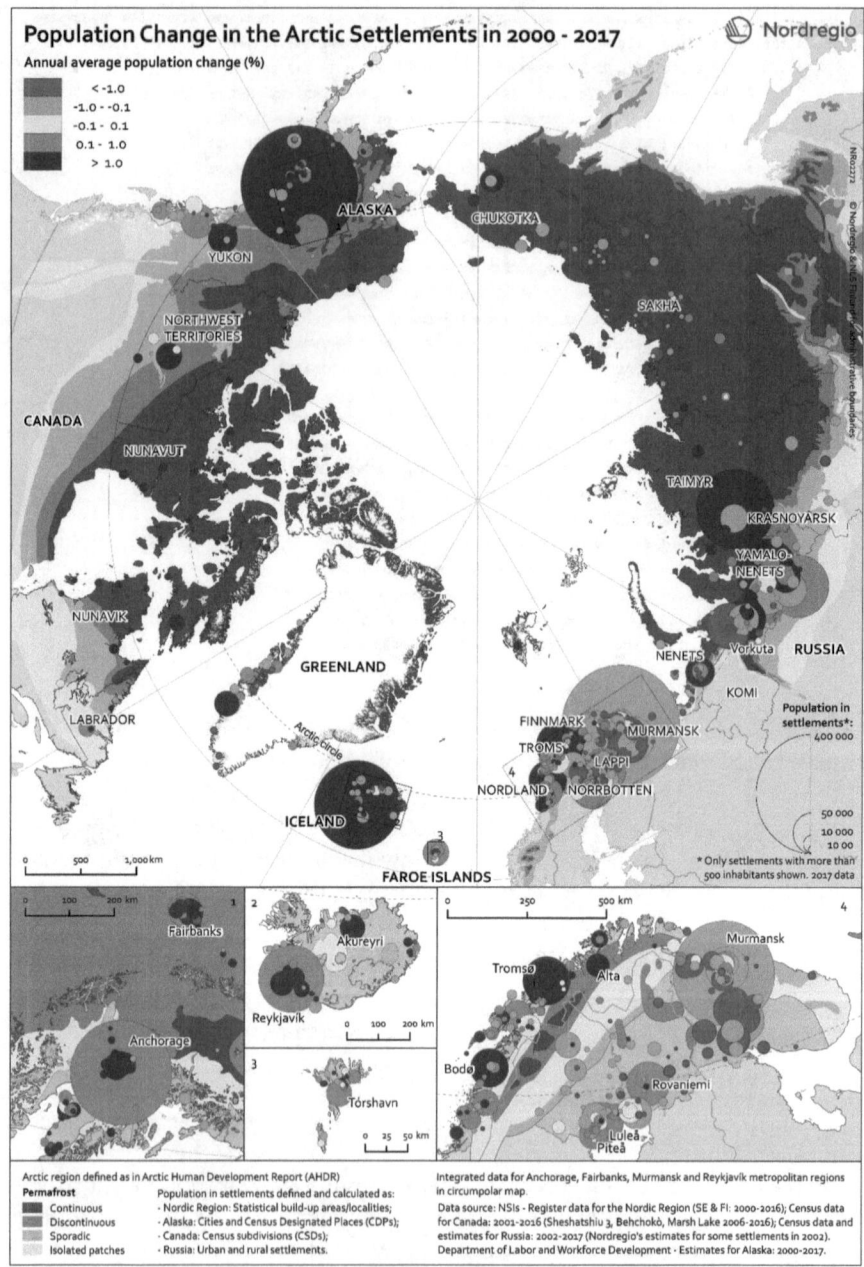

Fig. 4.3 Population change in the Arctic settlements, 2000–2017

the changes, with yellow showing places where little change has taken place, red indicating settlements with a declining population, and blue indicating a population increase. Alongside, there are four zoomed-in maps, showing Arctic Fennoscandia, Iceland, Faroe Islands, and Alaska, where the settlement density is high.

Alaska

Alaska is a non-contiguous state in the United States. It is the largest US state in terms of land area and is what makes the United States an Arctic country, a fact that escapes the attention of many in the country. Alaska was under Russian control until 1867 when it was purchased by the United States. At the time of the first contact with Russians and other Europeans, there were an estimated 80,000 Alaskan natives. Exposure to unknown diseases caused the population of Alaskan natives to plummet to a low of perhaps 6000 in the mid-1800s before recovering to 120,000 today. The demographic presence of Russians was never very significant and at its peak numbered only 900 (Levin 1991). The first of a number of waves of migration to the state occurred during the Alaska Gold Rush of 1897. There would be continued influxes of people from the lower 48 states associated with the discovery of new resources and the establishment of military bases during World War II and the subsequent Cold War.

Currently, the economy of Alaska is based largely on petroleum, centered on the large oil field in Prudhoe Bay on the North Slope (Glomsrød et al. 2017). The current population of 737,625 is unevenly spread with more than half the population residing in Anchorage and the nearby suburb of the Matanuska-Susitna valley. The three boroughs of North Slope, Nome, and Northwest Arctic are considered more truly Arctic because of their colder climates including extensive permafrost coverage. Many of the smaller settlements in these and other periphery regions are largely made up of Alaskan Natives and many are unconnected with the state-wide road system.

The reason for population decline in smaller settlements which are located far away from the two metropolitan regions, Anchorage and Fairbanks, is out-migration, which has canceled out the positive natural population growth. Almost all the settlements inside the two metropolitan regions have witnessed population growth, and their growth is even more significant than the two cities themselves. Proximity does matter when it comes to the migration of people out from the densely populated urban centers.

Arctic Canada

The economy of the three Northern Territories of Canada—Yukon, Northwest Territories, and Nunavut—has long been and is currently dominated by mining (Glomsrød et al. 2017). The Yukon Gold Rush of the late 1890s first brought large numbers of settlers to northern Canada. Currently, 95 percent of the value of diamonds in Canada originates from the Northwest Territories. Nunavut is the largest producer of gold among the three Northern Territories. Public administration is the largest sector in the economy of the three territories, and transfers from the federal government account for large portions of public revenues. The populations of the three Northern Territories reside in a quite limited number of settlements with large tracts of uninhabited land. Yukon's population resides in 25 settlements with 70 percent residing in Whitehorse. The population of the Northwest Territories resides in 33 communities with 48 percent in the capital of Yellowknife. The population of Nunavut resides in 25 communities. Nunavut has a deliberate policy of diffusing public-sector jobs to smaller communities outside the capital of Iqaluit, thus, the capital contains only 21 percent of the population.

The Canadian territories in general have high birth rates resulting in positive natural growth, which is the major contributor to the population growth, especially for small settlements in Nunavut, Nunavik, and Nunatsiavut. The phenomenon has resulted in population flow from small peripheral settlements to large centered settlements such as Yellowknife and Whitehorse.

Iceland

Iceland is the only Arctic region that is a completely sovereign state having become independent from Denmark in 1945. The economy was quite dependent on fisheries. After the banking crisis of 2008, the country started to aggressively promote tourism resulting in a huge influx. There was little immigration into the country for much of its history until the past few decades when the modernization of the Icelandic economy resulted in a significant rise in immigration, causing the percentage of foreign-born to increase from 4 to 11 since 1990 (Heleniak and Sigurjonsdottir 2018).

Greenland

Greenland is a self-governing region within the Danish kingdom. Fishing and related industries are the leading sectors (Glomsrød et al. 2017). Tourism is of increasing importance and there has been some oil and minerals exploration, though there is no significant production at present. The population of 55,877 (as of 1 January 2018) resides in 89 localities, including 17 towns, 54 settlements, 5 farms, and 5 stations. Nearly one-third of the population lives in the capital of Nuuk.

Faroe Islands

Like Greenland, the Faroe Islands are a self-governing region in the Danish kingdom. Like Iceland, the Faroes were completely uninhabited until the 800s. The economy is largely based on fishing, fish farming, and marine engineering (Glomsrød et al. 2017). Like other Arctic regions, tourism is a growing sector.

The rise and fall in fortunes of the seafood industry drive all other sectors and also migration to and from the country, mainly with Denmark. About 40 percent of the population resides in the capital of Torshavn. The rest of the population resides in a number of smaller coastal settlements on the 16 (of 18) inhabited islands. The government has long had a policy of linking all settlements via the national road system through a series of bridges and undersea tunnels to connect the entire population and reduce population decline in remote villages.

Arctic Fennoscandia

The Arctic regions of Norway, Sweden, and Finland have much milder climates than some of the other Arctic regions and have better transport connections to the southern regions of these countries. As such, they have more diversified economies and larger populations. In Arctic Norway, fishing, fish processing, and more recently aquaculture are the most important sectors (Glomsrød et al. 2017). The population of Arctic Norway was 487,000 in 2018 and includes several large cities including Tromsø (population 65,000) and Bodø (population 51,000). In Arctic Sweden, the manufacturing sectors, consisting of wood and metals processing and are significant as are the presence of several large universities (Glomsrød et al. 2017). The population of Arctic Sweden was 520,000 in 2018 including two large regional centers Umeå (population 85,000) and Luleå (population 77,000) located along the coast, where increasingly large portions of the populations reside. Arctic Finland differs from many Arctic regions in that it has a large manufacturing center based on electronics and other high-tech activities and several larger universities and research centers. The population of Arctic Finland was 660,000 in 2018 and consisted of several large cities such Oulo (population 201,000) and Rovaniemi (population 62,000).

Most of the smaller settlements in Fennoscandia have witnessed population decline during 2000 and 2017, with northern Norway being an exception. Though some smaller settlements experienced population decline, in the majority of the settlements in northern Norway—Norland, Troms, and Finnmark—the population has increased. The continuous migration inflow since 2010 has played a major role despite the fact that natural population growth has been negative for most of the small and medium-sized settlements. The relatively high birth rates in regional centers have resulted in even higher population growth such as Tromsø, Bodø, and Alta.

The dominating pattern in Fennoscandia is population growing in larger settlements and population shrinking in surrounding smaller settlements. This is similar to the pattern observed in the North Atlantic—Iceland, Greenland, and Faroe Islands. The capitals Reykjavik, Nuuk, Torshavn, and regional centers have been receiving inhabitants both domestically and internationally, while the settlements located in sparsely populated areas are losing their attractiveness. These population receivers are usually equipped with better education and healthcare facilities and have more opportunities in the job market. An interesting phenomenon for Reykjavik is that the settlements just outside the capital city are becoming more attractive than the capital city itself due to the crowdedness and high housing price inside the capital.

Arctic Russia

The manner in which the Soviet Union went about developing its Arctic and Siberian regions stands in sharp contrast to other Arctic countries which developed under market economy conditions (Heleniak 2009, 2010). Another distinction is that climatic conditions become progressively colder and harsher as one travels east rather than north as in the case of other Arctic regions. It was after the Bolshevik Revolution in 1917 and the institution of the centrally planned economy that the large-scale exploration of the Arctic and Siberian natural resources began. With the first Five-Year Plan in 1926, the Soviet Union sought to industrialize quickly and much of this was based on gold, oil, timber, and other resources found in the Arctic and northern peripheries. Economic planning decisions were made centrally by Gosplan (the State Planning Committee), rather than the market, which resulted in a non-market distribution of labor and economic activity (Hill and Gaddy 2003).

Development of Siberia, the North, and the Arctic was done in several over-lapping phases. The first was through gulag labor where large numbers of people were sent to establish labor camps and later to cities to extract necessary industrial resources. Later, a system of wage incentives and other bonuses was established to lure people to the Arctic. The Soviet economy became increasingly closed with the little movement of goods, information, or people across its borders. The northern wage increments, which often paid double or more for the same job in central Russia, were one way in which people could legitimately earn a high salary. Many people were induced by these incentives and migrated to the north for a period or a career.

Because labor was relatively undervalued as compared to in a market economy, the Soviet centrally planned economy put many more people and created much larger cities in the north and Siberia than had the region developed under market conditions. At the end of the Soviet period, there were many quite sizable Arctic and Siberian cities including Arkhangel'sk (417,000 in 1989), Murmansk (472,000), Petropavlovsk-Kamchatskiy (273,000), Noril'sk (180,000), Magadan (152,000), and Vorkuta (115,000) (Heleniak 2017).

When the Soviet Union broke apart and Russia made the transition to a market economy, including the liberalization of prices, the cost of the previous northern development system became burdensome and the wages and incentives for working in the north became worthless. There was large-scale out-migration from the north of one-quarter of the population, with larger portions from more distant regions such as the Magadan oblast which shrank in size by 60 percent and Chukotka which declined by three-quarters.

In post-Soviet Russia, two types of northern regions have developed, the oil and gas regions of the Khanty-Mansi and Yamal-Nenets okrugs and the rest. It is these two regions which are growing in population while most of the rest continues to decline slowly. Petroleum and other mining from these two regions make up more than half the northern economy (Glomsrød et al. 2017).

In the map, demographic change was analyzed in 177 Russian Arctic settlements and only in 41 of them the population was increasing. Consequently, over 75 percent of analyzed settlements were shrinking during the twenty-first century. The main reason for this decline was out-migration. In two regions of Nenets Autonomous okrug and Yamal-Nenets Autonomous okrug, positive natural growth was the strongest due to relatively low death rates and younger populations. They were also the only regions within the analyzed territory where the population was slightly growing. The population grew in Khanty-Mansi with about 14 percent during the analyzed time. These regions were not analyzed in the settlement map but are considered an Arctic territory within our definition. In these regions, there was growth in some small settlements, especially in big cities like Salekhard, Novy Urengoy, and Narian-Mar.

The most dramatic demographic change happened in the Komi region where the sub-region of Vorkuta is considered as an Arctic territory within our analysis. During the analyzed time, the region lost almost 40 percent of its inhabitants. The population declined in each of its settlements. In the more east side of Russian Arctic territory, the population was also decreasing.

Apart from the growth in the big city of Norilsk in Krasnoyarsk and the two fast-growing cities Anadyr and Egvekinot in the Chukotka Autonomous okrug, the population declined due to dominating out-migration. The population of Republic Sakha (Yakutia) region's population has been growing slightly due to high birth rates. However, within this map, we only analyzed the most northern part of Sakha, and as can be seen from the map the population was decreasing in almost all of its small settlements.

In the territories of Magadan Oblast and Kamchatka Kray, population decline was also remarkable. During this time, Magadan Oblast experienced both negative net migration and negative natural population change. In contrast, the continental region of Khanty-Mansi Autonomous okrug experienced both positive net migration and natural population growth.

ARCTIC POPULATIONS IN THE FUTURE

Identifying the broad demographic trends and the likely future size and distribution of the population of the Arctic is necessary for planning as well as for any possible interventions which governments wish to take to mitigate any negative socioeconomic consequences. All Arctic countries regularly do projections of their populations at both the national and regional levels (Heleniak 2019).

Global population growth in the past and into the future has had and will continue to have a profound impact on the Arctic and its population. Global population will continue to grow because of the momentum built into the global age structure, especially in the less developed regions. The global population is projected to reach 8 billion in 2023, 9 billion in 2037, 10 billion in 2055, and 11 billion in 2088 (United Nations 2017).

Between the time of statehood in 1959 and 1 July 2015, the population of Alaska grew from 224,000 to 737,625 (Alaska Department of Labor and Workforce Development 2016). The middle scenario, which calls for 0 percent net migration, projects a 22-percent population increase to 881,666 to 2040 (Fig. 4.4a).

According to the projections done for Arctic Canada by Statistics Canada, in Yukon, the population would increase slightly from 36,700 in 2013 to 43,100 in 2038 in the middle scenario. In some of the scenarios, Yukon's population in 2038 would be slightly lower than that observed in 2013, mainly because of interprovincial migration losses, which is the main driver of future population growth. The population of the Northwest Territories was 43,500 in 2013 and is projected to grow very slightly to 44,300 in 2038. Under several scenarios, population is projected to decline. Much of the variation in population change over the course of the next 25 years would depend on the nature of migratory exchanges with other parts of Canada. The population of Nunavut was 35,600 in 2013 and is projected to increase under all scenarios. The population is projected to increase to 46,600 in the medium scenario. Fertility and the youngest age structure in Canada are the key drivers of population growth in Nunavut. Its population would continue to increase despite losses in migration exchanges with the rest of Canada.

As noted above, there has been very little change in the population size of Greenland. In 1990, the total population was 55,558 and in 2016 was 55,847. The most recent set of projections done by Statistics Greenland go to 2040, in which the population of Greenland is expected to decline slightly to 52,207 by 2040 (Statistics Greenland 2017).

Iceland has had continuous population increase in recent decades which is expected to continue. The population is currently 348,450. Under the medium scenario, the population of Iceland is projected to grow to 405,338 by 2040 and 443,309 by the year 2065, a 33-percent increase over the next half-century (Statistics Iceland 2017).

The Faroe Islands has had slow population growth over the past 25 years with the population increasing by only 3 percent since 1990 and 2016. According to the middle value, the population is projected to slowly increase until a peak in 2029 of 50,941. It will then gradually decline until 2055 to 48,549. Overall, over the period 1990 to 2055, the population size of the Faroes is expected to change very little, reflecting the carrying and economic capacity.

The population of Norway is projected to increase by 21 percent between now and 2040. The Arctic regions are projected to increase but by less, Troms by 11 percent and Norland and Finnmark by 7 percent. Until the year 2040, the population of Sweden is projected to grow by 13 percent, but the Swedish

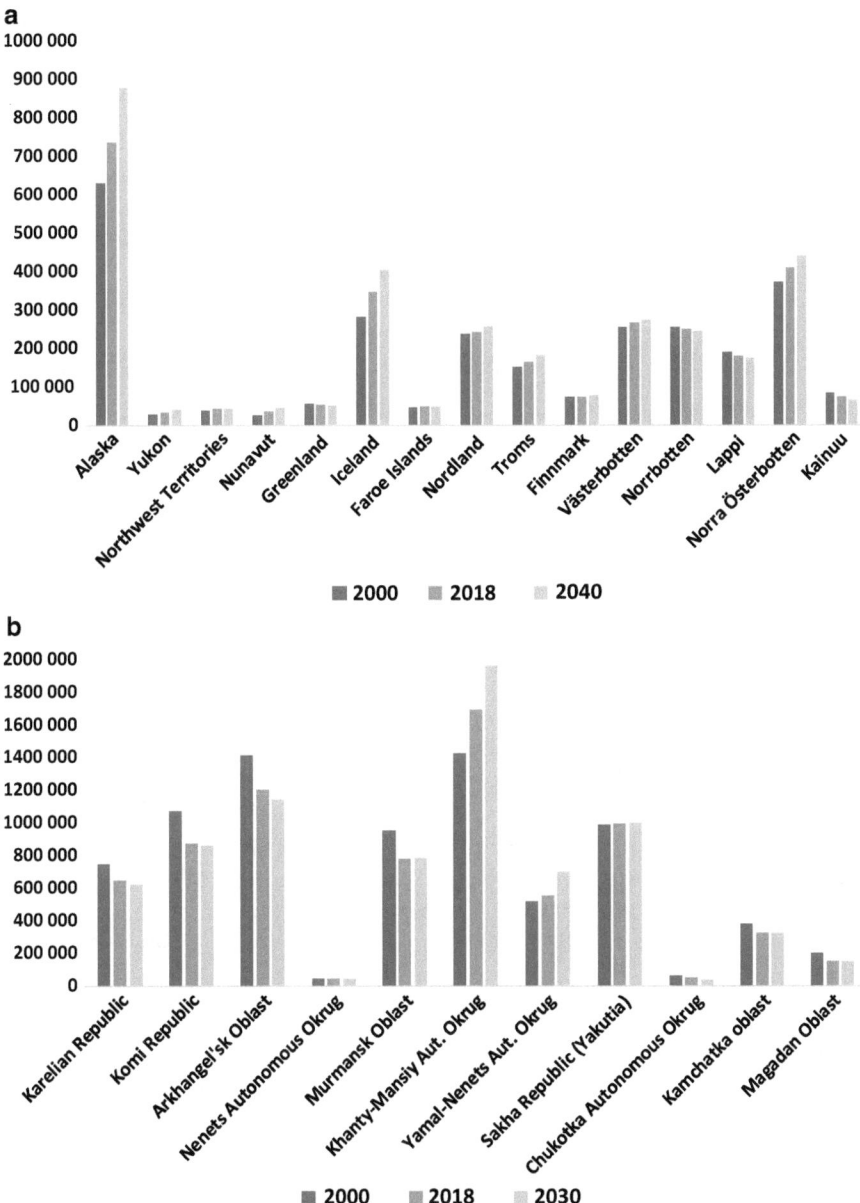

Fig. 4.4 Projected population of the Arctic. (**a**) Alaska, Arctic Canada, and Nordic Arctic. (**b**) The Russian Arctic. Sources and notes: National and regional statistical offices and Heleniak (2019)

Arctic regions will stay roughly the same with a 3-percent increase projected for Västerbotten and a 2-percent decline for Norrbotten. Until 2040, the population of Finland is projected to grow by 6 percent to 5.9 million. Most of this will be due to net immigration as natural increase is projected to be negative over much of the period. Pohjoil-Pohjanmaa (North Ostrobothnia) is projected to continue growing by 7 percent to 2040, mostly from natural increase. Kainuu is projected to continue to decline by 11 percent until 2040, mostly due to out-migration. Lappi is projected to have a moderate decline of 3 percent, mostly from out-migration.

Among Arctic Russian regions, the only three which are projected to increase are the Nenets, Khanty-Mansi, and Yamal-Nenets regions (Fig. 4.4b). All others are projected to decrease in size with Chukotka projected to have the largest percentage decline.

The Arctic regions will face both endogenous and exogenous demographic dilemmas in the future based on projections of current trends, some of which are already being addressed through different population policies (Nordic Council of Ministers). These include increased urbanization and the depopulation of smaller settlements for which Nunavut, Greenland, and the Faroe Islands have specific programs to counteract. The demographic challenge of aging is taking place globally and in all Arctic regions to differing extents. Higher out-migration of young women than from many smaller Arctic settlements is a trend which has been observed for quite some time but without a ready solution. Linked to population decline and aging is the challenge of international migration on which many Arctic regions have become quite dependent. Exogenous factors which will impact Arctic populations include climate change and permafrost thaw. Declining sea ice is already creating the need to relocate the populations of coastal settlements in Alaska. Thawing permafrost is impacting the infrastructure of settlements across the Arctic. There will be easier ocean transport across the Arctic brought on by less sea ice which could have both positive and negative impacts on Arctic port cities.

REFERENCES

Alaska Department of Labor and Workforce Development. 2016. *Alaska Population Projections 2015 to 2045*, April. http://live.laborstats.alaska.gov/pop/projections.cfm. Accessed 9 August 2017.

ArcticStat. http://www.arcticstat.org/.

Glomsrød, Solveig, Gérard Duhaime, and Iulie Aslaksen, eds. 2017. *The Economy of the North 2015*. Oslo-Kongsvinger: Statistics Norway.

Hassler, S., P. Sjölander, and U. Janlert. 2008. Northern Fennoscandia. In *Health Transitions in Arctic Populations*, ed. T.K. Young and P.B. Bjerregaard. Toronto, Buffalo, London: University of Toronto Press.

Heleniak, Timothy. 2009. Growth Poles and Ghost Towns in the Russian Far North. In *Russia and the North*, ed. Elana Wilson Rowe, 129–163. Ottawa: University of Ottawa Press.

———. 2010. Migration and Population Change in the Russian Far North During the 1990s. In *Migration in the Circumpolar North: Issues and Contexts*, ed. Chris Southcott and Lee Huskey, 57–91. Edmonton, Alberta, Canada: Canadian Circumpolar Institute Press, University of Alberta.

———. 2017. Boom and Bust: Population Change in Russia's Arctic Cities. In *Sustaining Russia's Arctic Cities: Resource Politics, Migration, and Climate Change, Chapter 4*, ed. Robert Orttung, 67–87. New York and Oxford: Berghahn Books.

———. 2019. *Polar Peoples in the Future: Projections of the Arctic Populations*. Nordregio Working Paper (forthcoming).

Heleniak, Timothy, and Hjördis Rut Sigurjonsdottir. 2018. *Once Homogenous, Tiny Iceland Opens Its Doors to Immigrants*. Migration Policy Institute, April 18. https://www.migrationpolicy.org/article/once-homogenous-tiny-icelandopens-its-doors-immigrants.

Hill, F., and C.G. Gaddy. 2003. *The Siberian Curse: How Communist Planners Left Russia Out in the Cold*. Washington, DC: Brookings Institution Press.

Hirsch, F. 2005. *Empire of Nations: Ethnographic Knowledge and the Making of the Soviet Union*. Ithaca and London: Cornell University Press.

Larsen, Joan Nymand, and Gail Fondahl, eds. 2015. *Arctic Human Development Report: Regional Processes and Global Linkages*. Copenhagen: Nordic Council of Ministers.

Levin, M.J. 1991. *Alaska Natives in a Century of Change*. Anthropological Papers of the University of Alaska, Vol. 23(1–2), Fairbanks, Alaska.

Statistics Greenland. 2017. *2017–2040 Population*. http://www.stat.gl/dialog/main.asp?lang=en&sc=BE&version=201702. Accessed 21 August 2017.

———. 2018. http://www.stat.gl/.

Statistics Iceland. 2017. *Population Projections*. http://www.statice.is/statistics/population/population-projections/population-projections/. Accessed 7 August 2017.

United Nations. 2017. *World Population Prospects: The 2017 Revision*. DVD ed. New York: Department of Economic and Social Affairs, Population Division.

Weeks, J.R. 2008. *Population: An Introduction to Concepts and Issues*. 10th ed. Belmont: Thompson Wadsworth.

PART II

Economic Development

State Expansion and Indigenous Response in the Arctic: A Globally Integrated Northern Borderland Emerges from the Historical Synthesis of Northern Frontier and Northern Homeland

Barry Scott Zellen

Since the state and its colonial proxies first encroached upon their homeland three centuries ago, natives of the North have continued to assert and defend their Aboriginal rights and cultural traditions, and have sought to preserve as much of the autonomy of their independent polities now threatened by state expansion as they could. As natives learned more about the many systems and structures of governance that were exported from Eurasia, from the commercial trading posts and early global networks they were part of, to the representative constitutional democracies that took root in their homelands, they found many new ways to reassert and, increasingly, restore their autonomy—through innovative domestic diplomacy, protracted (sometimes multi-decade) negotiations, and various forms of political protest and engagement (including the omnipresent threat of litigation to delay development projects proposed for their homeland).

This contrasted sharply with elsewhere in the Americas where the modern state collided more forcefully (and with greater kinetic energy) with the interests and sovereign aspirations of hundreds of indigenous empires, nations, and tribes from the late fifteenth century onward—the result of which was annihila-

B. S. Zellen (✉)
Center for Arctic Study and Policy (CASP), United States Coast Guard Academy, New London, CT, USA
e-mail: barry.s.zellen@uscga.edu

© The Author(s) 2020
K. S. Coates, C. Holroyd (eds.), *The Palgrave Handbook of Arctic Policy and Politics*, https://doi.org/10.1007/978-3-030-20557-7_5

63

tory warfare, genocide, forced migrations, and coercive assimilation policies reminiscent of what took place elsewhere in the world when states expanded onto indigenously self-governing lands—resulting in the general extinguishment of indigenous identity even as tribes brilliantly and heroically fought back against the state, but inevitably lost to the state's superior military power (Time-Life Books 1993; Todorov 1999; Utley and Washburn 1977).

Northward State Expansion: From Colonization to Increasingly Collaborative Northern Governance

In the Far North of the North American continent, the state collided with indigenous tribes much later in history, with economic contact, and later military interaction, starting in the seventeenth and eighteenth centuries. By the time the presence of the rapidly modernizing state began to be felt in the Arctic, its methods for asserting political control began to mellow, with hard power (as embraced by the Russians in their conquest of the Aleutians) shifting to soft(er) power, and treaty negotiation replacing conquest for the final integration of the Arctic territories into the American and the Canadian polities (Young 1998; Zellen 2008). With a "kinder, gentler" expansion and a less muscular conquest, northern natives did not feel compelled to engage the state frontally by force but began to see state expansion more symbiotically. Survival would require accommodation, rather than confrontation, from both sides (Figs. 5.1 and 5.2).

In 1867, America purchased Alaska from Russia, and with the purchase inherited the Russian-American Company's broad assertion of sovereignty over Alaska's interior tribes (even in the more remote and northern territories where the Russians never settled); because of Alaska's perceived harsh climate and remote location, most Americans thought Secretary of State William H. Seward was foolish to have spent $7 million on these frozen acres, dubbing the new territory "Seward's Ice Box" or "Seward's Folly" (Bockstoce 2009; Farrow 2016). Great Britain, and later Canada, similarly exchanged their way to sovereign expansion, not by purchasing the land from a competing power but by entering into a series of Numbered Treaties, nation-to-nation peace accords pledging friendship and mutual support (according to historical memory on the native side—even if the written versions of those treaties convey a "surrender" that natives today believe was never agreed to) that brought the western tribes into its expanding confederation (Fig. 5.3).

Thus, largely through negotiation between two unequal parties, tribe and state, the new territories of the northwest entered into southern control without, by and large, recourse to war—with exceptions including the Métis rebellion from 1871 through 1885—perhaps the greatest indigenous threat the young dominion would ever face—and in Canada's densely populated southern core, the far more limited (but no less intriguing) largely symbolic armed uprisings such as that which began at Oka, Quebec, in 1990 which led to the seizure of the Mercier Bridge connecting Montreal to the Southern Townships, a summer-long siege of a contemporary North American metropolis by an armed Mohawk movement of the type seen elsewhere in the world but which

PREPARING FOR THE HEATED TERM.

King Andy and his man Billy lay in a great stock of Russian ice in order to cool down the Congressional majority.

Fig. 5.1 Editorial cartoons at the time of the Alaska purchase were especially critical of Secretary of State Seward's "folly." Source: "PREPARING FOR THE HEATED TERM King Andy and his man Billy lay in a great stock of Russian ice in order to cool down the Congressional majority," *Frank Leslie's Illustrated Newspaper*, as cited at http://alaskaweb.org/history/hist4.html

caught many Canadians by surprise (York and Pindera 1991). (The bridge was believed to have been wired with explosives, but that proved to be a strategic deception.) Because the political integration of the predominantly native North, by contrast, was achieved largely without war, the preferred tools for reconciling the interests of tribe and state would remain predominantly non-violent, modeled on the treaty process, with negotiation helping to bring some balance to the many other asymmetries—such as economic and military power—that separated the indigenous tribes from the modern states laying sovereign claim to the North.

While the expansion of the modern state into the North did not require frontier warfare as experienced elsewhere in America's expansion, modern warfare nonetheless did have a profound impact on the relationship between natives and the modern state. This was most dramatically illustrated in June

"THE BIG THING."

OLD MOTHER SEWARD. "I'll rub some of this on his sore spot; it may soothe him a little."

Fig. 5.2 Editorial cartoons at the time of the Alaska purchase were especially critical of Secretary of State Seward's "folly." Source: Thomas Nast, "Map of the Russian Fairly Land," *Harper's Weekly*, April 20, 1867

1942 when Japan bombed Dutch Harbor and invaded the islands of Attu and Kiska in the Western Aleutians. With Japan's forcible resettlement of the surviving Aleuts from Attu to Hokkaido for the remainder of the war, Alaska natives quickly recognized that they too faced grave danger, and the crucible of war would help to tighten the bond between Alaska's indigenous peoples and the rapidly expanding modern state, which mobilized for war by building new airstrips, surging manpower, and cutting the Alaska Highway across 1400 miles of northern wilderness in 1942 (Chandonnet 1995).

While this rapid mobilization would create many stresses and strains on the long-isolated native population, including the painful odyssey of the remaining Aleut population as it was relocated outside the war zone to camps in Alaska's southeast (suffering significant losses to their population from the strains of displacement and internment), the wartime experience would in other ways help bring natives and newcomers closer together—as evident in the formation of the Alaska Eskimo Scouts in 1942, the famed "Tundra Army" organized by

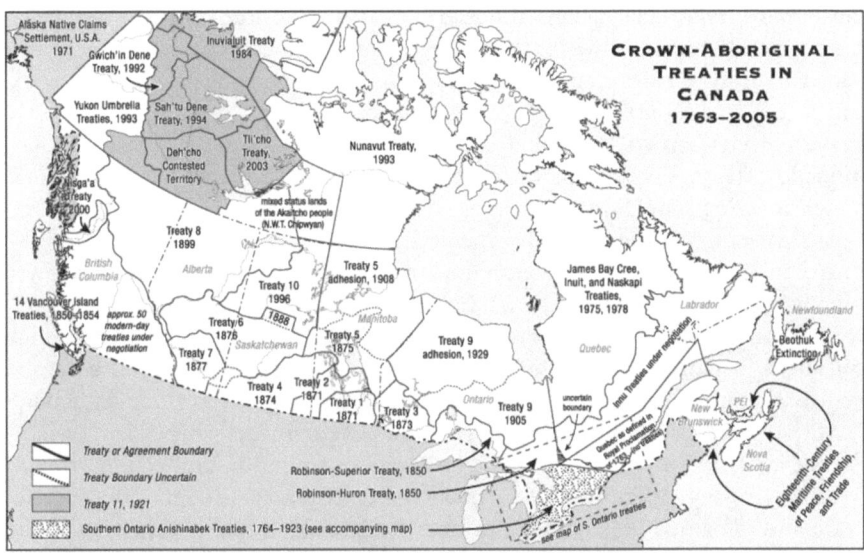

Fig. 5.3 Most of northwestern Canada was acquired through a series of numbered treaties negotiated during the late nineteenth and early twentieth centuries. Untreated lands to the north were later addressed through comprehensive land claims treaties negotiated in the late twentieth and early twenty-first centuries. Comprehensive land claims are now settled across the entire Arctic coast of Canada, from the 1984 Inuvialuit Final Agreement in the west to the 2005 Labrador Inuit (Nunatsiavut) Land Claim Agreement in the east. Source: Southern Chiefs' Organization Inc. website, http://scoinc.mb.ca/wp-content/uploads/2014/04/CrownAboriginalMap.jpg. As cited by "Treaties Recognition Week: First Full Week of November," Education Resources section of the Trent University Website (trentu.ca/education/resources), https://www.trentu.ca/education/sites/trentu.ca.education/files/documents/TrentUSchoolofEd_TreatiesWeek_ResourcePackage.pdf

Major Marvin "Muktuk" Marston, which would later become the Alaska Territorial Guard, with thousands of volunteers representing over 100 Aleut, Athabaskan, Inupiaq, Haida, Tlingit, Tsimshian, Yupik, and non-native communities (Marston 1969). In the high North Atlantic, the dual impact of the Battle of the Atlantic, and America's defense of Greenland and maritime Canada, would similarly bring modern state power into remote and traditional Inuit territories in Labrador, Baffin Island, and Greenland. Later, during the Cold War, the massive DEW (Distant Early Warning) Line Project and integration of the isolated Arctic coast into North America's air defense would have a similarly transformative impact, extending modern state power deeper into the homeland of the Canadian Inuit.

Native participation in the defense of Alaska would provide a powerful unifying force, stimulating the movement for native rights that culminated in the historic 1971 passage of the Alaska Native Claims Settlement Act, the pioneering land treaty transferring 44 million acres of land title and $1 billion in compensation to Alaska natives, a model embraced and later enhanced as Inuit land claims were negotiated across the entire North American Arctic, with Inuit gaining title to nearly one-tenth of their traditional land base, and new co-management structures enabling a joint approach to managing natural resources, land access, and economic development. By the time the Inuvialuit of Canada's Western Arctic settled their land claim in 1984, based on an Agreement-in-Principle (AiP) negotiated in 1978, they had greatly advanced the land claims model, favoring native traditions that ANCSA had overlooked. The Inuvialuit land claim entitled the 3000 Inuvialuit living in six communities to 35,000 square miles of land; co-management of land and water use, wildlife, and environmental assessment; wildlife harvesting rights; financial compensation of $45 million in 1978 dollars, inflation-adjusted to $162 million, for lands ceded to Canada; a share of government royalties for oil, gas, and mineral development on federal land; the formation of new national parks in their settlement area that further protect their land base from development, while allowing subsistence activities unhindered; and a commitment to meaningful economic participation in any development in their settlement area (Zellen 2008). This model has remained largely intact in later comprehensive land claims, showing great endurance as a model for northern development (Fig. 5.3).

A State/Tribe Synthesis Emerges: The Historic Reconciliation of the "Two Arctics"

The Inuvialuit land claim was the second comprehensive land claim in Canada, following the James Bay and Northern Quebec Agreement of 1975, which was extended in 1978 to include the Naskapi Indian Band. Just as the James Bay accord preceded major hydroelectric development in northern Quebec and ANCSA preceded the construction of the Trans-Alaska Pipeline System (TAPS), the Western Arctic (Inuvialuit) Claims Settlement Act preceded what was expected to be imminent, large-scale oil and gas development in the Western Arctic region. But Thomas Berger's famed Mackenzie Valley pipeline inquiry raised important social, political, and cultural questions, recommending that this imminent wave of exploration activity be frozen for a decade so that land claims could be settled in the North. In the Western Arctic, the pace of oil and gas exploration had been intensifying ever since the Prudhoe Bay oil confirmation in 1968: this caused numerous social tensions borne of the shock experienced when the two cultures collided suddenly. As environmental consultant Gary Wagner, a former resource staff member for the Inuvialuit Environmental Impact Review Board (EIRB), has recalled:

> On Banks Island in Canada's Northwest Territories in 1970, Inuvialuit hunters on the remote community of Sachs Harbour awoke one morning to find oil industry personnel off-loading drilling equipment from a large barge docked at the community wharf. None of the community members had heard that exploration activities were planned for the vicinity of Sachs Harbour. Hunting, fishing, and trapping—all essential to the Inuvialuit subsistence lifestyle—would be seriously curtailed by the work being done by the oil companies. To compensate,

Inuvialuit hunters were offered labourer jobs at about half the local government rate of pay. Set in the context of the assimilation policy of the federal government, and faced with deliberate and unannounced encroachment by developers on their traditional lands, the Bankslanders decided to oppose the cavalier way their government and the oil companies had treated them. If there was one catalyst that sparked the Inuvialuit land claim, perhaps the experience of the Bankslanders on that morning in 1970 was it. (Wagner 1996)

That 1970 morning, the Inuvialuit awoke to find a new, menacing manifestation of state expansion on their doorstep: a modern and industrial dynamo that differed profoundly from their traditional, pristine world. They thus pursued their claim to the newly contested lands of Banks Island and elsewhere throughout what would become the Inuvialuit Settlement Region 14 years later. For the Inuvialuit, the pursuit of their land claims—despite the central role of development and investment corporations in the land claims model defined by ANCSA that carried over into Canadian land claims policy—was paradoxically embraced as a shield to protect their homeland and hold this alien, industrial world at bay, even though in Alaska and northern Quebec land claims had paved the way forward for the rapid, large-scale development of the North. The end result of the Inuvialuit desire to slow the pace of Arctic modernization and the relentless pressure from Big Oil to develop northern resources was a complex compromise: a blueprint that sought to chart a middle course, balancing tradition and modernity. Land claims, and their corporate blueprint, would thus serve as a bridge, and not a barrier wall, between these two worlds, moderating the interface between tribe and state. And, with the passage of time and increased participation of northern natives in their emerging economic and political processes, land claims increasingly fostered a reconciliation of these two very different worlds.

In her history of Alaska's TAPS, Mary Clay Berry (1975) observed the existence of "two Alaskas" much as Richard Rohmer (1973) observed the existence of "two Arctics," and Thomas R. Berger (1985) described the North as both a "northern frontier" and a "northern homeland." Land claims, it appears, are borne of this inexorable synthesis of these two worlds. In her history, Berry cited long-time Alaskan leader Walter Hickel, who was quoted in *Time Magazine* in 1958 as saying "We're trying to make a Fifth Avenue out of the tundra!" (Berry 1975, 10). She also cited Alaska native Ewen Moses Laurnoff, who in 1968 said, "They tell me Russians sold our land to the Government. There were no Russians on our land. There were no white people. White people never came here. They never saw it. I think Government buy stolen property maybe. Tough luck for Government. Can buy whole world that way" (Berry 1975, 10). Berry, like Rohmer and Berger, identified the same compelling northern duality— both a natural North and a modern, industrializing North. Berry also observed both a non-native and a native Alaska, and the conflict between these two distinct groups of Alaskans in the 1950s as Alaska sought admission to the Union as a state (Berry 1975, 11). Berry recounted the native awakening and the struggle of native people against non-natives that culminated in the settlement of the Alaska native land claim (Berry 1975, 34–52), as well as the conflict between natural and industrial Alaska (Berry 1975, 53–84). The oil discovery at Prudhoe Bay and the TAPS proposal prompted a closer look at Alaska by the U.S. federal government: the courts, congress, and the executive branch were all drawn into the Alaska question. The push by industry for a pipeline from the North Slope to Valdez brought into focus these overlapping conflicts. Native people, environmentalists, industry, and government found themselves with a curious problem, and the solution to this problem required balance. This balance, of course, was ANCSA, which proved to be somewhat out-of-balance, leading to three subsequent decades of political conflict, litigation, lobbying, and state and federal legislation, all designed to redress the imbalance of a land claim that excluded subsistence. The two Alaskas were not equally satisfied with the Alaska land claim, its structure, or its impact on village life, and native dissatisfaction with the original land claim structure would come to dominate Alaska state politics well into the 1990s.

Rohmer observed the existence of the very same dichotomy in the Mackenzie Delta that Berry observed in Alaska. As he described it:

> It was all white. The frozen, undulating snow-covered tundra as it flashed by under the wings of our low-flying aircraft blended perfectly with an opaque overcast sky, a totality of whiteness, not a tree, not a building, not a road, not a moving thing. Suddenly straight ahead of us, at a point as far as they eye could see, was a tiny fleck of black, something foreign in this world of white. As we rapidly approached, it began to take the form of a black needle sticking up from the ground. And then closer its shape became the familiar silhouette of a drilling rig. Imagine! An oil rig sitting in one of the most desolate, remote, bleak, forbidding places in the world. What a feat of technical achievement that man could come here and discover from the surface of the earth where oil should be underground, bring his oil drilling equipment and supplies, pierce the ice-hard permafrost and then drill down through many thousands of feet into what at one time was a tropical jungle and what is now a pool of oil and gas. (Rohmer 1973, 72)

Before the oil boom worked its way into the Western Arctic, traditional economies, from trapping to hunting to whaling and fishing, dominated most peoples' work lives. The oil boom changed that. It brought the modern industry into an area long dominated by traditional subsistence. One result was the response by native people of the region, who sought greater control over their homeland. Indeed, just as in Alaska, the clash of cultures and the multi-party conflict of interests borne of oil exploration and discovery launched a grassroots movement for greater native control over the land and resources of the region. This was the birthing ground of Committee for Original People's Entitlement (COPE), which negotiated the

Inuvialuit land claim. It was in this maelstrom of conflict that "the local people [began] flexing their muscles and fight[ing] back at the great steam-roller" (Rohmer 1973, 79).

Rohmer described two incidents of local response to the oil companies and their intrusion into the Inuvialuit homeland: in Aklavik, where the people opposed further seismic exploration on their hunting and trapping grounds, and in Tuktoyaktuk where the people opposed exploration and development on traditional hunting lands at Cape Bathurst (Rohmer 1973, 78–80). Numerous examples of industry's insensitivity to local concerns spawned the movement by COPE for a land claim. The Alaska claim was a model of sorts, though COPE leaders learned from Alaskans that subsistence had better be protected or else traditional ways would not be preserved, something then NWT premier and later IRC chief Nellie Cournoyea told the Inuvialuit community corporations at their 1992 Annual General Meeting, to which Inupiat from Alaska were invited. The heart of Inuvialuit culture, like that of Inupiat, was and remains subsistence: hunting, fishing, whaling, and trapping. Unlike ANCSA, which did not protect subsistence, the IFA enshrined it, devoting great attention to the challenge of protecting the Western Arctic's environment and its wildlife resources. The goal of protecting the Arctic environment from the ravages of unregulated development and preserving Aboriginal subsistence as a cultural, economic, and environmental value was central to IFA; indeed, as the basis for subsistence, the environment is the very lifeblood of Inuvialuit cultural traditions and its protection is essential for the fulfillment of the IFA's objectives. Whereas the national interest has only in recent years included securing the health of the physical environment, the Inuvialuit tribal interest has long considered the environment sacrosanct, placing its protection at the very heart of their land claim—something the Alaska claim did not achieve.

Though the Inuvialuit felt compelled to protect their own regional biosphere, they did not do so at the expense of oil and gas development. As Gary Wagner has explained in his master's thesis on the implementation of the Inuvialuit land claim, "Somewhere between the extremes of unbridled industrial development and a return to a subsistence economy lies the balanced harmony of sustainable development" (Wagner 1996, 8). While the Inuvialuit land claim got its start, to a large degree, as an effort to stop the oil companies from coming into the Inuvialuit homeland, it evolved into a mechanism that transformed the Inuvialuit from virtual non-participants into a major economic player, with interests in oil and gas development, and in more recent years, even an equity stake in the resurrected, but ultimately unsuccessful, Mackenzie Gas Project. The emergence of such a strong, pro-development ethos in the Arctic has been evident across the region from Alaska to Nunavut. In Nunavut's first decade, there's even been a uranium rush in Nunavut as junior mining companies staked the new territory in search of this increasingly valuable metal. As one 2006 headline in *Nunatsiaq News* observed, "Uranium rush floods western Nunavut; Huge jump in price, demand has juniors racing to stake claims" (Minogue 2006). And, as *Arctic-Caribou.com*, the website of the Beverly and Qamanirjuaq Caribou Management Board (BQCMB), observed, "Uranium mineral exploration is feverish in Nunavut and NWT, which bear geological similarities to uranium-rich Saskatchewan.

Uranium has jumped to US$45.50 per pound from about US$11 per pound in late 2003, due in part to a huge increase in electricity needs from China and India" (Arctic-Caribou.com 2006b). On June 4, 2007, the Government of Nunavut announced that it would pursue a "balanced and sustainable approach" to uranium mining that included six principles announced by Economic Development and Transportation Minister David Simailak to "serve as a valuable guide for uranium mining in the territory," as "building a strong and sustainable economy is a key objective of our government and fostering a robust exploration and mining industry is a central part of that plan. As we move forward we must retain a balanced and sustainable approach and that is what we will be doing in the case of uranium" (Government of Nunavut 2007). While the Government of Nunavut announced it would be taking a "balanced and sustainable approach," the BQCMB, created in 1982 to "manage the more than half-million Beverly and Qamanirjuaq barren-ground caribou, which migrate in two herds across Manitoba, Saskatchewan, the Northwest Territories and Nunavut," and "whose migratory routes straddle two territories, two provinces, and four different native cultures" (as described on their website, Arctic-Caribou.com), disagreed.

As its website reported: "Alarmed by a pro-industry draft uranium policy proposed by land claim organization Nunavut Tunngavik Inc. (NTI), the BQCMB has lodged numerous concerns with NTI over its March 5, 2006 draft policy and consultation documents that favor uranium mining in Nunavut" (Arctic-Caribou.com 2006a). In a sidebar on the BQCMB website, titled "NTI think they know the answers," Baker Lake resident Joan Scottie, founder of the Baker Lake Concerned Citizens Committee that successfully opposed the Kiggavik uranium mine in 1990, described the response to her letter to the editor in *Nunatsiaq News* that criticized NTI for endorsing uranium mining without first consulting the Inuit:

> NTI responded by telling me that nuclear energy is used for peaceful and environmentally responsible purposes, that uranium mining in Nunavut brings significant economic benefits to the people of the local communities, to the region, to Nunavut and to Canada, that uranium mining is carried out in a manner that protects the health and safety of the workers and all Nunavummiut, that uranium mining will not cause significant adverse effects on the environment or wildlife, and that community members are given an opportunity for full and meaningful participation in both the environmental assessment process and the operations of uranium mining projects. (Arctic-Caribou.com 2006a)

As Scottie explained,

> The only message we are hearing loudly from our leaders is nothing but positive news about mining ... People eventually tend to believe and follow the leaders, such as NTI and Nunavut government officials, who make promises or positive guarantees to something that we have reservations about. Then there are the uranium mining companies that have been doing extensive public relations and have very strong input into the uranium policy. (Arctic-Caribou.com 2006a)

This may not have been the vision of the Arctic imagined back in 1970 when the Inuvialuit awoke to find oil drilling equipment being unloaded at the community dock in Sachs Harbour, but the lead role being taken by the native leadership in the development of their natural resource base is proof that land claims provided northern natives with unprecedented powers of regulation and self-governance. In 1970, without land claims at their disposal, the Inuvialuit were powerless to stop the oil companies and unlikely to benefit from any discoveries that might take place in their homeland. Fast forward to now, with natives now playing such a dominant economic role in the natural resource sectors in Alaska, the NWT, and Nunavut, it has become clear that land claims provided natives with powerful tools to benefit from northern development. They also provided northern natives with tools to regulate and protect their environment and to preserve their hunting traditions, but not at the exclusion of economic development. The concept that defined this balanced approach to developing the North's natural resources while protecting its environment so that traditional subsistence could continue was called co-management. It was this concept that bridged the "two Arctics," and moderated their intersection, so instead of collision, there was collaboration.

A new spirit of reconciliation between tribe and state thus emerged across the North, recognizing two fundamental truths on the ground: that the modern state had arrived and with it a preponderance of power but also that the indigenous peoples who had long been there, with their own traditions and cultures, still mattered greatly and contributed to the new, culturally diverse, polities taking shape. This reconciliation resulted in new governing institutions that moderated what might otherwise have been a tragic "clash of civilizations" along the last frontier, as new forms of local, regional, territorial, and even tribal governance took root—sometimes using a public governance model, while other times embracing a more traditional tribal model of governance. At the municipal level of government, there is the North Slope Borough in Alaska, a vast municipality that sustains itself through property taxation of the Prudhoe Bay oil facilities, a borough larger in size than the state of Massachusetts that governs a population of under 10,000—with hundreds of millions in petro-dollars to build world-class infrastructure and provide modern government services. At the territorial level, there is the vast Nunavut Territory, governing one-fifth of Canada's landmass, created through the peaceful secession of the Eastern Arctic from the old Northwest Territories, and home to some 30,000 people, almost entirely Inuit, scattered across 28 villages in an area three times larger than Texas —and a source of much of Canada's future natural resource wealth and strategic waterways. And at the tribal/indigenous level, there is the Inuit government of Nunatsiavut in northern Labrador, which has a unique Inuit constitution that governs its 2000 Inuit residents living in six villages in a traditional manner, rejecting a public governance model in favor of one more distinctively tribal in nature (owing to the demographic predominance of non-natives in Newfoundland and Labrador, making a public model inherently risky in contrast to Nunavut, which enjoys Inuit demographic predominance). As shown by these innovations in collaborative northern governance, indigenous culture has become increasingly recognized not as a fault line of conflict, but as a new and viable boundary line for political institutions, providing a foundation for political stability across regions, transcending national boundaries. This more benign form of state expansion, as witnessed across the North, suggests that with foresight and innovation and a willingness to redraw political boundaries to better preserve indigenous identity and autonomy, it is possible to create stable frontier regions free of war, and with effective mechanisms for mediating tribe-state disputes before they evolve into protracted, and seemingly perpetual, domestic or regional conflagrations.

THEORIZING THE NORTHERN BORDERLAND: AN EMERGENT— AND INCREASINGLY SALIENT—LEVEL OF ANALYSIS

My research on the Western Arctic region began during the final years of the Cold War as the "Ice Curtain" nominally dividing east and west across the North began its thaw (a full generation before the more literal thaw, catalyzed by Arctic climate change, began). Having studied International Relations (IR) theory with two pioneering political scientists, Kenneth Waltz and Ernst Haas, I naturally looked to IR theory for guidance in understanding this fascinating but relatively understudied corner of the Arctic, finding helpful insights in regional subsystems theory, which emerged in the early 1970s (Thompson 1973), and regime theory which soon followed (Krasner 1983; Young 1977, 1982, 1994, 1998). Regional subsystems theorists sought to fuse realism and structuralism with the diversity of regional politics around the world (adding some granular "reality" to an overly ethereal "neorealism") by drilling downward from the infamous and oversimplified third (systemic) toward a more complex and empirically authentic second (national) image, in search of their own distinct, hybrid level of analysis (Waltz 1959; Zellen 2018). With the rise of regime theory a decade later, we encounter another newly recognized structure between these same two levels, describing an analytical unit that is at once trans-state and sub-state and which can be useful to describe a multiplicity of collaborative and joint management efforts between states and/or regional associations of states in what some describe as distinct "borderlands" (yet another hybrid structure in world politics between the third and second images that would follow in more recent years with the advent of transnational studies). Such a nimble use of regime theory as a lens through which to understand what we now know as the northern borderland can be illustrated by the pioneering work of Oran Young, who wedded regime theory with the study of the Arctic and Subarctic in the 1970s and 1980s, and whose examination of Beringia as a regional subsystem can be viewed as a theoretical precursor to northern borderland studies, as can his broader work on the Circumpolar North which is, in essence, a circumpolar borderland that encompasses the boundaries of all the Arctic states (Osherenko and Young 1989; Young 1977; Zellen 2009a). Borderland theory has emerged more recently as a viable contender for this previously nameless (but nonetheless essential) structure in world politics where many northern intergovernmental, intertribal, and hybrid regimes (and other collaborative bodies, including co-management boards seen in the contemporary post-land claims settlement Arctic and Subarctic regions) operate. It appears well suited to the Western Arctic with room to incorporate not only the states that define the states-system that most IR theorists embrace but the many non-state entities (including stateless, sub-state, and trans-state indigenous peoples), which in many cases have been displaced or engulfed by state expansion, and yet who not only endure but continually push back against the very states which caused their initial displacement.

This is particularly evident in the Western Arctic region, and more generally across the Arctic, where states, when expanding into the North, asserted a lighter sovereign footprint than in other regions of colonization, and would over time increasingly turn to the indigenous peoples of the Arctic region for sovereign recognition through the mutuality of land claims negotiations, in which state and tribe recognized one another (with the state gaining much, in terms of legitimating its own claims to Arctic territories it barely occupied, from the process). It may well be indigenous recognition of the Arctic states' aspirations for sovereign recognition that has fueled the continuity of the Arctic land claims movement, bringing it from the very first Arctic land claims in 1971 (with the passage of the Alaska Native Claims Settlement Act) to the completion of the final Inuit land claim in North America in 2005 (with the Nunatsiavut Land Claim Agreement in Labrador).

In these transnational borderland regions, cultures, languages, identity, geography, and jurisdictional authority blend across borders, often echoes of salient structures from an earlier era that continue to informally shape the contemporary order, and which in many cases had established an underlying order that was later absorbed, often intact, by modern states during periods of state expansion. These underlying echoes of an earlier order can help bind borderlands together into distinct regions with an international presence—leveraging regional centripetal forces to offset the state's centrifugal forces. Instead of an anarchical realm of competing states, borderlands can assert an enduring regional order based on cross-border collaboration that precedes the formation of the very borders that now define them. By the time states turned to the Arctic region for their sovereign expansion, they did so less muscularly than they did in other parts of the world—guided by a globalizing vision of manifest destiny as illustrated by Secretary of State Seward's September 14, 1853, "Destiny of America" speech (in which he foretold of an America whose borders "shall greet the sun when he touches the Tropic, and when he sends his glancing rays towards the Polar circle") (Seward 1853) and a maturing (and increasingly inclusive) recognition of indigenous rights—which has enabled states to expand while integrating northern indigenous peoples largely intact into their evolving constitutional structures.

The northern tranquility long observed to define the Arctic region as a whole, and which is particularly salient in the Western Arctic, owes much then to the domestic reconciliation of tribe and state that has taken place during this historic and exemplary northward state expansion—dialectically catapulting the

Fig. 5.4 Justice Thomas R. Berger's 1977 report on the Mackenzie Valley Pipeline Inquiry (known throughout the North as the "Berger Inquiry") was appropriately titled "Northern Frontier, Northern Homeland," capturing the metaphorical tension between the "two Arctics" perfectly. Sustaining a synthesis of these two is the key to a stable, peaceful Arctic region. Source: Berger, Thomas R. *Northern Frontier, Northern Homeland: Report of the Mackenzie Valley Pipeline Inquiry, Berger Commission Report.* Ottawa: Minister of Supply and Services, 1977, cover

region beyond its earlier dichotomous categorization as either a northern "frontier" or "homeland," so that in synthesis it has evolved to become both frontier and homeland, and in this symbiotic fusion to emerge as something altogether new—a distinctive, globally integrated borderland unto itself (Fig. 5.4).

Acknowledgments The author would like to gratefully thank the Kone Foundation of Helsinki, Finland for generously funding my research on cross-border indigenous homelands and indigenous borderlands in 2016 and 2017, making research for this paper possible. The views expressed in this article are the author's own and do not necessarily represent the views or policies of the United States Coast Guard Academy, the United States Coast Guard, or any other branch or service of the United States Government.

REFERENCES

Berger, Thomas R. 1985. *Village Journey: The Report of the Alaska Native Review Commission.* New York: Hill and Wang.

Berry, Mary Clay. 1975. *The Alaska Pipeline: The Politics of Oil and Native Land Claims.* Bloomington: University of Indiana Press.

Beverly and Qamanirjuaq Caribou Management Board. 2006a. NTI Think They Know the Answers. *Arctic-Caribou.com.* http://www.arctic-caribou.com/news_july06.html.

———. 2006b. Pro-Uranium Position Elicits Grave Concerns. *Arctic-Caribou.com.* http://www.arctic-caribou.com/news_july06.html.

Bockstoce, John R. 2009. *Furs and Frontiers in the Far North: The Contest Among Native and Foreign Nations for the Bering Strait Fur Trade.* New Haven: Yale University Press.

Chandonnet, Fern. 1995. *Alaska at War, 1941–1945: The Forgotten War Remembered.* Anchorage: Alaska at War Committee.

Farrow, Lee A. 2016. *Seward's Folly: A New Look at the Alaska Purchase.* Anchorage: University of Alaska Press.

Government of Nunavut. 2007. *Balanced Approach for Uranium Mining in Nunavut.* Press Release, Government of Nunavut, Iqaluit, Nunavut, June 4. http://www.gov.nu.ca/Nunavut/English/news/2007/june/june4.pdf.

Krasner, Stephen D., ed. 1983. *International Regimes.* Ithaca: Cornell University Press.

Marston, Marvin "Muktuk". 1969. *Men of the Tundra: Eskimos at War.* New York: October House.

Minogue, Sara. 2006. Uranium Rush Floods Western Nunavut; Huge Jump in Price, Demand Has Juniors Racing to Stake Claims. *Nunatsiaq News,* April 28. http://www.nunatsiaq.com/archives/60428/news/nunavut/60428_05.html.

Osherenko, Gail, and Oran R. Young. 1989. *The Age of the Arctic: Hot Conflicts and Cold Realities.* Cambridge: Cambridge University Press.

Rohmer, Richard H. 1973. *The Arctic Imperative: An Overview of the Energy Crisis.* Toronto: McLelland and Stewart.

Seward, William H. 1853. The Destiny of America: A Speech of William H. Seward at the Dedication of Capital University in Columbus, Ohio, on September 14. *World Digital Library.* https://www.wdl.org/en/item/16955/.

Thompson, William R. 1973. The Regional Subsystem: A Conceptual Explication and a Propositional Inventory. *International Studies Quarterly* 17 (1, March): 89–117.

Time-Life Books. 1993. *The Way of the Warrior.* New York: Time-Life Books.

Todorov, Tzvetan. 1999. *The Conquest of America: The Question of the Other.* Norman, OK: University of Oklahoma Press.

Utley, Robert M., and Wilcomb E. Washburn. 1977. *The American Heritage History of the Indian Wars.* New York: American Heritage Publishing.

Wagner, Gary William. 1996. *Implementing the Environmental Assessment Provisions of a Comprehensive Aboriginal Rights Settlement*. Master's Thesis. Department of Environment and Resources Studies, University of Waterloo.

Waltz, Kenneth N. 1959. *Man, the State, and War: A Theoretical Analysis*. New York: Columbia University Press.

York, Geoffrey, and Loreen Pindera. 1991. *People of the Pines: The Warriors and the Legacy of Oka*. Boston: Little, Brown.

Young, Oran R. 1977. *Resource Management at the International Level: The Case of the North Pacific*. New York: Frances Pinter Publishers Ltd.

———. 1982. *Resource Regimes: Natural Resources and Social Institutions*. Berkeley: University of California Press.

———. 1994. *International Governance: Protecting the Environment in a Stateless Society*. Ithaca: Cornell University Press.

———. 1998. *Creating Regimes: Arctic Accords and International Governance*. Ithaca: Cornell University Press.

Zellen, Barry Scott. 2008. *Breaking the Ice: From Land Claims to Tribal Sovereignty in the Arctic*. Lanham, MD: Lexington Books.

———. 2009a. *Arctic Doom, Arctic Boom: The Geopolitics of Arctic Climate Change*. Westport, CT: Praeger.

———. 2018. Chapter Three: Crossborder Indigenous Collaboration and the Western Arctic Borderland. In *Part Two: Governance and Policy, The Networked North: Borders and Borderlands in the Canadian Arctic Region*, ed. Heather Nicol and P. Whitney Lackenbauer. Waterloo: Borders in Globalization/Centre on Foreign Policy and Federalism.

CHAPTER 6

The Economy of the Arctic

Joan Nymand Larsen and Andrey N. Petrov

General Characteristics of the Arctic Economy

The economy of the Arctic is composed of a diverse range of local, regional, and national economies of different size and resource geographies. These economies exhibit broad variations in the type and availability of economic and employment opportunities, and in the range of issues and challenges they face. While heterogeneity describes these economies, they also have many features in common, some of which are structural in nature and may include obstacles that interfere with the region's current and future economic development. Heterogeneity means that while local economies are subject to similar signals and disturbances from external economies, they respond differentially to regional and global changes. Differences in capacity to respond are linked to a broad diversity in physical, human, social, and natural capital (Duhaime 2004; Huskey et al. 2014; Larsen and Huskey 2015).

The economies of the Arctic share a number of distinct features that set them apart from non-Arctic economies. At a very general level, the Arctic economy can be divided into three separate sectors:

First, the formal market-based economy based on large-scale resource extraction that accounts for a large share of regional income generated, and related 'other' industries that often provide smaller, yet important, contributions to overall income;

J. N. Larsen
University of Akureyri and Stefansson Arctic Institute, Akureyri, Iceland

A. N. Petrov (✉)
University of Northern Iowa, Cedar Falls, IA, USA
e-mail: andrey.petrov@uni.edu

© The Author(s) 2020
K. S. Coates, C. Holroyd (eds.), *The Palgrave Handbook of Arctic Policy and Politics*, https://doi.org/10.1007/978-3-030-20557-7_6

Second, the traditional (subsistence-) based economy, including small-scale
family production, which is commonplace in smaller and more remote settings and in predominantly Indigenous communities; and

Third, the transfer (public) sector, which includes provision for both direct
transfers to households, public sector jobs, and provision of services in general (Larsen and Huskey 2010, 2015).

The importance and relative size of the market, subsistence, and transfer
sectors vary throughout the Arctic. The role and presence of large-scale capital
and skill-intensive industrial resource production is a common feature of the
formal and market-based economy, whereas the traditional, subsistence-based
economy is based on traditional pursuits that include hunting, trapping, gathering, fishing, and crafts, increasingly as part of a mixed economy setting with
close connections to the local market economy (Duhaime and Caron 2008;
Glomsrød et al. 2017; Petrov 2016; Poppel 2006). Formal, market-based
economies of the Arctic can be described by the importance of their international connections and their high share of the production in primary, especially
extractive, industries. Local economies tend to be more mixed with market and
non-market activities having an important role to play in providing for the
livelihoods of local communities. A range of economic and employment activities describe the economies of the Arctic, including from small-scale production for local consumption to large-scale natural resource production activities
for international markets. Sources of income include wage employment, traditional pursuits, and transfer income from government.

The Arctic region faces significant challenges related to regional and local
economic development, industrial production, and large-scale resource extraction activities, some of which are structural and persistent. This includes
remoteness and lack of accessibility, the high cost of production in the North,
the limited availability of human and other resources for large-scale industrial
projects, and the related challenge of addressing the growing demand for
skilled labor; the consequences of environmental impacts from industrial development and the negative spillover effects of industrial activity for local and
indigenous communities, culture, and tradition; and climatic change effects,
such as thawing permafrost and retreating sea ice (Larsen et al. 2014; Andersen
2015; Leadbeater 2009; Nielsen 2013; Nuttall 2013; Prowse et al. 2009).
These and other socio-economic and environmental challenges exert their
mark on the economic livelihoods of people in the Arctic. They play important
and growing roles in outcomes and decisions regarding resource allocation,
resource use, ownership, and control. They have significant consequences for
the structure, conduct, and performance of economic sectors and their options
for achieving economic sustainability.

The performance of the Arctic economy is also tied to key internal and external drivers, including the type and nature of large-scale resource production, the
price, supply, and demand conditions of these resources, and spending decisions
by local, regional, and national governments. A large share of the Arctic's
resources—such as oil, gas, precious metals, timber, and fish—are sold in exter-

nal markets, with decisions concerning these resources often being made by governments and businesses far outside the local area (Heininen and Southcott 2010). Aside from the prominent international aspects defining much of the formal economy, Arctic resources are also produced or harvested for a local market. This includes non-market-based traditional (subsistence) activities involving fishing, harvesting, hunting, and gathering. These activities are important in smaller and more remote and predominantly indigenous communities, where they provide an important source of income and food for local inhabitants.

Public spending creates jobs directly in government, as well as in the construction and service industries. Government also provides income through transfer payments, like public assistance or age-related support programs. In addition, the public sector contributes to residents' real income by providing services below their real costs; subsidized health care or housing, for example, adds to the residents' economic well-being. In more recent years, efforts have been geared toward developing alternative sources of income through a more diversified business structure, involving tourism, small-scale manufacturing, and the knowledge and cultural economy (Huskey et al. 2014; Petrov 2016; Larsen 2004a, b). However, efforts to achieve a higher level of economic diversification and a more stable economic setting are often impeded by structural obstacles involving a narrow range of marketable natural resources, small size, scattered population, remoteness and lack of accessibility to markets, and high costs of transportation.

General economic trends provide useful insights into where the future of the Arctic economy is heading. First, the pattern of resource development continues to be determined in large part by the development of natural resources destined for international markets. Second, structural and persistent factors, such as remoteness, lack of accessibility, long distance to markets and long supply lines, harsh climate, and an often-inhospitable environment, means that production costs remain exceptionally high in the Arctic and are unlikely to fall relative to costs outside the region. This leads to the high-grading of new renewable resources and prevents the commercial development of less impressive deposits. Third, natural resource production continues to be a driving force of the Arctic economy even when some project that climatic change could have a cost reducing effect. In fact, expectations of higher prices and lower costs for Arctic resource development may be overly optimistic, as the experience with oil and gas over the past decade demonstrates. Fourth, institutional arrangements that define the relationship between resource development and local communities, and in turn determine the degree of local control and ownership, continue to evolve. Done effectively, these arrangements contribute to an increasing share of the economic benefits from resource development that remain local. And fifth, the share of non-resource-related economic activity will continue to increase, including tourism, arts and crafts, electronics, and other trades, thereby providing the foundation for more of the economic value created to remain local (Huskey et al. 2014; Petrov 2016; Grimsrud 2017; Müller and Brouder 2014).

The Arctic Resource Economy

The formal and industrial sector of the Arctic economy is based on a relatively narrow set of commercial resources, including renewable and non-renewable natural resources such as hydrocarbons, metallic minerals, precious metals, precious and semi-precious stones, fish, and timber, all of which have created significant wealth (Aslaksen et al. 2015; Duhaime 2004). However, given the heavy geographic and commodity concentration of exports destined for international markets, the trade in these resources also tend to import a fair amount of economic volatility and earnings instability for industry and local economies. This, in turn, creates its own set of challenges where some economic settings are better equipped to respond and cope than others (Larsen 2010).

Production for external markets in the Arctic is supported by modern, large-scale, capital-intensive facilities. While small-scale and scattered communities may provide an efficient economic setting for traditional activities, the production of resources for export tends to be done at large-scale, concentrated production sites. Resource development, from exploration to extraction, is capital intensive and frequently requires the securing of import from outside the region. The scale, capital needs, and skills required when producing for an export market means that many of the human, technical, and financial resources used in production are brought in from outside the region or delivered to resource production sites by large international companies involved in the various stages of the resource industry (Duhaime 2004; Huskey et al. 2014; Duhaime et al. 2017). Thus, large-scale resource exploitation activities are frequently carried out to supply markets outside the Arctic using labor and capital inputs from outside the region. Therefore, primary resource production in the Arctic is associated with significant economic leakages, with potentially huge shares of income, profits, and resource rents accruing to owners of factors of production outside the region, with the result that a significant share of the wealth created through large-scale resource extraction often leaves the Arctic and, therefore, may not benefit the residents of the region. A reversal of these trends requires attention to questions related to the challenges of achieving a more diversified economy combined with changes to existing power relationships and governance structures via innovations in institutional arrangements (Duhaime 2004; Bjørst 2017; Fidler 2009; Halseth and Sullivan 2004).

With the major driver of the formal economy being the primary sector and with the international markets far from the Arctic, the prevailing economic trend is of the resources exploited in the Arctic leaving the Arctic. This contributes to a common phenomenon describing much of the Arctic, namely, that much of what is produced in the formal sectors of the Arctic economy is exported and a big share of what is consumed is imported. This has the effect of limiting the sector linkages, restricting opportunities for value-added developments, and dampening the local economic multiplier effects. Hence, while large-scale resource projects can be an important engine of growth in the Arctic, including contributing to economic diversification, transport systems, and regional infrastructure, there are also many potential drawbacks such as limited benefits remaining locally, pressures on facilities and infrastructure, and social problems associated with the influx of labor (Saxinger et al. 2016; Rodon and Levesque 2018; Larsen 2010).

The strength and importance of the Arctic's global connections resulting from the ever-growing force of globalization and the expanding economic integration across market and non-market economies have meant direct transmittal of global market volatility to the Arctic and the region's narrow resource-based local and regional economies. Impacts on employment opportunities, the distribution of income and wealth, the allocation of resources, and local community livelihoods related to these connections can be significant and they are amplified when Arctic local communities are situated in more resource-strategic positions (Larsen 2010). The economic future of the Arctic is increasingly tied to the direction of economic and global change processes and the ability to mitigate or successfully cope with the negative impacts of resource supply shocks, shifts in world prices for commercial resources, and the economic volatility associated with limited economic diversification and small regional multiplier effects (Larsen 2004a, b; Loeffler 2015; Petrov 2010).

The Arctic resource economy is organized in a variety of ways, including work located away from local communities, typically in fly-in/fly-out settings. These provide a way for local economies that supply the labor for resource projects to deal with challenges associated with economic volatility and external shocks of resource development (Storey 2010; Eilmsteiner-Saxinger 2011). The scale of activity and proximity to existing local communities are among the most important factors influencing the effects of fly-in/fly-out business on regions and the local communities providing the labor pool. While these settings may provide communities with opportunities for employment diversification and the expansion of the economic base, they may also prove destructive or harmful when they result in infrastructure and service demands that the communities cannot meet (Storey and Hall 2017).

A common tendency for limited economic net benefits accruing to regional economies from resource production has been explained by the familiar paradox of the 'resource curse' (also referred to as the 'Dutch disease'.) The basic theoretical elements of the resource curse refers to the case where economies with an abundance of natural resources tend to have less economic growth and development than those without resources. A resource boom becomes a 'curse', resulting in a falling economy. A boom leads to an expanding support sector that cannot be supported when the boom comes to an end, a situation only made worse when the downswing of an economy leaves excess capacity and debt (Huskey 2017). The loss of a region's comparative advantage is another element of the curse. Local labor is drawn into the resource industry away from their customary or historic activities, with nothing to return to when production comes to a close. Other aspects of the curse include the impacts associated with situations where policymakers and others forget to save during the upswing and boom phase of a resource development but instead focus their energy and resources on rent-seeking and wasteful spending. And finally, there is the problem associated with the 2q1 local costs, such as environmental damage or overutilization of local infrastructure, while there may be limited or no economic benefits to the region when unbalanced power relations are present between resource firms and the regional or local population (Huskey and Southcot 2016; Parlee 2015).

With the many challenges confronting Arctic economies, including limits on resource mobility and flexibility, constraints on finding and entering new foreign markets, and difficulties associated with a small and scattered population base that contributes to existing barriers to achieving economies of scale in domestic and local markets, a main concern for the Arctic is that of overcoming barriers to realizing long-run sustained growth while at the same time working on finding solutions to help minimize and eliminate the negative environmental and socio-economic impacts. There is a need to find and act on opportunities for sustainable economic development. With growing and tightening global connections and increased competition in key commodity and export markets, this will require a higher degree of mobility and flexibility in all aspects of resource development, including technological innovations and heightened efforts to address the persistent human and fiscal capacity challenges to be able to promote shifts into new export lines and favorable transitions into production for the domestic market, as well as the capacity to enter into new foreign markets.

BENEFIT SHARING: NEW COMMUNITY DEVELOPMENT OPPORTUNITIES IN THE ARCTIC EXTRACTIVE SECTOR

In recent years, a key question associated with the development of extractive industries in the Arctic relates to benefit sharing (Tysiachniouk and Petrov 2018). Benefit sharing is the distribution of benefits from extractive activity to different parties involved in or directly affected by this activity. Although extractive activities deliver economic gains (Loeffler 2015; Rodon and Levesque 2018), they often rest with the companies and other non-local actors, including national governments. The lack of localized benefits is generally explained by the dependence of local economies on external markets and actors, the effect of economic decoupling (where the resource sector is relatively detached from the local economy), 'resource curse' (with development work outpricing other forms of activity from labor market), uneven power relationships (local communities-state-companies), and lacking creative capital and competing technological paradigms (Bourne 2000; Huskey 2006; Petrov 2012). Consequently, benefit sharing presents a valuable way to ensure that more economic value stays in the Arctic. Most importantly, managing the benefit-sharing process and/or its co-management among companies, communities and the state becomes an important public policy issue in the Arctic.

The process of benefit sharing may be direct, in the form of revenues that are provided to individuals, or indirect, involving revenues negotiated and redistributed by intermediaries. Typical benefit-sharing options include employment, procurement, new infrastructure, community investments, government payments, and compensation for local disruptions (Wall and Pelon 2011). Community investments are voluntary contributions by companies beyond the scope of their normal business operations, while compensations are payments made by companies to mitigate the impacts of resource activity. Government payments may include taxes and royalties as well as other payment schemes intended for redistribution to communities (World Bank 2010, p. 3).

There are a number of dominant 'modes' of benefit sharing in the Arctic (Tysiachniouk 2016). The paternalistic mode is where the state is dominant and monitors and intervenes in the distribution of the benefits (Tysiachniouk et al. 2018). The company can add to or substitute for the state's efforts to provide support to local communities and Indigenous peoples who do not control the delivery of benefits. The corporate-oriented social responsibility (CSR) mode is pursued by companies to fulfill their CSR mandates through various arrangements with local communities, in which the company plays a defining role. The partnership mode, a form of tripartite partnership among the oil companies, government, and Indigenous communities, is a form unique to the Russian Federation (Wilson 2016). The beneficiary mode, common in Canada, is where companies work with community organizations and regional non-profit corporations and often conclude Impact and Benefit Agreements (Prno 2013). Benefits are managed and distributed by non-profit Indigenous corporations among eligible beneficiaries (i.e., community members). Finally, the shareholder mode, found in Alaska, emerges when Indigenous corporations with community members as shareholders deal directly with extractive companies and often themselves participating in extraction to gain direct profits. It is also important to differentiate the three core 'premises' of benefit sharing that shape the perspective from which the extractive industry stakeholders see the 'fundamental right' to give or receive the benefits: compensation (paying communities for damage or lost profit), investment (investing in community development), and charity ('giveaways' made at a giver's discretion).

The diversity of benefit-sharing arrangements in the Arctic generates variable experiences in local communities, with some benefiting more than others. Each of the modes has its advantages and problems, although the equity and voice of the local and Indigenous people are stronger in the partnership, beneficiary, and shareholder modes. Investments also tend to be more goal-oriented, long-term, and focused than are compensation or charity payments.

THE ARCTIC'S EMERGING POST-INDUSTRIAL ECONOMIES

Although the Arctic economy is still defined largely by the three 'pillars'—resource, public, and traditional sectors—there is a growing role of the Arctic's 'other' economies prompted by the shift to tertiary and quaternary sectors.

When compared to the staple sector, especially mining, some of the new industries have been growing faster and have demonstrated higher productivity (Glomsrød et al. 2017). These industries are prevalent in growing Arctic urban centers (Megatrends 2011). These post-industrial sectors are more advantageous for building a more sustainable economy in the Arctic.

'Other' economies (Petrov 2016) include a broad range of economic activities outside of resource-extractive, transfer, or subsistence sectors, although they may be connected to them. 'Other' economies tend to be more endogenous and post-industrial. As a result, they may have stronger internal linkages and multipliers and generate more local development. These economies are not solely local (Huskey et al. 2014), but are often a part of the international economy. Examples of 'other economies' include sectors such as knowledge-based industries, arts and crafts, small-case custom manufacturing, professional and technical services, tourism/recreation, and local retail trade.

There is a growing evidence that some Arctic communities are able to develop a diversified economy by engaging human capital (Beyers and Lindahl 2001; Boschma 2005; Gradus and Lithwick 1996; Selada et al. 2011). Local investment in human capital is an element in stimulating 'other' economic activities and diversifying local economies (Megatrends 2011; Petrov 2007, 2014). However, in order for this investment to work, there has to be a connection to localized knowledge and social capital that can be formed through institution-building and engagement with civic society (Aarsæther 2004).

A development-based, post-industrial economy is an emerging part of a larger sustainable development strategy, especially important for Arctic cities and towns (Pelyasov 2009). Bringing and sustaining new industries provides additional opportunities for northern urban communities to achieve diversification of their economic base, to break away from the boom-bust cycles, to reduce dependence on external economic and political actors, and ultimately to improve quality of life. The best cities for such economic diversification are administrative centers and urban economic hubs. Not all regions can develop strong 'other' economies, but it certainly may be an important ingredient necessary for achieving sustainable development in northern urban communities.

One of the examples of the post-industrial option is the *knowledge economy* (Petrov 2014) that is based on advanced skills in technology, culture, leadership, or entrepreneurship. There are 'hot-spot'-like concentrations of the labor force with occupations in applied and natural sciences, computer science, and engineering in the Arctic, notably in Alaska, Yakutia, and northern Scandinavia. People with knowledge economy occupations tend to locate in urban and more industrial areas. The Tech Pole Index (TPI) that looks at employment in high-tech industries (Florida 2002) estimates the volume of knowledge-based economic activity in the Arctic regions relative to the country's base. TPI is larger in the Northwest Territories, Yukon, and selected regions of Alaska. The index is much lower in northern Eurasia. Oil and gas-rich regions of the Russian Arctic have small numbers of high-tech employees. Most engineers and technology workers are employed in the extractive industry.

Innovation activity is another indicator of strengthening new economies. Patents are routinely utilized as a proxy to characterize the knowledge economy (Acs and Audretsch 1989; Feldman 2000). Patents are registered instances of innovation. The volume of patents registered to inventors in particular regions may be considered as closely related to the knowledge economy output in the area. The analysis of patents issued to Alaskans since 1976 (Zbeed and Petrov 2017) demonstrates that large concentrations of patents were in urban Alaska (Anchorage and Fairbanks). At the same time, many smaller areas had innovation activity, although many of them were highly specialized. In the non-metropolitan areas of the state, patented inventions more frequently resulted from the work of a few individuals. Most were also confined to the one or two main industries for which technologies were produced. Alaska inventors patented technologies or products pertaining to freezing, snow removal, winter sports, oil spill, outdoor activities, etc. Most patents registered to Alaska residents, especially in engineering and electronics, were prepared in cooperation with co-inventors from other states showing that Alaska inventors were involved in the external innovation networks. In sum, Alaska's knowledge economy gravitates to urban centers, demonstrates clustering of inventions and human capital, limited variety of produced knowledge, and strong differences.

The cultural economy is an important part of an emerging sector in the Arctic. Elements of traditional knowledge, such as arts and crafts, are not only important components of Indigenous culture but are also commodities that can bring economic profit. This economic sector is known as 'cultural economy'. Fragmentary evidence suggests (Hill Strategies 2010), however, that cultural economic activities are highly complementary to the traditional economy. It provides part-time or full-time work for thousands of northerners and brings millions of dollars. In Nunavut alone, it adds $30 million to regional GDP (Nordicity Group 2014). Most of the purchases are made by tourists and collectors of northern art. The main challenges for local artists in the Arctic are low and unstable income, difficulty in accessing external markets, lack of entrepreneurial skills and opportunities, and uncertain demand for their products.

THE ARCTIC SUBSISTENCE AND MIXED ECONOMY

The local economy of the Arctic is a mixed economy where market and non-market activities play important roles in supporting community livelihoods and contribute to the material well-being of local residents. Wage employment, traditional pursuits, and transfer income from government all provide important sources of income, with the relative size and importance of markets, non-market, and the transfer sector varying throughout the region. What emerges is a picture of a heterogeneous region affected by similar signals from regions beyond. Strategies for community economic development are many, and some include forming partnerships with outside actors in developing natural resources but also increasingly non-resource-related activities, combining subsistence activities with government or private sector employment and welfare in a variety of mixed economy settings, negotiating with governments for policies on regional development to create jobs and demand for local production, and using business and political networks within the context of growing regional and global connections to ensure access to international markets (Aarsæther et al. 2004; Tysiachniouk and Petrov 2018; Rasmussen et al. 2014).

The non-market part of the Arctic economy is described by subsistence (local production for local consumption) in the form of customary harvesting, hunting, fishing, and gathering, and this continues to play an important role in many parts of the region. The livelihoods of a significant number of indigenous people—including also some non-indigenous residents—continue to depend in many ways on harvesting and the use of living terrestrial, marine, and freshwater resources (Poppel 2006; Rasmussen et al. 2014; Myers 2000). Many of these resources are used as food, clothing, crafts, and other products, and make important contributions to the mixed cash economy of local communities and households. Local communities and indigenous people often mix formal sector activities (e.g., commercial fish harvesting, jobs in mining and other resource extraction, and eco-tourism ventures) with traditional or subsistence activities, which includes harvesting a variety of natural renewable resources to provide for human consumption and community livelihoods (Myers and Forrest 2000; Poppel 2006; Aarsæther et al. 2004). Productivity and economic outcomes in subsistence-based household production are closely linked to formal sector labor market participation as it requires substantial monetary investments to help purchase and maintain hunting and trapping equipment. The size and importance of these investments may continue to rise with the strengthening integration of market and non-market production, and as climate and environmental change may necessitate further travel to reach hunting grounds, thus, increasing the demand for modern harvesting and transport equipment.

Communities without strong market connections provide opportunities to combine subsistence activities and incomes from the public or corporate economies. Mixed income sources are often required to reach an acceptable level of living and to provide flexibility in household income generation. It is frequently necessitated by a number of critical factors including the small size of the local market economy; limited access to full-time, permanent, and well-paying modern-sector jobs; the high costs of doing business in the Arctic; and limited accessibility to markets and resources in general. This also helps explain why transfer income becomes an important source of household income for many northerners. The mix of market and non-market production and the close ties to outside markets define the local economy. The local mix of human, social, physical, and natural capital determines the outcomes of global change impacts (Larsen 2010; Hamilton et al. 2003).

While the contribution to income made by large-scale resource development in the Arctic is substantial and therefore also means that much control and influence is concentrated within that sector, the non-profit cooperative experience has been an important force in many local communities (Southcott 2010; Southcott et al. 2010). It provides an alternative economic solution that works well in mixed economy settings and can work alongside other economic models to help provide for the material well-being of local residents. Cooperatives have displayed entrepreneurial capacities on both local and regional levels, contributing to the financial, human, and social capital (MacPherson 2013). They have also supported the empowerment of Indigenous peoples, as illustrated in the case of Alaska where the non-profit sector has supported innovation and filled a critical gap between government services and community needs (McMillian et al. 2015). Similarly, Canadian non-profits and cooperative organizations have been a source of empowerment in the northern social economy, and have been important in their contributions to the development of social and human capital (Southcott and Walker 2009).

Local communities encouraged adaptation to new economic realities via the development of the mixed household economy. Mixed economies have components of both market exchange, subsistence activities, culture, and tradition (Burnsilver et al. 2016; Holen et al. 2015). Evidence suggests that the transition to a mixed economy has contributed to strengthening the viability of these local economic models. As early as the 1990s, Myers (1996) argued that despite predictions of the demise of the traditional economy, this sector of the northern economy has persisted for economic and cultural reasons. The mixed economy appears to be a fixture and is not as transitional as earlier commentators expected. As the mixed economy requires cash income to help support harvesting activities, more needs to be done to solve the lack of employment opportunities in the Arctic. Addressing the challenges of the mixed economy is an important step in finding workable solutions to a less volatile and more sustainable future for local economies.

Conclusion: Sustainable Economies and Policy Options in the Arctic

Socio-economic challenges related to global change can be expected to play a growing role in decisions on resource allocation, resource use, ownership, and control, as well as on emerging economic sectors, such as the post-industrial economy. There are important consequences for the Arctic economies and the region's economic sustainability. Sustainable development in the Arctic aims at improving the well-being, health, and security of Arctic communities and residents while preserving ecosystems' functions and resources (Petrov et al. 2017). Therefore, strategies for sustainable development and Arctic environmental protection need to consider the economic, social, and environmental linkages within the Arctic and between the Arctic and other regions of the globe. The future of the Arctic is connected to the growing role and economic dominance of global and transboundary connections, as well as the increasing presence and conduct of multinational corporations (Larsen 2010). At the same time, the sustainability of this economy in many ways depends on local entrepreneurship, creativity, social economy, development of the subsistence sector, and inclusion of the voice of stakeholders that are often unrecognized (Petrov 2016; Southcott 2010; Kruse 2010).

For many northern communities, the economic linkages between market sectors are limited. The production and use of resources may be less flexible and adaptable in northern local communities. There are many constraints related to the ability of resource product-mix to adapt to the realities of persistent economic volatility and the occasional resource supply shock effects. Many regional and local communities face a disparity between the structure of demand for goods and services and the use of natural resources. This increases dependence on external markets and limits the opportunities for regional multiplier effects and the creation of greater economic benefits.

Challenges to the viability and sustainability of Arctic economies are many and include remoteness, a narrow economic base, decisions made at a distance, and environmental change to list a few. Remoteness increases the cost of northern production because of the challenges of accessibility and the difficulties it presents when attempts are made to replace resource production with alternative economic opportunities. Also, a narrow resource base may result from high cost and raise the level of instability (Larsen 2004a, b). Furthermore, external decision making regarding production for external markets made by actors outside the region complicates the pursuit of regional and local development as they may reflect external rather than local conditions and visions. Climate change threatens the traditional economy and it is projected to have substantial future impacts on renewable resources, including fisheries, stock of marine mammals, terrestrial ecosystems, and agriculture. At the same time, it may reduce the opportunity to engage in traditional activities important to the identity and way of life of northern residents.

Economies of the Arctic are seeking solutions to challenges associated with the growing force of globalization and environmental changes. Innovative ideas are many. Among common options are finding and implementing measures to maximize local benefits and minimize negative impacts and costs. It involves spreading the burden of the costs among different economic sectors to raise the economic viability of individual sectors. Regional economies can substitute new activities that have fewer costs and are more sustainable. They can respond to changed economic incentives by reallocating the available natural, financial, and human resources toward new and more sustainable activities, including the post-industrial economy (Larsen and Huskey 2015).

Although life in the Arctic is increasingly shaped or influenced by events, decisions, and activities happening elsewhere, Arctic regions and communities are taking active steps toward positioning themselves to tackle the challenge and embarking on fostering alternative, locally embedded economic activities. These activities and steps forward include arts and crafts, tourism, small-scale manufacturing, and North-specific technological innovation, as well as addressing the coupling of traditional, market-based, and public and transfer economies. It also includes seeking new institutional arrangements that leave more control in the hands of locals. The Arctic is developing economic ways forward that may change the balance of power in the Far North.

Acknowledgments The authors are grateful to *Mr. Petr Grin* for his valuable assistance with preparing this manuscript. This work was partially supported by NSF #1532655.

REFERENCES

Aarsæther, N., ed. 2004. *Innovations in the Nordic Periphery*. Stockholm: Nordregio.

Aarsæther, N., L. Riabova, and J.O. Bærenholdt. 2004. Community Viability. In *Arctic Human Development Report (AHDR)*, 139–154. Akureyri: Stefansson Arctic Institute.

Acs, Z.J., and D.B. Audretsch. 1989. Patents as a Measure of Innovative Activity. *Kyklos* 4 (1989): 171–180.

Andersen, T.M. 2015. *The Greenlandic Economy – Structure and Prospects*. Economics Working Papers, 2015–14, Department of Economics and Business Economics, Aarhus University.

Aslaksen, I., S. Glomsrod, and G. Duhaime. 2015. ECONOR-The Economy of the North 2015. *Statistics Greenland*.

Beyers, W., and D. Lindahl. 2001. Lone Eagles and High Flyers in Rural Producer Services. *Rural Development Perspectives* 11 (3): 2–10.

Bjørst, L.R. 2017. *Uranium: The Road to "Economic Self-Sustainability for Greenland"?* Changing Uranium-Positions in Greenlandic Politics.

Boschma, R.A. 2005. Social Capital and Regional Development: An Empirical Analysis of the Third Italy. In *Learning from Clusters: A Critical Assessment from an Economic-Geographical Perspective*, ed. R.A. Boschma and R.C. Kloosterman, 139–168. Dordrecht: Springer.

Bourne, L. S. (2000). Living on the edge: conditions of marginality in the Canadian urban system. In Developing Frontier Cities (pp. 77-97). Springer, Dordrecht.

BurnSilver, S., J. Magdanz, R. Stotts, M. Berman, and G. Kofinas. 2016. Are Mixed Economies Persistent or Transitional? Evidence Using Social Networks from Arctic Alaska. *American Anthropologist* 118 (1): 121–129.

Duhaime, G. 2004. Economic Systems. In *Arctic Human Development Report*, 69–84. Akureyri: Stefansson Arctic Institute.

Duhaime, G., and A. Caron. 2008. Economic and Social Conditions of Arctic Regions. In *The Economy of the North 2008*, ed. S. Glomsrød and I. Aslaksen. Oslo: Statistics Norway.

Duhaime, G., A. Caron, S. Lévesque, A. Lemelin, I. Mäenpää, O. Nigai, and V. Robichaud. 2017. Social and Economic Inequalities in the Circumpolar Arctic. In *The Economy of the North 2015*, ed. S. Solveig Glomsrød, G. Duhaime, and I. Aslaksen. Oslo: Statistics Norway.

Eilmsteiner-Saxinger, G. 2011. 'We Feed the Nation': Benefits and Challenges of Simultaneous Use of Resident and Long-Distance Commuting Labour in Russia's Northern Hydrocarbon Industry. *The Journal of Contemporary Issues in Business and Government* 17 (1): 53.

Feldman, M.P. 2000. Location and Innovation: The New Economic Geography of Innovation, Spillovers, and Agglomeration. *The Oxford Handbook of Economic Geography* 1: 373–394.

Fidler, C. 2009. Increasing the Sustainability of a Resource Development: Aboriginal Engagement and Negotiated Agreements. *Environment, Development and Sustainability* 12 (2): 233–244.

Florida, R. 2002. The Economic Geography of Talent. *Annals of the Association of American Geographers* 94 (2): 743–755.

Glomsrød, S., I. Mäenpää, L. Lindholt, L., Mc Donald, H., Wei, T., & Goldsmith, S. (2017). Arctic economies within the Arctic nations. In: The Economy of the North 2015, ed. S. Glomsrød, G. Duhaime, and J. Aslaksen, Statistics Norway, Oslo.

Gradus, Y., and H. Lithwick. 1996. *Frontiers in Regional Development*. Lanham, MD: Rowman & Littlefield.

Grimsrud, K. 2017. Tourism in the Arctic: Economic Impacts. In *The Economy of the North 2015*, ed. S. Glomsrød, G. Duhaime, and J. Aslaksen. Oslo: Statistics Norway.

Halseth, G., and L. Sullivan. 2004. From Kitimat to Tumbler Ridge: A Crucial Lesson Not Learned in Resource-Town Planning. *Western Geography* 13/14: 132–160.

Hamilton, C.L., B.C. Brown, and R.O. Rasmussen. 2003. West Greenland's Cod-to-Shrimp Transition: Local Dimensions of Climatic Change. *Arctic* 56 (3): 271–282.

Heininen, L., and C. Southcott, eds. 2010. *Globalization and the Circumpolar North*. Fairbanks: University of Alaska Press.

Hill Strategies Inc. 2010. Artists in Small and Rural Communities in Canada. *Statistical Insights on the Arts* 8(2). http://www.arts.on.ca/AssetFactory.aspx?did=5773.

Holen, D., D. Gerkey, E. Høydahl, D. Natcher, M. Reinhardt Nielsen, B. Poppel, P.I. Severeide, H.T. Snyder, M. Stapleton, E.I. Turi, and J. Aslakse. 2015. Interdependency of Subsistence and Market Economies in the Arctic. In *The Economy of the North 2015*, ed. S. Glomsrød, G. Duhaime, and J. Aslaksen. Oslo: Statistics Norway.

Huskey, L. (2006). Limits to growth: remote regions, remote institutions. The Annals of Regional Science, 40(1), 147-155.

———. 2017. An Arctic Development Strategy? The North Slope Inupiat and the Resource Curse. *Canadian Journal of Development Studies* 39 (1): 89–100.

Huskey, L., and C. Southcot. 2016. That's Where My Money Goes: Resource Production and Financial Flows in the Yukon Economy. *The Polar Journal* 6 (1): 11–29.

Huskey, L., I. Maenpaa, and A. Pelyasov. 2014. Economic Systems. In *Arctic Human Development Report*, ed. J. Nymand Larsen and G. Fondahl. Copenhagen: Nordic Council of Ministers.

Kruse, J. 2010. Sustainability from a Local Point of View: Alaska's North Slope and Oil Development. In *Political Economy of Northern Regional Development*, ed. Gorm Winther, 55–72.

Larsen, J.N. 2004a. External Dependency in Greenland: Implications for Growth and Instability. In *Northern Veche, Proceedings of the Second Northern Research Forum, Veliky Novgorod, Russia*, September, 19–22, 2002, ed. Jón Haukur Ingimundarson and Andrei Golovnov. Akureyri: Stefansson Arctic Institute.

———. 2004b. Trade Dependency and Export-Led Growth in an Arctic Economy: Greenland, 1955–1998. In *Native Voices in Research, 2004*, ed. Jill Oakes. Aboriginal Issues Press, University of Manitoba.

———. (2010). Economies and Business in the Arctic Region. In Loukacheva, N. (ed) Polar Law Textbook. TemaNord. Nordic Council of Ministers, Copenhagen.

Larsen, J.N., and L. Huskey. 2010. Material Well-Being in the Arctic. In *Arctic Social Indicators*, ed. Joan Nymand Larsen, Peter Schweitzer, and Gail Fondahl. Copenhagen: Nordic Council of Ministers.

———. 2015. The Arctic Economy in a Global Context. In *The New Arctic*, ed. Birgitta Evengard, Joan Nymand Larsen, and Øyvind Paasche. London: Springer.

Larsen, J.N., O.A. Anisimov, A. Constable, A.B. Hollowed, N. Maynard, P. Prestrud, T.D. Prowse, and J.M.R. Stone. 2014. Polar Regions. In *Climate Change 2014: Impacts, Adaptation, and Vulnerability. Part B: Regional Aspects. Contribution of Working Group II to the Fifth Assessment Report of the Intergovernmental Panel on Climate Change*, 1567–1612. Cambridge, New York: Cambridge University Press.

Leadbeater, D. 2009. Single-Industry Resource Communities, 'Shrinking,' and the New Crisis of Hinterland Economic Development. In *The Future of Shrinking Cities: Problems, Patterns and Strategies of Urban Transformation in a Global Context*, ed. Karina Pallagst et al. Berkeley: Institute of Urban and Regional Planning, University of California Berkeley.

Loeffler, B. 2015. Mining and Sustainable Communities: A Case Study of the Red Dog Mine. *Economic Development Journal* 14 (2): 23–31.

MacPherson, I. 2013. *Cooperatives' Concern for the Community: From Members Towards Local Communities' Interests*, January 3. Euricse Working Paper No. 46/13.

McMillian, D., L. Wolf, and A. Cutting 2015. Alaska's Nonprofit Sector. *Economic Development Journal* 14 (2).

Megatrends. 2011. ed. R.O. Rasmussen. Nordic Council of Ministers, Copenhagen.

Müller, D.K., and P. Brouder. 2014. Dynamic Development or Destined to Decline? The Case of Arctic Tourism Businesses and Local Labour Markets in Jokkmokk, Sweden. In *Tourism Destination Development: Turns and Tactics*, ed. A. Viken and B. Granås, 227–244.

Myers, H. 1996. Neither Boom Nor Bust: The Renewable Resource Economy May Be the Best Long-Term Hope for Northern Economies. *Alternatives Journal*. 22 (4): 18–23.

———. 2000. Options for Appropriate Development in Nunavut Communities. *Etudes/Inuit/Studies* 24 (1): 25–40.

Myers, H., and S. Forrest. 2000. Making Change: Economic Development in Pond Inlet, 1987–1997. *Arctic* 53 (2): 134–145.

Nielsen, S.B. 2013. *Exploitation of Natural Resources and the Public Sector in Greenland*. Background Paper for the Committee for Greenlandic Mineral Resources to the Benefit of Society, Copenhagen.

Nordicity Group. 2014. *Needs Assessment: Arts Administration Skills and Resources in Nunavut's Arts and Culture Sector*. Report for the Canadian Center for the Arts. Nordicity Group. http://canadacouncil.ca/council/research/find-research/2014/needs-assessment.

Nuttall, M. 2013. Zero-Tolerance, Uranium and Greenland's Mining Future. *The Polar Journal* 3 (2): 368–383.

Parlee, B.L. 2015. Avoiding the Resource Curse: Indigenous Communities and Canada's Oil Sands. *World Development* 74: 425–436.

Pelyasov, A.N. 2009. *And the Last Become the First: Russian Periphery on the Way to Knowledge Economy* (И последние станут первыми: Северная периферия на пути к экономике знания). URSS. [In Russian].

Petrov, A. 2007. A Look Beyond Metropolis: Exploring Creative Class in the Canadian Periphery. *Canadian Journal of Regional Science* 30 (3): 359–386.

———. 2010. Post-Staple Bust: Modeling Economic Effects of Mine Closures and Post-Mine Demographic Shifts in an Arctic Economy (Yukon). *Polar Geography* 33 (1–2): 39–61.

Petrov, A.N. 2012. Redrawing the Margin: Re-Examining Regional Multichotomies and Conditions of Marginality in Canada, Russia and Their Northern Frontiers. *Regional Studies* 46 (1): 59–81.

Petrov, A. 2014. Creative Arctic: Towards Measuring Arctic's Creative Capital. In *Arctic Yearbook*, ed. L. Heininen, 149–166. Akureyri: Northern Research Forum.

———. 2016. Exploring the Arctic's Other Economies; Knowledge, Creativity and the New Frontier. *The Polar Journal* 6 (1): 30–50.

Petrov, A.N., S. BurnSilver, F.S. Chapin III, G. Fondahl, J.K. Graybill, K. Keil, et al. 2017. *Arctic Sustainability Research: Past, Present and Future*. London: Routledge.

Poppel, B. 2006. Interdependency of Subsistence and Market Economies in the Arctic. In *The Economy of the North, Chapter 5*, ed. Solveig Glomsrød and Julie Aslaksen. Oslo: Statistics Norway.

Prno, J. An analysis of factors leading to the establishment of a social licence to operate in the mining industry. Resour. Policy 2013, 38, 577–590.

Prowse, T., C. Furgal, R. Chouinard, H. Melling, D. Milburn, and S.L. Smith. 2009. Implications of Climate Change for Economic Development in Northern Canada: Energy, Resource, and Transportation Sectors. *Ambio* 38 (5): 272–282.

Rasmussen, R.O., G.K. Hovelsrud, and S. Gearheard. 2014. Community Viability and Adaptation. In *Arctic Human Development Report: Regional Processes and Global Linkages (AHDR-II), Chapter 11*, ed. Joan Nymand Larsen and Gail Fondahl. Copenhagen: Nordic Council of Ministers.

Rodon, T., and F. Levesque. 2018. From Narrative to Evidence: Socio-Economic Impacts of Mining in Northern Canada. In *Resources and Sustainable Development in the Arctic*, 98–116. London: Routledge.

Saxinger, G., A. Petrov, N. Krasnoshtanova, V. Kuklina, and D.A. Carson. 2016. *Boom Back or Blow Back? Growth Strategies in Mono-Industrial Resource Towns–'East' and 'West'*. Settlements at the Edge: Remote Human Settlements in Developed Nations.

Selada, C., I. Cunha, and E. Tomaz. 2011. Creative-Based Strategies in Small Cities: A Case-Study Approach. *REDIGE* 2 (2): 79–111.

Southcott, S. 2010. The Social Economy and Economic Development in the Canadian North: Constraints and Opportunities. In *Political Economy of Northern Regional Development*, ed. Gorm Winther, 55–72.

Southcott, C., and V. Walker. 2009. A Portrait of the Social Economy in Northern Canada. *The Northern Review* 30: 13–36.

Southcott, C., V. Walker, J. Wilman, C. Spavor, and K. MacKenzie. 2010. *The Social Economy and Nunavut: Measuring Community Adaptive and Transformative Capacity in the Arctic*.

Storey, K. 2010. Fly-in/Fly-out: Implications for Community Sustainability. *Sustainability* 2 (5): 1161–1181.

Storey, K., and H. Hall. 2017. Dependence at a Distance: Labour Mobility and the Evolution of the Single Industry Town. *The Canadian Geographer* 62 (2): 225–237.

Tysiachniouk, M., 2016. Benefit sharing arrangements in the Arctic: Promoting sustainability of indigenous communities in Areas of Resource Extraction. Arctic and International Relations Series, Fall 2016 (4): 18-21, https://jsis.washington.edu/arctic/research/arctic-and-international-relations-series/

Tysiachniouk, M.S., and A.N. Petrov. 2018. Benefit Sharing in the Arctic Energy Sector: Perspectives on Corporate Policies and Practices in Northern Russia and Alaska. *Energy Research & Social Science* 39: 29–34.

Tysiachniouk, M.; Petrov, A.; Kuklina, V.; Krasnoshtanova, N. Between Soviet Legacy and Corporate Social Responsibility: Emerging Benefit Sharing Frameworks in the Irkutsk Oil Region, Russia. Sustainability 2018, 10, 3334.

Wall, E. & Pelon, R. 2011. Sharing mining benefits in developing countries. Extractive industries for Development Series, 21, the World Bank, http://documents.worldbank.org/curated/en/359961468337254127/pdf/624980NWP0P1160ns00trusts0and0funds.pdf, (accessed 06.04.2017)

Wilson, E., 2016. What is the social license to operate? Local perceptions of oil and gas projects in Russia's Komi Republic and Sakhalin Island. The Extractive Industries and Society #3, 73-81

World Bank (2010) Mining Foundations, Trusts and Funds. A Sourcebook. The World Bank.

Zbeed, S., & Petrov, A. (2017). Inventing the New North: Patents & Knowledge Economy in Alaska. Arctic Yearbook 2017.

CHAPTER 7

Extractive Energy and Arctic Communities

Noor Johnson

INTRODUCTION

Development of oil and gas resources in proximity to Arctic communities has been happening since the nineteenth century (Avango et al. 2014). Large-scale projects were first developed in the 1960s in Russia (Komi Republic and Nenets Autonomous Okrug) and Alaska, where Prudhoe Bay, the largest field in the United States, is located (Hendersen and Loe 2014). Interest and investment in oil and gas in the Arctic has waxed and waned over the years, driven by global prices, domestic policy, and availability of oil and gas deposits in other regions. Given the long timeframe from exploration to development, oil and gas companies invest in speculative projects with the aim of generating additional investment through the promise of future return-on-investment for shareholders.

In the past decade, those with an eye on Arctic trends observed a cycle of interest, investment, and withdrawal. Interest increased prior to and following the release of a US Geological Survey report on offshore oil and gas potential (Dittmer et al. 2009); this was followed by a withdrawal of investment in 2015 when the global price of oil fell significantly. Offshore oil and gas exploration projects in Alaska, the Beaufort Sea region of Canada, Baffin Bay off of Canada and Greenland, and in the region of the Lofoten Islands of Norway led to expressions of concern from Arctic residents and concerted campaigns from Arctic communities and domestic and international environmental groups, such as Greenpeace's "Save the Arctic" initiative.

Futurists disagree about whether the 2015 withdrawal of oil and gas companies from planned projects signals an end to offshore interests or just a delay. Some predict that cheaper oil and gas reserves will continue to make the invest-

N. Johnson (✉)
National Snow and Ice Data Center, Cooperative Institute for Research in Environmental Sciences, University of Colorado, Boulder, CO, USA

© The Author(s) 2020
K. S. Coates, C. Holroyd (eds.), *The Palgrave Handbook of Arctic Policy and Politics*, https://doi.org/10.1007/978-3-030-20557-7_7

ment in Arctic sources prohibitive, and that meanwhile, renewable energy prices will continue to fall, taking the Arctic offshore out of the equation. Others argue that given the significant need for greenhouse gas mitigation to meet the globally agreed upon goal of limiting global warming to 2 degrees, Arctic deposits should be among those left in the ground. Others suggest that when the price of oil rises again, energy companies will once again turn to Arctic offshore deposits. Across the spectrum of possibilities, there is strong agreement that the hiatus of investment in the Arctic offshore offers an opportunity to establish frameworks and develop plans to mitigate risks posed by offshore development (Gulas et al. 2017; Osofsky et al. 2016; Sidortsov 2016).

While the Arctic has long been seen in the context of oil and gas development as a resource frontier, unlike many historical frontiers, the region is characterized by strong frameworks for international cooperation through the Arctic Council as well as bilateral agreements and other international policy instruments, and robust national legal and policy frameworks to guide development decision-making within Arctic states (Young 2005). In addition, Arctic policy discourse over the past decades has been characterized by increasing focus on ensuring that investments made in research and development benefit the region's residents, including and perhaps especially its Indigenous Peoples.

Depending on where you draw the boundaries of the Arctic, between 4 and 10 million people call the Arctic home; among them, 10 percent are Indigenous Peoples (Sidortsov 2016). Arctic residents, particularly those whose livelihoods are based on renewable resources, are arguably the most important stakeholders in oil and gas development projects because they are the ones whose landscapes and livelihoods stand to be disrupted by extractive infrastructure, demographic changes, and contaminants and hazards released through routine industry activity as well as oil spills. Often they are significantly affected by oil and gas exploration and infrastructure. In spite of this, they are sometimes overlooked in the discussions of "stakeholders" in oil and gas development.

Among Arctic residents, Indigenous Peoples have additional rights based on international agreements and domestic legislation. Indigenous Peoples have citizenship rights but in some cases also have additional constitutional or treaty rights, as well as rights based on international agreements, such as the UN Declaration on the Rights of Indigenous Peoples (Shadian 2017). These rights include the right to meaningful consultation on resource development projects, recognized as the right to "Free, Prior and Informed Consent" in the UNDRIP (Buxton and Wilson 2013). In addition, oil and gas companies have increasingly incorporated consultation and negotiation of impact and benefit agreements with Indigenous communities as part of corporate social responsibility.

This chapter offers a review of the current policy framework for extractive energy engagement with Arctic communities. I begin with a brief introduction to the various actors and policies at different policy scales that shape industry-community engagements, including the Arctic Council and its working groups,

Arctic states, non-Arctic states, NGOs, and industry. This is followed by a discussion of risks posed to Arctic residents and current strategies for risk mitigation. I then review impact and benefit agreements as reflecting the current state of practice in promoting benefits for Arctic residents from extractive development. In the conclusion, I discuss the challenge of climate change and the transition to renewable energy, with some reflections on how focusing on policy resources to support this transition may lead to greater sustainability and more opportunity for Arctic ecosystems and people.

Framing Community Experiences with Oil and Gas Development: The Role of International, State, Sub-National, and Civic Actors

The Arctic is a political space "inscribed with imaginaries of multiple interests and high stakes" (Dittmer et al. 2011, cited in Kristofersen and Langhelle). Recent debates about oil and gas development have illuminated some of these divergent perspectives and imaginaries. Here, I introduce some of the actors and policies at different scales of decision-making that shape the ways that oil and gas development projects impact communities as well as the possibilities for meaningful input into the process on the part of Arctic residents.

At the international level, organizing around Indigenous rights to self-determination has long focused on the challenges that extractive industries pose for Indigenous communities around the world. The United Nations Declaration on the Rights of Indigenous Peoples (UNDRIP) includes the right to Free, Prior and Informed Consent (FPIC), also recognized in the ILO Convention 169 (Wilson 2016). While only Norway and Denmark have ratified ILO Convention 169, all Arctic states have ratified UNDRIP. FPIC recognizes the rights of Indigenous Peoples to make free and informed choices about development projects on their lands, waters, and territories. FPIC aims to ensure that Indigenous Peoples are not coerced or pressured, that their consent is freely given prior to the start of a project, that they have full information about the scope and impacts of proposed activities, and that their choice to give or withhold consent is respected and upheld (Heinämäki 2015). FPIC has been unevenly implemented, however, with efforts by government and industry to weaken the meaning of "consent" to "consultation" (Heinämäki 2015). A key challenge around the implementation of FPIC is the question of who gives consent (and how to identify the right community members to speak for the collective), as well as whether or not "consent" includes the right to veto a project (Wilson 2016).

The main **regional forum** for negotiation of Arctic issues is the Arctic Council, which is comprised of eight member states, six "permanent participant" organizations representing Arctic Indigenous Peoples, and "observers" representing non-Arctic states and international organizations. The Arctic Council has two binding agreements that are relevant to oil and gas development. The first is a search and rescue agreement that delegates areas of responsibility among Arctic states in responding to emergencies (Arctic Council 2011); this agreement makes no particular mention of Arctic residents or Indigenous Peoples. The second, a marine oil pollution preparedness and response agreement, recognizes in its preamble threats to livelihoods of "local and indigenous communities" from marine oil pollution. The preamble also acknowledges that "indigenous peoples, local communities, local and regional governments, and individual Arctic residents can provide valuable resources and knowledge regarding the Arctic marine environment in support of oil pollution preparedness and response" (Arctic Council 2013). The Council's 2017 "Agreement on Enhancing International Arctic Scientific Cooperation" includes Article 9, "Traditional and local knowledge," which encourages the utilization of traditional and local knowledge in planning and conducting scientific activities in the context of "using the best available knowledge for decision-making" (Arctic Council 2017).

The Protection of the Arctic Marine Environment Working Group (PAME) has facilitated a number of initiatives relevant to oil and gas, including development of guidelines for offshore oil and gas drilling and exploration (1997, updated in 2002 and 2009) (Byers 2013), as well as additional guidelines for "systems safety management and safety culture" (PAME 2014). Of particular relevance to industry-community relations, PAME initiated a project in conjunction with the Social Development Working Group (SDWG) on "Meaningful Engagement of Indigenous Peoples in Marine Activities," which issued an initial report in 2015 but has not been completed due to a lack of dedicated funding. This effort was aimed at summarizing research and policies related to meaningful engagement but steered clear of issuing recommendations. The initiative struggled with lack of committed funding from Arctic Council states, suggesting that the topic did not rise to a high level of priority.

In 2011, the Inuit Circumpolar Council, an organization that represents 160,000 Inuit living in Alaska, Canada, Greenland, and Russia, issued the "Declaration on Resource Development in Inuit Nunaat" (ICC 2011) as a document to guide policy and decision-making; it emphasizes the rights of Inuit to be involved in all decisions that affect their lands and waters, and emphasizes the importance of sustainability and of decision-making that will allow hunting and harvesting activities, which are central to Inuit culture, to continue.

Arctic states play a particularly important role in shaping the policy framework that guides relations between industry and Arctic communities. States determine and oversee processes for licensing, set regulations and standards to promote industry safety, and create frameworks for environmental and social assessment processes. Arctic states are motivated by the need for resources to fuel economic growth as well as geopolitical concerns about energy access and national security. In the United States, Norway, and Russia, oil and gas contribute significantly to the national budget through taxes and revenues (Hansen et al. 2018). Ideally, state policy should seek to balance energy needs with an assessment of risks and impacts, including risks to residents and environmental policy goals. The frequency of lawsuits against states and energy companies brought by citizen groups and environmentalists, however, suggests that states often prioritize energy development over other interests.

While the development of oil and gas resources onshore lies clearly within the boundaries of domestic policy (although subject to international agreements on practices related to human rights, Indigenous Peoples' rights, and climate change policy), there are many international, regional, and binational mechanisms that contribute to governance of oil and gas decision-making in the offshore context.[1]

Sub-national governments (US states, Canadian territories, Greenlandic municipalities, Russian republics, and autonomous okrugs) play varied and, in some cases, significant roles in shaping oil and gas policy on their territories. The State of Alaska, for example, owns 28 percent of Alaska's land, including the North Slope oil fields where the Prudhoe Bay complex is located, and manages all licensing of oil and gas projects on those lands. Profits are invested in the Alaska Permanent Fund and used to support state spending, including public service spending, with a percentage of profits returned to state residents as dividends (Knapp 2012). Similarly, the Government of the Northwest Territories in Western Canada is responsible for onshore oil and gas policy, including exploration and development, benefits plans, and royalties. Offshore development in both Alaska (past 3 miles from shore) and Canada are overseen by federal government entities (in Canada, the National Energy Board, and in the United States, the Bureau of Ocean and Energy Management).

Native Corporations, formed as part of the Alaska Native Claims Settlement Act (ANSCA), manage oil and gas profits transferred from the state, manage royalties from projects on corporation-owned lands, and invest and manage other business ventures. Although these corporations create jobs and distribute benefits to shareholders, their creation has also led to increasing stratification due to the limited number of benefit-eligible shareholders (Tysiachniouk and Petrov 2018; Dombrowski 2007). In Canada, **co-management boards** established through regional land claim agreements oversee environmental impact review for all projects that may impact territories that are part of the land claim or on wildlife harvesting within the land claim region. **Land claim organizations** negotiate impact and benefit agreements on projects that will take place on lands held in trust on behalf of land claim beneficiaries.

While not a primary actor in oil and gas policy, **non-governmental organizations** (NGOs) have proven to be effective in lobbying for regulation, mobilizing widespread public awareness and protest of Arctic energy development, and working directly with Arctic residents and communities to organize resistance to extractive development and establish small-scale renewable energy demonstration projects. Greenpeace, for example, led a highly visible "Save the

---

[1] See Baker (2018) for a review of international legal frameworks for offshore oil and gas development.

Arctic" campaign against Arctic offshore drilling, targeting Shell in North America as well as industry activities in Greenland and Russia (Koivurova 2018).

Understanding and Mitigating Risks

People perceive risks through frames, which are simplified stories that we use to interpret or make sense of the phenomena (Callison 2014). Different frames represent strategic interests and positions of actors; for example, whether the Arctic is seen as a resource frontier or homeland (Avango et al. 2014). Koivurova (2018) recognizes four frames that represent different policy positions in relation to offshore oil and gas development: the "scramble for the Arctic" based on geopolitical conflict between states; Arctic oil and gas as a driver of domestic policy and energy security; Arctic oil and gas as contributing to climate catastrophe (which has animated "leave it in the ground" campaigns), and regional ecological risk from offshore oil. Arctic offshore oil and gas development has also been analyzed as an emergent "sacrifice zone," a term drawn from environmental justice that describes habitats destroyed or rendered uninhabitable through extractive practices (Reinert 2018). These frames capture distinct orientations toward the development of oil and gas resources, ranging from those that view these resources as belonging to the state and part of the natural extension of state power, to those that view them as part of a "commons" or public good that should be conserved. Missing from these frames is the perspective of Arctic residents, who are particularly attuned to not only environmental but also social risks stemming from development.

Differentiating Risks

Oil and gas development carries both ecological and social risks and hazards related to habitat fragmentation; disruption of migration from pipelines, roads, seismic testing, and marine vessel traffic; and increased pollution from oil spills. The Arctic is one of the world's remaining regions that still enjoys vast areas of undisturbed habitat, which is necessary for many Arctic animals, particularly those that migrate. Caribou are sensitive to disturbances from development activities and other alterations of their habitat (Dana et al. 2009). Marine species have high vulnerability to oil spill impacts, particularly those whose seasonal migrations mean that they are on the move in summer months when drilling takes place (Gulas et al. 2017).

Oil spills represent a "low probability, high impact" risk that carries a potentially devastating impact on local ecosystems, and is a common source of concern for coastal communities (McDowell and Ford 2014). Oil disperses and degrades very slowly at cold temperatures (Byers 2013). Studies have shown that two decades after the *Exxon Valdez* oil spill off Alaska's southern coast, oil from that spill persisted in the ecosystem (Guterman 2009, cited in Byers 2013). The *Exxon Valdez* led to a destruction of local herring and salmon stocks such that there was no herring fishing in the area for 15 years (Ravna and Svendson 2018).

Oil spills are difficult to clean up in the marine environment even in the absence of ice, as illustrated by the fallout from the Deepwater Horizon disaster in the Gulf of Mexico in 2010. The blowout, which occurred in the absence of a same-season relief well, released an estimated 3.19 million barrels of crude oil into the Gulf of Mexico. The cleanup effort released 1.4 million gallons of chemical dispersants, adding an additional environmental health concern (Ocean Portal n.d.). Subsequent analyses pointed to significant gaps in the regulatory system within the United States, as well as problems with the safety culture of the oil industry (Osofsky et al. 2013). Given that it took 89 days to cap the well in the warm and easily accessible Gulf waters raised questions about the industry's ability to safely pursue offshore drilling in the Arctic.

Sea ice in all its various forms adds significant complexity to the operational challenge of oil spill clean up (Wilkinson et al. 2017). Although large-scale spills are rare, smaller spills are a routine part of oil production and transportation. In the US Arctic, there were 10,985 oil spill incidents reported between 1995 and 2012 (mostly small); 67% of them were related to vessels (Gulas et al. 2017). In Russia, 1 percent or more of annual oil production is lost through leaks and spills (Byers 2013). In addition, other forms of contamination occur from normal industry operations, such as air emissions and chemical and oil residues from drilling (McDowell and Ford 2014). Some extraction methods are more polluting than others; for example, hydraulic fracturing (fracking), which can be conducted onshore and nearshore, includes a risk of seismic tremors and pollution of underground water reservoirs (Ellsworth 2013; EPA 2016). While fracking has not yet been undertaken on a large scale in the Arctic, in both Canada and Alaska, large-scale fracking projects are under consideration (Montgomery 2017; Struzik 2014; Toth 2018).

In addition to physical pollutants, acoustical disturbances from industry activity including seismic testing, drilling, and operation of ships, helicopters, and planes also affect Arctic wildlife. Underwater noise from a drillship operating off the coast of Greenland, for example, was detected at a level that would be audible to whales 38 kilometers away from the source (Kyhn et al. 2014). Noise from underwater seismic testing can travel 4000 kilometers, causing habitat avoidance of fish and marine mammals, as well as hearing impairment and even death in some species of marine invertebrates (Denchak 2018). Hunters and harvesters in Greenland have observed changes in the distribution of whale species during periods of testing in Baffin Bay (Nuttall 2017; McDowell and Ford 2014).

Oil and gas development is often accompanied by infrastructure development, which serves as an incentive on the part of the state to attract private-sector investment. New ports and roads bring increasing numbers of visitors and facilitate economic development through tourism and other forms of industry. These infrastructure projects, themselves, can have significant environmental and social impacts, both positive and negative. For example,

Canada invested in a road connecting the city of Inuvik in Canada's Northwest Territories to the small hamlet of Tuktoyaktuk on the Beaufort coast. Construction on the road project began in 2013 when there was industry interest in offshore development; after companies withdrew from the region following the decline in oil prices in 2015, the road was rebranded as tourism infrastructure (CBC 2017). Tuktoyaktuk residents have raised concerns, however, about the impact of the road's many culverts on fish migration, and of the easy access that it has provided for tourists and other Northwest Territories' residents to traditional fishing lakes.

In addition to ecological risks, resource development projects bring social risks related to demographic and economic changes at the community and household level. This may include an influx of new workers from the south, who may introduce different social norms and practices, or outmigration from smaller communities to regional centers set up to support industry practices (Ensign et al. 2014). Because many jobs associated with oil and gas development require technical expertise, only some residents will likely find employment, leading to economic stratification and growing inequality within and between northern communities (Arruda and Krutkowski 2017; Nilsen 2016). The increase in temporary workers who are flown in for short periods combined with the availability of cash from industry employment often increases the rates of alcohol and drug use in industry towns. These various changes are referred to as the "resource curse" (McCauley et al. 2016). On the other hand, from a social perspective, extractive resource development can bring benefits to communities. For example, residents of Hammerfest, Norway, the nearest city to Norway's large *Snøhvit* offshore natural gas field, report primarily positive impacts from industry on their city. These include a significant increase in jobs, younger people wanting to stay in the community, new houses and buildings built, and people moving there because of jobs and staying there (Loe and Kelman 2016).

Arctic Indigenous Peoples and residents whose ties to place span generations face additional cultural and social risks related to the erosion of livelihoods based on renewable resource use. Across the Arctic, resistance to oil and gas activities has often been led by Indigenous communities based on concerns for the continuity of hunting, gathering, and herding activities and the need for a healthy environment to support these practices. These practices are essential not only for food security and maintaining healthy local economies but also for cultural identity and a sense of connection to place. Risks that threaten these practices are, therefore, existential risks that are also viewed as threats to cultural resilience.

Many Indigenous Peoples across the Arctic continue to practice hunting, herding, fishing, and harvesting for subsistence or market-based livelihoods. Arctic Indigenous Peoples are integrated into the market economy and have been particularly affected by policies that limit access and rights to harvest animals. Throughout the history of the relationship between Arctic states and northern Indigenous Peoples, states have attempted to intervene into traditional subsistence practices, whether by encouraging market-oriented production or through policies of state-led, formal education in the Western tradition, which alienated Indigenous Peoples from land-based harvesting knowledge. In spite of this, in many parts of the Arctic, Indigenous communities continue to harvest and herd animals for household consumption; these practices are critical to food security for northern peoples, as store-bought food is expensive and often less nutritious (Arruda and Krutkowski 2017). The persistent importance of harvesting animals to northern peoples goes far beyond their role in providing nutrition; however, wild food collection, hunting, and herding are integral to Indigenous cultural resilience and continuity of language and knowledge traditions across generations.

Hunting, harvesting, and herding activities are impacted by energy extraction in a variety of ways. Oil and gas pipelines present barriers to migration and herding routes of caribou in North America and reindeer in Russia. Offshore oil and gas exploration and production pollutes the marine environment through noise from seismic testing, drilling, and shipping; oil spills are particularly harmful. Willox and colleagues identify a number of disruptions stemming from resource development on Indigenous lands, including disruptions to socio-cultural land activities, changes in individual and collective senses of place attachment (connection to one's home environment), and changes in mental conceptions of place (specific socio-cultural and psycho-social meanings attributed to specific areas) (2013).

Arctic residents must navigate and accommodate multiple sources of risk, including climate-related risks, such as increasingly unstable sea ice and stronger storms for coastal residents, or rain-on-snow events that affect reindeer herding communities, alongside the physical and social risks stemming from extractive development. Bennett et al. (2016) discuss both biophysical and socio-economic drivers of change, which can create a variety of exposures for coastal communities, from reduced sea ice and extreme weather (biophysical) to exposures stemming from demographic, economic, infrastructure and technology, governance and policy, or socio-cultural change. These multiple risks interact in uncertain ways, compounding a sense of uncertainty in relation to possible future harms, which McDowell and Ford suggest is, itself, a "key socio-ecological risk" (McDowell and Ford 2014: 105).

Risk Evaluation and Mitigation Tools

Risks to ecosystems and social systems are evaluated through a formal process of environmental and social impact assessment (EIAs). In some countries, these are separate processes, while in others, the social impact assessment is part of the environmental assessment. Impact assessments are one of the strongest existing tools for assessment, mitigation, and monitoring of impacts of resource development. They are also the primary mechanism for community consultation and engagement in decision-making.

Arctic countries have different requirements that affect how EIAs are conducted in practice. In some countries (Norway, Canada, Alaska), impact assessments are the responsibility of the project proponent (company or agency), and there is no requirement for direct government oversight until the review stage (Hansen et al. 2018). In other countries (Canada, Greenland), impact assessment is overseen by regulatory agencies. EIA processes contain an inherent contradiction in that they are funded and overseen by entities (industry, government) that have an interest in seeing development projects move forward (McDowell and Ford 2014).

For this reason, public participation in EIA processes is particularly important. While community engagement requirements vary, there are legal frameworks for consultation in place in all Arctic states (Newman et al. 2014). Canada has the most extensive framework, which is constitutionally mandated and backed by modern land claim agreements (Newman et al. 2014); Russia has the least robust system, with federal laws providing for Indigenous rights but lacking enforcement mechanisms (Wilson 2016; Yakovleva 2011). Minimally, meaningful engagement can be interpreted to mean that communities are engaged in planning and impact assessment, and that project proponents are open to making changes to their plans based on community input (Hansen et al. 2018). Meaningful engagement cannot be based on a single approach, however, because of the wide variety of cultural practices and ecological conditions across the Arctic (Newman et al. 2014). Newman et al. (2014) suggest that a spectrum analysis, which matches the depth of consultation required to the scale of impact of a project, is a "best practice" approach to consultation because it offers a way to align the interests of states and Indigenous Peoples. Their proposal suggests that those projects requiring deep consultation because of the potential for significant impact on Indigenous livelihoods may include a requirement of consent.

Project proponents benefit from conducting thorough and meaningful consultation processes in project assessment by gaining a "social license to operate," by learning from local and Indigenous knowledge relevant to project operation. An additional benefit is avoidance of costly legal fees incurred by proponents who fail to meet the minimum standards set out in national legislation or policy guidelines (Ehrlich 2010).

While EIAs provide important information, there are also many drawbacks to current methodological and interpretive approaches to impact assessment. Arctic residents have extensive knowledge about the region, such as historical conditions prior to resource development, as well as the behavior of ecosystems and geophysical phenomena (wind, direction of currents, etc.) (Ensign et al. 2014). Although some impact assessment processes attempt to document Indigenous knowledge, this process often emphasizes only Indigenous ecological knowledge relevant to oil and gas operations, rather than engaging Indigenous knowledge more holistically (Hansen et al. 2018). EIAs have been found to do a poor job of assessing cumulative impacts, including impacts occurring over time as a result of different phases of a development process (Kirkfeldt et al. 2017), synergistic impacts from multiple development projects (Hansen et al. 2018), and impacts related to climate change (McDowell and

Ford 2014). EIAs generally do a poor job of considering both biophysical and socio-economic drivers of change and the multiple exposures that result from them (Bennett et al. 2016). Finally, EIAs are only as good as the best available knowledge that they synthesize, and there are many aspects of Arctic social and ecological systems that are not well understood, especially in the context of rapid change.

While EIAs are primarily tools to evaluate risks associated with specific development projects, strategic environmental assessments (SEAs) can be used proactively to mitigate risks from development by considering the larger context (Gulas et al. 2017). SEAs are a "higher order" assessment tool conducted as a planning tool separately from and prior to considering specific project proposals (Noble et al. 2013). Among the Arctic nations, SEAs are mandated only in the Arctic region of Norway and in the United States only for actions that have the potential to significantly affect the environment. While not mandatory in Canada and Greenland, they are increasingly utilized (Kirkfeldt et al. 2017). Because they take into consideration multiple potential (and potentially conflicting) uses of the marine environment, SEAs are considered to be a better tool for assessing cumulative impacts than EIAs. The timing of SEAs, conducted prior to consideration of specific project proposals, can make them useful for informing regulatory decision-making (Noble et al. 2013).

In addition to planning and assessment tools, there are important regulatory approaches to risk mitigation. One policy option is the elimination of liability caps, which limit the amount of liability that companies must pay in the event of an accident. Liability caps can be viewed as a public subsidy on the oil and gas industry, and as such, policy analysts and environmental organizations have recommended their elimination (Byers 2013; WWF 2018) in favor of a "polluter pays" approach that they argue would also lead to a more realistic assessment of economic viability of Arctic offshore development. Another risk mitigation tool is the requirement for same-season relief wells, currently required in Canada, Norway, the United States, and Denmark, which facilitate a quick response in the event of a blowout (Gulas et al. 2017).

Thinking About Benefits

While there are many risks associated with oil and gas development, there are also examples of significant benefits to communities, including the potential for economic diversification for northern communities that often lack other good options for development. These benefits do not automatically accrue from development projects, however; rather, they are increasingly negotiated through impact and benefit agreements (IBAs). IBAs facilitate the distribution of benefits from resource extraction between companies, states, and local communities and help contribute to a "social license to operate" for energy companies (Tysiachniouk and Petrov 2018). The social license to operate reflects the growing awareness of oil and gas companies that successful operations require goodwill and support from at least some of the local population; this support must be sought through a process of relationship and trust building, encompassing both meaningful consultation and, in some cases, negotiation of IBAs.

IBAs can contain provisions for a certain percentage of contracts to go to local companies (such as contracts to provide goods and services to the industry); they can make stipulations for the number of local jobs they will create; and they can make arrangements for other social investments on the part of the company, such as new infrastructure investments or funding for education or cultural programming and activities (Cueva 2018). IBAs can also serve as a risk mitigation tool by requiring compensation in the case of a spill or accident, for example, for harvesting losses. This must be considered only a partial form of mitigation, however, since pollution can persist in the ecosystem for many years, often well beyond the requirements for compensation included within the agreement. Some IBAs include provisions for ongoing monitoring and assessment of environmental impact; however, these may or may not include provisions for oversight, since IBAs are usually negotiated separately from the EIA process (Cueva 2018).

IBAs are negotiated between different parties in different parts of the Arctic. In Canada, for example, they are negotiated between companies and landowners or organizations that hold land in trust, such as regional land claim organizations. In Greenland, IBAs are tripartite agreements involving the company, the Government of Greenland, and one or more municipalities (Mortensen 2018). While these agreements are often based on corporate social responsibility practices, they can be mandated through policy requirements. In Greenland, while IBAs are always required, the Mineral Resources Act states that they may in some cases be required by the government in order for a license to be issued (ibid.). Other requirements that are often negotiated through IBAs may include workforce and hiring, such as requiring a certain percentage of local sourcing of goods and services or a certain number of jobs allocated for residents. In the absence of such requirements, companies often pursue contracts with international suppliers, even for jobs that local companies are equipped to handle (Cueva 2018).

Even when there are policies in place that support negotiation of IBAs, oil and gas industry projects do not always support robust regional and local economic development. In recent exploration activities in Greenland and Canada, for example, Cairn Energy and Imperial Oil spent only approximately 11 percent of their total expenditures on locally based suppliers (Cueva 2018). Similarly, although IBAs often include hiring quotas, there are various obstacles to filling them. In particular, many jobs associated with oil and gas development require technical expertise in excess of what can be taught in relatively short-term job training programs. In reality, there is a limited number of high paying, year-round jobs associated with oil and gas development because of the technical expertise required. An analysis of the proposed Mackenzie Gas Project pipeline in Canada's Northwest territories suggested that only 50 people from the region would be employed in highly technical operating and maintenance jobs (Arruda and Krutkowski 2017: 281, citing Dana et al. 2008: 164). An offshore energy project run by Cairn Energy in Greenland in 2010 had only 7.7 percent local employ-ment. IBAs for recent onshore mineral development projects in Greenland had higher local hiring targets, ranging from 20–50 percent at the start of the proj-ect, but contained language softening the legal obligation of the companies involved to meet these targets (Mortensen 2018).

Tysiachniouk and Petrov (2018) identify four "modes" of benefit sharing: paternalistic, in which the state negotiates with oil companies on behalf of Indigenous communities without their participation; company-centered, in which a company plays the central role in setting up agreements based on glob-ally developed standards; partnership, in which energy companies, govern-ment, and Indigenous communities develop a tripartite relationship; and shareholder, in which communities become shareholders who are automati-cally allocated dividends from oil and gas sales. The native corporations of Alaska operate under the latter form.

There is a diversity of Indigenous perspectives and experiences with oil and gas development and its potential to contribute to regional and local economic development. Some communities, such as the Gwich'in who live in northeast-ern Alaska and in the Yukon territory of Canada, have consistently resisted extractive projects, raising concerns about the impact of proposed drilling in the Arctic National Wildlife Refuge (ANWR) on the Porcupine caribou herd (Arruda and Krutkowski 2017). Others have supported a variety of corporate arrangements and structures that facilitate the flow of benefits from oil and gas development, such as Alaska Native Corporations and, more recently, Arctic Inupiat Offshore (AIO). In July 2014, with the price of crude oil at more than $100 a barrel, Arctic Slope Regional Corporation and Shell announced the formation of Arctic Inupiat Offshore LLC, a joint venture that would provide royalties from offshore oil and gas profits to the Inupiat shareholders of ASRC. The partnership fell apart in 2015 when Shell announced that it was withdrawing from the offshore project after investing more than $7 billion in offshore exploration in Alaska. Arctic Inupiat Offshore subsequently purchased the offshore leases from Shell in anticipation of further exploration work and possible development of the offshore in the future (Bennett 2017).

In Western Canada, the Mackenzie Valley Pipeline project faced resistance from Indigenous communities; a public hearing resulted in a recommendation to halt the pipeline process until land claims were settled in the region (Berger 1988). The Inuvialuit Final Agreement was negotiated in response to oil and gas industry interest, with an eye toward ensuring that Inuvialuit were part of decision-making about oil and gas development and that they were able to benefit from development taking place, both on land and in offshore areas. After the IFA and other land claim settlements, representatives of the Invialuit, Gwich'in, and Sahtu Dene joined together to establish The Aboriginal Pipeline Group, a joint business venture that would have an ownership stake in the pipeline project (Dana et al. 2008; Nuttall 2009). As with Arctic Inupiat Offshore, however, the pipeline project was dissolved in 2017 by its main pro-ponent, Imperial Oil; the project would not have recovered the cost of investments due to market saturation by cheaply available shale gas from the United States and southern Canada.

In Russia, land rights are not settled for many northern peoples and national legislation addressing rights of Indigenous Peoples is not robustly enforced. There are 40 recognized Indigenous minorities living in the North, Siberia, and Far East, many of whom engage in reindeer herding, hunting, and fishing (Yakovleva 2011). State-led industrial development during the Soviet era was characterized by heavily polluting practices, many of which have continued into the present (Gulas et al. 2017). Overall, the oil and gas industry has been criticized for lack of attention to Indigenous rights, with industry practice that reflects a lack of understanding of Indigenous Peoples and their relationship with the land and animals (Stammler and Wilson 2006). In addition, EIA is not required for all projects in Russia, and when implemented, often fails to address social and cultural impacts (Yakovleva 2011).

More recently, some companies, recognizing the importance of public opinion in supporting a "social license to operate framework for consultation and benefit," have used a different approach, emphasizing communication with Indigenous communities and identifying mechanisms for benefit sharing (Wilson 2016). In the absence of a strong national framework for consultation and benefit sharing, international norms and pressure placed on extractive industry through standard setting, such as the Equator Principles, have helped shift corporate practice in recent developments. A comparative study by Wilson (2016) showed very different outcomes from negotiations between oil companies and communities in the Komi Republic and on Sakhalin Island, reflecting different cultures of extraction, with different norms and expectations, associated with Soviet/post-Soviet (reflected in the Komi Republic) and more recent cultural norms based on international standards (reflected in the Sakhalin project). In Sakhalin, local groups received information and support from a network of international NGOs, who provided important information about environmental monitoring and international standards. The use of strategic protests to gain international media attention led to shifts in proponents' willingness to address local concerns through compensation and benefit-sharing arrangements.

In summary, communities across the Arctic expect meaningful consultation on development projects. There are many different orientations toward development that exist between and within communities. Negotiation of IBAs that address the most significant risks to local livelihoods while guaranteeing some benefits to communities has proven helpful in the development of positive relationships between communities and extractive industry. At the same time, Arctic residents are aware that even in the best-case scenario, oil and gas projects bring social and environmental impacts, some of which cannot be anticipated ahead of time, and that the life span of these projects is finite. Part of the problem is the matter of limited options for remote communities where economies have historically been defined by natural resource use. If communities were given the choice between extractive industry focused development and more sustainable alternatives that would also create jobs and bring investment into their regions, interest in and support for extractive projects might well diminish.

CLIMATE CHANGE, ENERGY TRANSITIONS, AND SUSTAINABLE DEVELOPMENT

Across the Arctic and at different scales, northern economies remain dependent on oil and gas development. In Alaska, Norway, Greenland, and Russia, development of resources is directly linked to national and sub-national state wealth, which is tied to public investments in basic social services. With the development of Native corporations, the settlement of land claims in Canada and the devolution of Greenland from Denmark through self-government, Arctic Indigenous Peoples arguably have greater levels of self-determination now than at any other time in modern history. Sheila Watt-Cloutier, who chaired the Inuit Circumpolar Council from 2000 to 2005, has called on Inuit leaders to exercise leadership and responsibility in relation to development policy, stating "We must not permit the discussion of northern development to be conducted only in terms of sovereignty, resources, and economics. The focus must be on the human dimension, human communities and protection of human cultural rights" (Watt-Cloutier 2009).

Arctic states, too, have a responsibility to principles of sustainable development that have animated the discussion and work of the Arctic Council since its inception. These principles suggest responsibility not only to support Indigenous Peoples in maintaining their lands, waters, and cultural practices but also to recognize Indigenous sovereignty in decision-making about these territories and practices. Sustainable development commitments also require balancing future needs with the needs of the present. Given the rapid pace of climate change in the Arctic and the need to reduce emissions to limit climate change to a degree that will continue to support and sustain human and non-human life on Earth, this may require leaving some oil and gas reserves undeveloped, regardless of safety and technical capacity to do so (McGlade and Ekins 2015)

While Arctic states have relatively diverse economies, however, Arctic communities are much more constrained in their options. Investment in energy efficient building and alternative forms of energy including wind and solar power that can lower energy costs in remote communities can be a small part of the solution, and research suggests that residents are eager for these investments aimed at local and regional scale use, so long as they do not interfere with hunting and herding practices (McDonald and Pearce 2013).

At the same time, it is unrealistic to imagine that renewable energy can replace the role of extractive energy in Arctic economies single handedly. Economies in Norway, Alaska, and Alaska's North Slope, for example, are heavily invested in oil and gas, have continued to emphasize offshore exploration, even as some of their own residents and citizens voice values in line with those who believe that new reserves in the Arctic are best left undeveloped. From this perspective, policy pathways that emphasize integrated management and land use planning, including marine spatial planning, can facilitate the development of different sources of economic growth simultaneously, including fisheries development and tourism alongside extractive industry. Building more diversified regional and local economies will facilitate an easier transition to a post-petroleum scenario, whether by choice or due to market factors that make development of remote Arctic resources economically unattractive.

In the meantime, as this chapter has discussed, mechanisms to promote the participation of residents and Indigenous Peoples in decision-making about oil and gas development are unevenly developed and implemented across Arctic states. Greater involvement of residents can only improve safety and risk management through the utilization of Indigenous and local knowledge. Across the Arctic, there is a need for consultations that are meaningful and robust, and that identify pathways to mitigate risk and enhance benefit for affected communities. Even in countries where consultation is federally mandated, questions remain about how to implement FPIC in practice. Given the long history of extractive practices in the Arctic and the likelihood that they will continue for the foreseeable future, it is likely that these discussions will continue, even as elements are resolved, either through litigation or through adoption and utilization of guidelines endorsed by Arctic residents and Indigenous communities.

REFERENCES

Arctic Council. 2011. *Agreement on Cooperation on Aeronautical and Maritime Search and Rescue in the Arctic.* Nuuk, Greenland. May 12. https://oaarchive.arctic-council.org/handle/11374/531. Accessed 13 November 2018.
———. 2013. *Agreement on Cooperation on Marine Oil Pollution Preparedness and Response in the Arctic.* Kiruna, Sweden, May 15. https://oaarchive.arctic-council.org/handle/11374/529. Accessed 13 November 2018.
———. 2017. *Agreement on Enhancing international Arctic Scientific Cooperation.* Fairbanks, Alaska, May 11. https://oaarchive.arctic-council.org/handle/11374/1916. Accessed 13 November 2018.
Arruda, G.M., and S. Krutkowski. 2017. Social Impacts of Climate Change and Resource Development in the Arctic. *Journal of Enterprising Communities: People and Places in the Global Economy* 11 (2): 277–288. https://doi.org/10.1108/JEC-08-2015-0040.
Avango, D., L. Hacquebord, and U. Wråkberg. 2014. Industrial Extraction of Arctic Natural Resources Since the Sixteenth Century: Technoscience and Geo-Economics in the History of Northern Whaling and Mining. *Journal of Historical Geography* 44 (0305): 15–30. https://doi.org/10.1016/j.jhg.2014.01.001.

Baker, B. 2018. The Arctic Offshore Hydrocarbon Hiatus of 2015: An Opportunity to Revisit Regulation Around the Pole. In *Governance of Arctic Offshore Oil and Gas*, ed. C. Pelaudeix and E.M. Basse. London: Routledge.

Bennett, M. 2017. Alaska Native Corporation Pursues Offshore Oil. *The Maritime Executive*, July 23. https://www.maritime-executive.com/features/alaska-native-corporation-pursues-offshore-oil. Accessed 15 November 2018.

Bennett, N.J., J. Blythe, S. Tyler, and N.C. Ban. 2016. Communities and Change in the Anthropocene: Understanding Social-Ecological Vulnerability and Planning Adaptations to Multiple Interacting Exposures. *Regional Environmental Change* 16 (4): 907–926. https://doi.org/10.1007/s10113-015-0839-5.

Berger, Thomas R. 1988. *Northern Frontier, Northern Homeland: The Report of the Mackenzie Valley Pipeline Inquiry*. Vancouver: Douglas & McIntyre.

Buxton, A., and E. Wilson. 2013. *FPIC and the Extractive Industries: A Guide to Applying the Spirit of Free, Prior and Informed Consent in Industrial Projects*. London: International Institute for Environment and Development.

Byers, M. 2013. *International Law and the Arctic*. Cambridge: Cambridge University Press.

Callison, C. 2014. *How Climate Change Comes to Matter: The Communal Life of Facts*. Durham: Duke University Press.

CBC. 2017. New Tuktoyaktuk Road Life-Changing for Arctic Community. *CBC Radio "The Current"*, November 15. https://www.cbc.ca/radio/thecurrent/the-current-for-november-15-2017-1.4401756/new-tuktoyaktuk-road-life-changing-for-arctic-community-1.4401922. Accessed 13 November 2018.

Cueva, V.P. 2018. Impact Benefit Agreements and Economic and Environmental Risk Management in the Arctic. In *Governance of Arctic Offshore Oil and Gas*, ed. C. Pelaudeix and E.M. Base, 179–199. New York: Routledge.

Dana, L.P., A. Meis-Mason, and R.B. Anderson. 2008. Oil and Gas and the Inuvialuit People of the Western Arctic. *Journal of Enterprising Communities: People and Places in the Global Economy* 2 (2): 151–167.

Dana, L.P., R.B. Anderson, and A. Meis-Mason. 2009. A Study of the Impact of Oil and Gas Development on the Dene First Nations of the Sahtu (Great Bear Lake) Region of the Canadian Northwest Territories (NWT). *Journal of Enterprising Communities: People and Places in the Global Economy* 3 (1): 94–117.

Denchak, M. 2018. Ocean Pollution: The Dirty Facts. *Natural Resources Defense Council*, January 23. https://www.nrdc.org/stories/ocean-pollution-dirty-facts. Accessed 14 November 2018.

Dittmer, D.L. Gautier, K.J. Bird, R.R. Charpentier, A. Grantz, D.W. Houseknecht, T.R. Klett, and C.J. Wandrey. 2009. Assessment of Undiscovered Oil and Gas in the Arctic. *Science* 324: 5931.

Dittmer, J., S. Moisio, A. Ingram, and K. Dodds. 2011. Have You Heard the One About the Disappearing Ice? Recasting Arctic Geopolitics. *Political Geography* 30 (4): 202–214.

Dombrowski, K. 2007. Subsistence Livelihood, Native Identity and Internal Differentiation in Southeast Alaska. *Anthropologica* 49 (2): 211–229.

Ehrlich, A. 2010. Cumulative Cultural Effects and Reasonably Foreseeable Future Developments in the Upper Thelon Basin, Canada. *Impact Assessment and Project Appraisal* 28 (4): 279–286.

Ellsworth, W.L. 2013. Injection-Induced Earthquakes. *Science* 341 (6142): 1225942–1225942.

Ensign, P.C., A.R. Giles, and J. Oncescu. 2014. Natural Resource Exploration and Extraction in Northern Canada: Intersections with Community Cohesion and Social Welfare. *Journal of Rural and Community Development* 9 (1): 112–133.

EPA. 2016. *Hydraulic Fracturing for Oil and Gas: Impacts from the Hydraulic Fracturing Water Cycle on Drinking Water Resources in the United States* (Final Report). Washington, DC: U.S. Environmental Protection Agency. https://cfpub.epa.gov/ncea/hfstudy/recordisplay.cfm?deid=332990. Accessed 30 July 2019.

Gulas, S., M. Downton, K. D'Souza, K. Hayden, and T.R. Walker. 2017. Declining Arctic Ocean Oil and Gas Developments: Opportunities to Improve Governance and Environmental Pollution Control. *Marine Policy* 75 (October 2016): 53–61. https://doi.org/10.1016/j.marpol.2016.10.014.

Guterman, L. 2009. CONSERVATION BIOLOGY: Exxon Valdez Turns 20. *Science* 323 (5921): 1558–1559.

Hansen, A.M., S.V. Larsen, and B. Noble. 2018. Social and Environmental Impact Assessments in the Arctic. In *The Routledge Handbook of the Polar Regions*, ed. M. Nuttall, T.R. Christiansen, and M.J. Siegert, 389–399. London, New York: Routledge.

Heinämäki, L. 2015. Global Context – Arctic Importance: Free, Prior and Informed Consent, a New Paradigm in International Law Related to Indigenous Peoples. In *Indigenous Peoples Governance of Land and Protected Territories in the Arctic*, ed. T.M. Herrmann and T. Martin, 209–243. Cham: Springer.

Hendersen, J., and J. Loe. 2014. *The Prospects and Challenges for Arctic Oil Development*. Oxford Institute for Energy Studies. Working Paper 54.

ICC (Inuit Circumpolar Council). 2011. A Circumpolar Inuit Declaration on Resource Development Principles in Inuit Nunaat. https://iccalaska.org/wp-icc/wp-content/uploads/2016/01/Declaration-on-Resource-Development-A4-folder-FINAL.pdf. Accessed 29 July 2019.

Loe, J.S.P., and I. Kelman. 2016. Arctic Petroleum's Community Impacts: Local Perceptions from Hammerfest, Norway. *Energy Research and Social Science* 16: 25–34. https://doi.org/10.1016/j.erss.2016.03.008.

Kirkfeldt, T.S., A.M. Hansen, P. Olesen, L. Mortensen, K. Hristova, and A. Welsch. 2017. Why Cumulative Impacts Assessments of Hydrocarbon Activities in the Arctic Fail to Meet Their Purpose. *Regional Environmental Change* 17 (3): 725–737.

Knapp, G. 2012. *Alaska's Experience with Arctic Oil and Gas Development: History, Policy Issues, and Lessons*. Presentation at Energies of the High North-Arctic Frontiers, Tromso, Norway. January 25. https://scholarworks.alaska.edu/bitstream/handle/11122/3956/2012_01_25-Alaskas_Experience_with_Arctic_Oil_and_Gas_Development.pdf?sequence=1. Accessed 13 November 2018.

Koivurova, T. 2018. Framing the Problem in Arctic Offshore Exploration. In *Governance of Arctic Offshore Oil and Gas*, ed. C. Pelaudeix and E.M. Basse. London: Routledge.

Kyhn, L.A., S. Sveegaard, and J. Tougaard. 2014. Underwater Noise Emissions from a Drillship in the Arctic. *Marine Pollution Bulletin* 86 (1–2): 424–433. https://doi.org/10.1016/J.MARPOLBUL.2014.06.037.

McCauley, D., R. Heffron, M. Pavlenko, R. Rehner, and R. Holmes. 2016. Energy Justice in the Arctic: Implications for Energy Infrastructural Development in the Arctic. *Energy Research & Social Science* 16: 141–146.

McDonald, N.C., and J.M. Pearce. 2013. Community Voices: Perspectives on Renewable Energy in Nunavut. *Arctic* 66 (1): 94–104.

McDowell, G., and J.D. Ford. 2014. The Socio-Ecological Dimensions of Hydrocarbon Development in the Disko Bay Region of Greenland: Opportunities, Risks, and Tradeoffs. *Applied Geography* 46: 98–110. https://doi.org/10.1016/j.apgeog.2013.11.006.

McGlade, C., and P. Ekins. 2015. The Geographical Distribution of Fossil Fuels Unused When Limiting Global Warming to 2 Degrees. *Nature* 517: 187–190. https://doi.org/10.1038/Nature14016.

Montgomery, S.L. 2017. Large Scale Fracking Comes to the Arctic in a New Alaska Oil Boom. *The Conversation*, April 12. http://theconversation.com/large-scale-fracking-comes-to-the-arctic-in-a-new-alaska-oil-boom-75683. Accessed 14 November 2018.

Mortensen, B.O.G. 2018. Impact and Benefit Agreements in Greenland. In *Governance of Arctic Offshore Oil and Gas*, ed. C. Pelaudeix and E.M. Base, 199–210. New York: Routledge.

Newman, D., M. Biddulph, and L. Binnion. 2014. Arctic Energy Development and Best Practices on Consultation with Indigenous Peoples. *Boston University International Law Journal* 32 (Summer): 449.

Nilsen, T. 2016. Why Arctic Policies Matter: The Role of Exogenous Actions in Oil and Gas Industry Development in the Norwegian High North. *Energy Research & Social Science* 16: 45–53.

Noble, B., S. Ketilson, A. Aitken, and G. Poelzer. 2013. Strategic Environmental Assessment Opportunities and Risks for Arctic Offshore Energy Planning and Development. *Marine Policy* 39 (1): 296–302.

Nuttall, M. 2009. Energy Development and Aboriginal Rights in Northern Canada. In *Canada's and Europe's Northern Dimensions*, ed. A. Dey-Nuttall and M. Nuttall, 71–84. Oulu: University of Oulu Press.

———. 2017. *Under the Great Ice: Climate, Society, and Subsurface Politics in Greenland*. London, New York: Routledge.

Ocean Portal Team. n.d. *Gulf Oil Spill. Smithsonian Ocean Portal.* https://ocean.si.edu/conservation/pollution/gulf-oil-spill. Accessed 13 November 2018.

Osofsky, H., K. Baxter-Kauf, B. Hammer, and B.M. Mailander. 2013. Environmental Justice and the BP Deepwater Horizon Oil Spill. *NYU Environmental Law Journal* 99 (2179): 99.

Osofsky, H.M., J. Shadian, and S.L. Fechtelkotter. 2016. Arctic Energy Cooperation. *UC Davis Law Review* 1431 (1): 0–51.

PAME (Protection of the Arctic Marine Environment). 2014. *The Arctic Offshore Oil and Gas Guidelines: Systems Safety Management and Safety Culture Report.* Akureyri: Protection of the Arctic Marine Environment (PAME. https://oaarchive.arctic-council.org/handle/11374/418. Accessed 29 July 2019.

Ravna, Ø., and K. Svendson. 2018. Securing the Coastal Sámi Culture and Livelihood. In *Governance of Arctic Offshore Oil and Gas*, ed. C. Pelaudeix and E.M. Base, 153–166. New York: Routledge.

Reinert, H. 2018. Notes from a Projected Sacrifice Zone. *ACME: An International Journal for Critical Geographies* 17 (2): 597–6178.

Shadian, J. 2017. Reimagining Political Space: The limits of Arctic Indigenous Self-determination in International Governance? In *Governing Arctic Change: Global Perspectives*, ed. K. Stephen and S. Knecht, 43–57. London: Palgrave Macmillan.

Sidortsov, R. 2016. A Perfect Moment During Imperfect Times: Arctic Energy Research in a Low-Carbon Era. *Energy Research and Social Science* 16: 1–7. https://doi.org/10.1016/j.erss.2016.03.023.

Stammler, F., and E. Wilson. 2006. Dialogue for Development: An Exploration of Relations Between Oil and Gas Companies, Communities, and the State. *Sibirica* 5 (2): 1–42. https://doi.org/10.3167/136173606780490739.

Struzik, E. 2014. A New Frontier for Fracking: Drilling Near the Arctic Circle. *Yale Environment 360*, August 18. https://e360.yale.edu/features/a_new_frontier_for_fracking_drilling_near_the_arctic_circle. Accessed 14 November 2018.

Toth, K. 2018. NWT's Energy and Climate Plans Include Promoting Natural Gas. *CBC Online*, May 2. https://www.cbc.ca/news/canada/north/nwt-climate-change-natural-gas-1.4644221. Accessed 14 November 2018.

Tysiachniouk, M.S., and A.N. Petrov. 2018. Benefit Sharing in the Arctic Energy Sector: Perspectives on Corporate Policies and Practices in Northern Russia and Alaska. *Energy Research and Social Science* 39 (October 2017): 29–34. https://doi.org/10.1016/j.erss.2017.10.014.

Watt-Cloutier, S. 2009. Reclaiming the Moral High Ground: Indigenous Peoples, Climate Change, and Human Rights. *Nunatsiaq News Online*, December 21. https://nunatsiaq.com/stories/article/4567_reclaiming_the_moral_high_ground/. Accessed 29 July 2019.

Wilkinson, J., C. Beegle-Krause, K.U. Evers, N. Hughes, A. Lewis, M. Reed, and P. Wadhams. 2017. Oil Spill Response Capabilities and Technologies for Ice-Covered Arctic Marine Waters: A Review of Recent Developments and Established Practices. *Ambio* 46 (s3): 423–441. https://doi.org/10.1007/s13280-017-0958-y.

Willox, A.C., S.L. Harper, J.D. Ford, V.L. Edge, K. Landman, K. Houle, S. Blake, and C. Wolfrey. 2013. Climate Change and Mental Health: An Exploratory Case Study from Rigolet, Nunatsiavut, Canada. *Social Sciences and Medicine* 75 (3): 538–547.

Wilson, E. 2016. What Is the Social Licence to Operate? Local Perceptions of Oil and Gas Projects in Russia's Komi Republic and Sakhalin Island. *Extractive Industries and Society* 3 (1): 73–81. https://doi.org/10.1016/j.exis.2015.09.001.

Yakovleva, N. 2011. Oil Pipeline Construction in Eastern Siberia: Implications for Indigenous People. *Geoforum* 42 (6): 708–719. https://doi.org/10.1016/j.geoforum.2011.05.005.

Young, O.R. 2005. Governing the Arctic: From Cold War Theater to Mosaic of Cooperation. *Global Governance: A Review of Multilateralism and International Organizations* 11 (1): 9–15.

CHAPTER 8

Innovation, New Technologies, and the Future of the Circumpolar North

Heather M. Hall

INTRODUCTION

Regions throughout the Circumpolar North are experiencing unprecedented changes. For example, the impacts of climate change are more pronounced in the North (Hodge et al. 2016). This is placing new pressures on infrastructure and housing as permafrost melts (Lamb 2017). A decline in sea-ice is opening up new shipping routes as well as areas of exploration and resource development, creating economic opportunities but also environmental and geopolitical tensions (Jordans 2017; Dillow 2018). Across the North, there is also an increasing recognition of Indigenous rights, including land rights and a duty to consult (Josefsen 2010; Newman 2014). These changes are occurring alongside traditional challenges facing the North including the climate, low population densities, and remoteness. Innovation has the potential to counteract these pressing challenges facing the Circumpolar North; however, it could also deepen and present new challenges if it is not created with the North, for the North.

One of the most cited definitions of innovation is from the Oslo Manual, which defines innovation as "the implementation of a new or significantly improved product (good or service), or process, a new marketing method, or a new organizational method in business practices, workplace organization or external relations" (OECD 2005: 46). A more recent definition acknowledges that innovation "goes far beyond R&D" and "beyond the confines of research labs to users, suppliers and consumers everywhere—in government, business and non-profit organizations ..." (OECD 2015: online). This more holistic

H. M. Hall (✉)
University of Waterloo, Waterloo, ON, Canada
e-mail: h.hall@uwaterloo.ca

K. S. Coates, C. Holroyd (eds.), *The Palgrave Handbook of Arctic Policy and Politics*, https://doi.org/10.1007/978-3-030-20557-7_8

117

definition recognizes more incremental changes and the broader innovation ecosystem.

This chapter uses this broad view to explore innovation in the context of the Circumpolar North, including challenges and opportunities, while highlighting the importance of the innovation ecosystem. It also provides examples of new and adapted technologies that are being used in the Arctic to enhance traditional industries, promote social innovation, and encourage economic diversification. It concludes with a discussion of how to ensure that the development of new or improved innovative products, processes, and/or services occurs with the North, for the North.

Understanding Innovation in Circumpolar Regions

As Hall and Vodden (in press) (see also Hall and Donald 2009; Hall and Walsh 2013) argue, much of our academic and policy attention on innovation has focused on large-city regions. In fact, rural and northern regions are not typically cited in case studies on innovation and, perhaps more concerning, these regions are often discounted as "inauspicious" spaces for innovation (Johnstone and Haddow 2003). Coates and Poelzer (2014, 14) further contend that while much attention has been paid to finding innovative solutions to the challenges facing the Global South by governments, NGOs, and philanthropists, "no comparable effort is being made in the Far North."

There is, however, a small body of research emerging that is focused on understanding innovation in peripheral and northern regions (see, e.g. *Northern Review* 2017, Special Issue on Innovation in the Circumpolar North). Some of the insights from this body of literature are the importance of understanding the different types of innovation. For example, Isaksen and Karlsen (2010) explain how innovation in peripheral regions is typically focused on "doing-using-interacting" (DUI) versus "science, technology, innovation" (STI). In the DUI model, innovation is often more incremental in nature and might occur through in-house problem-solving by an individual or a group of workers or from addressing specific supplier, customer, or client group needs. This is in contrast to STI or more radical innovation that often occurs within large corporate R&D departments, research-intensive SMEs, and postsecondary institutions or other research centers. While the broader literature on innovation emphasizes the importance of institutions, including both "hard institutions" and "soft institutions" (i.e. social and cultural factors) (Harrison 2006; Amin and Thrift 1994, 1995), in peripheral regions, Tödtling and Tripple (2005) suggest three specific institutional issues: (1) thinness or low levels of clustering and a weak prevalence of institutions; (2) lock-in; and (3) fragmentation or a lack of interaction between institutional stakeholders.

Exner-Pirot et al. (2017) further discuss two barriers for innovation in the Arctic. First are the harsh environmental conditions, which means technologies developed in the South may be unreliable in the North. For example, Coates and Landrie-Parker (2016) discovered that some renewable energy projects in

Northern Canada used technology that was inappropriate for the climatic conditions and unique challenges facing northern communities. Especially problematic were equipment failures which required parts and labor to be flown in from the south at a significant cost that often took a substantial amount of time. The second barrier discussed by Exner-Pirot et al. (2017) is economies of scale. More specifically, the critical mass needed to secure an appropriate return on investment is often lacking due to the small, isolated populations across the Arctic. Hall and Donald (2009) also discuss the impacts of youth out-migration, access to postsecondary institutions, and infrastructure challenges on Northern innovation in Canada. While Coates and Poelzer (2014: 14) believe that a significant barrier to innovation is the fact that the Arctic is embedded in "rich-nations" and it "has fallen to these countries to take up the challenge of Arctic scientific and technological innovation and to develop innovative solutions to northern conditions." As a result, global philanthropists and NGOs turn their attention elsewhere.

Another important insight from the literature on innovation is the significance of the place or the geographical context (Hall and Vodden forthcoming). As Martin (2010: 20) argues "innovation is indeed often a highly localized phenomenon, dependent on place-specific factors and conditions." As such, it is imperative to recognize the vast differences between Circumpolar countries and within Circumpolar countries. Exner-Pirot et al. (2017) refer to this as the "many Arctics" highlighted by subregional geographical, economic, and cultural differences. They argue that the most significant cleavages are the rural-urban divide and geopolitical divide between Arctic nations. Many of the larger communities in the Circumpolar North (e.g. Whitehorse, Yukon or Tromsø, Norway) serve as regional service centers and have access to modern amenities like infrastructure connections and highspeed internet as well as economic opportunities, healthcare services, and postsecondary institutions. On the other hand, many rural communities, especially in Canada, face unique challenges. Some are off-grid and lack stable and affordable energy sources (Coates and Landrie-Parker 2016), while access to clean drinking water (Aiello 2017), internet and cellphone coverage (FCM 2017), and healthcare (Young and Chatwood 2017) are not available or exorbitantly expensive to access. Many rural northern communities in Canada can also only be accessed year-round by airplane. Ice roads are used by some communities in the winter months (MNDM 2017); however, the lack of reliable and affordable infrastructure and transportation options increases the costs of products and services prohibitively (Skura 2016; CBC 2018).

As Exner-Pirot et al. (2017) note, the second major cleavage is the geopolitical divide between Arctic nations. For example, northern regions in the Nordic countries (Norway, Sweden, and Finland) are comparatively more connected by land, air, and sea and have internet and cellphone coverage, access to Universities and research centers, and more economic development opportunities. A third significant distinction in the Circumpolar North is between Indigenous and non-Indigenous communities (Pigford et al. 2017).

Many Indigenous communities in the North are engaged in the social economy versus the market economy while preserving cultural heritage, traditional land, and traditional practices is vital. As a result of these variations across the Circumpolar North, innovation needs to be place-based whereby ideas and solutions reflect the unique challenges and opportunities of particular places. Put simply, what works in Finnmark, Norway might not work in Igloolik, Nunavut. Perhaps more importantly, innovation needs to occur in consultation with community stakeholders to ensure that the right solutions are applied to the needs of that particular community.

THE INNOVATION ECOSYSTEM: ARCTIC RESEARCH CENTERS

One of the most important arguments emerging from the innovation literature over the last several decades is the understanding that innovation is a social process (Wolfe 2009). Innovation, therefore, involves interaction between various economic actors or innovation stakeholders (e.g. firms, customers, postsecondary institutions, government agencies). A number of innovation models emphasize this social process from industrial districts (Becattini 1990), clusters (Porter 1990), and innovative milieus (see Proulx 1992) to learning regions (Morgan 1997), the triple helix/quadruple helix (Leydesdorff 2012), and regional innovation systems (Cooke and Morgan 1998). More recently, scholars and policymakers alike have gravitated toward the concept of the innovation ecosystem, which considers "not only firms, universities, colleges and polytechnics, but also a spectrum of intermediary players ... characterized by effective synergies, connections, and flows of knowledge and ideas" (Expert Panel on Federal Support to Research and Development 2012: 2–15). These intermediaries include tech transfer and applied research offices, incubators, public research institutes, and angels/venture capitalists. As Coates and Poelzer (2014) argue, developing innovation ecosystems in the North is essential to take advantage of technological advances. It is also imperative to ensure interaction and learning with innovation stakeholders in Northern regions, especially Indigenous communities to "reflect the values, interests and needs of Arctic communities" (Pigford et al. 2017: 3).

While innovation ecosystems are less advanced in the Circumpolar North, there are a number of Arctic Research Centres, postsecondary institutions, and alliances which are helping to ensure that new technologies, and more broadly economic/community development opportunities, are developed in the Arctic, with the Arctic, and for the Arctic. Postsecondary institutions include UiT– The Arctic University of Norway in Tromsø, the Luleå University of Technology in Sweden, the University of Oulu in Finland, and Yukon College, which is transitioning to become the first University in the Canadian Territories.

With regards to research centers, the Arctic Technology Centre (ARTEK) in Sisimiut, Greenland, provides engineering students from Greenland and Demark with the opportunity to learn about the Arctic and contribute to research on Arctic technologies. Areas of specialization include construction and physical environment; Arctic environmental engineering; buildings and energy in the Arctic; and planning sustainability and infrastructure (ARTEK 2018). Likewise, the Canadian government has recently constructed the Canadian High Arctic Research Station (CHARS) in Cambridge Bay, Nunavut. The facility includes a main research building and accommodation buildings for visiting researchers. There are also research labs as well as space for technology development, teaching, training, community engagement, and knowledge sharing (Government of Canada 2018).

Also in Canada, the Cold Climate Innovation (CCI) Research Centre at Yukon College, Whitehorse was created to develop, commercialize, and export sustainable cold climate technologies to subarctic regions. They provide partnership opportunities between applied science researchers, industry, and government to tackle the pressing cold climate issues impacting northerners around the world. The center provides funding, business mentoring and planning, assistance with prototype development, project management, marketing support, and patent advice. Projects supported by CCI have focused on alternative energy, building construction in a northern context, food security, and environmental remediation among others (Yukon College 2017a). One example is the partnership between CCI and the Tr'ondëk Hwëch'in to design and build a 3000-square foot community greenhouse to be used for production and teaching (Yukon College 2017b). In 2015, CCI started the Yukon Innovation Prize, which is awarded annually to a Yukon based entrepreneur. In 2017, it was awarded to Yukon River Skincare for their product innovation that includes birch sap and levan-based skincare products (Yukon College 2017c).

Similarly, in Fairbanks, Alaska the Cold Climate Housing Research Center (CCHRC) is a non-profit corporation created to "facilitate the development, use, and testing of energy-efficient, durable, healthy, and cost-effective building technologies for people living in circumpolar regions around the globe" (CCHRC 2018). Their research and testing facility is affiliated with the University of Alaska Fairbanks, which promotes collaboration between researchers, students, and faculty with the center. CCHRC has three main programs: building science research; sustainable northern communities; and policy research. Within the sustainable northern communities' program, for example, a Northern Shelter initiative was created in 2008 to build housing suited for the northern climate that also reflects the local culture, environment, and resources in a given community. The program has grown to include several prototype homes in over a dozen communities throughout Alaska (CCHRC 2018).

It is worth noting, however, that there are huge discrepancies between the innovation ecosystems in Scandinavia and North America. The innovation ecosystem in Northern Canada, for example, is institutionally thin and fragmented. In addition, many of the stakeholders in the innovation ecosystem, like research centers, universities, and investors have typically been located in the South (Pigford et al. 2017). However, there are signs of change exemplified by the transition of Yukon College to a University, the creation of a federal government economic development agency for the North (CanNor), and the creation of the Arctic Inspiration Prize by philanthropists Sima Sharifi and Arnold Witzig (Zilio 2018).

In the Scandinavian Arctic, on the other hand, there are a number of initiatives to support innovation by stakeholders in the North. For example, Hintsala et al. (2017: 83) describe the Oulu Innovation Alliance, which was created in 2009, to coordinate the efforts of education and research institutes, companies, and the public sector on agreed-upon innovation areas (e.g. Internet research, energy). It also promotes infrastructure investment and the creation and development of innovative tools that can be used by all innovation stakeholders. Another example is the "Arctic Valley" initiative to promote international business opportunities in the Arctic and to facilitate stronger cooperation between innovation stakeholders across Norway, Finland, and Sweden (Niemelä and Hintsala 2016). Likewise, in 2015, the Prime Minister of Finland commissioned a report to discuss how Norway, Sweden, and Finland could work together to promote sustainable growth in the North. The report focuses on four drivers of growth: cleaner energy, greener mining solutions, increased tourism, and world leaders in ice and cold climate solutions. The expert panel also recommends four instruments for achieving sustainable growth including having one voice, a long-term plan for transportation and infrastructure, and one pool of talent and labor.

Examples of New and Adapted Technologies in the Arctic

Recent attention on innovation has focused on new and adapted technologies which are ushering in an "age of disruption" (Deloitte 2015) that could fundamentally reshape the future of work and community development. According to Deloitte (2015), these technologies include artificial intelligence, collaborative connected platforms, advanced manufacturing, advanced robotics, and networks. These new technologies are reshaping industries and the relationships these industries have with people and places.

New and adapted technologies have the potential to be extremely beneficial to the North. Many of the challenges facing people and communities from clean drinking water to infrastructure, food security and access to healthcare will require innovative thinking and new or adapted technologies. As noted earlier, the Cold Climate Innovation Research Centre in Whitehorse, Canada is experimenting with different greenhouse designs to address food security while the Cold Climate Housing Research Centre in Alaska is developing new building methods and products to fit the needs of a northern climate. It is imperative that these innovations are place-based, the benefits are for the north, and they are created in consultation with stakeholders in the North. The following are some examples of where new and adapted technologies are enhancing traditional industries, promoting social innovation, and encouraging economic diversification in the Circumpolar North.

King Crab Fishery: Norway

Bugøynes, Norway is located in Finnmark county in northeastern Norway with a current population of roughly 230 people. In the 1980s, the community experienced a crisis in their cod fishery, their major industry, leading to a population decline of roughly 18%. A local action committee was created and they took a very bold approach to get the attention of policymakers and investors in Oslo. In 1989 they placed an ad in a leading newspaper in Oslo with the simple headline—*Will Someone Accept Us?* The ad went on to state:

> Is there a place in Norway that will welcome an increase in population of about 300 people? We ask as citizens of the fishing community Bugøynes in eastern Finnmark who now are fed up. The last few years have been a constant struggle to maintain the settlement. The reason is a fisheries policy that failed. Among other things, it led to the bankruptcy of the fish plant—our cornerstone firm—in 1987. In the two years since then no one has been able to get the plant started again. This is because of bureaucratic clutter and lack of will among bureaucrats and politicians. (In fact, they are still arguing about who owns the plant.) Now we feel it is time to put everything behind us, and start again somewhere else. We want to avoid becoming burned out and worn out in our struggle for existence for no purpose. We want to use our strength in a community where we can work for a future for ourselves and our children. We want to move together as a group—solidarity among the people of Bugøynes is strong! The adult part of the population has a mixed professional background. With our competence and go-ahead spirit we have much to give. We would be bringing 50 children. We are interested in moving south of Trøndelag. Even if there are difficulties in providing jobs there too, we won't let that scare us. We can help in creating new jobs. (as seen in Apostle et al. 1998: 297–298)

Their goal was to attract attention to the crisis facing their community and they were successful. Since the 1980s, the community and their fisheries industry have experienced ups and downs (see Apostle et al. 1998 for an overview), including the fish plant reopening and then going bankrupt again in 1996. However, the community is now a leader in catching and selling live king crab through innovative solutions that bring their product from fisherman to plate.

Norway King Crab has a facility in Bugøynes, which is where the fishery is located, and a crab hotel in Oslo, where live crabs are trucked to be flown all over the world. They are also integrating technology and research and development into their approach. For example, each king crab has a QR code, which when scanned by the consumer will provide information on size, catch date, a bio of the fisherman, and information about the catch site (Norway King Crab 2018a, b). The information about the fisherman and catch site uses high-quality video and imagery. This helps to establish a stronger connection between the consumer and the people and places producing the food that we eat.

SmartICE: Canada[1]

SmartICE is a community-government-university-industry collaboration that integrates adapted technology, remote sensing, and Inuit Traditional Knowledge to promote safe travel for all stakeholders in northern coastal environments. The founding partners include the Memorial University of Newfoundland, the Nunatsiavut Government, the Nain Research Centre, C-CORE, Ikaavik, Mittimatalik Hunters and Trappers Organization, the Canadian Ice Service, and the Hamlet of Pond Inlet Nunavut. Led by Dr. Trevor Bell, a geographer at Memorial University, SmartICE started as a social enterprise research project to monitor sea-ice conditions combining traditional Inuit knowledge about the ice with real-time satellite imaging and ice-sensing technology (Green 2016). Sea-ice across the North is vital for transportation between communities and to secure country foods for food security and cultural well-being. Arctic waterways are also increasingly being used for shipping. However, with climate change, this sea-ice is changing at an alarming rate and becoming less predictable for travel.

SmartICE was initially piloted in the Inuit communities of Nain, Nunatsiavut, and Pond Inlet, Nunavut. Nain had a population of 1125 people in 2016 (Statistics Canada 2017) and is the most northern community in the province of Newfoundland and Labrador on the east coast of Canada (Pitt and Pitt 2015). Pond Inlet had a population of 1617 people in 2016 and is located on the northern part of Baffin Island.

There are several components to the SmartICE technology, including (1) SmartBUOY, a stationary sensor, measures sea-ice thickness and transmits this information by satellite and (2) SmartQAMUTIK, a mobile sensor that is carried by *qamutik* (or sled) to measure sea-ice thickness along travel routes. This information is combined with traditional knowledge to generate maps that indicate Go, Slow, No-Go colour-coded travel zones (SmartICE 2018). The technology is made in the North for the North using adaptive production for cold Arctic temperatures including hard wiring, insulation, heaters, and interfaces that can easily be used with gloves (Canadian Northern Economic Development Agency 2017). As a social enterprise, there is also a youth-training component where at-risk youth in Nunatsiavut learn the technical skills needed to assist in the production of the SmartICE technology and monitoring systems (Green 2016).

After winning the Arctic Inspiration Prize in 2016, SmartICE was developed into a social enterprise and is expanding to communities across Northern Canada and around the world.

---

[1] Adapted with permission from Vinodrai and Hall (2018).

They have received funding from the Canadian federal government along with several other agencies and their partnerships have expanded across the North (Canadian Northern Economic Development Agency 2017). Going forward, their business model is focused on four pillars: improving predictability and planning for safe on-ice travel and shipping routes; expanding employment and training opportunities for local communities; exploring new economic prospects and strengthening existing markets; and creating an extensive bank of validated ice data for custom solutions (SmartICE 2018).

The Node Pole: Sweden

Luleå, Sweden, is located just under 100 kilometers south of the Arctic circle. Its population in 2016 was roughly 77,000 people, while the regional labor market (including the municipalities of Boden, Kalix, Piteå, Älvsbyn, and Luleå) had a population of 170,000 people. Traditional industries in Luleå include the fisheries, mining (particularly ironworks), and forestry. The community is also home to the Luleå University of Technology, which has a strong applied research focus through collaborations with industry in the surrounding region (Luleå Kommun 2017).

In 2013 Facebook opened its first datacenter or server farm outside the United States in Luleå. These server farms store massive amounts of data that are produced by users every day around the world. For Facebook, this is equivalent to 350 million photographs, 4.5 billion likes, and 10 billion messages per day (Harding 2015). They are often massive in size (e.g. 30,000 square meters) and they require huge amounts of electricity, accounting for roughly 2% of global power demand. As a result, they require strong, secure, and stable power and internet connectivity (Gregory 2013).

The company was attracted to Luleå by a number of place-based attributes including an abundance of cheap hydroelectric power remaining from Luleå's industry legacies in iron, steel, and paper. Perhaps, more importantly, Luleå marketed its cold Arctic temperatures as a natural coolant. According to Bickford et al. (2016), new technology is being used to cool the servers using outside air for roughly 8 months of the year while the remaining months use hydroelectric power. As a result, the Luleå datacenter is the "most energy efficient computing facility ever built" according to Facebook (Harding 2015). It is also 30% more cost-effective (Bickford et al. 2016) and a more environmentally friendly facility because it does not rely on the burning of fossil fuels (Gregory 2013).

Estimates suggest that roughly 4500 jobs will be created in a 10-year period related to the server farm (Bickford et al. 2016). Since opening, applications in computer science courses at the Luleå University of Technology have increased while five other companies have established datacenters nearby (Harding 2015). In addition, a number of new hotels and restaurants have opened in Luleå.

Innovation with the Arctic, for the Arctic

While these examples highlight the benefits of adapting new technologies, in some instances the social impacts can be quite pronounced. For example, Ticoll (2015) highlights the benefits and challenges of autonomation in the automotive sector. These include safety and environmental improvements as well as lower operating costs. In addition, some occupations will become redundant while new jobs with new skills will be needed. Likewise, Australian researchers have been exploring the social impacts of new technologies in the mining sector including autonomous vehicles and remote operations. They suggest a 30–40% reduction in overall employment, coupled with a significant shift in the types of skills needed in the industry (McNab and Garcia-Vasquez 2011). Other anticipated impacts include population and economic decline in communities dependent on the mining sector and an overall decline in community investment (Cosbey et al. 2016).

The North will not be immune to these changes. For example, mining company Agnico Eagle is exploring how to integrate automated technology at mine sites in the Kivalliq region in Nunavut, Canada. This would include driverless long-haul trucks and remote-controlled scoop loaders. The company is hoping that these new technologies will decrease production costs at their remote mine sites (Neary 2018). Likewise, in 2017 Statoil activated the Valemon control room in Bergen, Norway, which is the company's first fully automated offshore oil and gas platform operated from land. The platform is operated by 14 employees on seven shifts for a 4-week production period. This is followed by two weeks where the platform will be "manned" for maintenance and inspection. Initial estimates were that this 2-week period will only require about one-third of the normal crew (Wright 2017). In the Circumpolar North, these disruptive technologies will require a rethinking of employment opportunities, skills training, and community impact and benefit agreements (see Kielland 2015).

To understand the social impacts of new technologies in the mining sector, researchers in Australia have created a useful tool to assess technologies which could be applied in the Arctic context. Ideally, this technology assessment would be conducted during the design phase of a particular idea or before introducing it to allow for modifications. Table 8.1 outlines this approach developed by Franks and Cohen (2012) and McNab and Garcia-Vasquez (2011), with an added emphasis on the spatial impacts and unique place-based responses by Hall (2018).

Conducting a technology assessment will identify impacts (both positive and negative) on employment, education, training, business development, infrastructure and services, and community development. At the same time, it can determine possible options to mitigate or offset negative impacts and enhance opportunities. Ideally, this will ensure new technologies are reflective of the needs, realities, and values in the North versus southern solutions that have the potential to usher in new challenges for the North.

Table 8.1 Technology assessment framework

What is the technology?	Describe the technology and its purpose
	Identify the drivers (e.g. labor shortages, health and safety, environmental considerations, competitiveness)
	Current picture (e.g. under development, stage, in use)
	Identify stakeholders involved in its development and where it is being developed
Where will it be implemented?	Scope and profile current or anticipated geographical context
How will it affect employment?	Who and where will it affect?
	Scope and profile stakeholders
	How will it affect them?
	Forecast risks and opportunities
	Imagine possible and not impossible outcomes
	Identify knowledge gaps
How will it affect skills, education, and training?	Who and where will it affect?
	Scope and profile stakeholders
	How will it affect them?
	Forecast risks and opportunities
	Imagine possible and not impossible outcomes
	Identify knowledge gaps
How will it affect business development?	Who and where will it affect?
	Scope and profile stakeholders
	How will it affect them?
	Forecast risks and opportunities
	Imagine possible and not impossible outcomes
	Identify knowledge gaps
How will it affect presence effects?	How will it affect regional development opportunities? (e.g. infrastructure, population, services, etc.)
	Imagine possible and not impossible outcomes
	Identify knowledge gaps
How will it affect community investment?	Forecast risks and opportunities
	Imagine possible and not impossible outcomes
	Identify knowledge gaps
What can be done?	Identify options to mitigate, enhance, offset, or constraints on implementation
	Identify who can do what and where

Source: Hall (2018), adapted from Franks and Cohen (2012) and McNab and Garcia-Vasquez (2011)

CONCLUSIONS

This chapter has explored innovation in the Circumpolar North, which is experiencing rapid and unprecedented environmental, political, social, and economic shifts. As noted throughout this chapter, innovation has the potential to counteract the pressing challenges facing northern communities from food security, to healthcare, and economic diversification. However, innovation could also deepen and present new challenges if ideas are not created with the North, for the North. New and adapted technologies could, for example, lead to employment losses, shifts in the education and skills needed for traditional occupations, and broader community impacts. It is, therefore, imperative that ideas and technologies are assessed prior to their introduction in the North to mitigate and offset any negative impacts while enhancing opportunities. Involving regional innovation stakeholders, especially Indigenous communities, to recognize the unique needs, values, and opportunities in the north is also essential.

Scholars, policymakers, and innovators alike also need to recognize the importance of place. Multiple Norths exist throughout the Circumpolar North, both between countries but also within countries. As a result, one-size-will-not-fit-all. Related to this, innovation ecosystems need to be strengthened across the North at a subnational scale to reflect regional variations. This "local buzz" (Bathelt et al. 2004) or regional collaboration is a significant part of the innovation process. Likewise, "global pipelines" (Bathelt et al. 2004) or international collaborations are also essential to introduce new ideas, knowledge, and opportunities. Having both strong regional innovation systems and international collaboration will help ensure that solutions are appropriate for northern regions and that northern innovation stakeholders are engaged.

Finally, more research is needed on policies to support innovation and regional innovation ecosystems in the North. One area worthy of more attention is the potential of smart specialization approaches to identify regional competitive advantages and unite innovation stakeholders around a common vision (Healy 2017). More research is also needed that compares innovation ecosystems throughout the Circumpolar North to identify challenges and lessons learned. In addition, further empirical case studies on entrepreneurs across the Circumpolar North will provide insights on supporting innovation and economic development.

REFERENCES

Aiello, R. 2017. Can PM Trudeau Keep Drinkable Water Promise to First Nations? *CTV News*, December 28. https://www.ctvnews.ca/politics/can-pm-trudeau-keep-drinkable-water-promise-to-first-nations-1.3736954.

Amin, A., and N. Thrift. 1994. Living in the Global. In *Globalization, Institutions, and Regional Development in Europe*, 1–22. Oxford: Oxford University Press.

———. 1995. Institutional Issues for the European Regions: From Markets and Plans to Socioeconomics and Powers of Association. *Economy and Society* 24 (1): 41–66.

Apostle, R.A., G. Barrett, P. Holm, S. Jentoft, L. Mazany, B. McCay, and K. Mikalsen. 1998. Bugøynes: A Case Study of Community Resistance. In *Community, State, and Market on the North Atlantic Rim: Challenges to Modernity in the Fisheries*, 297–306. Toronto: University of Toronto Press.

ARTEK. 2018. *About ARTEK*. http://www.artek.byg.dtu.dk/english/about_artek.

Bathelt, H., A. Malmberg, and P. Maskell. 2004. Clusters and Knowledge: Local Buzz, Global Pipelines and the Process of Knowledge Creation. *Progress in Human Geography* 28 (1): 31–56.

Becattini, G. 1990. The Marshallian Industrial District as a Socio-Economic Notion. In *Industrial Districts and Inter-Firm Co-Operation in Italy*, ed. P. Pyke, G. Becattini, and W. Sengenberger, 37–51. Geneva: International Institute for Labour Studies (ILO).

Bickford, S.H., J.E. Krans, and N. Bickford. 2016. Social and Environmental Impacts of Development on Rural Traditional Arctic Communities: Focus on Northern Sweden and the Sami. *Journal of EU Research in Business* 2016: 1–11.

Canadian Northern Economic Development Agency. 2017. Getting Smart about Sea Ice. *News Release*, June 29.

CBC. 2018. Study Shows Cost of Food Varies Greatly Across Yukon. *CBC News*, March 29. https://www.cbc.ca/news/canada/north/yukon-food-cost-anti-poverty-coalition-1.4597639.

CCHRC. 2018. *Cold Climate Housing Research Centre – Programs*. http://www.cchrc.org/programs.

Coates, K., and D. Landrie-Parker. 2016. *Northern Indigenous Peoples & the Prospects for Nuclear Energy*. Saskatoon: International Centre for Northern Governance and Development.

Coates, K., and G. Poelzer. 2014. Arctic Innovation. In *Shared Voices*. Finland: UArctic International Secretariat.

Cooke, P., and K. Morgan. 1998. *The Associational Economy: Firms, Regions, and Innovation*. Oxford: Oxford University Press.

Cosbey, A., H. Mann, N. Maennling, P. Toledano, J. Geipel, and M. Dietrich Brauch. 2016. *Mining a Mirage? Reassessing the Shared-Value Paradigm in Light of the Technological Advances in the Mining Sector*. Winnipeg: International Institute for Sustainable Development.

Deloitte. 2015. *Age of Disruption: Are Canadian Firms Prepared*. Toronto: Deloitte. https://www2.deloitte.com/ca/en/pages/insights-and-issues/articles/future-of-productivity-2015.html.

Dillow, C. 2018. Russia and China Vie to Beat the US in the Trillion-Dollar Race to Control the Arctic. *CNBC*, February 6. https://www.cnbc.com/2018/02/06/russia-and-china-battle-us-in-race-to-control-arctic.html.

Exner-Pirot, H., L. Heininen, and J. Plouffe. 2017. Introduction – Change and Innovation in the Arctic. In *Arctic Yearbook 2017*, ed. L. Heininen, H. Exner-Pirot, and J. Plouffe, 11–18. Akureyri: Northern Research Forum. http://www.arcticyearbook.com.

Expert Panel on Federal Support to Research and Development. 2012. *Innovation Canada: A Call to Action. Review of Federal Support to Research and Development – Expert Panel Report*. Ottawa: Government of Canada.

FCM. 2017. *Northern Broadband*. https://fcm.ca/home/issues/northern-and-remote/northern-broadband.htm.

Franks, D.M., and T. Cohen. 2012. Social Licence in Design: Constructive Technology Assessment Within a Mineral Research and Development Institution. *Technological Forecasting and Social Change* 79 (7): 1229–1240.

Government of Canada. 2018. *The Canadian High Arctic Research Station (CHARS) Campus*. https://www.canada.ca/content/canadasite/en/polar-knowledge/CHARScampus.html.

Green, J. 2016. Nobel of the North. *The Gazette*, December 9. Memorial University of Newfoundland.

Gregory, M. 2013. Inside Facebook's Green and Clean Arctic Data Centre. *BBC News*, June 14. https://www.bbc.com/news/business-22879160.

Hall, H.M. 2018. Remote Controlled: Technology in the Mining Sector & the Future of Development in Peripheral Regions. *SSHRC Insight Grant*.

Hall, H.M., and B. Donald. 2009. *Innovation and Creativity on the Periphery: Challenges and Opportunities in Northern Ontario*. Working Paper Series: Ontario in the Creative Age. REF. 2009-WPONT-002.

Hall, H.M., and K. Vodden. In Press. Learning, Knowledge Flows and Innovation in Canadian Regions. In *Regional Development: A Critical Review of Theory, Practice, and Potentials in the Canadian Context*, ed. K. Vodden, D. Douglas, S. Markey, and B. Reimer. London: Routledge.

Hall, H.M., and J. Walsh. 2013. *Knowledge Synthesis. Advancing Innovation in Newfoundland and Labrador Project*. St. John's: Harris Centre.

Harding, L. 2015. The Node Pole: Inside Facebook's Swedish Hub Near the A Circle. *The Guardian*, September 25. https://www.theguardian.com/technology/2015/sep/25/facebook-datacentre-lulea-sweden-node-pole.

Harrison, J. 2006. Re-Reading the New Regionalism: A Sympathetic Critique. *Space and Polity* 10 (1): 21–46.

Healy, A. 2017. Innovation in Circumpolar Regions: New Challenges for Smart Specialization. *Northern Review* 45: 11–32.

Hintsala, H., S. Niemelä, and P. Tervonen. 2017. Arctic Innovation Hubs: Opportunities for Regional Co-Operation and Collaboration in Oulu, Luleå, and Tromsø. *Northern Review* 45: 77–92.

Hodge, G., H.M. Hall, and I.M. Robinson. 2016. *Planning Canadian Regions*. 2nd ed. Vancouver: UBC Press.

Isaksen, A., and J. Karlsen. 2010. Different Modes of Innovation and the Challenge of Connecting Universities and Industry: Case Studies of Two Regional Industries in Norway. *European Planning Studies* 18 (12): 1993–2008.

Johnstone, H. & Haddow, R. (2003). Industrial Decline and High Technology Renewal in Cape Breton: Exploring the Limits of the Possible. In D.A. Wolfe (ed.), Clusters Old and New: The Transition to a Knowledge Economic in Canada's Regions, 187-212. Montreal & Kingston: McGill-Queen's University Press.

Jordans, F. 2017. Battle for Arctic Resources Heats Up as Ice Recedes. *Global News*, August 23. https://globalnews.ca/news/3690400/arctic-resources-shipping-routes/.

Josefsen, E. 2010. *The Saami and the National Parliaments. Resource Centre for the Rights of Indigenous Peoples*. Geneva, New York: Inter-Parliamentary Union and United Nations Development Programme.

Kielland, N. 2015. Supporting Aboriginal Participation in Resource Development: The Role of Impact and Benefit Agreements. Papers in the Library of Parliament's. In *Brief* series. Publication No. 2015-29-E. Ottawa: Library of Parliament.

Lamb, D.M. 2017. 'It Scares Me': Permafrost Thaw in Canadian Arctic Sign of Global Trend. *CBC News*, April 17. https://www.cbc.ca/news/canada/north/it-scares-me-permafrost-thaw-in-canadian-arctic-sign-of-global-trend-1.4069173.

Leydesdorff, L. 2012. The Triple Helix, Quadruple Helix, …, and an N-Tuple of Helices: Explanatory Models for Analyzing the Knowledge-Based Economy? *Journal of the Knowledge Economy* 3 (1): 25–35.

Luleå Kommun. 2017. *Short Facts About Luleå and Luleå Municipality*. https://www.lulea.se/download/18.58e5fa4715e2f9269a28cee/1505286210691/korta%20fakta%20engelsk%20nr%204%202017%20webb.pdf.

Martin, R. 2010. Roepke Lecture in Economic Geography—Rethinking Regional Path Dependence: Beyond Lock-in to Evolution. *Economic Geography* 86 (1): 1–27.

McNab, K.L., and M. Garcia-Vasquez. 2011. *Autonomous and Remote Operation Technologies in Australian Mining*. Prepared for CSIRO Minerals Down Under Flagship, Minerals Futures Cluster Collaboration, by the Centre for Social Responsibility in Mining, Sustainable Minerals Institute, The University of Queensland, Brisbane.

MNDM. 2017. *Northern Ontario Winter Roads.* https://www.mndm.gov.on.ca/en/northern-development/transportation-support/northern-ontario-winter-roads.

Morgan, K. (1997). The Learning Region: Institutions, Innovation and Regional Renewal. Regional Studies, 31(5), 491-503.

Neary, D. 2018. Agnico Eagle Prepares for Automated Kivalliq Mine Sites. *Nunavut News*, April 2. http://nunavutnews.com/nunavut-news/agnico-eagle-prepares-for-automated-kivalliq-mine-sites/.

Newman, D.G. 2014. *Revisiting the Duty to Consult Aboriginal Peoples.* Saskatoon: Purich Publishing Limited.

Niemelä, S, and H. Hintsala. 2016. Arctic Business Potential from Oulu Region's Perspective – Opportunities and Obstacles. *ePOOKI*, March 3. http://www.oamk.fi/epooki/2016/arctic-business/?ccm_paging_p_b1802=2.

Northern Review. 2017. *Special Issue – Innovation in the Circumpolar North* 45.

Norway King Crab. 2018a. *Norway King Crab.* https://nkc.no.

———. 2018b. *Tracking.* https://nkc.no/track/.

OECD. 2005. *Oslo Manual: Guidelines for Collecting and Interpreting Innovation Data.* OECD Publishing.

———. 2015. *Defining Innovation.* https://www.oecd.org/site/innovationstrategy/defininginnovation.htm.

Pigford, A., G.M. Hickey, and L. Klerkx. 2017. Towards Innovation (Eco)Systems: Enhancing the Public Value of Scientific Research in the Canadian Arctic. In *Arctic Yearbook 2017*, ed. L. Heininen, H. Exner-Pirot, and J. Plouffe, 24–49. Akureyri: Northern Research Forum. http://www.arcticyearbook.com.

Pitt, R.D., and J.E.M. Pitt. 2015. Nain. *Historica Canada.* https://www.thecanadianencyclopedia.ca/en/article/nain/.

Porter, M.E. 1990. *The Competitive Advantage of Nations.* New York: Free Press.

Proulx. (1992). Innovative milieus and regional development.Canadian Journal of Regional Science, 15(2), 149–154.

Skura, E. 2016. Food in Nunavut Still Costs Up to 3 Times National Average. *CBC News*, June 24. https://www.cbc.ca/news/canada/north/nunavut-food-price-survey-2016-1.3650637.

SmartICE. 2018. *Technology.* https://www.smartice.org/technology/.

Statistics Canada. 2017. Nain, T [Census subdivision], Newfoundland and Labrador and Newfoundland and Labrador [Province] (table). Census Profile. 2016 Census. Statistics Canada Catalogue no. 98-316-X2016001. Ottawa.

Ticoll, D. 2015. *Driving Changes: Automated Vehicles in Toronto.* Discussion Paper. Toronto: Innovation Policy Lab, Munk School of Global Affairs.

Tödtling, F., and F. Tripple. 2005. One Size Fits All? Towards a Differentiated Regional Innovation Policy Approach. *Research Policy* 34: 1203–1219.

Vindorai, T., and H.M. Hall. 2018. *Innovation-Led Growth and Economic Development Beyond the Metropolis: A Review of Best Practices.* Prepared for the Federal Economic Development Agency for Southern Ontario (FedDev Ontario).

Wolfe, D.A. (2009). 21st Century Cities in Canada: The Geography of Innovation. The 2009 CIBC Scholar-in-Residence Lecture. Ottawa: Conference Board of Canada.

Wright, S. 2017. Statoil Activates Valemon Automated Rig Control Room. *InnovOil*, Issue 59. https://www.innovoil.co.uk/single-post/2017/11/29/Statoil-activates-Valemon-automated-rig-control-room.

Young, T.K., and Susan Chatwood. 2017. Delivering More Equitable Primary Health Care in Northern Canada. *Canadian Medicinal Association Journal* 189 (45): E1377–E1378.

Yukon College. 2017a. *Cold Climate Innovation.* https://www.yukoncollege.yk.ca/innovation/cold-climate-innovation.

———. 2017b. *Tr'ondëk Hwëch'in Cold Climate Greenhouse.* https://www.yukoncollege.yk.ca/innovation/trondek-hwech-in-cold-climate-greenhouse.

———. 2017c. *Innovation Prize.* https://www.yukoncollege.yk.ca/innovation/yukon-innovation-prize.

Zilio, M. 2018. Vancouver Couple Donate $60-Million to Esteemed Arctic Inspiration Prize. *The Globe and Mail*, January 31. https://www.theglobeandmail.com/news/national/vancouver-couple-donate-60-million-to-esteemed-arctic-inspiration-prize/article37812955/.

CHAPTER 9

Arctic Advanced Education and Research

Anne Husebekk, Kenneth Ruud, Sveinung Eikeland, and Geir Gotaas

The basic premise for a discussion of the role of universities in development processes and development models is that universities, supplemented by university colleges and vocational education programs, are key actors in educating a skilled labor force and in engaging in research that is relevant to the regions they serve, and to the international research community, and, thus, that they are instrumental for the development of a region.

For a university to succeed in this role, certain basic prerequisites must be met, such as infrastructure, a certain degree of urbanization and a long-term political commitment to building a strong research-based university—a university that has a regional identity, but that is also globally relevant.

Infrastructure, be it buildings, means of communication (high-speed internet connection or easy access to the rest of the world through travel) or a stable and reliable power supply, are things we tend to take for granted, but they are, nevertheless, key requirements for a well-functioning academic institution—be it a university, a university college or a research institute. Institutions like these are, in turn, key contributors to the overall local and regional infrastructure, and they are in many cases drivers of infrastructure development. In this capacity (separate from, and in addition to the services they provide), academic institutions can have significant societal impact. Research stations are examples of a different kind of infrastructure that can be extremely valuable in a regional context, both because they (often) are closely linked to certain natural characteristics (and therefore cannot be located in any other place), and because they attract leading scholars from across the world engaged in

A. Husebekk • K. Ruud • S. Eikeland • G. Gotaas (✉)
University of Tromsø, Tromsø, Norway
e-mail: anne.husebekk@uit.no; kenneth.ruud@uit.no; sveinung.eikeland@uit.no;
geir.gotaas@uit.no; geir.gotaas@npolar.no

133

K. S. Coates, C. Holroyd (eds.), *The Palgrave Handbook of Arctic Policy and Politics*, https://doi.org/10.1007/978-3-030-20557-7_9

research and education linked to that particular set of natural characteristics. They may also have significant societal impact, though this depends on the nature of the research conducted at the station. However, research stations do not offer the broad academic scope and the education opportunities for undergraduate students that universities or university colleges can provide, and consequently their importance in contributing to regional development is much more limited.

In order to deliver high-quality research and education, and deliver it sustainably (i.e. over an extended period of time), a university relies on a certain "critical mass" of people, competence and resources. This critical mass makes a university attractive for students and qualified staff, but it is also important if the university's potential for contributing to regional development is to be realized; there must be a public and private sector for the university to interact with, or at least a clear political intention to support the development of these sectors.

The importance, relevance and "value" of a university may be easiest to identify at the regional level, but every single university is part of a global knowledge network, and as such contributes to the overall global capacity building—each university bringing to the knowledge commons their particular (regional) perspectives.

Arctic research is a prime example of the value that this total global knowledge production brings to the table. The Arctic is a key priority for both Arctic and non-Arctic states, primarily because the changes in the climate—already very visible in the Arctic—affects the whole world and creates challenges and opportunities of global interest. The global reach of climate change means that a global and holistic approach is necessary in order to understand the mechanisms at play, and to try to find ways to mitigate the changes that we see coming. At the same time, the most pronounced effects of climate change are observed in the Polar Regions, and this means that some of the most pressing research questions raised by the global changes can only be answered by carrying out studies in the Arctic and in the Antarctic. We believe that universities located in the Arctic are uniquely qualified and equipped to engage in research aimed at providing answers to these pressing questions—answers that must be at the core of the international Arctic policy development.

Policy development is one area where universities can provide valuable background information and insights—that is, provide the foundation needed to make knowledge-based decisions. Commercialization and business development is another area where universities play a similar role in that research-based innovation within a university or a research institute frequently serves as a starting point for business development outside the institution. Frequently, such activities rely heavily on interactions with existing industries. In regions where the industrial base is weak, this presents the research institutions and the governmental support structures (in the case of Norway, e.g. Innovation Norway—innovationnorway. no) with a considerable challenge. One report (Gjelsvik 2015) comparing UiT's innovation activities, entrepreneurship output and interaction with local businesses with those of other young universities and university colleges in Norway (the University of Stavanger and Nord University in particular) found that the absence of a strong research-based industry in the region around Tromsø has made UiT adopt unique strategies. Rather than focusing on existing industries, it has built on the comprehensive strength of its academic profile and—to a much larger extent than the other two universities—focused on creating new businesses and start-up companies. This has led to a diversification of local businesses in Tromsø, and the city today is home to a very strong biotechnology cluster as well as a number of small and medium-sized companies in the information technology sector. With the increasing focus on personalized medicine and health technology, and with strong research groups in the health sciences and in big data analyses at UiT, the region has a potential for further growth in this business area.

UiT The Arctic University of Norway: "Ordinary" but Unique

The idea of a university in Tromsø was first put forward in 1918, when the visionary businessman Hans A. Meyer from Mo i Rana in Northern Norway wrote an op-ed in the national newspaper "*Tidens Tegn*". He argued that a university in the North was necessary to secure a work force that both had an in-depth understanding of the region, and that could contribute to its development. At that time, people with higher education would typically come from the south, stay and work for a short while in the north, and then leave—having a limited understanding of the region when they arrived, and to a very limited extent contributing to its development.

However, 50 years passed before the Norwegian parliament in 1968 decided to establish a university in Tromsø—after intense deliberations and against considerable skepticism. At the time, few believed that a university so far north could attract academic staff, or that the goal of 2000 students was attainable.

A lack of health care professionals in the north of Norway was one of the most important arguments for establishing of a university in Tromsø. However, while a medical faculty was a central part of the new university (and still is), the Norwegian parliament decided to establish the University of Tromsø (UiT) as a classical, comprehensive university. Other core activities at the new university were based on research at Tromsø Museum (from 1872), and at the Geophysical Institute in Tromsø (from 1918), and linked to financially important businesses, such as fisheries (and later, aquaculture). Thus, UiT, from 1972, when the first students were matriculated, was given a research and education profile comparable in breadth and scope to that found at the existing universities in Bergen and Oslo. We believe that this bold and foresighted decision by the Norwegian parliament in 1968 has been central to the success of the University.

The new university did stand out in one area, though; it was given a particular, national responsibility for research and education related to Sami language, history and culture. This is a responsibility that is still very much a part of UiT's identity, and it is a prioritized area also in our most recent strategy.

It is safe to say that Tromsø (and Northern Norway) of 1972 was somewhat unprepared for a university, and for the political discussions (in many ways quite radical) that staff and students brought to the region. The university was met with considerable skepticism—also locally. However, 50 years later, the importance and significance of a regional university, with national and international prominence in a number of academic fields, is very much appreciated. The university has delivered doctors and dentists, physics teachers and philosophers, engineers and economists, both to the region and to the nation. Furthermore, research at the university has had a profound impact in that it has improved socioeconomic characteristics, public health, industrial development, and—perhaps more difficult to measure, but no less important—the self-esteem of people in the north. Instead of being the subject of research, the people in the region now define and engage in their own research agenda. The university has brought attention to the region; it has brought visitors and new permanent residents, and it has made Northern Norway more international. The city of Tromsø alone, with a population of 76,000, has citizens from more than 140 countries around the world—many of them attracted by opportunities at the university, and at businesses and research institutes that have originated from or collaborate closely with the university.

UiT is the northernmost university in the world. After merging with four university colleges since 2008, UiT now has main campuses in all three counties in Northern Norway (i.e. Nordland, Troms and Finnmark), and in Longyearbyen on Svalbard. The university hosts more than 16,600 students, and a staff of almost 4000. The "internationalization" seen in Tromsø after the establishment of a university in the city is evident also in the other campus cities—particularly in Narvik, Harstad and Alta, which were host cities for university colleges prior to the mergers of these institutions with UiT.

In fulfillment of the political ambitions as they were formulated in 1968, UiT of today is a comprehensive university engaged in education, research and knowledge dissemination in a wide range of fields—as are the other comprehensive universities in Norway—thereby contributing to the global academic community and to the national and the regional economy. However, UiT is also characterized by an integration of classical university disciplines (such as mathematics, linguistics and political science) with vocational study programs addressing needs and demands of specific professions, both in the public and the private sector. This interaction between "pure" academic disciplines and broader study programs is found at our Faculty of Health Sciences Educating Physicians, Dentists and Nurses, at our Faculty of Humanities, Social Sciences and Teacher Training Educating Teachers, Social Workers and Arctic Adventure Tourism Guides, at our Faculty of Engineering and Technology, and our Faculty of Science and Technology—both of them educating engineers, and at our Faculty of Biosciences, Fisheries and Economics Educating Economists, Marketing Specialists and Masters in Aqua-medicine.

The fact that UiT is a comprehensive university is also reflected in its research priorities, which include epidemiology (linked to a series of population studies in Tromsø—the first of which was conducted in 1974); development of vaccines against infectious diseases in salmon, which has paved the way for the multibillion dollar industry that today is Norwegian aquaculture—with next to no use of antibiotics; Sami history, culture and language; remote sensing—where sensors and algorithms that will be key to our monitoring and understanding of the impact of climate change on the Arctic environment are being developed; and law of the sea research related to the sustainable and equitable utilization of marine resources, and marine environmental protection.

These examples from our education and research portfolio clearly show that UiT has an Arctic focus and that our research projects have an impact on Arctic communities. In addition, UiT has a large number of projects not directly related to the Arctic—exemplified by a center of excellence related to pathogenesis, diagnosis and treatment of thrombosis, a center of excellence in theoretical chemistry and a highly rated research group in theoretical linguistics.

Since 1968, more than 58,000 candidates have graduated from UiT, and more than 70% of them have remained in Northern Norway (UiT 2018). Every single municipality in the region employs graduates from the university. The same is true for most enterprises. Furthermore, a number of new companies have been established based on ideas from UiT employees and graduates.

The level of education among the inhabitants close to a university campus is in general above the mean level for Norway, but the level decreases proportionally with the distance from a university campus (UiT 2018). The exceptions from this general rule are a few important administrative centers without a university campus, but with other key public or private institutions or businesses. A prime example is Karasjok, in the center of Sápmi and host municipality for the Sami parliament. A majority of the employees in the parliament administration are university graduates, many of them UiT alumni.

The importance of a university (or university college) campus can be illustrated by the situation in Alta (in Finnmark). In a report from Statistics Norway (2013), Alta was shown to be one of the Norwegian municipalities in which a university college (Finnmark University College), and later a university (UiT) has had the greatest impact as an attractor for students and for employees with a university degree—regardless of their line of work or their employer. While the (then) university college recruited 33% of its students from Finnmark county, 66% of their students sought employment in Finnmark after graduation—contributing to a net influx of highly qualified persons to the county. However, the general picture is still that a majority of municipalities in northern Norway are lagging behind the rest of the country when it comes to their ability to attract university graduates, and the formal qualification level of the general work force.

The demographic trends in Northern Norway largely mirror the trends in the rest of the world; elderly people make up the majority of the population in rural and remote communities, young women tend to seek higher education opportunities and move to cities, while young men tend to stay behind—uneducated (or under-educated)—in the countryside (Megatrends 2011). Like many other sparsely populated regions, Northern Norway has a lower proportion of youth finishing their upper secondary level education than the rest of the country. But at the same time, Northern Norway has a low level of unemployment and is, in fact, dependent on an immigrant work force in order to fill vacancies.

Following the merger of The University of Tromsø with Finnmark University College in 2013, the new, resulting, multi-campus university was given a new name by the government: University of Tromsø—The Arctic University of Norway. The addition of "The Arctic University of Norway" to the name was in part a recognition of the fact that UiT was (and still is, in 2018) the leading Norwegian university in polar research (Norwegian Polar Research 2017) (both in terms of the number of researchers engaged in Polar research, and in the research output) and in part an encouragement from the government to the university to focus even more on this aspect of our education, research and outreach. In response to this, UiT published a revised institutional strategy in 2014, with the overarching goal of "Developing the High North". While this was not a radical change, but rather a natural continuation of the first 50 years, the awareness of being an Arctic university was emphasized and communicated more clearly—internally as well as externally.

As part of this sharpened focus on Arctic issues, a key point in UiT's revised strategy from 2014 was to increase the involvement in all Arctic policy spheres. With the introduction of the Sustainable Development Goals by the United Nations in 2015, global perspectives have been interwoven with the Arctic frame of reference, and this has made it even more important for UiT to develop its key role in the university landscape in the Arctic region—both through collaboration with other institutions, and through strategic research priorities.

THE UNIVERSITY LANDSCAPE IN THE CIRCUMPOLAR NORTH

Through its presence in 10 cities—most of them north of the Arctic Circle—UiT The Arctic University of Norway is delivering both on the task that it was given 50 years ago of building competence in Northern Norway, and on a key undertaking for all universities—engaging in cutting edge research that contributes to regional and national development. With the new national and international interests in the Arctic, and the societal changes that follow from long-term trends of urbanization, globalization and demographic changes, the experience that UiT has gained over the last 50 years in developing higher education and research in North Norway is relevant in a broader Arctic and northern context. Similarly, Norway can benefit from adopting best practices from other Arctic countries, and from learning how other countries have faced challenges that are similar to the ones we face in "our" North.

Collaboration with other universities both in the Arctic and outside of the region is key to the success of UiT in this learning process, and in adopting best practices.

Among the Arctic countries, Sweden and Finland are perhaps the ones that are most similar to Norway in terms of the challenges we face. Consequently, UiT is collaborating closely with Umeå University and Luleå Technical University in Sweden, and with Oulu University and University of Lapland in Finland in a network that we have labeled "Arctic Five". Through this network we have identified six areas that are of strategic importance to all our universities, and where we all have set aside strategic funding to promote collaboration. Energy, health, mining, tourism, teacher training and indigenous issues are areas where we see similar opportunities for knowledge development and capacity building, while we at the same time acknowledge that Norway, Sweden and Finland address these issues from different starting points, and that—as a consequence—we can learn from each other.

While Norway, Sweden and Finland (and to a certain extent Iceland) have many similarities when it comes to factors such as infrastructure, demographics, climate and so on, some of the challenges that need to be addressed in the other four Arctic states—the United States, Canada, Greenland and Russia—are quite different.

With a population density of 0.028/sq. km, Greenland ranks as the most sparsely populated country in the world. Large distances, poorly developed infrastructure and a small population base (56,000) make it extremely challenging to build and maintain higher education and research at a level that can contribute to local/regional development. Today, many Greenlandic youths get their higher education at Danish universities, and as we know from a Norwegian context, where you study has an impact on where you subsequently choose to live and work. This brain drain needs to be tackled, and Greenland is taking concrete steps toward expanding the breadth and scope of the education programs offered at Ilisimatusarfik (University of Greenland).

While Canada, the United States of America and Russia all have world-class universities, they face many of the same challenges as Greenland when it comes to providing their northerners with higher education, and building and maintaining research institutions that can contribute to regional development: vast areas, few people, and extreme infrastructure challenges. Broadly speaking, the university structure in the three countries are quite similar, with a number of very highly ranked universities that excel also in Arctic research, but which are typically located far south of the Arctic. This is obviously an oversimplification, and particularly so as far as Alaska goes, where the three universities in the University of Alaska system (Fairbanks, Anchorage, and Southeast) each have multiple campuses serving local communities—also in the north of that state.

In Canada, the Northwest Territories (NWT), Nunavut and Yukon all have plans for strengthening their education and research infrastructure, albeit through somewhat different approaches. While NWT is planning a transition of its Aurora college to a polytechnic university, Nunavut is working toward establishing a formal partnership between its Nunavut Arctic College and a southern university, and Yukon is in the process of re-shaping its Yukon College to Yukon University—starting this year by offering three bachelor's programs in Indigenous governance, business and northern studies. Full university status is expected to be in place by the spring of 2020.

In other words, Arctic states have chosen different strategies, but with the same end goal; strengthening culture, self-determination and governance in the North through education and research.

Other, more southern universities in the United States, Canada and Russia obviously face many of the same challenges, with long distances, sparse population and infrastructure challenges, but it is fair to say that these obstacles are particularly evident in the northern parts of all these three countries, as they are in Greenland, Iceland, and the northern parts of Norway, Sweden and Finland.

Along with common challenges, these regions have common opportunities; an abundance of natural resources, which in turn attracts the interest of actors from outside the region. The question then is how local communities in the north can take an active role in securing a sustainable exploitation of these resources when the external actors that they have to negotiate with might not always have regional development at the top of their list of priorities. This is where universities can play a key role, providing the tools that ensure local capacity building and competence development.

In addition to the efforts Arctic universities (as individual institutions) make to provide individual northerners and local communities with the tools needed to build capacity, this issue is typically high on the agenda in bilateral and multilateral university collaborations. Furthermore, it was a key concern in the discussions that were initiated by the Arctic Council in the late 1990s—discussions that led to the establishment of the University of the Arctic (UArctic) in 2001.

This university network was originally set up to facilitate cooperation between universities established and working in the Arctic region, focusing on capacity building in local communities with limited access to higher education programs. Distance learning was a core element of the activities initiated and promoted by UArctic, in particular through the development of a bachelor of circumpolar studies, offered by several member institutions, both on campus, and as a distance learning program.

Today, UArctic is a cooperative network of universities, colleges, research institutes and other organizations concerned with education and research in the North. UArctic builds and strengthens collective resources and collaborative infrastructure that enables member institutions to better serve their regions. Through cooperation in education, research and outreach, UArctic enhances human capacity in the North, promotes viable communities and sustainable economies and forges global partnerships. Almost 180 institutions are members, mostly from Arctic Council member states, but also from other countries. While the circumpolar study programs are still offered by several member universities, the main focus today is on collaboration through thematic networks, which cover a wide range of topics; health, education, natural sciences, engineering, technology, humanities, art, business, politics and law. In addition, UArctic has established four institutes; the latest addition is a Science and Research analytics institute.

Into an Uncertain Future

The Arctic is a hot topic also outside of the circumpolar north. The region is rich in natural resources such as fish, minerals, oil and gas. Tourists are visiting the region in larger numbers than ever before—paradoxically many of them come to experience the pristine wilderness and the quiet of the North. Thinner and less extensive sea ice cover means that maritime activity is on the increase, and moving further north than ever before. Today's sailings through the Northern Sea Route may in our lifetime be replaced by voyages straight across the Arctic Ocean.

The global climate change—more clearly manifested in the Polar Regions than anywhere else—might have a profound impact on flora and fauna on land and in the ocean, it might make natural resources more accessible, and it might force indigenous and non-indigenous people in the region to change their traditional way of life. The impact of these changes on the culture of indigenous peoples in particular is hard to assess today, but given the long history of these peoples' interaction with the natural environment, the impact is likely to be dramatic.

In order to prepare for, and adapt to an uncertain future we, as global citizens, need to make use of the combined strengths of research-based knowledge and traditional knowledge. In this process, universities in the north must play a key role. We as university communities have an obligation to monitor the changes carefully and to help people in the north understand how best to adapt to the changing environment. In fulfilling these obligations, universities in the circumpolar north will help promote sustainable development in the communities they serve, while at the same time contributing in a meaningful way to our common objective of reaching the 17 UN Sustainable Development Goals—thereby helping secure the best possible Arctic and global future.

References

Gjelsvik, M. 2015. *University-Firm Linkages as Drivers of Innovation.* https://www.forskningsradet.no/prosjektbanken/#/project/NFR/212275.

Megatrends. 2011. *TemaNord.* Nordic Council of Ministers, 527. www.norden.org/en/publications/publications/2011-527.

Norwegian Polar Research & Svalbard Research: Publication Analysis. 2017. ISBN 978-82-327-0260-2. https://www.nifu.no/publications/1476382/.

Statistics Norway. Report 6/2013 (In Norwegian). ISBN 978-82-537-8590-5. http://bit.ly/2CYHDY9.

UiT. 2018. *In House Analysis.* UiT, The Arctic University of Norway.

CHAPTER 10

Circumpolar Business Development: The Paradox of Governance?

Andrei Mineev, Elena Dybtsyna, and Frode Mellemvik

INTRODUCTION

The purpose of this chapter is to describe the state-of-the-art in commercial activity and raise stakeholders' awareness of challenges and opportunities for business development in the Arctic using comprehensive, comparable and regular socio-economic information.

In the past ten years, the Arctic regions with their abundant natural resources have attracted a lot of attention among nation states, global businesses and international policy-makers. Challenges and opportunities for sustainable socio-economic development in the Arctic were addressed by international co-operation institutions such as The Arctic Council, The Arctic Economic Council, The Nordic Council of Ministers, The Barents Euro-Arctic Council, the OECD, the World Economic Forum and also governments and organizations in the Arctic and non-Arctic states.

The coming years will bring many changes in the Arctic—changes in its economy, population, climate and environment (AMAP 2017). The Arctic is characterized by significant and increasing geopolitical interest as well as major opportunities for business development and value creation. At the same time, there are serious constraints and challenges, and indeed considerable knowledge gaps regarding the development of business and socio-economic life in the various Arctic regions.

A. Mineev • E. Dybtsyna • F. Mellemvik (✉)
High North Center for Business and Governance,
Business School, Nord University, Bodø, Norway
e-mail: andrey.mineev@nord.no; elena.dybtsyna@nord.no; frode.mellemvik@nord.no

© The Author(s) 2020   143
K. S. Coates, C. Holroyd (eds.), *The Palgrave Handbook of Arctic Policy and Politics*, https://doi.org/10.1007/978-3-030-20557-7_10

As the circumpolar areas belong to different national regimes, the information on social and economic issues and business development has been dispersed and not easily available.

Business Index North (BIN) is a project co-ordinated by the High North Center for Business and Governance[1] aiming to contribute to sustainable development and value creation in the Arctic. The overall goal of BIN is to set up a recurring, knowledge-based, systematic information tool for stakeholders and interested parties in the Arctic such as companies, governments, regional authorities, academia and media (Business Index North 2017, 2018). The BIN report gives both an overview and a detailed picture of the socio-economic development and business opportunities within the BIN area.

The first "Business Index North" periodic analytical report (Business Index North 2017) focused on socio-economic developments in eight northern regions of Norway (Finnmark, Troms and Nordland), Sweden (Norrbotten and Västerbotten) and Finland (Lapland, North Ostrobothnia and Kainuu). The second "Business Index North" report (Business Index North 2018) included two more regions—Murmansk Oblast' and Arkhangelsk Oblast' in North-West Russia.

These 10 ten regions are referred to collectively as the "BIN area" (Fig. 10.1). The definition of the BIN area correlates with the EU concept of a macro-region—an area including a territory from a number of different Member States or regions associated with one or more common features and challenges. The BIN area runs across national borders and although there are differences between the nations, it has common characteristics, challenges and indeed opportunities, not least when it comes to business development. The aim of the third BIN report (forthcoming in 2019) is to increase the geographic scope by including territories of US Alaska and the Canadian High North, as well as more regions of the Russian Arctic.

According to Arctic Business Analysis (2018) supported by the Nordic Council of Ministers, there is a need to strengthen and promote the collection and dissemination of Arctic-specific data, and the BIN project is mentioned as an important contributor to ensuring knowledge development and raising the awareness through data of the situation prevailing in the Arctic.

In this chapter we present the findings of the BIN report on its six major topics: (1) People, Life and Work; (2) Business Activities and Innovations; (3) Maritime Transportation by the Northern Sea Route; (4) Connectivity; (5) Renewable Energy; and (6) Cross-Border Co-operation. Each topic outlines the state of-the-art and key trends in the area and provides salient information to be born in mind in discussions on policies for socio-economic development

---

[1] The High North Center for Business and Governance, established in 2007 and located at Nord University, is a national center for research, education and policy development. The center develops and communicates knowledge to contribute to innovation, business creation and societal development in the Arctic or High North as we call this geographical area. The High North Center for Business and Governance focuses on assisting companies, organizations and public institutions to increase both awareness of and commitment to the opportunities in the High North.

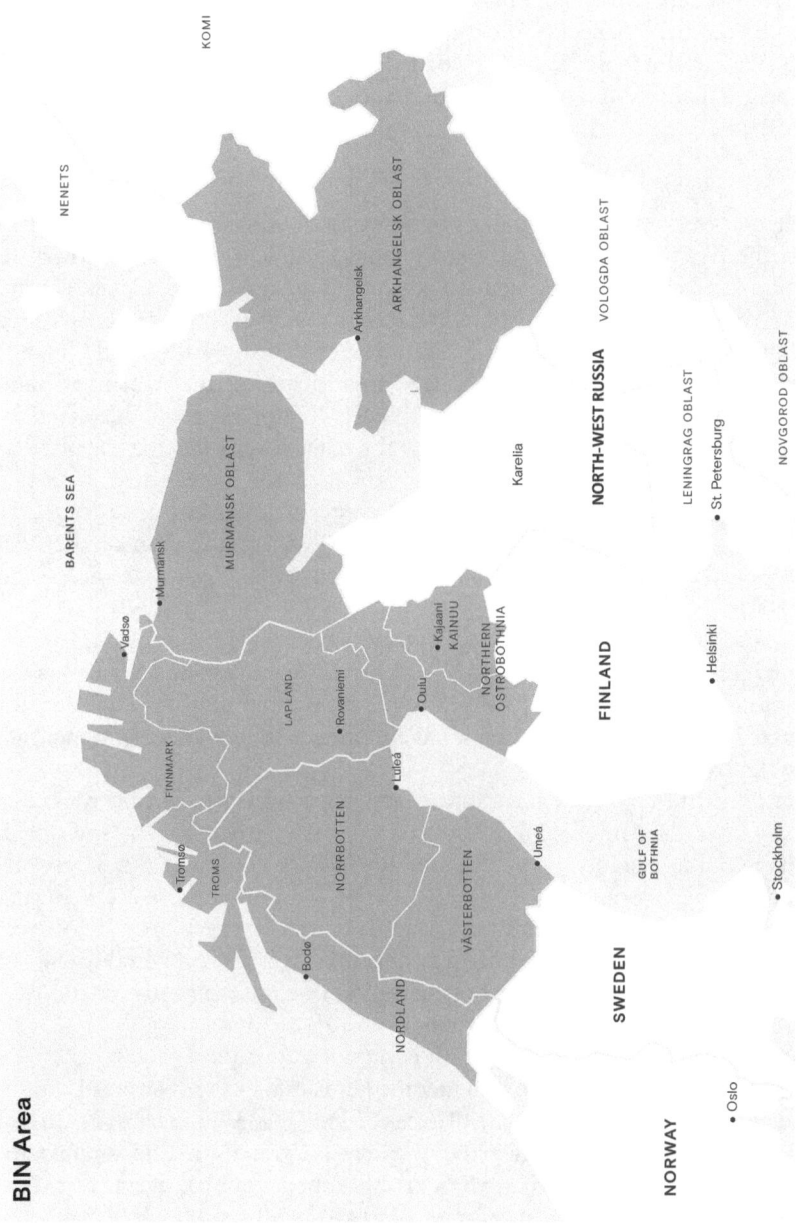

Fig. 10.1   The BIN Area—10 regions included in the BIN report (Business Index North 2018)

in the Arctic. The chapter moreover outlines several paradoxes and dilemmas that may affect the future development of business in the Arctic.

The geographic area considered in this chapter comprises Northern Norway, Northern Sweden, Northern Finland and North-West Russia—the area so far covered by the BIN project. Other territories of the circumpolar Arctic including US Alaska, Northern Canada, North-East Russia, Iceland and Greenland are to be gradually added in the coming issues of the BIN report.

PEOPLE, LIFE AND WORK

Circumpolar business development needs people. Are there enough people in the North? What about their education and health? Do they earn enough to reside and live happily in the Arctic? These issues are addressed in this section.

The BIN reports show that the population growth in the Nordic BIN area is 2.7 times slower than in the Nordic countries (Norway, Sweden and Finland) as a whole. The population in the BIN area including the Murmansk and Arkhangelsk (without the Nenets Autonomous Region) regions decreased by 3.1% in the period 2007–2016. However, the main concern is that the population of the BIN area is aging. The population aged 65+ grew by 13.3%, while that aged 0–19 declined by 7.5%. When it comes to the most productive age group, the situation is not so encouraging either, since population aged 20–39 declined by 6.8% and that aged 40–65 declined by 3.7% during the period 2007–2016.

The BIN results show that life expectancy at birth in the Nordic BIN area is higher by 13.6 years for males and 7.6 years for females than in the Russian BIN regions. In the Russian BIN regions, life expectancy at birth is 65 years for males and 76 for females, although it has a tendency to increase at a higher rate than in the Nordic countries.

Regarding the level of education among people in the Nordic BIN area university education among 20–59 year old males is fairly high, lagging only 5 percentage points behind the average for the Nordic countries and 3 percentage points for females in that age group. At the same time, there is a worrying disproportion of highly educated men and women in the Nordic BIN area: only 26% of all adult males hold higher education degrees compared to 36% of all adult females. This may cause girls with higher education to leave the BIN area, either for family reasons or for work.

How much do people in the North earn from their jobs? The recent BIN report shows that disposable annual income per capita in the Nordic BIN is on average 22,700 euros and in the Russian BIN area 5990 euros in 2016.[2] However, this means that in the whole BIN region, the disposable annual family income is 4–10% lower than the corresponding country average (except Murmansk region, which outperforms North-West Russia by 14%). Income inequality in all 10 BIN regions (measured as the ratio of the income level of

---

[2] This difference looks much smaller if income figures are considered in terms of purchasing power parities for the respective national currencies.

10% of the richest and 10% of the poorest) is lower than their corresponding country averages. This creates conditions for more coherent societies in the North.

There is a favorable employment development in the Nordic BIN area compared with the Nordic nations, and the Nordic BIN area started to catch up during the period 2015–2016 while lagging behind the national average during the period 2011–2014. The BIN report shows that the employment development in the Russian part of BIN area is affected by loss of jobs in the sectors of wholesale and retail trade, mining and quarrying, transport and communication for Murmansk region, and in manufacturing, forestry and agriculture, transport and communication for Arkhangelsk region. However, both regions show growth in the number of workplaces in the accommodation and restaurants, real estate and professional services sectors, as well as in the mining industry in the Arkhangelsk region.

In light of the information presented in the BIN reports it is important to contemplate the depopulation in the BIN area, especially the lack of growth in young population. A prosperous socio-economic development in the Arctic demands young and well-educated people earning incomes being at least similar to those they can earn in the capital areas of their countries. These findings presented in the BIN report are in line with those presented in other studies (see e.g. Olsen et al. 2016), and more coherent policies in education, work, living conditions, quality of life and infrastructure including transport and digital infrastructure are needed in the North.

Business Activities and Innovations

The BIN area is a shining example of a region, probably beyond most people's expectation, where companies are able to grow despite limited access to financing and human resources, especially compared to companies in the capital areas. In this area, we find successful companies with high growth opportunities, good value performance yet, a less aggressive approach to innovative competitiveness.

The businesses in the BIN region have already developed a significant innovation potential in terms of clusters, brands and successful companies—an issue often overlooked when the region is viewed in terms of its natural resources. Many innovative businesses create brands by emphasizing and building upon their identity with Northern lifestyle and values.

The most successful businesses in the BIN area[3] are in Northern Norway—aquaculture firms and real estate developers, in Arkhangelsk Region—businesses related to the emerging mining industry, in Murmansk Region—traditional mining companies, and in general, in the Northern Nordics—manufacturing based on electric energy. An 87% increase in turnover from 2008 to 2016 and 18% from 2012 to 2016 in the BIN area con-

[3] Companies providing headquarter services, for example, oil and gas companies, banks, and subsidiaries in BIN regions were not included in the current study.

tributed to the area's turnover that today exceeds 90 billion Euro in 10 BIN regions altogether.

One of the essential drivers of economic development in most of the BIN regions is tourism. This industry contributes to the promotion and formation of a positive image of the whole BIN area and the European Arctic.

Being innovative often requires ample capital, R&D and knowledge-intensive production. The BIN report shows that the intensity of patenting activity in the Nordic BIN area is 2.5 times lower than for the Nordic countries' average. However, three regions within the BIN area (Northern Ostrobothnia, Norrbotten and Västerbotten) demonstrate relatively high patenting activity. The level of patenting activity in the five other Nordic BIN regions (Nordland, Troms, Finnmark, Lapland and Kainuu) is rather low.

The opportunities for the future in the BIN area are prosperous in economic terms. At the same time, there is continued loss of jobs in traditional BIN industries such as mining, quarrying and manufacturing, agriculture, forestry and fishing. To be able to realize the economic potential, competitive advantage has probably to be developed both through automation and digitalization and infrastructure investments.

Maritime Transportation by the Northern Sea Route

The economics of the circumpolar regions needs good transportation infrastructure. Internal Russian traffic and traffic between Russian ports and non-Russian ports are the most common means of transportation on the Northern Sea Route (NSR). The total volume of cargo transported along the NSR increased from 7.5 million tons in 2016 to 10.5 million tons in 2017. Among all regions, the south-western part of the Kara Sea had the highest traffic density on the NSR in the period 2016–2017. Altogether, 129 shipping companies were operating on the NSR in 2016, of which 75 were Russian companies and 54 non-Russian. The majority of non-Russian shipping companies operating on the NSR in 2017 were Norwegian, with 11 vessels making 92 separate voyages.

Exploitation and transport of natural resources out of the Arctic to markets in Europe and North-East Asia is the main driver of the increased shipping.

The business opportunities afforded by the Northern Sea Route influence transport infrastructure development as a whole in the BIN area. This includes, for instance, the new Finnish railway project[4] and digital infrastructure projects for strengthening the transportation infrastructure between the northern parts of Norway, Sweden and Finland (roads, railways, flight routes, etc.) with further extension to Russia and improved mobile broadband coverage in some unpopulated areas.

---

[4] The Finnish Ministry of Transport and Communications has announced that a railway route to the Arctic Ocean via Oulu, Rovaniemi and Kirkenes is the one that will be examined further (Press release dated 09.03.2018, https://www.lvm.fi/-/study-on-the-arctic-rail-line-completed-kirkenes-routing-to-be-examined-further-968073).

CONNECTIVITY

The BIN area is a huge territory with sparsely populated areas. Basic fixed broadband is currently available to 95% of households in the Nordic BIN regions and to 75% of households in the Russian BIN regions. There are, however, some regional differences in the availability of fixed high-speed broadband (speeds over 100 Mbps): Troms, Nordland in Norway and Norrbotten in Sweden lag behind their country averages by 8 and 7 percentage points respectively, while the Finnish regions of Northern Ostrobothnia, Kainuu and Lapland outperform Finland's average by 8 percentage points.

On the positive side, mobile broadband coverage (3–4G) is good over all the populated places in the BIN area. In terms of territorial coverage in 2016, the BIN regions in Norway had the best coverage lagging behind the national average by only 3 percentage points. The Swedish BIN regions lagged behind by 14 percentage points and the Russian BIN regions[5] lagged 21 percentage points behind their respective national averages.

Future opportunities for business development in the BIN region demand data cable connections. The BIN area currently has no direct connection to Europe and North America via subsea data cables. A number of landing points of data cables to Europe are on the coast of South Norway, South Sweden and South Finland. North-West Russia has one subsea data cable to Finland. Direct trans-Atlantic data traffic between Europe and North America proceeds through 12 subsea cable systems and lands in Denmark, the UK, the Netherlands, Germany, France, Spain and Portugal.

While basic fixed broadband accessibility is good, better access to high-speed speed internet is needed to increase economic growth in the BIN area. Active involvement of the national governments and strong consortia in subsea fiber cable projects are required for development of such infrastructure.

RENEWABLE ENERGY

The Nordic BIN area is a substantial provider of renewable energy and accounts for 25% of the hydropower production and almost 40% of the wind power production in the Nordic countries. There are three regions with a surplus of renewable energy: Västerbotten, Norrbotten and Nordland. These regions are also increasing their renewable energy production.

Within the Nordic BIN area, several new wind farms are under construction, thus enhancing the BIN area's position as an important renewable region in the North.

The BIN regions in Finland have to rely on energy import. The planned Hanhikivi 1 nuclear power plant in Northern Ostrobothnia will change the energy balance in the region since at present the main exporter of the electricity to Northern Finland is Sweden.

---

[5] The North-West Federal District of Russia is used as a reference comparison for the Russian BIN regions.

CROSS-BORDER CO-OPERATION

Successful cross-border co-operation in the Arctic requires openness and new ways of thinking and acting. The mental boundaries (especially for those living far away from cross-border regions) associated with borders between countries could easily inhibit co-operation between neighboring areas. In the current geopolitical situation, it is important to continue people-to-people co-operation and secure flows of knowledge, information, goods, workforce and youth.

The harsh and demanding Arctic environment puts pressure on the people and businesses operating there, making them creative and motivated out of necessity, for co-operation with each other. Thus, there are examples of companies successfully co-operating across borders. These companies do not consider borders as obstacles but as opportunities for building comparative advantages. For example, Kimek companies, one of the largest northernmost mechanical environments in Norway, built their business on access to highly qualified workforce in neighboring Russia. Such international strategic thinking is a specific feature of several innovative organizations highlighted in the 2017 Business Index North report.

The existing transport infrastructure and resource flows in the north of the Nordic countries are developed along a north-south dimension. This represents a challenge for developing cross-border co-operation in the east-west direction. The railway corridor "Ofoten Railway" from Norway to Sweden, the so-called Iron ore railway with further connection to Finland, is one of the few examples of such an east-west transportation corridor.

There is potential for future business opportunities in cross-border co-operation between small- and medium-sized enterprises (SMEs) and in the creation of new industries in the Arctic, for example, industries utilizing minerals and local suppliers to larger infrastructure projects. Furthermore, there are good co-operation opportunities for SMEs in the BIN regions in telecommunications, software and computer technologies, health care solutions, engineering and processing industries.

Circumpolar cross-border business co-operation can contribute to cost-effectiveness in public and private investments and strengthen economic complementarities (see e.g. Olsen et al. 2016). However, this co-operation must be stimulated with coherent policies and requires joint efforts at local, regional and national levels in the Arctic countries.

THE PARADOX OF CIRCUMPOLAR BUSINESS DEVELOPMENT

The Arctic climate is changing and this requires adaptation actions (AMAP 2017), including developing innovations that improve co-operation and co-ordination across governance levels. As the BIN reports have shown, the High North regions are attracting increasing attention due to their abundance of natural resources (e.g. fish, oil, gas, minerals, tourism potential, increasingly transport solutions by sea), creating opportunities to contribute to economic growth of benefit to the whole world. However, an important challenge is how to ensure the long-term sustainable business development of the local communities of the High North based on those resources, which are quite sensitive to changes in prices and market regulations.

The BIN area currently has about 3.56 million people, which is about 10% of the total population of Norway, Finland, Sweden and the North-West of Russia, accounting for 7% of the total economic value creation for these countries (for Russia—North-West Russia is included as reference for comparison). However, this area is characterized by two trends: steady economic growth and depopulation—especially the loss of young people. Consequently, depopulation challenges the future of the people of the High North and the safeguarding of local communities' future health and welfare. At the same time, there is economic growth in traditional industries and industries based on natural resources. Paradoxically this goes hand in hand with a decline in the number of jobs in these industries. Another important feature is the lack of innovative capacity in the northern regions; for example, the innovation levels are lower than the averages for the respective countries. With the present rate of depopulation and the lower innovation capacity, long-term sustainable economic development of the High North is problematic and threatened.

One key problem in ensuring the sustainable development of the High North regions may be the so-called governance paradox, indicating tension between local and global interests. Governance of the High North communities is increasingly addressed from the local perspective, with the promise to consider the values/interests of inhabitants/population of the High North and its sustainable development (e.g. Russian Strategy of Arctic Development, Norway's Arctic Strategy, Finland's strategy for the Arctic Region, Sweden's Strategy for the Arctic region, etc.)

Norway is one of the Arctic nations that for the longest time has been actively working with strategies for ensuring sustainable development in the Arctic or High North (e.g. documents issued in 2006, 2013, etc.) and here we will refer to the recent Arctic Strategy of Norway (2017) in discussing this paradox.

The Norwegian Arctic Strategy states that the vision for *the Arctic is to be a peaceful, innovative and sustainable region* (Norwegian Ministries 2017). To contribute to this, the Norwegian government is working toward an integrated strategy that incorporates both foreign and domestic policies. In particular, the government addresses five priority areas that are of importance not only for Norway but also for all countries with interests in the Arctic. The priority areas are: international co-operation, business development, knowledge development, infrastructure, environmental protection and emergency preparedness.

International Co-operation and Development of Infrastructure

The principles and ambitions for international co-operation in the Arctic are more or less clear and being widely addressed (see e.g. Arctic Council, Arctic Economic Council, etc.). But how do we ensure that the discussions and the political ideas will lead to successful actions and real economic co-operation and social development? The BIN reports stress that one of the first things to be addressed by policy-makers and business leaders is development of infrastructure.

Business and economic co-operation will not be able to grow unless the issue of Arctic infrastructure is placed high on the international agenda. Building infrastructure requires long-term commitment, co-ordinated effort at the top political level and a huge amount of joint investments. Both digital and transport infrastructure are of great importance. Subsea data cables (such as Arctic Connect) and new data centers to connect the Arctic to the world are important initiatives. At the same time, there is a need for further development of East-West transport corridors, improvement of existing transport infrastructure and a possible connection to the Northern Sea Route. In this respect, the plans of the Finnish Government to build a new railway route to the Arctic Ocean via Oulu, Rovaniemi and Kirkenes appears promising. The plans of major actors such as Maersk to use the Northern Sea Route may serve as a catalyst for other shipping companies, thus making the Northern Sea Route more attractive and internationally oriented than it is today.

Business Development

While regional development policies in the Arctic tend to address the development of human capital and sustainable communities, one cannot disregard the global trends associated with the increased use of information and communication technologies and robotization. Even today, we can already see the growing business efficiency of companies in the North in spite of job losses in traditional occupations. Business owners tend to follow this trend with increased pressure for more efficient, machine-based and less risky production.

The Arctic is diverse, and it is doubtful for at least some areas where business develops whether there is the need and opportunity to invest in local communities. Remote resource areas that lack infrastructure and have harsh natural conditions, for example offshore oil developments, would likely involve more and more technology and reduce the number of people involved in production, and so reduce the risk of accident for human beings. In such areas, the resources would likely be explored and exploited by workers commuting in and out, meaning that the workers do not have to live in the High North.

Technologization, robotization and commuting could increase depopulation if sustainable communities are not empowered by policy measures. The spread of high-speed internet and the creation of so-called smart communities may present an opportunity for people living in the North, both by making them globally connected and allowing them to live in innovative local communities. However, such a development in the High North will demand true engagement and real investment from central governments. Without strong measures from the central governments, local communities in the High North will increasingly be at risk of a dramatically deteriorating demographic situation.

Knowledge Development

Universities in the circumpolar areas are in a unique position to contribute to regional development due to their proximity to local businesses and their understanding of the issues as well as of the High North wider societal context and global development trends. Partnerships between universities and industry to develop innovations seem to be an important driver for industrial development. University-based research on Arctic issues benefits from international co-operation, rigor in analytical approaches as well as academic independence from more political or business interests. The universities in the High North are important players in the implementation of Arctic policies as they connect the views and interests of various stakeholders in the Arctic. Many politicians and business people who have contributed much to the development of both new universities in the Arctic as well as to funding of the established universities are also aware of this.

However, the current trend is that the focus in Arctic research is mostly on natural sciences and not so much on the social sciences. The latter includes socio-economic development issues, business development and innovations. We hope that a more balanced, inter-disciplinary research approach would in future attract the attention of policy-makers and research funders. This would enable more socially sustainable business development building on knowledge from a wide range of scientific fields. The Arctic region has much more to offer than natural resources. More business and management studies are needed to highlight the Arctic as an area for innovations.

Environmental Protection and Emergency Preparedness

Due to the natural and geographic conditions, the Arctic has the potential to become a huge supplier of renewable, environmentally friendly energy. Therefore, policy-makers may stimulate the development of new power-intensive industries. Instead of focusing mainly on the extraction and export of natural resources, some of these resources could be processed in the Arctic regions. For example, more spare parts for the automobile industry can be produced from high quality steel in the Northern Nordics. Data storage centers and servers for global providers are another example of a new industry being established in the North. The Northern Nordic region has the potential to become a future base for green power intensive industries. The North-West of Russia has great improvement potential as it has not only huge natural resources but also many people living in the area. It is important to create new and attractive jobs to encourage people not to move away. Emergency preparedness is an issue of the utmost importance and again, it cannot be resolved only on national basis.[6]

A WAY FORWARD

In summary, it is easy to conclude that it is important to balance national (and global) and local interests when it comes to strengthening business development in the Arctic. For national and global interests it is crucial to arrive at a sustainable way of using the resources of the Artic. Because these resources are vast and most people in all the Arctic nations do not work and live in the Arctic, the decisions about major projects with significant economic effects for the nations and big companies involved may easily be taken without much influence from those who live and work in the Arctic. Historically, the structure of production in the Arctic has meant few linkages between resource production and the communities of the north, resulting in most of the potential benefits flowing out of the northern regions (Huskey and Southcott 2016). In some cases, strategic and financial planning remains far from the 'local' High North, rather being guided by decisions by central governments, big corporations or global institutions (Bourmistrov et al. 2017). This creates a paradoxical situation: while the region's popular image is one of rich resources and many opportunities for development, the local perspective is dominated by views that emphasize a lack of resources and services and regions struggling to benefit from regional development (Tennberg et al. 2014).

The Norwegian Arctic Strategy of 2017 created a new political forum in an attempt to reduce or at least address some of the challenges connected to the "governance paradox". Norway has established "a regional forum for systematic dialogue at political level" between the national government, the three regional counties of North Norway and the *Sameting/Sámediggi*. Other key players in the Arctic, such as the business sector, academia and so on are also invited to take part in the forum, when appropriate. This kind of governance dealing with the regional driving forces, solutions and consequences of circumpolar business development makes it possible to operationalize investments in human and social capital, transport and ICT infrastructure that can fuel sustainable economic growth and a good quality of life for the people working and living in the Arctic.

---

[6]An example of successful international co-operation in this respect is the MARPART project network (Maritime Preparedness and International Partnership in the High North) led by High North Center for Business and Governance. The MARPART Consortium consists of 13 universities and research institutes that focus on emergency management and crisis preparedness in the Arctic.

REFERENCES

AMAP. 2017. *Adaptation Actions for a Changing Arctic: Perspectives from the Barents Area*. Arctic Monitoring and Assessment Programme (AMAP), Oslo, Norway.

Arctic Business Analysis: PPPs and Business Cooperation. 2018. Copenhagen. https://doi.org/10.6027/ANP2018-707.

Bourmistrov, A., I. Khodachek, and E. Aleksandrov (eds). 2017. *Budget Developments in Russia's Regions: New Norms, Practices and Challenges*. FoU rapport nr. 18, Nord Unviersitet, Bodø.

Bullvåg, E., A. Mineev, P. Pedersen, A. Hersinger, O. Pesämaa, M. Johansen, S. Ovesen, A. Middleton, and J. Simonen. 2017. *Business Index North – A Periodic Report with Insight to Business Activity and Opportunities in the Arctic*, Issue No. 1, Bodø, Norway.

Huskey, L., and C. Southcott. 2016. "That's Where My Money Goes": Resource Production and Financial Flows in the Yukon Economy. *The Polar Journal* 6 (1): 11–29.

Middleton, A., A. Hersinger, A. Bryksenkov, A. Mineev, B. Gunnarson, E. Dybtsyna, E. Bullvåg, J. Simonen, O. Pesämaa, P. Dahlin, S. Balmasov, and S. Ovesen. 2018. *Business Index North – A Periodic Report with Insight to Business Activity and Opportunities in the Arctic*, Issue No. 2, Bodø, Norway.

Norwegian Ministries. 2017. *Norway's Arctic Strategy – Between Geopolitics and Social Development*. https://www.regjeringen.no/contentassets/fad46f0404e14b2a9b-551ca7359c1000/arctic-strategy.pdf.

Olsen, L.S., A. Berlina, L. Junsberg, N. Mikkola, J. Roto, R.O. Rasmussen, and A. Karlsdottìr. 2016. *Sustainable Business Development in the Nordic Arctic*. Nordregio Working Paper 2016:1, Stockholm, Sweden.

Tennberg, M., J. Vola, A.A. Espiritu, B.S. Fors, T. Ejdemo, L. Riabova, E. Korchak, E. Tonkova, and T. Nosova. 2014. Neoliberal Governance, Sustainable Development and Local Communities in the Barents Region. *Barents Studies: Peoples, Economies and Politics* 1 (1): 41–72.

CHAPTER 11

Multinational Corporations in the Arctic: From Colonial-Era Chartered Companies to Contemporary Co-management and Collaborative Governance

Barry Scott Zellen

An imperial crossroads of global—and in particular, economic—importance for centuries, the Western Arctic region remains largely underdeveloped with a population less than 20,000 people from Alaska's North Slope to Canada's Mackenzie Delta. It's a vast, single region spanning two countries, encompassing one state and two territories that converge along a boundary once separating two global empires, Russia's and Britain's, at the zenith of their territorial breadth—until the consequential sale of Russian-America to contain British North America, and thereby protect Russia's northeastern flank. (Moscow's logic would hold true only until after the War of 1812 ended, paving the way for America's own imperial rise and eventual rivalry with Russia during the Cold War period.) (see Fig. 11.1). The Western Arctic's relative underdevelopment has resulted in a small influx of settlers, despite the active role of the world's first multinational corporations (MNCs), the crown-chartered companies of the colonial era, leaving the indigenous peoples of the region with a substantial and sustained demographic majority that contributed to their recent, and historic, re-empowerment.

As the Western Arctic region evolved from the age of empire to the post-land claims settlement era, this re-empowerment positioned the indigenous peoples of the region to be masters of their own fate in an increasingly globalized and

B. S. Zellen (✉)
Center for Arctic Study and Policy (CASP), United States Coast Guard Academy, New London, CT, USA
e-mail: barry.s.zellen@uscga.edu

© The Author(s) 2020   157
K. S. Coates, C. Holroyd (eds.), *The Palgrave Handbook of Arctic Policy and Politics*, https://doi.org/10.1007/978-3-030-20557-7_11

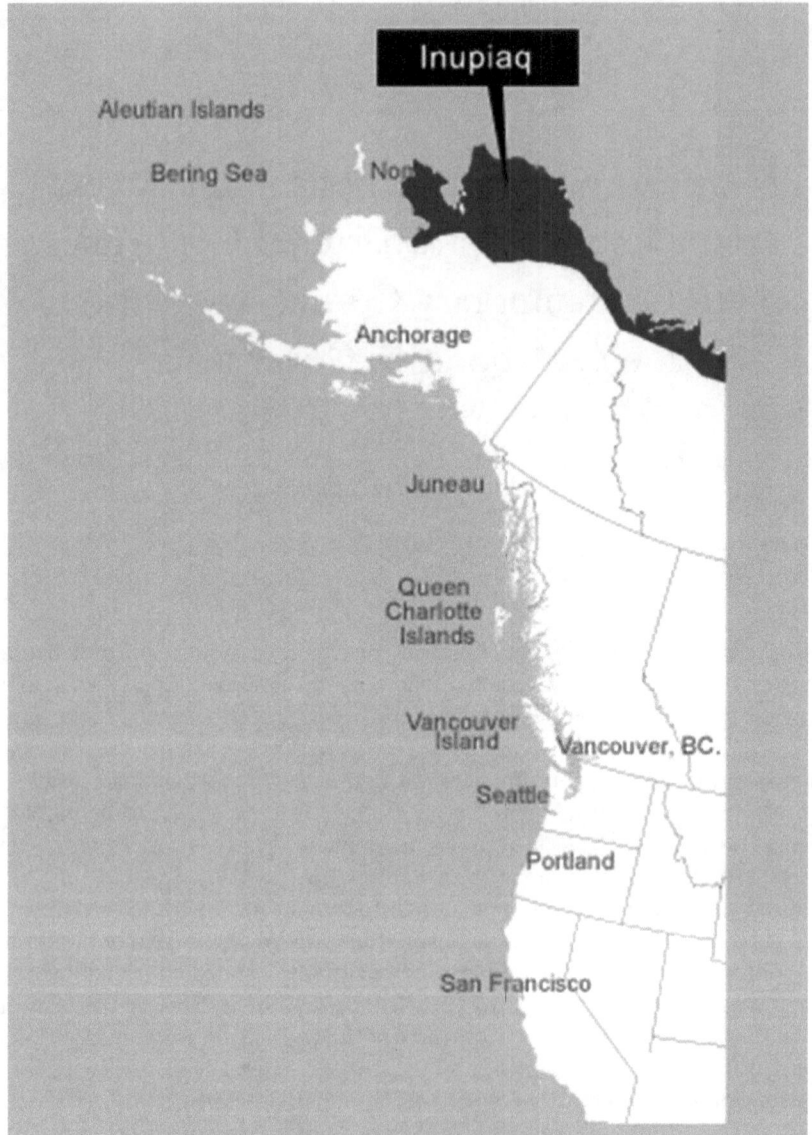

Fig. 11.1 One language, two countries, three territories: Inupiaq language map. Source: https://stoningtongallery.com/wp-content/uploads/CU-map_inupiaq_VIV_1.jpg

economically integrated part of the world. Even after the famed oil strike at Prudhoe Bay in the 1960s, with the rapid expansion of multinational oil companies to the North Slope (and with it, a surge in exploration that extended deep into Canadian territory)—and despite the construction of the Dalton Highway in 1974 linking the Alaska Highway to the Arctic coast and the subsequent opening of the Dempster Highway soon thereafter (1978) under Canada's

"Roads to Resources" development policy—the pace of external settlement has remained relatively slow. Inuit demographic predominance continues to define the coastal region. While an influx of settlers further south during the Klondike Gold Rush, permanently rebalanced the demographics of the Yukon Territory, and to a lesser degree the Mackenzie River valley and Great Slave Lake basin, the Western Arctic region overall attracted little settlement (Coates 1985; Easton 2016). The oil town at Deadhorse/Prudhoe Bay on Alaska's North Slope is a rare exception to this pattern, as was the whaling community at Pauline Cove on Herschel Island just off the Yukon North Slope. The latter occurred a century before during the final years of commercial whaling, in an earlier era of multinational interest in the Arctic by the global whale oil industry, a precursor to the modern multinational petroleum industry. (For more on the impact of the oil boom on the Inupiat, see Jorgensen 1990.)

While Russian-America, through the Russian-American Company which operated from 1799–1867 and British North America—and in particular, the North-Western Territory (see Figs. 11.2 and 11.3) adjacent to Rupert's Land, through both the Hudson's Bay Company (HBC)'s activities in its Mackenzie

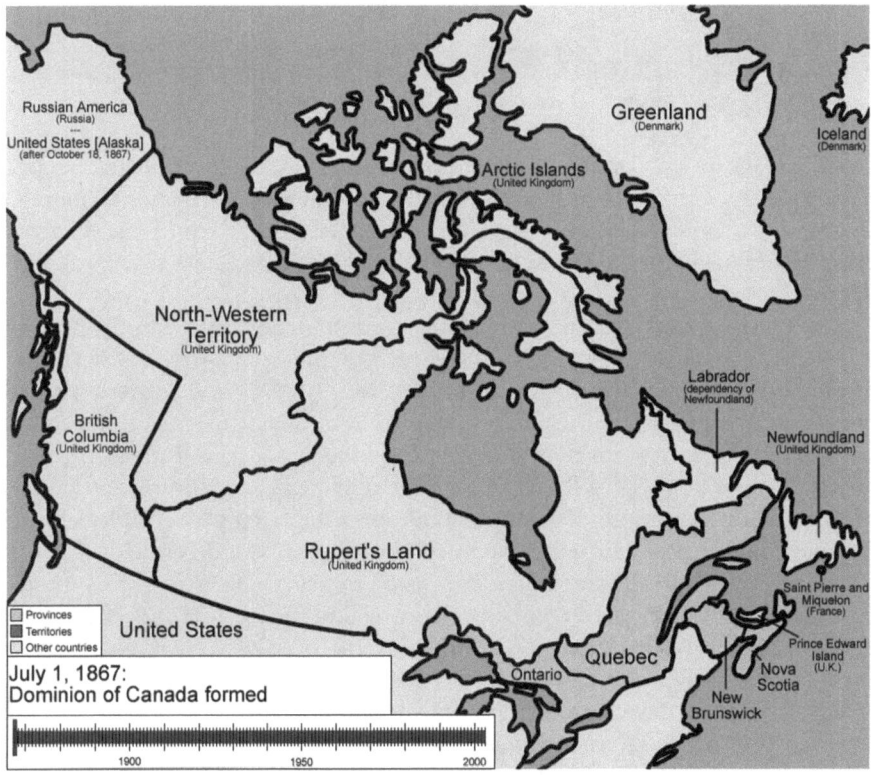

Fig. 11.2 Russian America, the North-Western Territory, and Rupert's Land as of July 1, 1867. Source: https://upload.wikimedia.org/wikipedia/commons/9/9d/Canada_provinces_evolution_2.gif

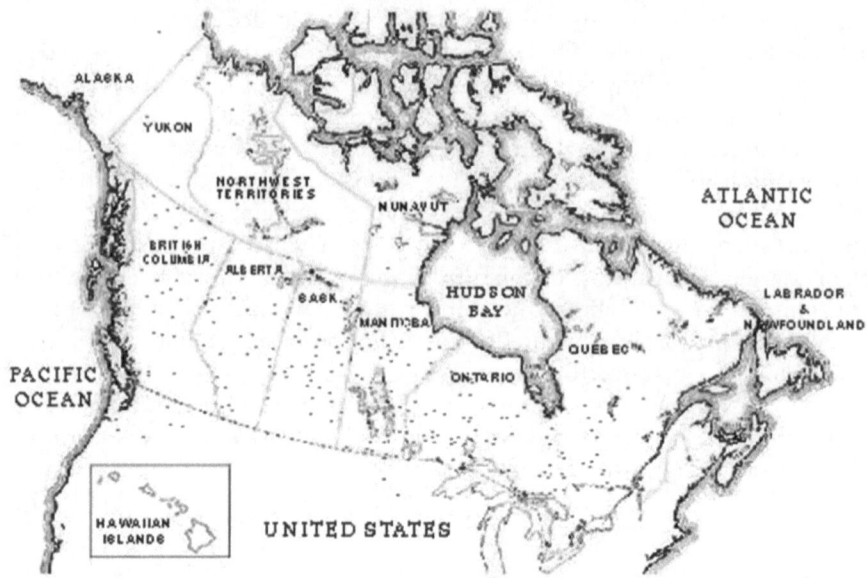

Fig. 11.3 HBC: Map of trading post locations. Source: https://www.gov.mb.ca/chc/archives/hbca/resource/cart_rec/postmap/hbc_c.html

and Great Slave Lake districts—asserted quasi-sovereign control over the region through Crown-chartered trading companies, defining an international boundary that, to a large degree, has remained unchanged as the world transitioned from the era of chartered companies into the modern era of state sovereignty in the Arctic.

This boundary—forged in colonial times and subdividing the Inupiat homeland today—emerged from the cauldron of expanding empires, marking the frontiers of the international fur trade powered by a network of indigenous hunters and trappers on both sides. Some like the Aleuts (Unangan) were virtual slaves, displaced from their homelands by the conquering Russians, while others, like the trappers of Rupert's Land, maintained their cultural and much of their political autonomy even as they became integrated with the globalized British economy. Both fur empires, in spite of their vast differences in governance and respect for indigenous traditions, remained united in their mutual decision to only lightly settle the region, protecting the fur-bearing ecosystems upon which they depended by holding back the pace of colonial settlement (Arnold 2011; Bockstoce 2009; Newman 1985, 1998). John Bockstoce (2009) observed that the northern fur trade was already part of a well-established inter-indigenous/international trading network linking Alaska natives to Chukchi traders across the Bering Sea, who in turn traded with both China and Russia, when Russia expanded across the Bering. The pre-existence of such a network connecting northern furs to Eurasian markets, and its continuity under direct Russian control after its colonial expansion to Alaska, suggests a continuous and enduring globalized political economy linking the self-governing era of pre-contact indigenous polities to the colonial era, when the earliest MNCs reached into the Arctic.

But even a light demographic intrusion can prove calamitous, particularly as waves of new diseases carried by representatives of these many early modern multinationals, whether fur traders, whalers, or miners, devastated long-isolated native populations at great immunological risk. Indeed, epidemics decimated local populations of Mackenzie Inuit during the early years of the twentieth century, exposing the region to a demographic upheaval (McGhee 1976). Ironically, the fluidity of crossborder migration by the Inupiat, drawn in part by economic opportunities presented by the integration of their homeland with these expanding global trade networks preserved their demographic predominance, while it was a crossborder migration by non-native settlers during the Gold Rush that would transform the Klondike. This challenges many of our preconceptions about settlers and about what constitutes indigeneity; that both were in flux at this time gives the Western Arctic region a particularly dynamic nature. As recounted by Robert McGhee (1976, 144),

> By 1910, the Mackenzie Eskimos were reduced to a few score survivors scattered among the more numerous Alaskan Eskimo immigrants who flooded into the Delta in the company of European whalers and traders.

As McGhee further described:

> After the appearance of the American whaling fleet along the Mackenzie Delta coast in 1889, and with the increasing association between the indigenous population and the whalers wintering at Herschel Island and elsewhere, the effects of disease and the disruption of aboriginal social patterns accelerated rapidly. The population was subjected to two devastating measles epidemics in 1900 and 1902. By this time, according to police reports, the Mackenzie Eskimo population had declined rapidly from an estimated 2500 people in 1850 to about 250 in 1905 and under 150 in 1910. At the same time as Eskimos were being decimated by disease, local aboriginal culture was being submerged beneath a wave of American and Alaskan Eskimo introductions. (McGhee 1976, 144)

The influx of Inupiat settlers from Alaska in the years that followed ensured continued Inuit demographic predominance, and in many ways helped to solidify the cultural and linguistic cohesiveness of the region, imbuing it with enduring qualities that continues to reach across the international boundary.

The endurance of indigenous polities and their demographic predominance assured that as the very first corporate entities—the Crown-chartered companies that launched the globalization of the world economy—came north, their imperial ambitions would be moderated from the land and power grabs experienced elsewhere in the colonial world, most famously with the East India Company, which many scholars describe as the world's first MNC (Robins

2004). Indeed, as Nick Robins (2004) has argued, "the East India Company, romantic as it may seem, has more profound and disturbing lessons to teach us. Abuse of market power; corporate greed; judicial impunity; the 'irrational exuberance' of the financial markets; and the destruction of traditional economies (in what could not, at one time, be called the poor or developing world): none of these is new. The most common complaints against late twentieth- and early twenty-first-century capitalism were all foreshadowed in the story of the East India Company more than two centuries ago."

Instead, the early northern chartered companies would—though not without exception, as evident in the case of Russia's more heavy-handed imperial expansion across Siberia to Alaska—collaborate with the indigenous peoples whose lands and resources they coveted. In time, this collaboration evolved into formal co-management, and later, even joint ventures and native managerial training positions with newly formed native corporations that arose from the land claims process. That is why perhaps even today, the HBC retains its popularity across the North, as does the ubiquitous HBC store and its descendants—first as "The Bay," and later among the 178 former North West Company stores, as the "Northern" store (until being sold off by HBC in 1987, to form a separate company that would resurrect The North West Company name—causing much grumbling in the northern communities when it was learned the stores could no longer be called the Bay) (Burns 1987). As reported by the *New York Times*: "To Indians, Eskimos and other Canadians living in communities served by the stores, from Labrador in Newfoundland to Whitehorse in the Yukon Territory, and in dozens of settlements in between, it will mean an end to trips to 'the Bay,' as the company's stores are universally known. It has been enough to stir protests in some of the affected communities, where the stores have become a symbol of contact with a distant and more comfortable world" (Burns 1987).

Indeed, the sale of the HBC's northern stores also riled many northern experts, as the *New York Times* further noted: "Many enthusiasts for the old Hudson's Bay find the situation hard to accept, believing that the sense of abandoning history will be felt well beyond the company. One of them, the Canadian writer Peter Newman, at work on a multi-volume history of the company, demanded that the parchment carrying the Royal Charter, housed in offices adjacent to the Bay store at Yonge and Bloor streets in Toronto, be transferred to the northern stores' new owners. 'They're the ones who are fulfilling the charter'" (Burns 1987). Newman went on to publish his three-volume history of HBC—*Company of Adventurers* (1985), *Caesars of the Wilderness* (1987), and *Merchant Princes* (1991). *Up Here* magazine similarly turns to Newman to capture the grand historical legacy of the HBC: "Historians speak in lofty terms of the grand old company's early days—no mere mortal men, the fur traders and explorers were titans, kings, Caesars in birch bark canoes. Through cunning, ambition, skirmish and diplomacy, the HBC gained sovereignty over a third of what is now Canada. It was a trade monopoly, selling furs and other resources back to Mother Britain, but to those living and working under the company's lordship, there was no greater authority, at least on this side of the ocean. 'We know only two powers—God and the Company,' an old HBC chief factor is quoted as saying by journalist Peter C. Newman.

Even under its new ownership, however, the former HBC stores remained vital to the communities they serve, and a welcome and important part of the northern business community, in marked contrast to how the liberated, post-colonial citizens of India felt about the British East India Company, or the displaced Aleut (Unangan) community felt about the Russian-America Company after their departures. Indeed, Dorothy M. Jones (1981) characterized Russia's treatment of the Aleuts as "brutal" (though she is equally critical of US treatment after the Alaska purchase), writing: "In the first fifty years of Russian occupation, the free trade period, Russian fur hunters brutally mistreated the Aleuts and at the same time commanded their labor. The Russians stole the Aleuts' wives, slaves, and possessions, and slaughtered any who resisted their domination. They sent Aleut men on long sea hunting expeditions from which many never returned and during which many women and children, left alone in the villages, suffered severe deprivation. This mistreatment, combined with the diseases the Russians introduced, nearly decimated the Aleut population. In the first thirty years of Russian contact, the Aleut population declined from an estimated 12,000 to about 1900." She further chronicled:

> Eager to prevent ruinous competition between its companies, to regulate traders' treatment of the natives, and most importantly, to protect and expand its sovereignty in Russian-America (its first overseas colony), the Russian government in 1799 granted a monopoly to a private firm, the Russian-American Company. The government gave the company not only a monopoly on trade but authority to govern and garrison the new territory. Apparently the Russians applied the experience of the British East India Company in using a private business as an instrument of government. The establishment of an outright Russian government administration in North America might have provoked conflict with the United States and Great Britain, which was averted by establishing a company administration. The first charter granted to the Russian-American Company contained no definite regulations about the status and treatment of natives other than an injunction to treat them amicably and convert them to Christianity. The second and third charters, in 1821 and 1844, however, specified natives' political status. Aleuts and other natives under company administration were declared Russian subjects: 'Tribes inhabiting the places administered by the company are ... Islanders, Kurils, Aleuts, and others.... As Russian subjects they shall conform to the general laws of the empire and shall enjoy the protection thereof.' The protection was hardly forthcoming. (Jones 1981)

The absence of sovereign protection and the painful experience of subjugation was not unique to Russian-colonized Alaska, but instead was illustrative of Russia's centuries-long imperial expansion extending its formal sovereignty across Siberia from the early sixteenth century on through a highly oppressive system of taxation, forced tributary payments, and virtual enslavement—paving the way for the subsequent mistreatment of the Aleuts by Russia's maritime fur traders across the Bering Sea. And just to the south of Russia, on the island of Hokkaido and along the Kuril island chain (which both Japan and Russia had coveted), the Ainu were similarly denied sovereign protection while being quite brutally subjugated by the northward-expanding Japanese state, which in a manner reminiscent of the Russian expansion across Siberia would ultimately expropriate Ainu lands as well as their trading networks, reducing the hitherto independent Ainu to a condition of near slavery previously comparable to the Aleuts in Russian-America.

In marked contrast, the HBC provided the natives of what would later become Canada's northern territories a more protective, mutually beneficial, and sustainable relationship based on the enduring, multigenerational reciprocity of its commercial relationship with the hunters and trappers of the Arctic and Subarctic. In many ways, the HBC seeded the north, where subsistence and regional trading networks, which already interconnected with an indigenous international trade system that stretched across the Bering Sea to Asia, as described by Bockstoce (2009), facilitated a less intrusive form of colonization as compared to the Russian-America and British East India models— which more aggressively interwove the formal sovereign powers of state with the corporate prerogatives of the chartered companies—requiring instead not the subjugation, displacement, or enslavement of the native population, and much more than the mere survival of the indigenous people of the North: namely their sustained and supportive participation in the newly globalized commercial activities. The experience of natives in Rupert's Land thus differed greatly from the much harsher Aleut experience in Russian-America, whose population was forcibly displaced from their traditional homeland and virtually enslaved by the Russian invaders—a tragic experience that still casts a painful shadow across Southwest Alaska, where uninhabited Aleutian islands stand empty to this day, and impoverished Aleuts in continuing urban exile serve as a reminder of how close the North came to the more exploitative form of both colonialism and multinationalism.

Nearly a century after the Russian displacement and enslavement of the Aleuts, when oil was found in commercial quantities on the North Slope, an exploration boom that would extend far across the Western Arctic region, with seismic survey crews entering native lands. The threat to native stewardship of the North presented by petroleum MNCs helped catalyze a movement across Alaska and the Western Arctic for the preservation of native rights, traditions, and lands—culminating in the historic land claims movement. Curiously, it was not just the threat to natives from Big Oil that precipitated the pioneering Alaska Native Claims Settlement Act in 1971; it was also the threat of years of native litigation against Big Oil and its plan to pump oil out of the North Slope to southern markets via the Trans-Alaska Pipeline System (TAPS) that worried the state of Alaska and the federal US government (Zellen 2008). It was thus the historic convergence of oil, money, land, and power that brought together Big Oil, the Alaska natives, and both the state and federal levels of government to conclude the unprecedented Alaska Native Claims Settlement Act of 1971 (ANCSA). It was these four pillars of the emerging political economy of the North ("Oil, Money, Land, and Power") that became the title of the undergraduate honors thesis of Alaska lawyer and then-Harvard student Clifford John Groh II (1976). And despite ANCSA's many original flaws, the Alaska land claims nonetheless spared a wave of claims settlements that would ripple across the North to include, the Western Arctic, Nunavut, Nunatsiavut in northern Labrador, the Mackenzie River valley and to the communities governed by the Council of Yukon Indians (CYI), now known as the Council of Yukon First Nations (CYFN).

As the land claim model transformed unsettled lands to the east and south of Alaska's North Slope, it would evolve and expand to include new models of self-governance that radically transformed the land claims model from its original design to primarily economically integrate the North. Many critics felt ANCSA was designed to fail, and thus precipitate a transfer of lands, wealth, and corporate control from the native community back to the non-native majority as several Alaska native corporations (ANCs) succumbed to bankruptcy. That's because ANCSA only protected native ownership and control for 20 years, and did not guarantee "new natives" (young Alaska natives younger than 18 at the time of ANCSA's signing as well as all those born later) shares in the new native corporations and thus in the governance of their lands and resources. This situation was described as the "1991 Time Bomb," named for the year when the Alaska Native Claims Settlement Act was widely expected to collapse (Worl 1988).

But in the 20 years that followed ANCSA, two things happened: first, land claims spread across the border via the Inupiat, whose original excitement (and newfound wealth) inspired the Inuvialuit, primarily descendants of the Inupiat from an earlier migration into Canada, persuading them to quickly settle their own land claim. They broke with the Inuit of Nunavut, whose lands were not yet under major development pressures. The Inuvialuit were able to improve upon the land claims template dramatically, ensuring all "new natives" were automatically enrolled as shareholders and preventing shares from ever being sold to non-natives (ensuring continuity of native corporate control), while also preserving native subsistence and protecting the land and its wildlife (Zellen 2008). Alaskan natives recognized the original ANCSA structure that was agreed to in 1971 was seriously flawed—and thus worked to improve it, taking inspiration from their Canadian relatives, who showed that an improved model could be agreed upon with government. The "1991 Time Bomb" was thus defused (Worl 1988) and the ANCs were reborn. They are now important stakeholders in Alaska's economy, comparable to the largest native corporations in Arctic Canada.

All this took place because, in large measure, it was the corporations that came north before the state, in some cases, centuries ahead of the state's formal arrival to the North; these crown-chartered companies brought quasi-colonial ambitions and quasi-sovereign responsibilities, serving as imperial proxies but under the administration of corporate employees rather than political leaders and appointees. Because the Arctic did not have forests to cut, or plantations to plant, it did not need an influx of farm laborers to up-end the indigenous majority. The political economy that emerged was based on the fur trade, which required native participation and continued stewardship over their lands. And so the chartered companies of the North, notably the HBC and the North West Company, would become collaborative partners in the modernization of the North, bringing northern natives along with them into a shared future.

The corporate model would thus adapt and respond to the unique conditions of the North, allowing the corporation to become the foundational unit of governance for northern natives, the gateway toward true self-government rather than a vehicle designed to abolish self-government (as many believe the original ANCSA was intended to do, most notably Thomas R. Berger, former B.C. Supreme Court Justice and now legendary figure in the history of Canadian Arctic land claims) (Berger 1985). The native corporations introduced the northern native leadership to the experience of governing—and pave the way toward a more balanced form of co-management of the North.

Land Claims in the Contemporary Arctic: Embracing the North's Corporate Legacy for Achieving Collaborative Governance and Co-Management

Today's Western Arctic presents a fascinating set of settled land claims, dynamically evolving systems of indigenous and regional governance, distinctly indigenous and collaborative international diplomacy, and flexible balancing of subsistence culture with economic modernization and development. The system blended two worlds—one traditional, one contemporary—and presented a compelling example of enduring order in the absence of strong state institutions. The North provides compelling evidence that not all regions of the world are defined by international anarchy, nor dominated by armed conflict and political violence. Some have found their own ways to mitigate regional conflict and to foster peaceful and collaborative interaction across borders, sometimes borrowing ideas and emulating policies for application from adjacent areas. The Western Arctic is just such a place.

Here is an alternate historical narrative, defined by an historic reconciliation of tribe and state, a restoration of indigenous land and cultural rights, and a rise in Indigenous regional participation in international relations. Here, ideas and insights from the Alaska land claims process of the 1970s flowed across the international boundary and into the Western Canadian Arctic where they were re-thought, refined, revised, and re-applied—resulting in a stronger, more resilient, and ultimately more scalable model for northern development (Zellen 2008). This reflected a deep and enduring commitment to collaborative cross-border management, intergroup (and international) partnerships, and constructive transboundary relationships that present a compelling model for how the world could be governed. All this has its roots in the first wave of northern MNCs, and the blueprint for northern development crafted by the HBC. Because of the distinct challenges of the northern landscape, corporate survival required native resilience; and corporate success ultimately depended upon native success. The HBC-administered north was no Shangri La; but it was a far cry from the excesses witnessed in Russian-America. Its legacy was the enduring collaboration between the indigenous peoples of the North and the governing entities that asserted sovereignty over the North. It was not always frictionless collaboration, since there were times and issues where interests could and did clash. But despite these very real and recurring collisions of values between native, environmental, settler, and resource-extractive interests—the oil strike in Prudhoe Bay catalyzed the rapid formalization of the Alaska Native Claims Settlement Act (ANCSA) in 1971—collaborative efforts between natives and northern governments, and between neighboring native communities that stretch across the border, have remained ongoing, helping to mitigate inevitable conflicts.

When ANCSA was enacted, it sought to quickly bring Alaska natives into the modern economy, and at the same time to clarify the limits of aboriginal title, making it possible to fully develop the state's natural resources and in particular to build the Alaska pipeline (Naske and Slotnick 1979; Zellen 2008). Because these objectives were largely economic, its corporate model became its defining and most transformative characteristic—not without controversy, since the corporate model was rightly viewed with some skepticism by indigenous leaders as a tool of assimilation. There remains a debate over the appropriateness of the corporate model to the indigenous north (Berger 1985). ANCSA formally extinguished aboriginal rights, title, and claims to traditional lands in the state, while formally transferring fee-simple title to 44 million acres—or some 12 percent of the state's land base—to Alaska natives, with $962.5 million in compensation for the lands ceded to the state, $500 million of which was to be derived from future oil royalties. Over half of the "compensation" was to be derived from resources extracted from the Inupiat homeland—an irony not missed by Alaska natives (Tundra Times 1969; Zellen 2008). ANCSA also created 12 regional native corporations (and later a 13th for non-resident Alaska natives), and over 200 village corporations to manage these lands and financial resources (see Fig. 11.4). These new corporate structures introduced a new language and culture, as well as a new system of managing lands and resources that seemed at variance with the traditional cultures of the region and their traditional subsistence economy. The early years of ANCSA were famously described by Thomas Berger as dragging Alaska natives "kicking and screaming" into the twentieth century. Many native corporations approached the brink of bankruptcy, forced to monetize their net operating losses in a last desperate bid to stay in business (Berger 1985). A new cottage industry of northern investment, legal, and policy advisors emerged—sometimes to the benefit of their clients, but often not—a problem that would recur in each new region of settled land claims.

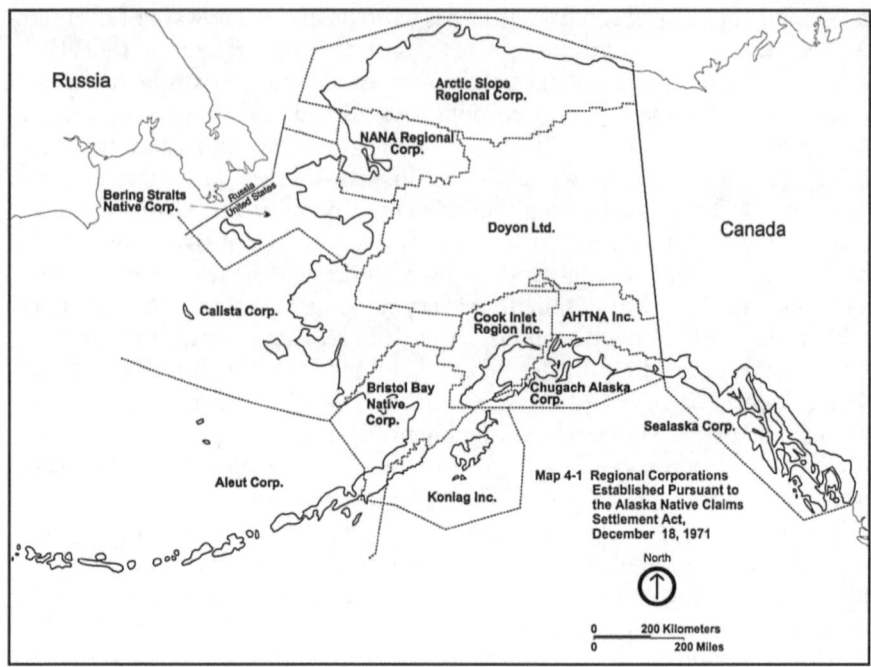

Fig. 11.4 Regional Native Corporations Created by ANCSA in 1971. Source: https://www.nps.gov/parkhistory/online_books/norris1/images/map4-1.jpg

In addition to the *corporatization* of village Alaska, ANCSA's original design also had structural flaws that also nearly proved fatal to the land claims experience, including its 20-year moratorium on transferring shares in native corporations to non-natives, which many feared would inevitably result in the dilution of native ownership (Worl 1988; Sykes et al. 1985a). While critics of the land claims process are correct to point out these original structural flaws and the assimilating pressures introduced by contemporary corporate structures, the land claims model has nonetheless proved resilient and adaptive. Native corporations matured and their boards, managers and shareholders found ways to better balance traditional and modern values, learning capitalism as they went. Today, the native corporations represent a huge economic force in the state of Alaska (Sykes et al. 1985b).

On the Canadian side of the international boundary that divides the Western Arctic, lessons from the Alaska land claims experience and its initial structural flaws were closely studied. This crossborder flow of ideas and insight influenced a new model for land claims settlements that ensured native lands and corporations would remain in native hands, that young natives would be automatically enrolled as shareholders upon adulthood, and that subsistence would forever be protected on both native-owned lands as well as adjacent government lands. The Alaska experience thus proved critical in guiding Canadian natives forward in their quest to assert, and protect, their Aboriginal rights. Just across the border from Alaska, the Inuvialuit of the Western Canadian Arctic—many of whom were descendants of early twentieth-century Inupiat settlers—followed ANCSA developments closely, and were impressed by all the money that was flowing north, as well as the new corporate structures created and the sizeable land quantum formally transferred to Alaska natives. But they also took note of the continuing threat to indigenous culture, and the lack of adequate protections of subsistence rights, traditional culture, and environmental protection, and were determined to do better. When the Inuvialuit negotiated their landmark 1984 Inuvialuit Final Agreement (IFA), the land claims model became significantly enhanced. In addition to creating new native corporations, the IFA also made an equal institutional commitment to the preservation of native culture and traditions, to preserve the land and the wildlife, and to empower not just new corporate interests but also traditional cultural interests as well, by creating new institutions of co-management and more powerful hunter and trappers' committees. They also made sure all Inuvialuit became shareholders, and that no non-Inuvialuit ever could, learning from the Alaskan experience. The Inuvialuit thus successfully modified the land claims concept, so that its structure included a natural institutional balancing that has enabled a greater commitment to cultural and environmental protections (Zellen 2008) (see Fig. 11.5). But one issue that was not yet on the table in the late 1970s and early 1980s when the Inuvialuit chose to pursue their own regional land claim—and thereby gain some control over the intense oil boom in their homeland—was the establishment of new institutions for autonomous, meaningful, effective aboriginal self-government, something that the Inuit of the central and eastern Arctic—the future Nunavut territory—desired. The Inuvialuit felt they did not have the luxury of time given the intense pace of oil and gas exploration in their lands. But Nunavut remained more isolated than the Western Arctic and under much less external pressure to develop, thus providing more time to re-think and renegotiate the land claims model.

As the states that now assert sovereignty over Arctic North America expanded into the northernmost reaches of our planet, they could have done what the Russians did when they crossed the Bering Sea, or what the Japanese did when they set their sights on the island of Hokkaido, or even what a younger and more imperial United States did when it began its own westward expansion centuries ago—simply crushing and displacing the tribal entities they found in their path. But by the time they turned to the Arctic region for sovereign expansion, they did so more gently and less muscularly than their counterparts in other parts of the world—guided not only by their still maturing recognition of indigenous rights which has enabled them to expand while integrating northern indigenous peoples largely intact into their constitutional structures, but also to the enduring, resource-based northern political economy long in place, having been nurtured by the chartered companies which enjoyed their quasi-sovereign dominion over their northern territories for so many generations. The northern stability observed by many in Arctic North America as a

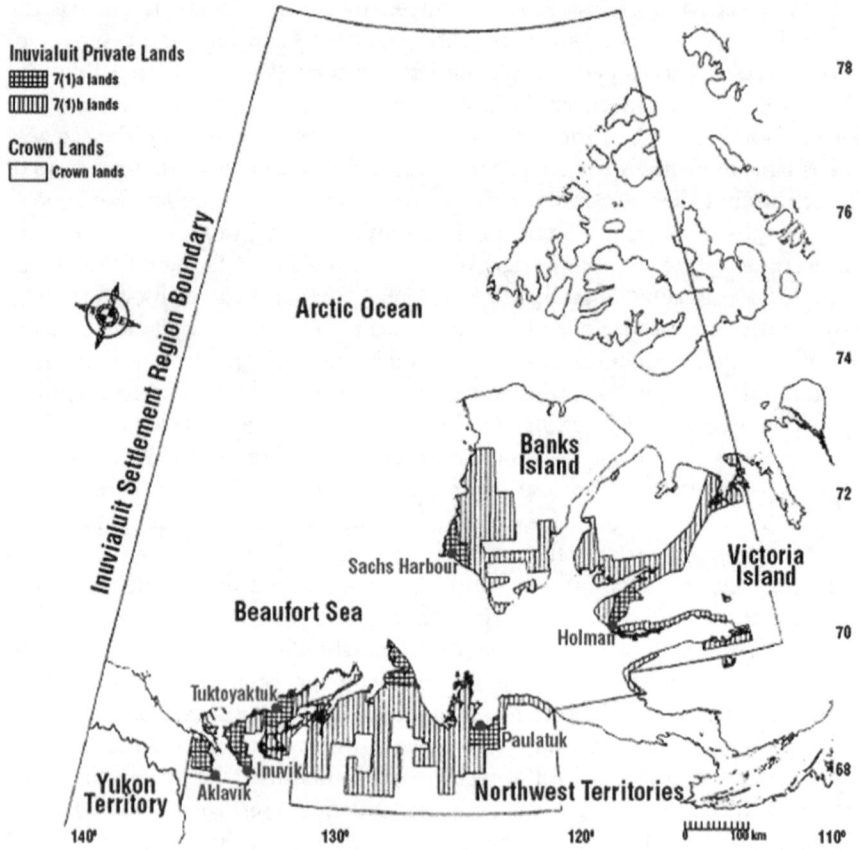

Fig. 11.5 The Inuvialuit settlement region created by the Inuvialuit Final Agreement of 1984. Source: https://www.aadnc-aandc.gc.ca/DAM/DAM-INTER-HQ/STAGING/ images-images/al_1100100031068_eng.gif

whole owes much then to the mutual reciprocity of commerce embraced by the HBCs.

The Inuvialuit land claim thus presents a substantial evolutionary leap beyond the Alaska land claim which inspired it, with many prescient and endur- ing advances in collaborative management and stronger protections of native lands and traditions missing from the Alaska claim. Had the Inuvialuit not so enthusiastically embraced and constructively improved the land claims model, with native corporations at its core, the many structural weaknesses of the Alaska land claim—as described by Thomas Berger (1985) and Edgar Blatchford (2009, 2013)—might well have doomed the model altogether. Blatchford noted how the land claim model that transformed the political economy of Alaska, Yukon, NWT, Nunavut, and northern Labrador would ultimately be rejected by Indian Country in the "lower 48" as a flawed model, where corporations are viewed with much more skepticism and as contrary to native values. But in the Arctic, it has become a central and evolving blueprint for strengthening the bond between First Nations and the state, and a defining feature not only of the Western Arctic but the entire Arctic littoral of North America, an increasingly strategic region of not only the North, but of the world. This continuing process of reforming the land claim model as it flowed from the Inupiat to the Inuvialuit and on to Nunavut and Nunatsiavut is a reflection of the collaborative mechanism that defines the Western Arctic, reminiscent of the integrative and symbiotic dynamic of HBC and North West Company trading posts, and reflective of the very dynamic and integrative "Dene/Inuit interface" of the Mackenzie Delta as described by William C. Wonders (1989).

The Western Arctic contains many of the same ingredients found in more contested regions of the world, including intense pressures of militarization and geopolitics from the time of the fur empires up through to the Cold War, with World War II's interstate violence reaching right up to Alaska's shores and resulting in the militarization of much of the region. And yet, from this process emerged a strikingly collaborative, crossborder dynamic on both sides of the Western Arctic. The region's propensity for accommodation and reconciliation, for welcoming cultural diversity and inclusion, and for collaborative problem-solving would continue to re-assert itself. This would reflect the enduring legacy of the early pioneers in northern economic development, and some of the world's very first multinational corporations.

This is collaborative spirit is evident today in the close collaborative relationship between the Inupiat and the Inuvialuit, who have partnered on numerous crossborder issues—including the Inuvialuit-Inupiat Polar Bear Management Agreement in the Southern Beaufort Sea, and the Inuvialuit-Inupiat Beaufort Sea Beluga Whale Agreement. The collaboration extended to the resumption of bowhead whale harvesting by the Inuvialuit during the 1990s, when community-to-community exchanges ensured the transfer of traditional knowledge required for a successful and safe restoration of bowhead hunting. It is equally evident in the complex (and occasionally prickly) relationship between the Inuit and the Dene along the "Dene/Inuit interface" (Wonders 1989) a history that includes intertribal warfare as evident at the tragedy of Bloody Falls during Samuel Hearne's fateful expedition (Newman 1985, 1998) but which also includes efforts (ultimately unsuccessful) across decades of negotiation and collaboration, to form a "Western Arctic Regional Municipality" to jointly govern Dene and Inuit in the Western Arctic (Zellen 2008).

The result is the emergence of a diverse, inclusive, and fascinating political culture across the Western Arctic which has embraced a deep and enduring commitment to collaborative governance and crossborder co-management, intergroup (and international) partnerships, and constructive transboundary relationships that present a compelling model for how the world and Indigenous-government relationships specifically can (and one might argue, should) be governed. It is not always frictionless collaboration, as residents of the Mackenzie Delta well know, since there will inevitably be times and issues where interests will again clash—but it nonetheless presents our world us with an intriguing model for crossborder collaboration worthy of emulation. Indeed, this model has now been copied across the North, as is evident in many of the more recent land claim settlement areas, including Nunavut and Nunatsiavut to the east, and the many Dene and Yukon First Nations settlement areas to the south. In these adjacent regions, many of the collaborative structures pioneered by the Alaska land claim and refined by the Inuvialuit land claim have now been copied and adapted—rooted in the shared, symbiotic relationships not only between neighboring natives (such as the Inuit and Dene) but also between natives and settlers. The latters' interrelationships have evolved over time to become increasingly mutually beneficial, fostered by the reciprocal and collaborative foundational culture of the HBC and Northwest Company trading posts that dotted the northern landscape, seeding the emergence of today's municipalities and regional governing structures and embracing the same mutual commitment to collaboratively co-managing the North, much as has been done since the very first northern trading posts were formed.

Acknowledgments The author would like to gratefully thank the Kone Foundation of Helsinki, Finland, for generously funding my research on crossborder indigenous homelands and indigenous borderlands in 2016 and 2017, making research for this chapter possible. The views expressed in this chapter are the author's own and do not necessarily represent the views or policies of the United States Coast Guard Academy, the United States Coast Guard, or any other branch or service of the United States Government.

REFERENCES

Arnold, Carlene. 2011. *The Legacy of Unjust and Illegal Treatment of Unangan During World War II and Its Place in Unangan History.* University of Kansas, Department of Global Indigenous Nations Studies, Master's Thesis.

Berger, Thomas R. 1985. *Village Journey: The Report of the Alaska Native Review Commission.* New York: Hill and Wang.

Blatchford, Edgar. 2009. *The Unintended Consequence of the U.S. Congress and the Alaska Native Claims Settlement Act: The Demise of Corporate Democracy and the Threat of Native Ownership of the Land Base.* Dartmouth College Master's Thesis.

———. 2013. *Alaska Native Claims Settlement Act and the Unresolved Issues of Profit Sharing, Corporate Democracy, and the New Generations of Alaska Natives.* University of Alaska, Fairbanks Doctoral Thesis.

Bockstoce, John R. 2009. *Furs and Frontiers in the Far North: The Contest Among Native and Foreign Nations for the Bering Strait Fur Trade.* New Haven: Yale University Press.

Burns, John F. 1987. Farewell to the 'Northern Stores'. *New York Times,* March 29. https://www.nytimes.com/1987/03/29/business/farewell-to-the-northern-stores.html.

Coates, Kenneth S. 1985. *Canada's Colonies: A History of the Yukon and Northwest Territories.* Toronto: Lorimer.

Easton, Norman. 2016. *History of the Border in White River.* Presentation to the 2016 Borders in Globalization Summer Institute, Yukon College, Whitehorse, Yukon, June 21.

Groh, Clifford John, II. 1976. *Oil, Money, Land and Power: Passage of the Alaska Native Claims Settlement Act of 1971.* Cambridge MA: Harvard University/Harvard College Honors Thesis.

Jones, Dorothy M. 1981. *A Century of Servitude: Pribilof Aleuts Under U.S. Rule.* Lanham, MD: University Press of America. http://arcticcircle.uconn.edu/HistoryCulture/Aleut/Jones/jonesindex.html.

Jorgensen, Joseph G. 1990. *Oil Age Eskimos.* Berkeley: University of California Press.

McGhee, Robert. 1976. The Nineteenth Century Mackenzie Delta Inuit. In *Supporting Studies, Vol. 2 of Inuit Land Use and Occupancy Project Report,* ed. Milton Freeman, 141–152. Ottawa, ON: Supply and Services Canada.

Naske, Claus-M., and Herman E. Slotnick. 1979. *Alaska: A History of the 49th State.* 1st ed. Grand Rapids: William B. Eerdmans.

Newman, Peter C. 1985. *Company of Adventurers.* Markham, ON: Viking/Penguin.

———. 1998. *Empire of the Bay.* Markham, ON: Penguin Books Canada.

Robins, Nick. 2004. The World's First Multinational. *The New Statesman,* December 13. https://www.newstatesman.com/politics/politics/2014/04/worlds-first-multinational.

Sykes, Jim (Producer), Bill Dubay (Writer), Sue Burrus (Writer and Editor), and Frankie Burrus (Researcher). 1985a. Part 10: The Newborns—Left Out of ANCSA. *Holding Our Ground: A Radio Documentary.* Anchorage: Western Media Concepts, Inc.

——— (Producer), Bill Dubay (Writer), Sue Burrus (Writer and Editor), and Frankie Burrus (Researcher). 1985b. Part 11: From Hunter, Fisher, Gatherer to Corporate Director. *Holding Our Ground: A Radio Documentary.* Anchorage: Western Media Concepts, Inc.

Tundra Times. 1969. Bulk of the Monetary Settlement May Come from Navy Petroleum Reserve No. 4. *Tundra Times,* March 14, as reprinted in *A Scrapbook History: Alaska Native Claims Settlement Act.* Anchorage: Tundra Times and the Alaska Federation of Natives, 1991, 34.

Up Here. 2005. The Far Flung Corners of the Empire. *Up Here,* July 3. https://uphere.ca/articles/far-flung-corners-empire.

Wonders, William C. 1989. The Dene/Inuit Interface in Canada's Western Arctic, N.W.T. In *For Purposes of Dominion: Essays in Honor of Morris Zaslow,* ed. Kenneth S. Coates and William R. Morrison. Toronto: University of Toronto Press.

Worl, Riccardo. 1988. The 1991 Time Bomb Defused. *Alaska Native Magazine,* No. 1 (June), 1.

Zellen, Barry Scott. 2008. *Breaking the Ice: From Land Claims to Tribal Sovereignty in the Arctic.* Lanham, MD: Lexington Books.

CHAPTER 12

The Future of Work in the Arctic

Ken S. Coates

Innovation enthusiasts have been commenting for years on the work-saving, environment-helping and society-changing elements of scientific and techno-logical transformation. A smaller number of cautionary critics, starting with Jeremy Rifkin's prescient *The End of Work*, worried about how rapid and largely unregulated technological change would alter patterns of work and employ-ment opportunities (Rifkin 1995). The standard analysis, captured in Erik Brynjolfsson and Andrew Mcafee's *The Second Machine Age*, is that robotics, digitization and artificial intelligence will collectively eliminate millions of highly paid, stable jobs (2012, 2014). While there will be expanding opportu-nities in some sectors, particularly in high technology areas, the consensus is that many manual, repetitive jobs will disappear in a wave of workplace innova-tion. Higher end positions, including in business operations, health care and governance, are also at risk. But as Laurence Smith argues, northern innova-tion realities will not follow international patterns (2010):

> We know that innovation will not unfold evenly. There is already a substantial "digital divide" between those areas with ready access to the Internet and digital technologies and those without. We can now see the development of an "innova-tion divide," between those parts of the world with the key elements of a modern scientific and technological society and those who lack infrastructure, educational facilities, economies and scale and the other critical elements required for global technological competitiveness.

As Heather Hall has shown (in this collection), Arctic regions struggle to keep up with global technological change. Poor Internet connections, par-ticularly in the North American Arctic and Russia, inhibit technological

K. S. Coates (✉)
Johnson-Shoyama Graduate School of Public Policy, University of Saskatchewan, Saskatoon, SK, Canada

© The Author(s) 2020
K. S. Coates, C. Holroyd (eds.), *The Palgrave Handbook of Arctic Policy and Politics*, https://doi.org/10.1007/978-3-030-20557-7_12

development. But companies and governments continue to explore financially and commercially viable applications as they seek to reduce the high costs of operating in remote, Arctic and sub-Arctic conditions. There are many areas requiring innovative solutions—developing appropriate responses to winter, darkness, distance, diseconomies of scale and other Arctic realities—and comparatively little effort being made to come up with new technologies or digital processes that address specific Northern conditions.

But in almost all parts of the Arctic, employment has long been insecure, tied to commodity markets, seasonal fluctuations in work and episodic surges and declines in government spending. Most Arctic regions are more government-dependent than resource-dependent, despite the prevailing image of the northern regions as centers of mining, forestry, and oil and gas extraction. Government jobs—teaching, nursing, medicine, the military, infrastructure maintenance, social welfare management, environmental monitoring and the like—may be less subject than most others to technological displacement. The maintenance of such positions is, moreover, highly political and subject to considerable public backlash if positions are pulled out of a community or region. Even here, however, major shifts are possible. In Finnmark, Norway, for example, experiments have been undertaken with automated snow removal at airports. If these systems work, hundreds of employees could be displaced through new technologies, all the while providing better, more reliable and cost-effective service to the region.

It is important to remember that in the small and remote communities that dominate the Arctic social network, a small number of government jobs have an out-sized impact on the local economy. These positions are, in contrast to most posts (other than high-end jobs in the natural resource sector) stable, well-paid and reliable. These workers are primary customers for local businesses and services. When a handful of government jobs leave a small town, the financial shockwaves can be considerable. In many parts of the North, to a degree that northerners do not fully acknowledge, government is the primary economic engine at the community and regional level.

But even government positions are not guaranteed. Government work is as susceptible to the impact of technology as any other kind of activity. Web-based services, digital program delivery, video-conference based consultations, long distance digital surveillance and monitoring are but a few of the current technologies that are altering the need for regular, paid government employees across the North. If major investments are made in tech-based service delivery, government can and will save a substantial amount of money by switching away from regular employees. The future of work is intricately linked to the sustainability of Arctic communities and the viability of Northern businesses and regional economies (Arctic Human Development Report 2015).

The impact of these transitions is complex and multi-directional. There may be more jobs, or less work. Technologically enhanced changes in the coming years could be dramatic. Major advances are possible in the resource sector, bringing new mines, oil and gas properties and other developments into viable operations. Older properties could be reopened due to improved extraction systems and more favorable market conditions. Activities that were commercially impractical only a few decades ago could now become practical. New businesses can emerge and old ones close. Few of the transitions can be anticipated in full and the implications are often not recognized until it is too late to change course.

There are countless examples of community-level changes based on emerging technologies. The introduction of radial tires, improved gas mileage and more effective paving systems introduced over the past 40 years resulted in the closure of hundreds of roadside garages and hotels along northern highways. Improved communication systems clearly helped many parts of northern life, but Internet-delivered entertainment accelerated the closure of northern movie theaters and had considerable impact on community social life. Automated teller machines, now available in even isolated locations, brought great convenience but resulted in the closure of many small bank and credit union branches. Online shopping, particularly access to digital mega-stores like Amazon and Alibaba, undercut the viability of hundreds of Arctic businesses. The potential for large-scale transformation holds, too, through tele-medicine, tele-health and tele-education. In each case, new technologies could improve services and consumer opportunities for Arctic residents but could, in the process undercut current patterns of employment.

On a grand scale, it is not clear that the North has an obvious role in the emerging, technology-based economy. The region has few post-secondary institutions, mostly colleges and a few small universities. While there were some successes built around universities (Umeå in Sweden, University of Alaska-Fairbanks, UiT in Tromsø, Norway, and Oulu, Finland), the simple truth is that the biggest research universities and the vast majority of the researchers are in the south. The big cities and southern areas have other advantages, including access to investment capital, a ready supply of trained workers (often called Highly Qualified Personnel), proximity to commercial markets and greater support from national and regional politicians. The long-standing challenges of northern economic development now face the additional barrier of overcoming the innovation and technological deficits of the Far North. While optimists argue that innovation could and would occur where entrepreneurship and new ideas came together—a substantial truth—the reality is that the Far North is far from competitive nationally and internationally.

The new economy presents some unique and puzzling challenges, and none are greater than the future of work in the North. New technologies typically eliminate or transform jobs. Some new positions open up—coding and the management of big data and control of autonomous vehicles were not significant occupations 20 years ago—but in most areas the job losses mount up faster than the employment gains. Salaries are one of the major inputs for any operation, from a factory to a government office or from a drilling platform to deep ocean fishing. Eliminating jobs through labor-saving devices and technologies can make companies viable, produce greater profits and reinforce business stability. But labor savings mean fewer jobs. Fewer jobs mean fewer opportunities. Fewer opportunities create economic uncertainty, if not panic, in the local and regional population. Without good and sustainable jobs, people leave communities, often heading for cities or larger urban centers which are perceived to offer greater opportunities. Towns and villages suffer economic decline and an overall downward spiral in population and community development. This has become the reality for much of the rural world over the past generation, with the likelihood of an acceleration of these challenges in the near future.

The story is far from one-sided. New technologies could and often do eliminate difficult and dangerous jobs, ones that companies face difficulties filling. This is particularly important in areas of extreme Arctic weather. New technological systems could and will make many commercial operations more viable. Improvements in mining operations, for example, have moved people away from a lot of the dangerous underground work. This is good news, unless you are an underground miner with few prospects for new employment in the new economy. Further, and as Rifkin argued years ago, one of the goals of humanity overall has been to reduce the amount of work and to replace work with labor-saving devices (formerly machines and now, increasingly, robots and digitally controlled systems) particularly in the case of repetitive, dangerous, simple and low-skilled work (Ford 2015).

Consider one example: Many communities have re-sorting stations for recycled materials. The work is not exciting, but provides solid work for people of limited skill and with a decent work ethic. In some communities, particularly in Canada, this work has been set aside for people with intellectual disabilities. But these jobs are, in technical terms, already largely redundant. The workers could be replaced with machines that could handle the sorting. The machines would do it faster, more accurately and likely more cheaply. Local authorities face the choice of eliminating jobs in the name of efficiency. Sometimes, saving money is not the only thing that matters. This issue has been addressed directly by many Indigenous companies, which have established community well-being as a higher priority than the production of profits. In these companies, hiring and retaining workers has high social value; laying people off and creating economic despair is not viewed as acceptable. (This approach, incidentally, is not unique to Indigenous peoples. Japan and other Asian countries have long had comparable commitments to maintaining employment, although these systems have been under pressure in recent years.)

Of course, the new economy is supposed to create new jobs and new opportunities. That is true, but only in a limited way. First, many aspects of the new economy require fewer jobs to provide comparable services or produce similar copies of a particular product. Google, one of the wealthiest companies in the world, has only 20,000 employees, 1/20th the number of automaker Toyota Corporation. Second, the new economy places a premium on high-end computing, engineering, medical, mathematical or scientific skills. These skills are already hard to acquire and in great demand, particularly in the Arctic where inconsistent and often inadequate educational services track well behind competitive international norms. Technological skills will become more important and more scientifically rooted in the future. Equally important, people who are displaced from work in the traditional industrial economy can only rarely shift into the high technology jobs that are being created.[1] The new economy requires, in other words, different people with different skill sets. Truck drivers, to use one example, cannot easily be retrained or repositioned as systems designers or electrical engineers.

The major problem facing workers in the North is simple. The historic prosperity of many northern communities rested on the availability of substantial numbers of comparatively low skill, high wage work. In most areas, the construction workers, nurses, miners, transportation workers and others were imported from the South, save for in Scandinavia, which did a good job of training local residents for emerging jobs. In both Scandinavia and North America, this conjunction of employment and income was largely due to the presence and authority of unions, which forced up wages and improved working conditions. The result was generally egalitarian northern societies, with substantial numbers of working people earning decent incomes while doing work that did not require advanced education or highly specialized training.

But this pattern has changed substantially in recent years. New machinery, digitization, robotics and automated systems are replacing the hardy northern miner, logger, fisher and construction workers of regional lore. The new technologies are expensive to implement but comparatively inexpensive to operate. When selected and used properly, the new approaches are productive, efficient and cost-effective. The forest industry has already experienced massive work place reform, with new harvesting machinery undermining historical logging work. Mechanized and computerized sawmills and pulp and paper operations have been displacing tens of thousands of workers around the world. Northern Scandinavia, particularly Sweden, has been a world leader in the advanced mechanization of forest operations, leading a technological revolution that globally displaced workers while maintaining, if not expanding, wood production.

The possibilities for the technological displacement of workers are substantial. Here are some of the examples that are currently possible (and note that the major dislocations of workers are in the near to medium-term future as technological capacity improves):

- **Remote Operation of Mines, Including the Operation of On-Site Machines and Related Transportation Systems**: These labor-saving systems have been implemented in large-scale operations in Sweden and Northern Canada, and are further advanced in remote regions of Australia, where the major impediment facing the workers is extreme heat rather than frigid Arctic winters. The changes generally result in increased ore production, much lower labor costs and other improvements. The massive

---

[1] Rick Miner, Jobs Without People, People Without Jobs.

Kiruna iron ore mine in northern Sweden has reached an agreement with its workers that will allow for rapid automation, with income and job protection for existing employees. When these systems are implemented in full, the mining industry in the Far North will need hundreds if not thousands fewer workers.

- **Long Distance Truck Driving**: Automated vehicles are being tested extensively around the world and have already been introduced to a number of northern mining operations. The safe and secure operation of automated vehicles requires advanced communications systems (i.e., cell phone and wireless coverage), which is not common place in the Far North. But trucking firms appreciate the considerable advantages that would go along with this technology. The ice road trucking that is common in remote parts of the Canadian North would, if the communications challenges of working across vast, sparsely populated areas can be worked out, benefit greatly from replacing human drivers with advanced automated driving systems. If these systems work—and they are being tested more on southern urban roads than Arctic highways—there could well be a reduction in hundreds of Arctic truckers.

- **Internet-Based Surgery and Other Medical Services**: Many parts of the Arctic struggle to provide basic health care. Maintaining proper preventative health care coverage is difficult in areas without reliable medical or nursing care. Northern areas have long been innovators in this area, using everything from Internet-based doctor's "visits" to remote assessment of medical images, such as X-rays. Newer innovations include remote surgery (involving the manipulation of medical devices over the Internet), interactive video-conferencing for psychiatry, robot-based nursing education and a growing variety of monitoring systems that provide for easy, inexpensive and reliable evaluation of personal health conditions. Once a full suite of these technological solutions are in place—and developments in this area are moving fast—northern communities could well need fewer nurses, fewer doctors (or doctor's visits in places where physicians are not permanently based in the communities), better preventative medicine, closer monitoring of the health Arctic residents and, proponents anticipate, substantially better health outcomes.

- **E-Teaching and e-Learning**: The digital education industry is attracting billions of dollars in annual investment. Schools, educational companies, colleges and universities have created enormous quantities of educational material, much of it highly sophisticated and pedagogically sound and creative. Research and teaching efforts have included major improvements in digital learning systems, tele-conference-based classrooms and new systems for monitoring and evaluation of students. Some of the systems, using what promoters call the twenty-first century classroom method have produced impressive results, with higher test scores, lower teaching costs and the replacement of some teachers with digital instruction. Because of limitations of northern Internet, many of the communities that could benefit the most from e-education approaches do not have access to the best technologies and applications, adding to the educational and training divide. In the near future, the emphasis of this educational effort will shift from high school and post-secondary education and will focus instead on twenty-first century employment, scientific and technological skills. These technologies could level the educational playing field between North and South, but it is easy to see that a displacement in Arctic teachers is possible, although more technicians and digital specialists will be required to monitor the equipment and ensure the effective operation of the classrooms.

- **Drone Technologies**: The rapid development of drones has been significant across the North. They have been used for remote rescue operations, often in extremely hostile conditions, and are becoming commonplace in mineral exploration and soils testing. They help with wildlife monitoring, assist environmental oversight and, as they get larger and more dependable, will be contributing significantly to the delivery of supplies to mining camps, remote settlements, scientific stations and the like. These are all exciting, cost-saving and helpful innovations. But in each instance, they involve either major changes in existing work or the elimination of current activities. The expanded use of drones in mineral exploration, for example, has lessened the demand for summer field workers, many of them Indigenous people in recent decades. The drones are much more efficient and less expensive than traditional techniques, but as has become commonplace in recent years, they also eliminate jobs.

- **Shopping and Northern Stores**: Fully digital grocery shopping, going beyond well-known current systems to more advanced technologies that allow for personalized clothes shopping and enhanced consumer control of the marketplace, are coming onstream. Consumers around the world are capitalizing on unprecedented access to a wide range of shopping opportunities. For residents of remote northern communities, typically served by a small local store and occasional trips to larger centers, online shopping has reduced prices, expanded choice and often improved service. The expansion of digital shopping, however, erodes the viability of northern stores and will, in many instances, result in fewer retail workers and even the closure of Arctic businesses, perhaps while increasing jobs in the delivery sector. Northern content producers, of course, can use the same technologies to reach consumers and markets around the world. This has already proven beneficial for Arctic artists and performers—who can sell their songs, movies, artwork and designs around the world—and for small tour operators who can reach potential clients that otherwise would never learn of their services. As with many of the digital disruptions, it is too soon to tell if, on balance, this works to the benefit or cost of the Arctic economy.

- **Digital Surveillance**: Remote monitoring of wildlife, environmental changes and other natural phenomenon has been a growth sector in recent decades, accelerating recently due to the attention devoted to climate change. Rapidly developing sensors and quantum computing, combined with digital communication systems, have the potential to expand oversight of natural and human phenomenon across the Arctic. These devices could provide comprehensive coverage of the natural environment and climate change while reducing dramatically the demand for field workers, many of whom are northern residents specifically trained by Arctic scientists to provide on-the-ground specialists.

- **Professional Services**: The service economy has been growing apace in recent years, covering a wide array of financial, legal, health and other supports. Regional centers across the North have found the service sector to be an important source of economic growth. Emerging technologies have automated many such services—smart phone-based banking systems being the best example—resulting in the displacement of numerous workers, including in the North. Outsourced professional work, including the reading of medical charts, legal work and accounting services, are both enhancing Arctic services and reducing the need for northern workers. Of course, one of the most creative and disruptive elements in the digitally enabled economy is that Internet-based services work in all directions, allowing northern-based service providers to market their professional services around the world.

- **Additive Manufacturing or 3D Printers**: The transformation of work is not all in one direction. The development of additive manufacturing—which works on the extrusion of various materials (ranging from paper and ceramics to concrete and metal)—has considerable potential in the Far North. 3D printers can produce everything from homes and buildings to replacement parts for machines and equipment. They will enable users in the most remote locations to download (Internet-permitting) coded instructions to be printed and for items to be produced on site. This fast-moving development has the potential to disrupt a great deal of work; if 3D printed homes work in the North (not assured), hundreds of seasonal home builders and carpenters will lose their jobs, while a smaller and indeterminant number of 3D printing companies and workers will create new businesses and find new work.

The list could go on. The transformations based on technological changes are not unique to the Arctic. They are, in fact, becoming increasingly common across the industrial world and, increasingly, in the developing world (Hirsch-Kreinsen 2016). On a broader scale, the technological transitions are substantial, in both scale and reach. Some of the other transitions of work in the North include:

- Artificial intelligence systems that could replace human decision-makers in fields such as finance, insurance and, eventually, government services;

- Robot-based manufacturing, which is already displacing millions of workers world-wide and which permit automation of a great deal of processing work, a small amount of this in the Far North;
- Online systems for downloading movies, books and magazines, technologies that have already disrupted the content sector (news and entertainment) around the planet.
- Small scale commercial innovations, such as self-service gas stations, online check-in for airlines, electronic ticket purchases and self-checkout at retail stores, have found substantial following among consumers, with quite significant job losses in standard retail operations;
- Automated payment systems, including for government services such as income tax filings, which substantially reduce the need for human interventions.

These systems, and many others, are changing the foundations of the economy and upsetting established patterns of work, commerce and government services. Technology emerges rapidly. Some flounder and disappear. Others grow exponentially and become globally significant. The essence of digital technologies is that they use new systems to replace large numbers of workers, while creating new sectors and producing incremental work in targeted and emerging areas. The main elements here are uncertainty, unpredictability and potential disruption. The new technologies do allow for work to proceed in multiple directions. Work does not automatically relocate to the South and could, in fact, be based in the North if there is appropriate infrastructure, sufficient entrepreneurial capacity and technical abilities.

New technologies, including such now ubiquitous services as GoToMeeting, Skype and Zoom, have tackled distance and the need for in-person meetings. At one extreme, this permits highly skilled professionals to live wherever they wish; the Government of Yukon refers to these highly desired, high-income people as "lone eagles," and pursues a strategy of building the local economy "one (highly paid professional) job at a time." Northerners who wish to stay in the Arctic but whose clients are located elsewhere now have the potential to connect digitally and to maintain global professional engagements. The same works in reverse. Where the Arctic has long received thousands of professional visitors per year, including consultants, government officials and professional service providers, many of these individuals have the option of conducting their business online. The systems may lack the immediacy and flexibility of in-person engagement, but these digital connections are inexpensive and reduce time-consuming travel from the South to the North. While the benefits in terms of immediacy and cost are evident, the shift from on-site meetings to digital connectivity will also reduce the demand for air travel, hotel rooms and associated taxi rides, restaurant meals and the like. The balance in terms of work and employment remains unknown.

At present, the global economy is in the process of eliminating hundreds of thousands of low skill, decently paid work and a growing number of white-collar and professional positions as well. The net effect could be net positive for the global economy as a whole. The transition is working particularly well for people of significant technical skill and substantial wealth. For small towns and remote regions, the likely effects are less promising, even if costs for services and products could go down and speed and efficiency could improve. Currently, high quality medical services, like radiological assessments, are being provided in South Asia for North American doctors and patients. In time, this could apply to surgical procedures as well as more interventionist treatments like psychiatric care. Purchasing even high-end professional services will be cheaper and more efficient as they are done remotely, with work relocated to low wage, decent education societies like India, the Philippines and Pakistan. People will not protest if they are able to get a full medical checkup without leaving their homes, and most would appreciate digital health oversight that caught acute attacks and monitored chronic illnesses. Society will, in the abstract, benefit from enhanced, less expensive and more readily available digital and technological services.

The question arising out of these workforce transitions, however, is fundamental and critical: what will northerners do for work in the near and distant future? A job-less society is not in sight, and there may actually be more work, albeit episodic, seasonal and perhaps less remunerative. Certain jobs, particularly those requiring high levels of scientific knowledge and/or technical skill, will grow in number and, likely pay. Service jobs will remain strong but, in the North, often seasonal due to the reliance on the tourism industry. The long-standard northern resource jobs—comparatively low skilled and high waged—are likely to decline substantially. For governments, the top issue is that work will change, perhaps dramatically, and that people and educational institutions need to prepare for an uncertain future.

Looking to the future–not the distant future but the next 10 to 15 years–suggests that even more changes are coming down the line. Algorithms (mathematical programs that replicate human thinking and decision-making processes) are becoming incredibly sophisticated. Research shows that digital solutions are more accurate at even some fundamentally human tasks: selecting the best candidates for a job, making a medical diagnosis, managing airport and highway traffic, completing a psychological and psychiatric assessment, among many others. People will resist some of these and accept others. All we know for certain is that the future of work is very uncertain. And for the Far North, where the vulnerabilities are greater than the obvious opportunities, there is particular reason for the concern.

POLICY RECOMMENDATIONS

Adjust to the uncertainties of the future of work will require considerable foresight and policy innovation. In recent decades, many Northern governments have had challenges connecting education, training and workforce preparation to the job opportunities associated with Arctic infrastructure and resource development. This has led, in turn, to a reliance on imported and fly-in fly-out workers and to a receptiveness to technological solutions. Across much of the North, with a marked exception of Scandinavia, Indigenous peoples are particularly disconnected from many parts of the high wage workforce, despite concerted and now long-term educational and training efforts. Approaching the future of work, in an era of accelerated and unpredictable changes, will require new and creative processes of workforce development. Several key considerations stand out.

Anticipating the Future of Work

Planning for the future of work will not be a one-time evaluation and planning exercise. Conditions change rapidly and preparing individuals, families and communities for the challenging years ahead will require diligence and unusual foresight. Governments and the education and training industries will have to engage the public in their research and contemplation of what lies ahead. Regional authorities will have to monitor changes closely and think carefully about the workforce needs and training requirements. At present, parents and young adults live in an age of 'career paranoia,' where there is a broad commitment to advanced education and skills training but great uncertainty about the best career paths for young people. Across the Western industrial nations, there are distressing patterns already evident: a rise of contingent labor and the development of the "gig" economy, high student debt loads in North America, substantial underemployment of college and university graduates, outmigration for northern communities in search of work and the rapid expansion of job opportunities in major urban centers. At times, and this appears to be one of them, there are reasons for paranoia. In these circumstances, governments and educational institutions have to make it clear that they are contemplating seriously to continued transformation of the workforce, that they are focused on career-readiness, and that they are developing strategies for long-term viability. At this juncture, few people expect official prescience and a comprehensive understanding of the evolving nature of work. Citizens do, realistically, expect that their governments, colleges, universities and skills development programs will be alert to coming changes and engaged with preparing young people for the jobs of the future.

Trainee Preparation

There is an equal onus on people entering the job market to be alert to future prospects and the changing nature of the workforce. This is not a current strength of the post-secondary and training system. In the advanced industrial nations, there is a growing and significant avoidance of physical and outdoor labor, with a strong preference for office and creative work. In many areas, including across much of the North, this resulted in shortages of skilled trades people and underemployment for non-trades personnel.

In much of the world, there are few direct controls on student choices for college, university or trades programming. This often results in dramatic mismatches between the number of graduates from specific programs, particularly in lifestyle offerings like outdoor recreation and forestry and wildlife management. Many areas of strong employer demand, from the skilled trades or advanced data management and artificial intelligence, are not yet connected to the number of trainees. Put directly, young people hoping to have secure, reliable employment with decent wages are strongly encouraged to align their studies and training with workforce requirements, with the vital caveat that the job opportunities of the present may not hold longer-term.

Educational and Training Institution Responsibilities

High schools, colleges, universities, apprenticeship programs and other training centers have a clear obligation to attend to two distinct, but overlapping markets: individuals preparing for the contemporary job market and those planning for the work of the future. At the elementary and high school level, this requires accelerated attention to scientific and technological training, ensuring that many more students have foundations in Math, Computer Science, Physics and Chemistry and are given access to experiential learning opportunities which better prepare them for participation in the paid workforce. Colleges and polytechs, particularly those programs that combine practical and applied elements, focus on preparing graduates for the world of work. Even here, though, the emphasis on current needs has often resulted in programs that are not attuned to the largest technologies and techniques and that may not provide full preparation for anticipated changes in the workplace. Universities and research institutes, which play an important role in pushing the technological agenda, have a wide range of successes and failures. Some elements—engineering, computer science, health systems and so on—are at the cutting edge of the new economy. The social sciences, humanities and fine arts have extremely valuable contributions to make, but these fields can be both conservative in their approach to advanced education and troubled by the growing evidence on the employability of graduates.

There is a good chance that the dictates of the new economy will require a substantial transformation of advanced education and training. Some of the shifts will focus on delivery and assessment systems; online and distance courses, programs and degrees will become ever more commonplace. More effort will be made to produce "just in time" instruction, including for company employees and individuals and communities facing dramatic workforce changes. Modularized programs, competency-based learning, simulator instruction and the use of a variety of gamification and visualization technologies may make more responsive and creative programming close to the norm, pushing the more standard diploma and degree formats into the background. More of this work will be done in the workplace, as part of ongoing employee upgrading processes and will be connected closely to emerging companies and technologies.

A major advantage rests with communities, regions and companies that get these connections between education, training and work right. Arctic regions, many with underdeveloped elementary, secondary and post-secondary education systems, find themselves at a disadvantage. Even areas like northern Scandinavia, with robust, high quality institutions with strong connections with regional employers, find that it is difficult to match prospective workers with employment opportunities. This does not suggest that there are major institutional shortcomings but rather that the comparatively small size of emerging businesses and job markets rarely produce sufficient student demand to justify program expansion. Colleges, which can open and close programs with relative ease are more responsive than universities which emphasize longer time horizons and more substantial investments.

Employer Engagement

Companies have, in recent decades, taken a back seat in the development of new and existing employees. That is changing rapidly, as firms realize the need to train, retrain and retain their workers. Larger firms have generally done better than smaller companies in responding to the preparatory work associated with technological and work place change. In the new economy, it is vital that companies provide accurate and thoughtful estimates for evolving workplace needs. This is highly sensitive information, for current workers and workers' associations/unions, who may be concerned about job loss or substantial retraining obligations. (Some companies, like the managers of Sweden's Kiruna iron ore mine, handled these transitions with minimal conflict and assurances of long-term positions for existing workers.) They need to work with educational institutions and governments to anticipate job losses and the training needs of new employees. Countries, as in Scandinavia, where employers generally collaborate closely with communities and governments, will likely do much better than those regions without forward-looking collaborations. Major companies and sectoral associations need to collaborate on the development of long-term national and regional labor force requirements and training strategies, no easy assignment in the best of circumstances and a serious challenge at times of intense change.

Preparing Entrepreneurs

The intersecting elements of the gig/contract economy, the global reach of portions of the service, culture and entertainment sectors and rapid technological change have recreated the entrepreneurial landscape. While the live anywhere, work anywhere expectations of the early years of the Internet did not transpire as anticipated, the reality is that digital and other technologies have redrawn the fundamental elements of entrepreneurship. Collaborations of local businesses, government agencies and post-secondary institutions are required to develop and support a new generation of Arctic-centered, technologically proficient and globally engaged business people. Entrepreneurs are typically in short supply. They cannot be trained or produced in, say, the same manner as skilled trades people or medical professionals. Producing a steady stream of entrepreneurs requires the creation of a substantial and creative innovation eco-system, the development of which is typically associated with places like London, Silicon Valley, Waterloo (Canada), Oulu (Finland) and Tokyo. Efforts have started toward building an Arctic eco-system, as through university-commercial collaborations in Oulu, Finland and Bodo, Norway. Arctic governments are investing heavily in various incubator and start-up facilitation efforts, but so, too, are governments around the world. There is no easy path to the creation of an Arctic entrepreneurial culture, but it helps to discuss openly the need for such an effort and to build Circumpolar connections to facilitate Arctic-based commercial innovation. The Far North needs more and larger companies based in the region, if it is to produce more and better paid jobs.

Income and Transition Support

The insightful study, *The Second Machine Age*, ends with the startling declaration that perhaps 25 to 50% of North American jobs are at risk in the coming decades due to digital and other technological innovations. This raised, for the authors Erik Brynjolfsson and Andrew McAfee and readers alike, the central question: what are governments to do to prepare for such a change? The book shied away from clear and definitive policy suggestions save for one: prepare new systems of income support to ensure that a large portion of the population does not fall into poverty and long-term unemployment. They favored, in particular, a generous form of basic income guarantee or a comparable income substitution program. Setting aside the fact that recent pilots in Finland and Ontario, Canada, were abandoned mid-stream there is little evidence of widespread public acceptance for such an initiative or a scalable funding model. The problem, however, is that governments may not really have a choice, if thousands of workers lose their jobs and are without hope for a quick return to the workforce. In the North American Arctic, of course, long-term unemployment or underemployment has been endemic and there is considerable evidence that dependence on government transfers does not address the social and cultural need and desire to have meaningful employment (Caputo 2012).

Internationally, commentators worried about the future of work have called for more aggressive and progressive tax systems, taxes on robots and other job-killing machines, comprehensive taxes on digital companies and other revenue producing measures. It is difficult to ensure that these taxes, if forthcoming and sustainable, will filter properly into northern jurisdictions, for the dislocation of work will likely be equally pronounced in more densely populated non-Arctic regions.

Regional governments will have to be alert to the scale and intensity of job location. They can count, based on historical patterns, on the out-migration of many displaced workers, who came North for the jobs and, typically, high incomes. This would alleviate the pressure substantially, while undercutting the North's economy generally. But if other revenue streams, from resource royalties, for example, stay strong, governments could be in a good position to sustain and expand other forms of work. Encouraging a shift to high content fields—creative production (art, film, music, theater), service work delivered globally, and culturally focused activities with Indigenous organizations—could create sustainable opportunities that are much less vulnerable to technological displacement. Indeed, focusing on such emerging fields could provide a strong economic alternative, with considerable commercial upside, to the traditional economies of the Arctic, a reality long-ago demonstrated by the emergence of Inuit art in the Canadian Arctic.

At this juncture, the primary government obligation is to take seriously the prospect of substantial displacement of traditional workers and the prospect of there being growing numbers of unemployed or underemployed workers in the Arctic. This would be on top of the current and often large cohort of Arctic people already in such situations. Exploring new economic models, ranging from job sharing, the development of cooperatives, reduced work weeks, lifelong training systems and various tax regimes (such as jointly funded sabbaticals from work) that could create more work or distribute existing work more equitably. Perhaps most importantly, the emphasis in government planning could shift from producing higher incomes to highlighting improvements in the quality of life that can come from new approaches to work and income generation.

Anticipating a World with Less Work

Evaluators of the future of work argue that major transformations are in the offing—with the important caveat that technological implementations rarely unfold simply or as expected. Technological and societal changes are messy affairs, with many instances of political and social resistance, economic disruptions and protests. There is no technological nirvana in the near future, nor will there be a rapid descent in a world of economic chaos and social unrest. What is clear is that many of the less attractive, difficult and dangerous jobs will be eliminated or substantially modified. Many routine white-collar jobs are well on the way to be replaced. In the near future, artificial intelligence and related digital systems will allow numerous jobs with higher order thinking—medical assessments, psychological assessments, staff recruiting and even basic journalism and coding—will be handled by machines of one type or another.

No scenario comes remotely close to describing a world without work and, indeed, standard economic concepts suggest that money and spending displaced in one area will produce consumer demand and therefore employment in another. The likelihood does exist that the world, including in the Far North, will have considerably less work available, which poses challenges and opportunities of their own. Myriad options for adapting to these emerging realities: reducing the number of jobs within a family, moving to a three or four day work-week, reducing the work day, connecting continuous education and training with paid employment, and the like.

The problem does not lie in capitalizing on the freedom from intense work and filling the available hours. Rather, the challenge rests with providing citizens with a high enough income to provide a desirable quality of life. New technologies will, of course, create new employment opportunities, but of indeterminant nature and salary at present. None of the available models for state intervention in a world of less work appear to be workable or financially sustainable. The problem is particularly acute for people of low or medium skill who are not well situated for retraining into higher end, more technologically sophisticated jobs.

Northern governments have more experience than most regional authorities in working with such work and employment environments, with the standard exception of northern Scandinavia. Across the region, unemployment rates are high, seasonal work is commonplace and the threat of technological displacement is quite significant. The strategies employed across the North American North—welfare payments, make-work projects, multiple retraining programs, employment insurance—have generally not been effective. These approaches have typically entrenched existing workforce problems rather than solving them, and have not addressed the underlying social, cultural and personal challenges associated with being unemployed and without a clear work function within the family and community. In particular, current northern government initiatives have struggled to connect with Indigenous values, cultures and adult responsibilities (Friedman et al. 1998).

The likely evolution in the world of work generates numerous policy challenges for Arctic governments. The region stands to experience considerable workplace dislocations based on technological change. Preparing for the future requires a modernization and intensification of education, potentially radical adaptations in post-secondary education and workplace training, adjustments in government income support systems, new approaches to the development of entrepreneurs and the incubation new firms, and the creation of an Arctic-centric, future-oriented strategy for economic development and engagement. Making these transitions will require a substantial amount of money, institutional flexibility, government prescience, private sector engagement and Circumpolar collaboration.

The task of responding to the future of the work in the Arctic is much the same as for other parts of the world. In this global competition, to secure a place in the future economy, the Arctic has few natural or regional advantages and significant regional liabilities. The Arctic requires substantial educational improvements, major infrastructure upgrades, commercial creativity and coordinated and sustained government commitments. Northern regions will do so while all other countries are making comparable investments and pursuing similar strategies, often with significant advantages and substantial head-starts. It is quite possible, perhaps even likely, that Arctic regions will lag behind other jurisdictions, enduring significant out-migration and economic decline. Given all of the others issues on the Arctic agenda, the slow-drip of the erosion of northern work has the potential to exacerbate existing vulnerabilities and draw attention to regional shortcomings. It clearly remains to be seen if Arctic governments and peoples can respond effectively to the challenges posed by the future of work.

REFERENCES

Brynjolfsson, Erik, and Andrew McAfee. 2012. *Race Against the Machine: How the Digital Revolution Is Accelerating Innovation, Driving Productivity, and Irreversibly Transforming Employment and the Economy.* Brynjolfsson and McAfee.
———. 2014. *The Second Machine Age: Work, Progress, and Prosperity in a Time of Brilliant Technologies.* New York: WW Norton & Company.
Caputo, Richard, ed. 2012. *Basic Income Guarantee and Politics: International Experiences and Perspectives on the Viability of Income Guarantee.* New York: Springer.
Ford, Martin. 2015. *Rise of the Robots: Technology and the Threat of a Jobless Future.* New York: Basic Books.
Friedman, Stewart D., Perry Christensen, and Jessica DeGroot. 1998. Work and Life: The End of the Zero-Sum Game. *Harvard Business Review* 76: 119–130.
Hirsch-Kreinsen, Hartmut. 2016. Digitization of Industrial Work: Development Paths and Prospects. *Journal for Labour Market Research* 49 (1): 1–14.
Larsen, Joan Nymand, and Gail Fondahl, eds. 2015. *Arctic Human Development Report: Regional Processes and Global Linkages.* Nordic Council of Ministers, Chapter 4.
Rifkin, Jeremy. 1995. *The End of Work: The Decline of the Global Labor Force and the Dawn of the Post-Market Era.* New York: GP Putnam's Sons.
Smith, Laurence C. 2010. *The World in 2050: Four Forces Shaping Civilization's Northern Future.* New York: Penguin.

PART III

Policies of Arctic Nations

CHAPTER 13

Russia's Arctic Regions and Policies

Gail Fondahl, Aileen A. Espiritu, and Aytalina Ivanova

Introduction

The North has been crucial as a material, mental and identity resource for
Russia, since its origin as a state. Some trace Russia's strong connection to the
North to the Middle Ages, during which Pomor fishermen with Viking connections established settlements along the shores of the European Arctic Ocean
and traded with both indigenous herders and hunters and international mer-
chants. From the sixteenth century on, the North and Siberia were colonized,
and became a crucial resource base for the Russian Empire, first as a source of
hard currency income via the fur-trade, then as a source of non-renewable
resources (oil, gas, minerals) during the Soviet period. In the post-Soviet
period, Russia's Arctic is of critical importance economically as a resource
hearth, while its role in nuclear deterrence is also fundamental. Thus, Russia
has a strong awareness of the importance of its North—economically, militarily
and as a symbol of its unique global position and historical roots—an awareness
that drives its policy (Medvedev 2018). If the attention paid to this region has
waxed and waned in response to other geopolitical developments, the growing
global interest in the Arctic as a potential source of natural resources and as a
possible shipping route has motivated Russia's renewed focus on the North

G. Fondahl (✉)
University of Northern British Columbia, Prince George, BC, Canada
e-mail: gail.fondahl@unbc.ca

A. A. Espiritu
University of Tromsi - Norway's Arctic University, Tromsø, Norway
e-mail: aileen.a.espiritu@uit.no

A. Ivanova
North Eastern Federal University, Yakutsk, Russian Federation

© The Author(s) 2020
K. S. Coates, C. Holroyd (eds.), *The Palgrave Handbook of Arctic Policy
and Politics*, https://doi.org/10.1007/978-3-030-20557-7_13

once again. This heightened interest is evident in a number of policy documents produced in the last decade, produced in order to more effectively guide the development of this rich (and diverse) region. The state has recently designated its Arctic as a specific region deserving its distinct policy.

This policy, domestic and foreign, is communicated chiefly through four key documents. We organize our discussion around these pronouncements, and the priorities they identify for Russia's Arctic in the twenty-first century. After a brief overview of the geography of Russia's Arctic, we examine the four documents: Russia's 2008 "Foundations of State Policy of the Russian Federation in the Arctic to 2020 and Beyond" (Osnovy 2008; henceforth referred to as 2008 Foundations), its 2013 "Strategy for the Development of the Arctic Zone of the Russian Federation and Guaranteeing National Security to 2020" (Strategiya 2013; henceforth 2013 Strategy), the State Program "Social-Economic Development of the Arctic Zone of the Russian Federation" (Gosudarstvennaya 2014; henceforth 2014 State Program), and a draft law "On the Arctic Zone of the Russian Federation" (AZRF 2017; henceforth Draft AZRF Law). We then discuss the main priorities articulated in these documents. These include the Arctic's role in providing strategic resources for the development of Russia as a whole; maintaining the Arctic as a zone of peace and cooperation; preserving the Arctic ecosystem; and developing the Northern Sea Route as a key transportation-communications system (Osnovy 2008). We also discuss the dearth of policy pronouncements regarding Indigenous peoples of the Arctic in these key policy documents, which we identify as a glaring shortcoming.

Given the excellent expert analyses of the 2008 Foundations document, and more generally, the pre-2014 situation, that is available to readers of English (e.g. Carlsson and Granholm 2013; Laurelle 2014; Zysk 2015), we focus somewhat greater attention on the 2014 State Program and on the draft AZRF law, and what these indicate about shifts in policy direction—as not much has been written yet in English on these documents. The draft law itself has evolved over the past half-decade, as described below. Geopolitical events, such as Russia's annexation of Crimea and involvement in Syria, have notably affected Russia's Arctic policy and the policy directing its development. We touch on other key legal acts (laws, decrees, presidential edicts) that address Russia's Arctic region. Indeed, more than 500 legal acts currently govern the Arctic (Makova 2018). As Heininen notes (2018), Russia's Arctic policy is part of a larger modernization plan for the country; a corollary is that national policies and legislation of the Russian Federation not specifically focused on the Arctic continue to often affect the Arctic.

While the documents that we review address policy dimensions of military security and issues pertaining to Russia's work on defining its continental shelf, discussion of these issues can be found elsewhere in this volume (chapters by Huebert and Koivurova et al., respectively; see also chapters by Laukenbauer, Bankes and Pincus), rather than in this chapter. We also note that our focus is mainly limited to federal policies, pronouncements and actions, although

numerous Russian northern federal subjects[1] have generated their own policies, which in some instances diverge from federal policy (Zysk 2015; see
Ivanova and Fondahl n.d., for some discussion of regional policies).

RUSSIA'S ARCTIC REGIONS

Russia's Arctic comprises a huge area of land and sea. The 2008 Foundations
document draws on a late-Soviet period definition (Decision of the State
Commission under Council of Ministers of USSR, of 22 April 1989) and an
early Soviet decree (Decree of the Presidium of Central Executive Committee
of USSR, 15 April 1926), the latter naming the islands belonging to the
USSR. The 2008 document also calls for legal acts that would determine the
geographical boundaries of the AZRF including its southern zone (Osnovy
2008, Chapter 4, Article 9c), a call that was met by two acts from 2014 (Putin
2014; Gosudarstvennaya 2014).

The State Program on the "Socio-economic development of the Russian
Arctic Zone until 2020" (Gosudarstvennaya 2014), defines a somewhat narrower 'Arctic Zone of the Russian Federation', as does a presidential edict of
May 2014 (Putin 2014). By these definitions the Arctic Zone comprises the
jurisdictions along the Russian coast of the Arctic Ocean: Murmansk Oblast,
Nenets, Yamal-Nenets and Chukotka Autonomous Okrugs in their entirety, as
well as the Taymyr (Dolgan-Nenets) municipal areas of the Krasnoyarsk Kray,
several 'counties' (*rayony/ulusy*) of the Sakha Republic (Yakutia) (Allaykhovsk,
Anabar, Bulun, Nizhnekolymsk and Ust-Yana) and of the Arkhangelsk Oblast
(Mezen, Onezhsk and Primorsk). The zone also includes the urban districts of
Arkhangel'sk, Noril'sk, Novodvinsk, Severodvisk and Vorkuta. In 2017, a
related presidential edict expanded the Arctic Zone to include three additional
counties, the Lousk, Kemsk and Belomorsk municipal districts of Karelia (Putin
2017). These territories together equal over 20% of the land territory of the
Russian Federation. Over 40% of the Arctic's land as a whole falls within Russia,
and over 40% of the Arctic's population is also found here.

While, given its span, Russia's Arctic zone is characterized by significant
geographical variation, some general characteristics commonly influence any
policy directed toward it. The region's challenging climatic conditions, including severe, prolonged winters and short cold summers, long polar nights and
frequent inclement weather and the presence of permafrost shape its development potential. Geophysical conditions, such as elevated electromagnetic noise
and the challenges to using traditional geo-positioning systems in the Far
North, confound communications and navigation. The environment is fragile:
with limited biodiversity, it is especially susceptible to disturbance (including
anthropogenic) and slow to recover. With climate warming more pronounced

---

[1] 'Federal subjects' refer to the various administrative units such as republics, territories (*kray*),
regions (*oblast*) and districts (*okrug*).

in the Arctic (the past three decades have witnessed a 2 °C rise in average temperature), permafrost degradation is expanding across northern Russia.

Russia's Arctic is homeland to a dozen Indigenous peoples (Sami, Nentsy, Entsy, Evenki, Dolgany, Eveny, Yukagiry, Chukchi, Chuvantsy, Eskimosy, Kereky, Koryaki), although numerically, Russians and other nationalities predominate. Much of the region is sparsely inhabited, with great distances between settlements. Russia's Arctic experienced a massive out-migration in the post-Soviet period, of 810,000 people from the 7 Arctic regions (oblasts, autonomous okrugs, republic and kray, noted above) between 1900 to 1990 (Tabata and Tabata 2018).[2] Development of the region is very costly, as most goods have to be shipped great distances, and infrastructure of all kinds is limited. For instance, Russia's Arctic zone in 2016 had less than 5700 km of paved roads (Rosstat 2016).

At the same time, the Russian Arctic is rich in natural resources, including hydrocarbons (on- and off-shore), and minerals. For instance, in 2016, the Russian Arctic zone produced over 89% of the country's natural gas, almost 25% of its petroleum gas and 100% of its apatite (Rosstat 2016). With less than 2% of Russia's total population, the Arctic zone produces about 12–15% of the Russian Federation's GDP, and provides about one-quarter of its exports (ACPOL 2016).[3]

The western part of Russia's Arctic (west of the Taymyr peninsula) is significantly more developed than its eastern sector, with cities of more than 100,000 inhabitants (Arkhangel'sk, Murmansk, Noril'sk), and significant infrastructure, including rail and road networks as well as port facilities. In the eastern Russian Arctic, the largest town (Tiksi) has just over 5000 inhabitants. Less is known about the resources of the eastern Russian Arctic (Antrim 2017).

Soviet Policy Toward the Arctic

The Soviet Arctic was defined by consolidation of Soviet power, nation-state building and Cold War politics. This section focuses primarily on the Soviet state's policies toward Arctic Indigenous peoples and the regime's drive to remake them into Soviet citizens. First, we outline how Marxist-Leninist theory formed the basis for nationality policies as they shaped the social and political culture of Soviet citizens, notably indigenous Northerners.

Second, we discuss the history of the Committee of the North, followed by its dissolution and Josef Stalin's intensive industrialization policies beginning with the First Five-Year Plan in 1928. Third, we show how these nationalities and industrialization policies had both beneficial and detrimental effects on the

---

[2] The figures include the populations of the regions as a whole, not just their Arctic counties, due to the available statistics.

[3] Other sources suggest a smaller share for these resources in Russia's Gross Regional Product: for instance, Tabata and Tabata (2018) give 8.1% (for 2015). They note that the Arctic fishing accounts for over 28% of Russia's fishing GRP and almost 26% of its subsurface (hydrocarbon and mineral) GRP comes from the Arctic (Tabata and Tabata 2018, 17).

native peoples of the North and Siberia, eventually coming to fruition in the Gorbachev era and after, as indigenous elites used what they learned from the Soviet system to gain political power within the Soviet political infrastructure.

Lenin's Nationalities Policies and the Soviet Arctic

Lenin and his party were pragmatic. They knew that they had to consolidate power, including in the borderlands, where many local leaders were trying to establish governments independent of the old Russian regime. In order to do so, the Bolsheviks declared support for self-determination and secession so that they could gain support from the borderland populations against the encroaching armies of the counter-revolutionary Whites.

The Bolsheviks were aware of the inherent contradictions of self-determination as the basis of nationality policies. They were cognizant that self-determination and secession were real possibilities that oppressed nations would take advantage of. Lenin and the Bolsheviks continued the rhetoric of self-determination while retaining the power to deny independence to any nationality or ethnic group if doing so would be deemed as harming the Soviet project. Moreover, "the party need only proclaim that separation was not in the best interests of the toiling masses, in order to brand any move in that direction counterrevolutionary" (Connor 1992, p. 19). With the discussion of self-determination came the promise of a federal structure. Federalism, much like the right to self-determination, carried a two-pronged utility for the Bolsheviks: centralization and advancing the power of the Communist Party within a federal structure (Pipes 1964, p. 242).

The effects of Soviet rule on the Arctic populations of the Russian Federation were profound. There were two corresponding policies that changed the traditional organization of the country and its people: Lenin's nationality policy and the industrialization policies advanced by the Soviet regime to bring its mostly peasant and rural population into the modern and industrial era, which changed their lives and livelihoods almost completely. To conform to Marxism-Leninism, it was necessary for the Bolsheviks to create a proletariat, urban or not. Both nationality and industrialization policies intertwined, in Bolshevik form.

Within the confines of the highly centralized Communist Party, the regime subjected Indigenous Arctic peoples to the same nationality policies as applied to other nationalities and ethnic groups such as Ukrainians, Kazakhs or Belorussians. While centralization and federalism were contradictory, so too were the policies and practices that emerged from these ideas. For many of the non-Russian minorities in the Soviet Union, federalism under a strong all-encompassing communist party meant that certain ethnic and/or national expressions were tolerated, even promoted so long as Marxist-Leninist ideology, as it would be determined by the leadership, was maintained and followed. For example, indigenous peoples could pursue their traditional economies under a collectivized setting, but the teaching of their respective ethnic languages would quickly fall by the wayside in favor of the dominant Russian language.

Asserting the ideals of an egalitarian society, Stalin moved to apply central-ization policies in order to eliminate backwardness giving non-Russians the opportunity to catch up with the rest of Russia (Gleason 1992). These policies were used by Stalin as the rationale for modeling economic, political and cul-tural development along the same lines as European Russia. Stalin made allow-ances for the demands of the ethnic minorities so that they supported the aims of the central government based in Russia. For example, the educational poli-cies formulated by Stalin tolerated the right to native language training and usage. It was clear, however, that these concessions were temporary until the "backward nationalities" had been industrialized and educated enough to be able to defend socialism (Gleason 1992).

The Committee of the North

The intellectuals attempting to better the lives of the Natives debated these issues while the Bolshevik regime implemented policies through the Committee of Assistance to the Peoples of the Northern Borderlands (*Komitet sodeystviya narodnostyam severnykh okrain*, abbreviated as the Committee of the North). It was established in 1924 and replaced the work of Narkomnats. The Committee was initiated by intellectuals exiled to Siberia under the Imperial regime. The Committee of the North answered the Communist Party's Central Executive Committee and determined that "the small peoples of the North" were at the stage of primitive communism, and thus had to be brought rapidly toward capitalism. The mandate of the Committee was very similar to the Northern department under Narkomnats—to help the Natives help themselves. The Committee proposed to help the Natives by attempting to provide such basic services as health care and education.

It was also at this time that the state created national districts for different ethnicities and nationalities. The national districts would serve the administra-tion well as it brought [Russian] culture, politics and economic development to the population. But the divisions of territories were not determined by the users: that is, the "national" territorial units were roughly what Native peoples would have designated for themselves but were not a precise representation of their traditional land use. Some of the ramifications of these divisions are only now being resolved as issues of self-governing territories (*obshchiny*) and eco-logical preservation are being debated among Native leaders, the government and development interests.

The Committee of the North could see that the growing proportion of Slavic Europeans in Siberia threatened the demarcation of national districts; thus they worked quickly to establish them. They feared that without the national districts, the degradation of cultures and numbers of the Indigenous peoples of the North would continue unchecked. Ironically, while there was conflict between the Committee and the regime regarding how to bring Natives to civilization, the two agreed that Natives had to be "civilized" and that it was their responsibility to "assist the small peoples in their difficult climb up the evolutionary ladder. Cultural progress meant getting rid of backward-ness, and backwardness, in the very traditional view of the committee members, consisted of dirt, ignorance, alcoholism and the oppression of women" (Pika and Prokhorov 1994).

The Committee promoted political development and awareness among Indigenous groups through the kul'tbazy, concerning itself with formalizing and transforming native traditional organizations into functional political structures. The Marxist-Leninist political education of Indigenous peoples was meant to make Natives aware of the struggle of creating a socialist state and their place in it. The Committee of the North aimed to educate the Northern peoples in their own Native languages until 1937. It was cumbersome to teach the Native languages in Latin script and Russian in Cyrillic, thus the govern-ment opted to eliminate the Native language publications and classes in favor of advancing Russian among non-Russians (Forsyth 1994).

This policy coincided with the dissolution of the Committee of the North in 1934, and its replacement with the *Glavsevmorput'* (the Central Agency for the Northern Passage). *Glavsevmorput'* did not want to have anything to do with northern peoples after 1938, concentrating primarily on the management of northern waterways and transportation. The peoples of the North, therefore, were left with no administrative institutions until 1957. For Stalin and his regime, the northern Natives had sufficiently bridged the centuries of develop-ment to be treated like other citizens of the Soviet Union and should contrib-ute the necessary means to rapidly industrialize the country, especially the North and Siberia (Grant 1993; Slezkine 1992).

Collectivization of Traditional Economies: Making Natives into Proletariats

While the economic and political environment that resulted soon after Stalin's rise to leadership of the Soviet Union was severe and brutal, early in the 1930s, the central government was compelled to create policies sensitive to the ethnic minorities under Soviet rule. The vastness of Soviet territory and the multiplic-ity of ethnic groups and nationalities made for difficult and intricate economic policy-making processes because the central government was obligated to structure economic policies that accounted for the uneven development from one part of the Soviet Union to another. The Stalin government also accounted for the differences in political outlook while at the same time making sure that the centrally planned economic policies were being implemented.

The primary aim was to bring industrialization to the masses, but in order to appease the minority ethnic voices, collectivization and industrialization were permitted to take on traditional ethnic characteristics. This was true for the Northern peoples, whose traditional economies were collectivized (par-tially depending on accessibility), progressing toward Stalin's strategy of transforming the northern peoples into proletariats and bringing socialism to every corner of the Soviet Union.

By the time Stalin's policies of collectivization were implemented in the North and Siberia, there were already many Natives from West to East who accepted the promises of the Soviet state and the benefits that they would be granted if they were dutiful members. Willing participation in the collectivization and industrialization process meant that some Natives would eventually gain status among the Russians who outnumbered them in the large urban centers and the villages (Tabelev 1936). There were more opportunities for all to engage in the system as other citizens would, by joining the pioneers, komsomols and the ranks of the working class. The regime argued that these activities were important, not just for the development of the Soviet Union as a nation, but also so that Natives could cultivate their own middle-rank technical personnel and their own intelligentsia (Stalinskaia Konstitutsiia i rabota na Krainem Severe 1936). This was also a convenient way for the government to settle Northern Indigenous peoples more centrally making it expedient (theoretically) to deliver social services such as medical care, education and cultural socialization. It was also an important way to engage in politicization, agitation and policy implementation in remote regions of the North and Siberia (Balzer 1983).

Industrialization

Also of concern to policy makers was the establishment of an industrialized proletarian population of the North with the aim of training Northern Natives on how to be and act like good Soviet citizens. For the Soviet Union of the 1930s, especially in the latter part of the decade, this was exemplified by rapid development of people and resources. By the mid-1930s, the Stalin government was becoming intolerant of national and ethnic differences leading to Russification of the education system and to the training of local cadres in political work along Soviet lines.

The most significant event in Siberia during World War II and after was the massive in-migration of Slavic populations into the region. The evacuation of families to Siberia during the war, the need for industrial workers for the armaments industry, and a population of prison inmates meant that the non-Native population of Siberia increased tremendously leading to the complete immersion of the Native population into the greater European Slavic population. For example, in Northwest Siberia, the non-Native population increased as much as five times from the end of the 1950s up to the early 1990s (Dienes 1987).

Intensified industrial development enlarged and maintained this non-Native population growth after World War II and particularly after the early years of economic recovery. Economic development between 1960 and 1980 was unequaled in any other phase in the country's history (Aganbegian 1984). While collectivization of Native traditional economies was already well established in many communities or continued to be a major policy goal by the time of Stalin's death in 1953, industrial and resource development in the

Russian North and Siberia initiated by Nikita Khrushchev and intensified under Leonid Brezhnev would modernize, urbanize and Sovietize the society, economy, landscape and culture of Soviet citizens from the Kola Peninsula to Sakhalin.

Whether it was nickel mining on the Kola and Norilsk, oil and gas development in Northwest Siberia, or diamond mining in Yakutia, industrial development took precedent over concern for the land and environment. The impact of flagrant industrialization under the Five-Year economic plans coincided with the nation-state building of the Soviet Union and, after World War II, the Cold War competition with the West. Caught in the middle were Native and Indigenous peoples of the North and Siberia, their land and traditions. By the time Nikita Khrushchev came to power in 1956, the Soviets were able to declare that the country was industrialized and by 1960, that a major portion of its population urbanized, though uneven in development.

Policies of industrialization and infrastructure development would continue throughout the Brezhnev era funded by the finding, exploitation and export of oil and gas from Northwest Siberia beginning in the early 1960s. Oil became the most valuable commodity for bringing hard currency to the Soviet Union, but its rapid extraction exerted a heavy toll on the environment and culture of Indigenous peoples in Northwest Siberia. However, by the early 1980s, just like the Soviet economy, the oil industry was in trouble and many of the existing fields were producing less and less. In the last years of Brezhnev's reign, the emphasis was on extracting as much as possible out of already developed fields, rather than exploring for new sources of oil or conservation. For the Soviets, the strategy was to expand the efforts to extract as much oil out of already producing fields. Thus unlike in the West, there was no collapse of the oil industry in the USSR in 1978 or even the beginning of the 1980s (Ebel 1994). The Brezhnev government's strategy was to invest as many resources as possible into increasing production, and Gorbachev continued to practice this same policy throughout his leadership.

Despite policies to export more resources for hard currency revenues, by the time Gorbachev came to power in 1985, the Soviet economy was failing and the system was not producing enough consumer goods to satisfy the market, leading to a deeply dissatisfied population. In 1986, Gorbachev introduced *perestroika* and *glasnost* with a measure of democratization. A year later, in 1987, Gorbachev would make his "Murmansk Initiative" (Gorbachev 1987) speech that would change the thinking and attitudes toward the Arctic and how it would be governed. In the "Murmansk Initiative," Gorbachev famously argued to make the Arctic a "zone of peace" and into a nuclear free zone (Åtland 2008). For its neighbors, it was a signal that the Soviet Union wanted to engage in open dialogue, at least on Arctic issues. Arguably, this would become the foundation for the eventual creation of the Arctic Environmental Protection Strategy (AEPS) that included all 8 Arctic states and would eventually become the Arctic Council in 1996, and would become a foundation for Arctic policy within the Arctic Council member states.

THE POST-SOVIET PERIOD: RUSSIAN ARCTIC AS A FOCUS OF STATE POLICY

The 1990s

With the collapse of the Soviet Union, ensuing economic challenges, and cessation of the Cold War, the Russian government turned its attention away from the Arctic. The area was perceived as more of a burden than a source of wealth, given the cost of supporting remote populations with food, fuel and other basic needs. Indeed, some would characterize the Soviet development of its northern regions as overextension (Hill and Gaddy 2003). The state considerably cut support for deliveries of supplies. Industrial sites were abandoned. A mass out-migration ensued, transforming once well-provided-for, thriving communities into ghost towns. The remaining population experienced increasing isolation (Hill and Gaddy 2003; Vitebsky and Alekseyev 2000), and a heightened dependence on local resources (Pika and Prokhorov 1994). During this period, the Russian state also abandoned much infrastructure along the shores of the Arctic Ocean and in its North more generally: meteorological stations were closed, while ports, airports and waterways received no support. The governance of the area was left, to a much larger extent than before, to the regional governments, leading to a wider variety of regimes of governance and development trajectories (see Ivanova and Fondahl n.d.).

The 2000s: About-Face

Climate change, including decreasing ice in the Arctic Ocean, and various globalization trends, increased international interest in the Arctic, including in its hydrocarbon and mineral resources, and in its possibility of serving as an alternate marine transportation route between Asia and Europe (30% shorter than that currently used). Russia refocused on its North, now emphasizing much more strongly on the Arctic (rather than the much larger "North and regions equated with the North"). It submitted its claim regarding its continental shelf to the Commission on the Limits of the Continental Shelf in 2001. The planting of the Russian flag on the seabed at the North Pole (while collecting additional information to support its continental shelf claim) provided another definitive moment in indicating that the Arctic was very much reviving as a priority region for the country. In 2008, Russia adopted its first post-Soviet policy regarding development of the Arctic, "Foundations of the State Policy of the Russian Federation in the Arctic, to 2020 and Beyond" (Osnovy 2008). Russia was the second Arctic country to adopt such a policy, after Norway (in 2006); since then the other six Arctic countries, as well as several non-Arctic countries, have also adopted such policies.

2008: "Foundations of the State Policy of the Russian Federation in the Arctic, to 2020 and Beyond"

Russia's 2008 *Foundations*, which deals mostly with its domestic Arctic, identifies four key priorities:

- securing the Arctic as a strategic resource base for the development of the entire country
- preserving the Arctic as a zone of peace and cooperation
- preserving the unique ecosystems of the Arctic
- using the Northern Sea Route (NSR) as a unified nation-wide transport infrastructure (Osnovy 2008, Chapter 2, Article 4).

The 'goals for socio-economic development' as referred to in this document mainly focus on extracting Arctic hydrocarbon and mineral resources to satisfy Russia's needs as a whole, for both internal use and export. The 2008 Foundations document proposes that the Arctic Zone is to become Russia's 'leading strategic resource base' by 2020 (Osnovy 2008).

2013: "Strategy for the Development of the Arctic Zone of the Russian Federation and Guaranteeing National Security to 2020"

The 2013 "Strategy for the Development of the Arctic Zone of the Russian Federation and Guaranteeing National Security to 2020" (Strategiya 2013) updates the 2008 Foundations policy, and elaborates on its objectives. It outlines risks and threats to developing the (Russian) Arctic, identifies priority development directions in more detail than in the 2008 document and fleshes out some of the means and mechanisms for achieving objectives. A pronounced goal of the 2013 Strategy is to create the conditions of achieving 'sustainable development' in the Arctic, through multi-level governance between the Federal, regional and municipal levels (Section 2).

The 2013 Strategy also highlights security more than the 2008 Foundations policy does: the term indeed is introduced in its title. It is worth noting that military security and the protection of the Russian state borders in the Arctic comes as the last of six points of priorities, after (1) socio-economic development, (2) science and technology, (3) creation of a unified information space, (4) environmental security and (5) international cooperation (Section 7). As expressed in this 2013 document, security appears to be perceived by key Russian policy makers as much in terms of 'soft' determinants, such as the social and environmental conditions of the region, as in terms of defense/military conditions. Military security, while being mentioned in the 2013 Strategy, is not prominent: military security and measures relating to the protection of state borders in the Arctic figure only in Section 18, after environmental security and international cooperation measures.

Post-2014 Developments

Several developments during and after 2014 have altered some of the aspects of state policy toward the Arctic. These include Russia's annexation of Crimea, its heightened role in the Syrian conflict and the resulting economic sanctions imposed by Western countries. The sharing of key western technologies for deep-water and shale hydrocarbon extraction, critically needed by Russia to develop its Arctic resources, was now prohibited (Motomura 2018). This disrupted progress toward achieving the goals and objectives articulated in the 2008 Foundations and 2013 Strategy policy documents.

The end of a run of higher oil prices (peaking in 2008, then drastically declining in 2014) further stalled the development of (expensive) Arctic hydrocarbons, especially those found off-shore. Continuing political friction with the West has encouraged Russia to look increasingly to eastern countries both as markets and as investors in Arctic mega-projects.

Russia has also increasingly prioritized military security once again. It claims that its military expansion is of a defensive nature, and is in many cases just restoring what was lost in the 1990s (e.g. continuous radar coverage), with many new or restored installations serving both a civilian and military role (Antrim 2017). Justification also involves the need for creating a safer environment in the Arctic through heightened capabilities for search and rescue operations, and responding to environmental emergencies (e.g. oil spill). Growing interest in the Northern Sea Route, especially from Asian states, may have enhanced Russia's attention to this priority. A new version of Russia's military doctrine, adopted in late 2014, for the first time (in peacetime) assigns protection of Russia's national interests in the Arctic to Russia's armed forces (Voennaya… 2014), while a new national security strategy, approved in 2015, identifies the possibility of international competition for Arctic natural resources as an issue of national security (Putin 2015). This policy also underscores the importance of the Northern Sea Route to Russia's economic security (Ibid.). Nevertheless, the Russian government concurrently has continued to highlight the Arctic as an area of international cooperation (see below).

The State Program "Social-Economic Development of the Arctic Zone of the Russian Federation"

Adopted in 2014, the State Program "Social-Economic Development of the Arctic Zone of the Russian Federation" is supposed to be the main mechanism for implementing the 2013 Strategy (Gosudarstvennaya 2014). However, initially no funding was provided to pursue activities prioritized under the program. A new version of the program was adopted, extending to 2025 (removing the date from the title, and focusing on the creation of 'support zones' (*opornye zony*)) as a means of developing the AZRF, as well as developing the NSR and facilitating the extraction of hydrocarbons. This state program identifies 13 other programs that specifically refer to the Arctic (e.g. on energy, fishing, ship-building, environmental protection), as well as another 7 governmental programs that concern it and 9 federal program targets.

Draft Law "On the Development of the Arctic Zone of the Russian Federation"

Initiated under President Boris Yeltsin, a draft law governing Russia's Arctic Zone was discussed in the late 1990s. Discussion regarding the need for such a law revived in 2013, with the development of the policy documents discussed above (Nifontova 2015). In 2015, Russia's Expert Council of the Arctic & Antarctic of the Federation Council concluded that such a law was needed. The Ministry of Economic Development has been developing a draft over the past five years. The idea is to bring "to a common denominator scattered, fragmentary and non-systematic regional and federal legislation in this area, whose foundations in most cases were laid in the Soviet times" (Tsybul'skiy 2015: 11), and to provide one comprehensive and cohesive law to govern the multi-faceted development of Russia's Arctic. As of this writing, the draft law still awaits revision and adoption by the Russian parliament. Input from 'stakeholders' across the Arctic has been solicited, during the drafting and re-drafting process, including from experts within the Russian North such as the Faculty of Law of the University of Yakutsk. This confirms that there is still room for regional agency in legal processes (Stammler and Ivanova 2016) despite the centralization trends of power in Russia.

In its 2016 version, the draft law more or less evinced the spirit of the 2008 and 2013 policies described above, as demonstrated by the fact that economic development, environmental protection and international cooperation remain important foci, while Arctic security and defense are listed as the last of ten articulated goals. Economic development remains front-and-center, a situation unsurprising in that the Ministry for Economic Development was the principal agency responsible for drafting the law. The 2017 version focuses much more conspicuously on creating 'support zones', the topic of the draft law's second chapter (the first being devoted to definitions). As of mid-2018, it was unclear whether and in what form this law would be adopted by the Russian parliament, but its frequent discussion in the press suggests that it is still seen as an important and necessary instrument for laying out the governance of the Arctic in Russia.

Other Policies/Developing the Arctic to Assert Russia's Geopolitical Position

While the draft law on the Arctic Zone does not significantly diverge from former policy documents (2008 and 2013), the change of the geopolitical climate may have influenced another field of legal reform in Russia with far reaching consequences for the Arctic. In 2014, Russia updated its military policy (Voennaya 2014) and a year later, its marine policy (Morskaya 2015). In these two documents, the focus is clearly on the increased importance of traditional security considerations. Military capacities in the Arctic are key to delivering on the two principal goals of Russian State policy, namely the maintenance of its sovereignty and the protection of its borders. Experts from the Centre for Political Conjecture explicitly acknowledge that this is a 180° turn in Russian Arctic policy, and that these strategic goals now outweigh previous priorities of international cooperation and internal Arctic development (Rossiskaya strategiya nd).

Key Domestic Arctic Policy Priorities
of the Russian Federation

The Arctic as a Resource Warehouse

Current Russian Arctic policy prioritizes the development of its natural resources.[4] The development of these is seen as a means to support the economic development of Russia as a whole, and to assert its position as a great power. Arctic policy statements as well as Russia's 2016 National Security Strategy underscore the need for Russia to (continue to) lead in the development of Arctic resources. Special emphasis was on hydrocarbons, with Russia's Energy Strategy of 2009 envisaging the large-scale development of off-shore resources. The development of the port of Sabetta as an LNG terminal provides just one example of some 15 mega-projects, underway or planned for the Arctic (11 related to oil and gas, four to ore and coal; Radushinsky et al. 2017). However, the deterioration of political relations with the West, sanctions imposed on the import of critical technologies for such development and the decline in world oil prices have caused some shift in focus to land-based mineral extraction. A re-orientation to Asian countries, as providers of investment, technology and as markets, has also characterized the years since 2014.

Introduced in the 2014 State Program and developed in the Draft AZRF Law is the concept of "support zone." These zones of intensified economic development are seen as the key mechanism for developing the Arctic. Over the Draft AZRF Law's evolution, each version has amplified the crucial role of such zones. Some experts have criticized such a focus, noting that a broader, more holistic approach to Arctic development is needed, and that Russia's current Arctic policy is colonial in its extractivist focus on accessing Arctic resources for the benefit of the non-Arctic populations, with little attention to improving the lives of local inhabitants (e.g. Zhukov 2018). The need to diversify economic activities in the Russian Arctic, including a move from the overwhelming dependence on extraction and exporting of raw natural resources to the processing of at least some of these locally, has been argued (e.g. Tsybul'skiy 2015).

---

[4] Heininen (2018) notes that this focus on economic development of natural resources parallels similar developments at the level of the Arctic Council.

Policies prioritizing the development of natural resource extraction have focused minimally on the issue of necessary human and social capital in the North to achieve these goals, though labor shortages may seriously constrain such development (Pavlenko et al. 2017). The outflow of population has intensified challenges of having adequate numbers of qualified workers to achieve the proposed development objectives. Russia's Arctic territories lost around 30% of the working age population in the 1990s (Pavlenko et al. 2017). Some feel that efforts should be made to build capacity among the Arctic's population to meet these needs, rather than continuing to depend on labor imported to the region or, increasingly, on shift-workers (V zakone 2016). Recent discussions about the justification, level and targets of northern increment (*severnaya nadbavka*) payments reflect the tension between a difficult economic situation that constricts these and the need to ensure the necessary conditions to retain workers who have taken up jobs in the North because of the benefits. This goes as far as the Russian labor ministry saying that that system is *archaic* while on the other hand refusing to change it (Topilin 2017).

The Northern Sea Route as Global Transport Corridor

A second major thrust of Russia's Arctic policy is the development of an integrated, national unified communications and transportation system, of which the Northern Sea Route (NSR) would play a central role (Osnovy 2008; Strategiya 2013). Currently the NSR is the only commercial shipping route in the Arctic. While it has served as a route for Russia for more than eight decades (Antrim 2017), a warming climate and concomitant declining sea ice has heightened international attention to this route. Development of the NSR as a commercial, international trade route has geopolitical as well as economic implications for Russia. Russian Arctic policy and practices have increasingly underscored the need to protect state sovereignty and enhance state security (in the more traditional meaning of the term) in this.

The NSR is touted as a route that would decrease transport and overhead costs of moving cargo between Asia and Europe. Yet numerous uncertainties thwart its development: the necessity of an icebreaker escort, variability of ice cover and preponderance of floating ice, the unreliable length of the navigable season, to name a few. Shipping through the route is likely to be limited to summer months in the near future. Yet, while some claim that the main obstacle to NSR navigation is climate conditions (e.g. Radushinsky et al. 2017), others argue that it is economic conditions and political uncertainty that more likely will limit shipping than natural conditions (e.g. Otsuka et al. 2018).

Policy to develop Russia's Arctic resources for both domestic and international markets also in part depends on the NSR. The NSR at present remains mostly a domestic transport route, used for supplying Russia's northernmost communities, and transporting some minerals out of the Arctic. The off-shore hydrocarbons, the development of which has been halted, were to be a large share of the future freight traffic along the route, bound for export markets.

Since relations with the West deteriorated, Russia has been wooing Asian investors to help move the route from domestic to international profile (Radushinsky et al. 2017).

In prioritizing the development of the NSR for international as well as domestic maritime trade, recent Russian policy identifies the need for the protection of this trade, including the upgrading of military infrastructure. The state has articulated plans to upgrade and expand coastal infrastructure (ports, hydrometeorological facilities, etc., as well as 20 border guard stations along the Arctic). With this new attention to the Arctic and the NSR in particular, some experts have recommended reviving an administrative unit for governing the whole Arctic littoral area, as existed for three decades during the Soviet period (*Glavsevmorput*; see above) (Zhukov 2018).

A Zone of International Cooperation and Peace: A Continued Commitment

"Preservation of the Arctic as a zone of peace and cooperation" is one of the four main "national interests in the Arctic" identified in the 2008 Fundamentals document (Osnovy 2008). Commitment to this principle is underscored in Russia's other key Arctic policy documents as well as in its foreign policy discourse (Tynkkynen 2018), in which the Arctic is often described as a 'territory of dialogue and cooperation' (e.g. Barbin 2015). International cooperation is cited as critical to addressing both Russia's national and international interests and commitments, such as sustainable Arctic development. Russia acted as a co-lead with the United States to developing an "Agreement on Enhancing International Arctic Scientific Cooperation," which was adopted by the Arctic Council at its 2017 Ministerial meeting (Agreement 2017). It had previously co-chaired the Arctic Council's Task Force on Search and Rescue, which developed an "Agreement on Cooperation in Aeronautical and Maritime Search and Rescue in the Arctic," signed into effect in 2011. In addition to these Arctic Council activities, the 2014 State Program notes the importance of bi-lateral agreements and calls out the Barents Euro-Arctic Council as a key example of such cooperation (Gosudarstvennaya 2014). In the economic sphere, after 2014, Russia has been increasingly engaging with China in the development of LNG, as well as with other Asian partners (including India and Vietnam; Barbin 2015).

Yet a strong dualism is evident here: while we see a serious commitment to fostering and supporting international cooperation in the Arctic, other policies emphasize the building of military might (Baev 2018), and foreigners are portrayed as intent on pillaging Russia's North for its wealth (Tynkkynen 2018).

Protecting the Arctic Environment

Seemingly incompatible with a focus on the rapid development of non-renewable resources, the majority of which contribute directly to GHG loads is a strong declaration in the policy documents on the importance of protecting and in cases restoring the environment of the Russian Arctic. With 15% of the

AZRF contaminated by industrial and military activities, the Russian state has declared the pollution a security issue. A program was initiated under the auspices of the Russian Geographical Society, with thousands of tons of waste being removed from several of Russia's Arctic islands in the past five years. It is expected that the program will run until 2025, though a shortage of funding, as well as other constraints, have caused delays.

One key environmental issue in a warming Arctic is that of permafrost degradation. The thawing of permafrost can cause great damage to extant infrastructure; any new industrial development in the Arctic needs to consider its potential impact on permafrost, as well as the influence future climate warming may have, regardless of measures taken to protect against industrial thawing. Some have called for a special federal law on the protection and rational use of permafrost; a regional law of the Sakha Republic (Yakutia), "On protecting Permafrost," was adopted in 2018 (Ob okhrane 2018). The law strongly emphasized the need to monitor more closely the thawing of permafrost, for which there should be a state agency established. In the explanatory text to the law, the importance of permafrost throughout Russia is highlighted, as it underlays 65% of the country's territory. Correspondingly, regional policy makers consider the country's first law on permafrost a milestone (Vasilieva 2018).

Indigenous Northerners: Forgotten? What's Missing? Indigenous Rights

Notable in its lack in Russia's current Arctic policy are concrete provisions for the improvement of the situation of Russian's Arctic peoples. A *Concept Paper on the Sustainable Development of Indigenous Peoples of the North, Siberian and the Far East of the Russian Federation*, published in 2009, outlined the social and economic challenges facing Indigenous Northerners and the need to improve their living conditions and other facets of their socio-economic development (Konseptsiia 2009). It called, inter alia for the adoption of legislation to ensure their priority hunting and fishing rights (in the face of increasing competition from industry). However, little action has come of this and earlier concepts, and indeed, subsequent legislation has eroded Indigenous rights to subsistence activities.

The adoption of legislation in Russia involves opportunities for interested parties to comment on drafts of the legislation and propose amendments and changes. In 2016, the head of the Russian Association of Indigenous Peoples of the North (RAIPON), as well as a deputy of the Federal State Duma[5] in his capacity as the latter criticized the current (2016) version of the Draft AZRF law, for its dropping of several clauses that would protect the rights and recognized interests of Indigenous northerners. RAIPON made eight suggestions for changes to the draft law in relation to Indigenous northerners, including several that would strengthen protection of their lands, and require

---

[5] The lower house of the Federal Assembly (national legislature) of Russia.

their input on development projects (Ledkov n.d.). Yet the latest draft AZRF law (as of this writing, from 2017) limits its identification of objectives for the Indigenous peoples of the Arctic to improving educational programs and preserving their cultural heritage and language, and folk arts and crafts. The draft law provides no legal mechanism for protecting lands used by Indigenous northerners for their "traditional activities" (reindeer husbandry, hunting, fishing, gathering), in the face of a strong policy focus on resource extraction. Lands used by Indigenous groups for centuries, and the resources lying within/under these, can be transferred to industrial companies on the basis of other legal acts. Indigenous leaders in Russia note that this approach fails to meet the international responsibilities that Russia has to protect the rights of Indigenous peoples.

CONCLUSION: THE ARCTIC IN RUSSIAN POLICY

Russia's recent Arctic policy in many ways beckons back to the Soviet period, with an increased focus on viewing the region as a storehouse of natural resource, to be used to support the economic growth of the country, and to bolster its reputation as a 'great power'. If in the 1990s we saw a decline in attention to the Arctic at the national level, in some areas (e.g. the Sakha Republic (Yakutia)), increased attention was paid to Indigenous rights. Today, while the policies give some attention to sustainable development, the consideration of Indigenous rights seems to be eroding. The increased focus on, and legal provisions facilitating, resource extraction is accompanied by a failure to facilitate any real move toward encouraging the development of the knowledge-economy in the Russian North. The strong focus on hydrocarbons and coal preempts any real push to development of renewable energy in the Arctic, thus giving endorsements of sustainable development a hollow ring. Minimal attention is paid to improving the capacity and capability of northern populations to contribute to northern development, another necessity of sustainable development in the North. And the strong articulation of the need to maintain the Russian Arctic as a zone of international peace and cooperation sits in distinct tension with what appears to be an increasing practice of asserting sovereignty in ways often interpreted in the West to be more than just 'defensive'. While certainly paying increased attention to its Arctic, Russia's view of this region seems to be reverting once again to an extractivist, colonial standpoint.

Acknowledgments The authors gratefully acknowledge support for this research from the Social Sciences and Humanities Research Council of Canada (N. 435-2016-0702), the Norwegian Research Council (NORRUS Project N. 257644), and the Russian Fund for Fundamental Research (N. 18-59-11001)

REFERENCES

ACPOL. 2016. *Natsionalinyy obshchestvenny standart "Ecologicheskaya Bezopasnosti Arktiki" Proekt.* Moscow: Libri Plus. http://arcticas.ru/docs/2016/Broshura_Arctica.pdf.

Aganbegian, A.G. 1984. *Zapadnaia Sibir' na rubezhe vekov* [Western Siberia at the Turn of the Century]. Sverdlovsk: Srednie-Ural'skoe knizhnoe izdatel'stvo.

Agreement. 2017. Agreement on Enhancing International Scientific Cooperation. https://oaarchive.arctic-council.org/handle/11374/1916.

Antrim, C. 2017. The Next Geographical Pivot: The Russian Arctic in the Twenty-First Century. *Naval War College Review* 63 (3): 15–37.

Åtland, K. 2008. Mikhail Gorbachev, the Murmansk Initiative, and the Desecuritization of Interstate Relations in the Arctic. *Cooperation and Conflict* 43 (3): 289–311.

AZRF. 2017. O razvitii arkticheskoy zony Rossiyskoy Federatsii. Rossiyskaya Federatsiya. Federal'nyy Zakon. Proekt. [On the Development of the Arctic Zone of the Russian Federation. Russian Federation. Federal Law. Draft]. November 2017 version.

Baev, P. 2018. Examining the Execution of Russian Military-Security Policies and Programs in the Arctic. In *Russia's Far North. The Contested Energy Frontier*, ed. V. Tynkkynen, S. Tabata, D. Gritsenko, and M. Goto, 113–125. London: Routledge.

Balzer, Marjorie. 1983. Ethnicity Without Power: The Siberian Khanty in Soviet Society. *Slavic Review* 42 (4): 633–648.

Barbin, V.V. 2015. Russia Is Committed to Cooperation in the Arctic. *The Arctic Herald* 3 (14): 12–17.

Carlsson, M., and N. Granholm. 2013. Russia and the Arctic: Analysis and Discussion of Russian Strategies. FOI Report, March 2013.

Connor, W. 1992. The Soviet Prototype. In *The Soviet Nationality Reader*, ed. R. Denber, 15–33. Boulder, CO: Westview Press.

Dienes, L. 1987. *Soviet Asia: Economic Development and National Policy Choices.* Boulder, CO: Westview Press.

Ebel, R.E. 1994. *Energy Choices in Russia.* Washington, DC: Center for Strategic and International Studies.

Forsyth, James. 1994. *A History of the Peoples of Siberia: Russia's North Asian Colony 1581–1990.* Cambridge: Cambridge University Press.

Gleason, G. 1992. The Evolution of the Soviet Federal System. In *The Soviet Nationality Reader*, ed. R. Denber, 107–120. Boulder, CO: Westview Press.

Gorbachev, M.S. 1987. The Speech in Murmansk At The Ceremonial Meeting on the Occasion of the Presentation of the Order of Lenin and the Gold Star Medal to the City of Murmansk, 1 October 1987. Moscow: Novosti Press Agency, pp. 23–31. https://Www.Barentsinfo.Fi/Docs/Gorbachev_Speech.Pdf.

Gosudarstvennaya. 2014. Gosudarstvennaya programa Rossiyskoy Federatsii "Sotsial'no-ekonomicheskoe razvitie Arkticheskoy zony Rossiyskoy Federatsii na period do 2020 goda" [State Program of the Russian Federation "Socio-Economic Development of the Arctic Zone of the Russian Federation for the Period up to 2020"]. http://government.ru/media/files/AtEYgOHutVc.pdf.

Grant, B. 1993. Siberia Hot and Cold: Reconstructing the Image of Siberian Indigenous Peoples. In *Between Heaven and Hell: The Myth of Siberia in Russian Culture*, ed. G. Diment and Yu. Slezkine, 227–253. New York: St. Martin's Press.

Heininen, L. 2018. The Twofold Development of the Arctic. Where do the Arctic States Stand? In *Russia's Far North. The Contested Energy Frontier*, ed. V. Tynkkynen, S. Tabata, D. Gritsenko, and M. Goto, 84–95. London: Routledge.

Hill, F., and C. Gaddy. 2003. *The Siberian Curse. How Communist Planners Left Russia Out in the Cold*. Washington, DC: Brookings Institution Press.

Ivanova, A., and G. Fondahl. n.d. Legal Reform, Governance and Security in the Russian Arctic (Manuscript Under Review).

Konseptsiia. 2009. Kontseptsiia Ustoychivogo Razvitiya Korennykh Malochislennykh Narodov Severa, Sibiri i Dal'nego Vostoka Rossiyskoi Federatsii [Concept for the Sustainable Development of Small Indigenous Population Groups of the North, Siberia and the Far East of the Russian Federation], 4 February 2009. http://docs.cntd.ru/document/902142304.

Laurelle, M. 2014. *Russia's Arctic Strategies and the Future of the Far North*. Armonk, NY: M.E. Sharpe, Inc.

Ledkov, G. n.d. Predlozheniya deputata Gosudarstvennoy Dumy Federal'nogo Sobraniya Rossiyskoy Federatsii, prezidenta Assotsiatsii korennykh malochislennykh narodov Severa, Sibiri I Dal'negog Vostoka Rossiyskoy Federatsii G.P. Ledkova v proekt federal'nogo zakona "O razvitii Arkticheskogo zony Rossiyskoy Federatsii".

Makova, E. 2018. Arkticheskoe zakonodatel'stvo segodnya, Tsentr Informatsionnogo i pravovogo obespecheniya razvitiya Arktiki. http://arctic-centre.com/ru/analitika/item/324-arkticheskoe-zakonodatelstvo-segodnya.

Medvedev, S. 2018. Simulating Sovereignty. The Role of the Arctic in Constructing Russian Post-Imperial Identity. In *Russia's Far North. The Contested Energy Frontier*, ed. V. Tynkkynen, S. Tabata, D. Gritsenko, and M. Goto, 206–215. London: Routledge.

Morskaya. (2015). *Morskaya Doktrina Rossiyskoy Federatsii* [Maritime Doctrine of the Russian Federation], 26 July 2015. http://docs.cntd.ru/document/555631869.

Motomura, M. 2018. Perspectives on Oil and Gas Development in the Russian Arctic. In *Russia's Far North. The Contested Energy Frontier*, ed. V. Tynkkynen, S. Tabata, D. Gritsenko, and M. Goto, 27–42. London: Routledge.

Nifontova, M.S. 2015. Russia Needs the Law on the Arctic. *The Arctic Herald* 3 (14): 18–25.

Ob okhrane. 2018. Ob okhrane vechnoy merzloty v Respublike Sakha (Yakutia) Yakutian Regional Law, N 2006-3 N 1571-V, 22 May 2018.

Osnovy. 2008. *Osnovy Gosudarstvennoi Politiki Rossiiskoi Federatsii v Arktike na Period do 2020 Goda i Dal'neishuiu Perspektivu* [Foundations of the State Policy of the Russian Federation in the Arctic Up to and Beyond 2020]. http://government.ru/info/18359/.

Otsuka, N., T. Tamura, and M. Furuichi. 2018. In *Russia's Far North. The Contested Energy Frontier*, ed. V. Tynkkynen, S. Tabata, D. Gritsenko, and M. Goto, 43–63. London: Routledge.

Pavlenko, V.I., A.A. Dregalo, V.I. Ul'yanovskiy, S.Yu. Kutsenko, K.O. Malinina, and S.M. Balitskaya. 2017. Metodologicheskie ossobennosti sotsial'no-ekonomicheskikh issledovaniy arkticheskoy zony Rossiyskoy Federatsii [Methodological Features of Socio-Economic Studies of the Arctic Zone of the Russian Federation]. *Izvestiya Komi nauchnogo tsentra UrO RAN* 1 (29): 109–114.

Pika, A.I., and B.B. Prokhorov. 1994. *Neotraditsionalizrn na Rossiskom Severe* [Neotraditionalism in the Russian North]. Moskva: Institut narodnokhoziaistvennogo prognozirovaniia Tsentr demografii i ekologii cheloveka RAN, Mezhdunarodnaia rabochaia gruppa po dWelam korennogo naseleniia (IWGIA).

Pipes, R. 1964. *The Formation of the Soviet Union: Communism and Nationalism, 1917–1923*. Cambridge: Harvard University Press.

Putin, V. 2014. Ukaz Presidenta Rossiyskkoy Federatsii "O sukhoputnykh territoriyakh Arkticheskoy zony Rossiyskoy Federatsii" [Edict of the President of the Russian Federation "On land territories of the Arctic Zone of the Russian Federation"] N°296, 2 May 2014, with modifications, Ukaz N° 287, 27 June 2017. http://www.kremlin.ru/acts/bank/38377.

———. 2015. Ukaz Presidenta Rossiyskoy Federatsii "O strategii natsional'noy bezopasnosti Rossiyskoy Federatsii" [Edict of the President of the Russian Federation "On the strategy for national security of the Russian Federation"], No.638, 31 December 2015. http://www.consultant.ru/document/cons_doc_LAW_191669/61a97f7ab0f2f3757fe034d11011c763bc2e593f/

———. 2017. Ukaz Presidenta Rossiyskoy Federatsii "O vnesenii izmeneniy v Ukaz Presidenta Rossiyskoy Federatsii ot 2 Maya 2014 g. N° 296 "O sukhoputnykh territoriyakh Arkticheskoy zony Rossiyskoy Federatsii"" [Edict of the President of the Russian Federation "On the Introduction of Changes to the Edit of the President of the Russian Federation "On Land Territories of the Arctic Zone of the Russian Federation""] Ukaz N° 287, 27 June 2017. http://www.kremlin.ru/acts/bank/42021.

Radushinsky, D., A. Mottaeva, L. Andreeva, and G. Dyakova. 2017. The Evaluation of the Modernization Cost of the Transport Infrastructure of the Northern Sea Route in the Arctic Zone of the Russian Federation. *IOP Conference Series: Earth and Environmental Science* 90: 012137. https://doi.org/10.10888/1755-1315/90/1/012137. (7 pp.).

Slezkine, Yu. 1992. From Savages to Citizens: The Cultural Revolution in the Soviet Far North, 1928–1938. *Slavic Review* 51 (1): 52–76.

Stammler, F., and A. Ivanova. 2016. Resources, Rights, and Communities: Extractive Mega-Projects and Local People in the Russian Arctic. *Europe Asia Studies* 68 (7): 1220–1244.

Strategiya. 2013. *Strategiya Razvitiya Arkticheskoi Zony Rossiyskoi Federatsii i Obespecheniya Natsional'noi Bezopasnosti na Period do 2020 Goda* [The Strategy for the Development of the Arctic Zone of the Russian Federation and Ensuring National Security for the Period to 2020]. Approved by President Vladimir Putin on 20 February 2013. http://government.ru/info/18360/.

Tabata, S., and T. Tabata. 2018. Economic Development of the Arctic Regions of Russia. In *Russia's Far North. The Contested Energy Frontier*, ed. V. Tynkkynen, S. Tabata, D. Gritsenko, and M. Goto, 11–26. London: Routledge.

Tabelev, V.F. 1936. Na Yamalskom Severe [In the Yamal North]. *Sovetskaya Arktika* 7: 11–23.

Topilin, M. 2017. Severynye nadbavki i rayonnye koeffitsienty trogat' ne nado. Press Release of the Russian Labour Minister, 10 October 2017. https://rosmintrud.ru/labour/relationship/305.

Tsybul'skiy, A. 2015. Speaking of the Arctic We Speak of Integrated Development of the State. *The Arctic Herald* 3 (14): 6–11.

Tynkkynen, V. 2018. Introduction. Contested Russian Arctic. In *Russia's Far North. The Contested Energy Frontier*, ed. V. Tynkkynen, S. Tabata, D. Gritsenko, and M. Goto, 1–8. London: Routledge.

V zakone. 2016. V zakone ob Arktike zabyly pro lyudey [The Law of the Arctic Forgot About People]. *Parlamentskaya Gazeta*, November 29. https://www.pnp.ru/social/2016/11/29/v-zakone-ob-arktike-zabyli-pro-lyudey.html.

Vasilieva, T. 2018. V Yakutii prinyat zakon ob okhrane vechnoy merzloty [In Yakutia a Law on the Protection of Permafrost Is Adopted]. Press Release of "Il Tumen", Sakha (Yakutian) Parliament, 25 May 2018. http://iltumen.ru/content/v-yakutii-prinyat-regionalnyi-zakon-ob-okhrane-vechnoi-merzloty.

Vitebsky, P., and A. Alekseyev. 2000. Coping with Distance: Social, Economic and Environmental Change in the Sakha Republic (Yakutia), Northeast Siberia. Report on Expedition Funded by the Gilchrist Educational Trust in Association with the Royal Geographical Society, 1999. Cambridge: Scott Polar Research Institute. Unpublished Report.

Voennaya. 2014. *Voennaya Doktrina Rossiyskoy Federatsii* [The Military Doctrine of the Russian Federation], 26 December 2014. http://static.kremlin.ru/media/events/files/41d527556bec8deb3530.pdf.

Zhukov, M. 2018. Ob ocherednom proekte federal'nogo zakona "O razvitii Arkticheskoy zony Rossiyskoy federatsii" (Interview with M. Zhukov). *Redkie Zemli/The Rare Earth Magazine*, 3 March 2018. http://rareearth.ru/ru/pub/20180313/03767.html.

Zysk, Katarzyna. 2015. Russia Turns North, Again: Interests, Policies and the Search for Coherence. In *Handbook on the Politics of the Arctic*, ed. L.C. Jensen and G. Hønneland, 437–461. Cheltenham: Edward Elgar.

CHAPTER 14

Government, Policies, and Priorities in Kalaallit Nunaat (Greenland): Roads to Independence

Adam Grydehøj

LOOKING TOWARD THE FUTURE

Kalaallit Nunaat (in English, 'Greenland') may be regarded as an autonomous subnational island jurisdiction (SNIJ) of the metropolitan state of Denmark. Alternatively, Kalaallit Nunaat may be regarded as one of the three constituent nations of the Danish Realm (*Rigsfællesskabet*), alongside the Faroe Islands and Denmark proper. When appearing on the global stage, Kalaallit politicians often take a third approach, highlighting that Kalaallit Nunaat is the only Indigenous Arctic territory with a legally established roadmap toward independence. Such statements point to Kalaallit Nunaat's exceptionalism relative to both Arctic Indigenous territories without autonomy (e.g. Sápmi) and Arctic Indigenous territories with autonomy but without a straightforward method of attaining political independence (e.g. Nunavut).

For the past 70 years, the people of Kalaallit Nunaat have engaged in negotiations of various kinds at various levels with the people of Denmark, resulting in immense societal change occurring alongside gradual increases in jurisdictional capacity (the de facto or de jure ability to formulate and execute policy). These increases in jurisdictional capacity have ultimately amounted to a territory that can in many ways be regarded as genuinely autonomous, and Kalaallit Nunaat (together with the Faroe Islands) is in the almost unique position of having been granted by its metropolitan state the ability to politically separate itself from that metropolitan state at a time of its own choosing. Despite strong ethnic and political tensions between Kalaallit and Danes,

A. Grydehøj (✉)
University of Prince Edward Island, Charlottetown, PE, Canada
e-mail: agrydehoj@islanddynamics.org

K. S. Coates, C. Holroyd (eds.), *The Palgrave Handbook of Arctic Policy and Politics*, https://doi.org/10.1007/978-3-030-20557-7_14

and despite widespread dissatisfaction on both sides regarding aspects of the relationship between these two nations within the Danish Realm, there is a general understanding that, because Kalaallit Nunaat controls the timing of the decision to become independent, the territory is being given the opportunity not simply to gain independence but also to lay the economic, political, and social groundwork for a *successful* independence. Kalaallit Nunaat has the ability to prepare for independence at its own pace, to retain its annual block grant from Denmark while creating the conditions for survival without the block grant, to make full use of its bridges before burning them. It is uncertain whether independence will come in three years, within the lifetimes of today's leading political figures, or within their children's or grandchildren's lifetimes, but whatever the decision, it will be made by the people of Kalaallit Nunaat.

So goes the narrative, for those who are so inclined.

There is an undeniable appeal to the process of extended negotiation and peaceful engagement between an Indigenous territory and its metropolitan state, between an Indigenous territory and its former colonizer. And it is undeniable that this process has presented certain benefits for the people of Kalaallit Nunaat, especially seen in light of the conditions reigning in many former island colonies that were either pushed into or demanded independence on a short timescale. Nevertheless, this process has resulted in a situation in which Kalaallit Nunaat is perpetually stuck in a neither/nor position, continually in the midst of jurisdictional transformation, always preparing for a future that never quite arrives. For while Kalaallit Nunaat modernizes its economy, raises the welfare of its citizens, and hones its powers of statecraft, the goalposts keep moving: The economy keeps changing, societal expectations keep evolving, and the political demands that are placed on states of all sizes grow ever more complex. How can Kalaallit Nunaat look toward its future without neglecting the needs of the present? And equally, how can efforts to cope with the needs of the present avoid succumbing to the power of inertia and the siren call of perfectionism? No one wants to live in an independent Kalaallit Nunaat that is beset by problems, yet equally, if states needed to be perfect to be sovereign, then there would be few sovereign states in the world indeed.

This chapter provides an overview of Kalaallit Nunaat's political development over time and its policy priorities today. This chapter does not advocate any particular policy with regard to Kalaallit Nunaat's political connection with Denmark, but it does bear in mind that, for many decades, a large proportion of Kalaallit Nunaat's population has expressed a desire for independence or at least a radically altered relationship with Denmark. Those who research Kalaallit Nunaat's politics and economy often take their point of departure from a conviction that independence is financially unfeasible or is otherwise a bad idea—and then develop their analyses from there. Whether or not this conviction is correct is, however, to some extent immaterial, for a scholarly perspective that does not take into account the desires of the Kalaallit will not do much good for the people of Kalaallit Nunaat. As such, the present the chapter seeks to offer something approaching a Kalaallit perspective on Kalaallit Nunaat, grounded in the awareness that the author, who is neither Kalaallit nor Danish, can never gain full understanding or awareness of local worldviews.

HISTORICAL OVERVIEW

Kalaallit Nunaat consists of a large central island ringed by numerous small islands, with a human population of 55,847 (Naatsorsueqqissaartarfik 2016).

Norse immigrants from Iceland settled Kalaallit Nunaat's hitherto-unpopulated southwest coast in the late 900s CE. The Thule people, ancestors of today's Kalaallit, arrived around 1200 CE, eventually spreading from northern Kalaallit Nunaat across the western, southern, and eastern coastal regions. By the 1400s, the long-declining Norse settlements had disappeared completely (Gulløv 2008), with the result that there is no continuity between Kalaallit Nunaat's Medieval Norse population and the ethnic Danes resident in Kalaallit Nunaat today. Although the Kalaallit had contact with European whalers in the 1600s, European imperialism first came to the archipelago in 1721, with the arrival of a combined religious and trade mission led by the Norwegian pastor Hans Egede, supported by the Danish-Norwegian crown. Over the following decades, this developed into a full-fledged colonial project, with Danish missionaries and administrators establishing tiny 'colony towns' up and down the coasts to lead Christianization efforts and engage in trade (primarily in seal skins) with the Kalaallit.

This was never a settler colonialism project, and Danish colonialism in Kalaallit Nunaat was not grounded in physical violence or forced labor. Colonialism nevertheless exerted social violence. Christianization disrupted Indigenous epistemologies and power structures, while the introduction of market economies led to the exploitation of surplus labor in what had previously been a nomadic subsistence economy. Similarly, the establishment of a Danish trade monopoly and the colonial desire for marketable products led to the encouragement of seal hunting at the expense of other traditional economic activities (Rud 2014), ultimately serving to entrench the colonial division of labor.

When seal populations collapsed in the early 1900s, the colonial administration encouraged a transition to another export-oriented primary sector activity, namely fishing. Unlike seal hunting, however, the fishing industry required the Kalaallit to live in permanent settlements and could not be the sole basis for a subsistence economy. It is thus from the 1920s that most Kalaallit finally became truly *dependent* on the Danish colonial administration not just for luxury items, schooling, and other services, but for their very survival. Following World War II, and facing global demands for decolonization, the Danish government sought to remove Kalaallit Nunaat's status as a colony by integrating the territory into the Danish state as a municipality. This occurred in 1953, following limited consultation with the educated Kalaallit elite.

THE SELF-GOVERNMENT SYSTEM

The 1960s and early 1970s witnessed rising Kalaallit nationalism, and an increasing demand for political authority led to the creation of the Home Rule system in 1979, following a referendum in Kalaallit Nunaat (70.1% in favor of

Home Rule). The advent of Home Rule was both complicated and influenced by Denmark's joining of the European Economic Community (EEC) following its own referendum, at a time when Kalaallit Nunaat was still treated as a Danish municipality. A new referendum in Kalaallit Nunaat in 1982 decided narrowly against continued EEC membership (53% opposed to remaining). In the debate regarding EEC membership, fisheries policy was a major issue, but there was also a sense that tying themselves to Europe represented a step backward in a time when the Kalaallit were otherwise at last being given control over their own territory. It is thus that, in 1985, Kalaallit Nunaat became the first (and so far—pending the exit of the UK—only) territory to leave what would later develop into the European Union.

In terms of de jure distributions of authority, the Home Rule system granted Kalaallit Nunaat the potential for significant autonomy. The Kalaallit Nunaat state received from Denmark an annual block grant, which increased as the state took on responsibility for additional policy areas from Copenhagen. In the event, however, the Kalaallit Nunaat government never exploited the potential powers under Home Rule in full, and a feeling emerged within Kalaallit Nunaat that Home Rule was neither satisfactory as the end point in the political decolonization process nor as the final way station before eventual independence. Significantly, the Home Rule system did not grant Kalaallit Nunaat ownership over subterranean natural resources, and it did not recognize Kalaallit Nunaat's rights as a 'nation' (folk) under international law, specifying instead that "Greenland represents a special national community within the Kingdom of Denmark" (Statsministeriet n.d.; translation my own). This led to negotiations between the governments of Denmark and Kalaallit Nunaat in the 2000s, resulting in a 2008 referendum in Kalaallit Nunaat regarding the introduction of a new and more powerful Self-Government system (75.5% in favor of Self-Government), followed by the Danish parliament's adoption of the Self-Government Act in 2009 (Lov om Grønlands Selvstyre 2009).

A key aspect of the new Self-Government system was that it set forth a process by which Kalaallit Nunaat could become independent from Denmark at a time of the territory's own choosing. Whereas the 1979 Home Rule system had involved an annual block grant that increased as responsibility for additional policy areas were transferred from the Danish state to the Kalaallit Nunaat state, however, the 2009 Self-Rule system froze the block grant at then-current levels (adjusted for inflation, 3.6 billion Danish kroner, around 483,700,000 million euro) so that the devolution of new responsibilities would not result in increased monetary contributions from Denmark. It was hoped that this would be offset by the ability to profit from Kalaallit Nunaat's subterranean resources. Indeed, the Self-Government Act ties the block grant directly to subterranean resource extraction, stating that when Kalaallit Nunaat achieves profits from these activities, the subsequent year's block grant will be reduced by an amount equal to half the level of profits exceeding 75 million Danish kroner. In the event that these reductions ever equal or exceed the block grant, then the Danish and Kalaallit Nunaat governments must negotiate "future economic relations between the Kalaallit Nunaat Self-Government and the [Danish] state, including regarding distribution of income from natural resources activities in Greenland" (Lov om Grønlands Selvstyre 2009, §4.10; translation my own).

These legal mechanisms would neither force political independence upon an economically independent Kalaallit Nunaat nor deny the people of Kalaallit Nunaat the ability to opt for political independence in the absence of economic independence. However, one result of the way in which the Self-Government Act was constructed has been to create a conceptual association between independence and the extractive industries (particularly concerning precious metals, oil, gas, and jewels). Over the past decade though, development in these industries has not met expectations. Active mining projects remain scarce, and in terms of long-term planning, the Committee for Greenlandic Mineral Resources to the Benefit of Society (2014) has dashed ambitions for a rapid, mining-fuelled transition to independence.

Kalaallit Nunaat thus remains highly dependent on Denmark. The block grant represents around 25% of Kalaallit Nunaat's GDP (Naatsorsueqqissaartarfik 2017, p. 2). Furthermore, although all of Kalaallit Nunaat's nationally elected politicians are Kalaallit, the bureaucracy, educational system, and business community remain highly dependent on imported Danish skilled labor, leading to a prevalence of Danish policy solutions to Kalaallit problems.

Kalaallit Nunaat's Self-Government system is in many respects modeled after the Danish system of government, with a strong central government in the capital (Nuuk) and the distribution of various responsibilities to five municipalities. The Kalaallit Nunaat government has de jure jurisdictional capacity over most policy areas excluding defense, foreign policy, immigration, monetary policy, and specialist legal areas. This means that the Self-Government can engage in genuinely meaningful decisions regarding welfare, services, education, industrial policy, transport, and other crucial areas. As shall be discussed below though, not only do the responsibilities still held by Denmark continue to place some de facto and de jure constraints on Kalaallit Nunaat's jurisdictional capacity, but the Self-Government at times acts beyond its formal capabilities, for instance, by engaging in paradiplomacy and informal diplomacy.

Party Politics

Politics in Kalaallit Nunaat are shaped by a complex party political system that has developed within what is in many respects a remarkably simple electoral framework. Elections under the Home Rule system (beginning in 1979) involved the election of 21 representatives, split between single-person constituencies in the relatively peripheral regions of North Kalaallit Nunaat and East Kalaallit Nunaat as well as multi-member constituencies in the West Kalaallit Nunaat region. This system favored the dominant political parties, which at the time were Siumut (a social democratic party that had been at the vanguard of the independence movement) and Atassut (at that time, a conservative, unionist party). Although Siumut and Atassut would battle over first and second place status in elections throughout the subsequent two decades, Siumut was able to maintain its position at the head of the government, due both to its strength in the small towns and villages in the north and the east and to its capacity to occupy Kalaallit Nunaat's political center, well placed to collaborate with all other parties and to build upon its reputation as the stalwart of Kalaallit autonomy. Over time, Inuit Ataqatigiit (a socialist, pro-independence party) displaced Atassut as Siumut's primary competitor, and Atassut itself transitioned to a pro-independence stance.

Today, Kalaallit Nunaat's parliament of 31 representatives is elected through nationwide proportional representation, a system that is favorable to the formation and maintenance of small parties. The 2018 parliamentary elections, for example, saw the voting in of seven parties, with Siumut winning nine seats on 27.2% of the vote, Inuit Ataqatigiit winning eight seats on 25.5%, and Demokraatit (an economic liberal party founded in 2002) experiencing a surge in support and winning six seats on 19.5%. Remarkably though, Siumut has managed to maintain its hold on power at all but one election since 1979, with Inuit Ataqatigiit leading the government from 2009–2013. One source of Siumut's resilience may be its capacity (paradoxically bolstered by the proportional representation system) to weather splits that lead to the formation of new parties, the voters for which seem particularly likely either to eventually be reabsorbed by Siumut or to prove equally problematic for other parties. Thus, despite having emerged as a breakaway party from Siumut, Demokraatit is battling for votes primarily with Inuit Ataqatigiit and the newly formed Samarbejdspartiet/Suleqatigiissitsisut (another economic liberal party and the only current party to take a clear unionist stance), which likewise have their strongest appeal in Nuuk, among people with high levels of education, and among ethnic Danes. Whereas Siumut is a 'big tent' party that retains a core vote in the small towns and settlements that is sufficiently robust to withstand periodic disruption by breakaway parties laying claim to similar emotional and ideological ground, Kalaallit Nunaat's more cosmopolitan voters are increasingly fragmented. The continual decline in Siumut's vote share must thus be assessed in light of equally worrisome developments for its key political opponents.

The result of all this is that Kalaallit Nunaat's highly fluid party political ecology in some respects conceals a great deal of stability. Kalaallit Nunaat's governments have shared many of the same overarching political objectives and have engaged in and built upon the same national narratives for most of the period since 1979. This is important to bear in mind when considering the major political priorities and key policy areas in Kalaallit Nunaat today.

LINKS BETWEEN ECONOMIC AND POLITICAL INDEPENDENCE

A poll undertaken in late 2016, based on a representative sample of 708 adults resident in Kalaallit Nunaat, found an overwhelming desire for political independence among Kalaallit Nunaat's public: Asked whether eventual independence was important to them, 34% of respondents said it was very important, and 25% said it was somewhat important, versus 16% who said it not very important, and 8% who said it was not at all important while 12% of respondents replied that they did not know (Skydsbjerg and Turnowsky 2016). That is, relatively strong desire for independence was nearly 2.7 times higher than weak or no desire for independence. The poll was undertaken at a time when all the parties in the Kalaallit Nunaat parliament supported independence to some degree, and once 'don't knows' were removed, it was only among supporters of the relatively small party Atassut that a majority of respondents did not regard independence as important.

A follow-up poll carried out in early 2017 provided further nuance to these results. Whereas 11% of respondents desired independence no matter what, and 12% desired independence even if it meant a small drop in living standards, 44% stated that they only desired independence if it did not lead to a drop in living standards, while 27% preferred a form of strengthened autonomy that fell short of independence (Turnowsky 2017). Although this again indicates 67% support for independence, it highlights the conditional nature of much of this support and explains why the heavily pro-independence Kalaallit political establishment has not thus far grasped the opportunity to declare independence from Denmark.

Gad (2016) has theorized this situation as being intimately linked to conceptions of Kalaallit identity discourse. Analyzing political discourse of the early 2000s, Gad identifies three 'basic positions' underlying the apparent consensus in favor of independence:

> One according to which self-governance was a precondition for [economic] self-support and another according to which self-support was a precondition for self-governance. Between these positions, a third and more complex construction of 'self-supportedness' as a joint project shared by Greenland and Denmark meant that self-government would remain a gradual affair. (Gad 2016, p. 56)

Key here is the idea of economic self-support and jurisdictional capacity as intimately related, an idea that may explain in part why Kalaallit Nunaat's present situation of having its economy buttressed by Danish money is widely regarded as unacceptable in the long run. From a long-term perspective, economic dependence on Denmark is seen as inimical to political independence *even if this were to prove acceptable to Denmark*. The political tumult in Kalaallit Nunaat in September 2018, when the strongly pro-independence Partii Naleraq quit the government in part in protest over the prospect of Danish investment in the expansion of some of Kalaallit Nunaat's airports (Turnowsky 2018), is indicative of this tension: Some political actors see the airports as a means of helping Kalaallit Nunaat create global connections and a stronger economy, but for other political actors the involvement of Danish money would by its very nature preclude the airports from representing a step toward greater economic or political independence.

This conception—which exists among the publics of both Kalaallit Nunaat and Denmark—that political and economic self-sufficiency are mutually dependent has fostered a gradual, incremental independence-seeking process in Kalaallit Nunaat. Although pro-independence rhetoric can take strong forms, there is little desire among Kalaallit politicians for political independence in the short term. This patience or willingness to wait is no doubt encouraged by the Self-Government Act's legally established roadmap toward independence. Many independence movements around the world get caught up in either overcoming resistance from the metropolitan state (e.g. the battle for referenda in Catalonia and New Caledonia) or become limited by electorally determined deadlines and timeframes (e.g. the pressure on the Scottish National Party to call a referendum after its electoral victory in 2011). In contrast, Kalaallit Nunaat is capable of laying the groundwork to ensure perfect conditions for independence. As Gad (2016, p. 51) argues though, this renders the conception of Kalaallit Nunaat itself fundamentally unsettled:

> In Greenlandic identity discourse, the national principle is what ties aboriginality and modernity together: Greenland *ought to be* an independent state to allow Greenlandic culture to flourish within a welfare society. Some deem this impossible, some view it as a perspective far away on the horizon, while yet others would declare independence soon. So in terms of Greenlandic discourse, the present situation is not as it ought to be: Greenland is not an independent nation-state. Hence, Greenlandic political identity is transitional: Greenland sees itself as somewhere on the way from colonial submission to future independence.

Policymaking in Kalaallit Nunaat is always looking to the future. Domestic and foreign policy cannot be separated, and the most mundane of policy areas can take on immense significance in narratives of nationality.

Policy Priorities

Political and economic independence may be the long-term goal for all but one of Kalaallit Nunaat's current political parties, but the means of reaching this goal differ, both among and within parties. Although there are areas of shared concern, such as health, poverty, low educational outcomes, child abuse, alcoholism, and violent crime, the present overview will focus on areas of political tension that have significance with regard to divergent visions of Kalaallit Nunaat's future.

Population concentration policies play a major role in the narrative of Kalaallit identity and political development. Prior to the 1920s, Kalaallit Nunaat's towns were oriented around supporting the Danish colonial project, in terms of trade, religion, and education. Crucially, the focus on seal skin exports meant that there was a colonial desire to maintain and even strengthen the fundamentally nomadic seal hunting culture (Rud 2014). The advent of commercial fishing, however, led to a focus on establishing small-scale permanent settlements that could serve as bases and processing centers for the fishing industry.

The formal decolonization of Kalaallit Nunaat in the 1950s and the desire to raise the territory's standards of living to those of other Danish municipalities once again brought about new demands. Although focus did not shift from the importance of fishing, politicians and administrators in Copenhagen began pursuing assimilationist policies. Given the vast distances between Kalaallit Nunaat's settlements, Danish-style public services in health care, education, and other areas required the concentration of the population around centers of service provision. Although the Kalaallit politicians of this era were not seeking assimilation, there was a widespread feeling that the path toward greater Kalaallit self-government was paved with Danish-language skills and Danish professional qualifications.

It was in this period, in the 1950s–1970s, that Kalaallit Nunaat's large towns emerged from among the various other 'colony towns' through the introduction of mass housing and neighborhood construction in the form of blocks of flats and open spaces that, at the time, represented the cutting edge in Nordic urban design (Grydehøj 2014). There was a state-sponsored movement of people from the periphery to the new centers. Although this period has come to be regarded as a late-colonial attempt to disrupt traditional Kalaallit culture, the reality is perhaps more complex. Thus, for example, the most notorious example of the population concentration policy is often said to be the town of Qullissat, which was closed in 1972, leading to the forced resettlement of its population in towns across Kalaallit Nunaat. Qullissat was not, however, a hunting or even a fishing settlement but instead a coal mining town that had only been founded in 1924 and was one of Kalaallit Nunaat's largest towns by the 1960s (Andersen et al. 2016). The experiences from Qullissat nevertheless speak to a wider and legitimate complaint regarding the limits of Danish paternalism and the historical tendency for Kalaallit Nunaat's 'development' to occur on Danish terms. The efforts to establish and maintain permanent settlements in East Kalaallit Nunaat in the early twentieth century were partially a result of attempts to reinforce Danish sovereignty over the region in the face of Norwegian territorial claims. In the early 1950s, the Danish state carried out the forced removal of the population of Pituffik in North Kalaallit Nunaat to clear space for the establishment of the USA's Thule Air Base. It is thus that population movements in general—whether forced or merely encouraged by lack of services, employment, or opportunities—have come to be regarded with suspicion in today's Kalaallit Nunaat.

There is a growing divide in opinion between the residents of Nuuk in particular and those of Kalaallit Nunaat's other towns and settlements. The attractions of life in Kalaallit Nunaat's largest town and the growth in business and administrative jobs located there have served to pull migrants from elsewhere in Kalaallit Nunaat, resulting in a consistently strong demand for new housing, despite significant expansion in Nuuk's housing mass over the past decade and a half. As a result, Sermersooq Municipality, the local government body responsible for Nuuk, is pursuing ambitious urban expansion plans. In February 2017, plans were presented within the Sermersooq Municipal Council for the Siorarsiorfik Nuuk City Development project, which seeks to extend the existing Qinngorput neighborhood (constructed in the late 2000s and early 2010s) and link it with a new neighborhood called Siorarsiorfik. In the first instance, the plan seeks to construct 1600 residences to accommodate 5000 people over the next 6–8 years. This expansion is moreover regarded as a precondition for longer-term efforts to extend Nuuk to the south, out across the fjord (Siorarsiorfik Nuuk City Development 2017). Seen in a positive light, this urban expansion project has the potential to promote the tertiary sector, decrease the population's distance from services, boost employment and consumer spending during the construction phase, make Kalaallit Nunaat more attractive for foreign direct investment, and help attract and retain skilled and educated residents (Grydehøj 2018, p. 82).

This plan, which has cross-party support within Sermersooq Municipality, is significantly less popular among both politicians and the public outside of Nuuk, given that Nuuk's gain in population is to a great extent the reminder of Kalaallit Nunaat's loss. Further centralization is primarily supported by those already at the center or who see themselves as somehow belonging to the center. Hendriksen (2017) has called for nationwide planning, to ensure that local interests do not end up overriding national ones, yet even such a system would be unable to prevent local-national conflicts of the sort raised by projects that, like Siorarsiorfik, are pursued at the municipal level rather than the national level.

As noted above, Kalaallit Nunaat's national politics have long been dominated not by Nuuk but by sentiments in the smaller towns and settlements. It is thus that the Kalaallit Nunaat government is itself engaged in a major airport expansion and construction project, which will boost mobility to and from Nuuk but also seeks to buttress the economy and standards of living on the periphery.

Conceptions of Kalaallit identity remain rooted in the small towns and settlements dotted along the coast and in the hunting and fishing livelihoods that these promote. In the minds of many, Nuuk and urbanism in general continue to be associated with Danishness, resulting in a valorization of the periphery and less willingness to embrace dual Kalaallit-Danish identities (Grydehøj 2016). This is not to be confused with nostalgia or a wish to turn back the clock; indeed, the two political parties with the most assertive stances on independence and the most radical policy programs (Partii Naleraq and Nunatta Qitornai) have their support bases on the periphery. Partii Naleraq in particular stands out for its emphasis on forging global partnerships as a means of bypassing economic dependence on Denmark. So strong is this emphasis that Partii Naleraq withdrew its support from and caused the collapse of a Siumut-led coalition government in September 2018, when it came to light that Kim Kielsen (Kalaallit Nunaat's prime minister) was entering into an agreement with Lars Løkke Rasmussen (Denmark's prime minister) to secure Danish investment in the airport expansion project (Turnowsky 2018).

Center-periphery divides in visions for Kalaallit Nunaat's future may nevertheless come into play in overarching tensions between the ideal of centralization as a means of achieving economies of scale versus the ideal of decentralization as a means of supporting peripheral livelihoods. There has thus, for example, been a long-running debate over a fishing industry law, with clashes between the desire to support industry-wide competitiveness versus the desire to redistribute fishing quotas to a greater number of smaller actors. These political processes are, however, complex. State-owned enterprises (SOEs) play a prominent role in the Kalaallit Nunaat economy and are a vital mechanism by which the state can distribute resources and employment (Grydehøj 2018, pp. 76–77), yet SOEs may in some senses be perceived as fundamentally centralizing and anticompetitive. It is thus that the two political parties that are furthest to the right on economic matters (Demokraatit and Samarbejdspartiet/Suleqatigiissitsisut) combine criticism of the economic dominance exercised by SOEs (the very incarnation of centralized power) while also having their support bases in the center, in Nuuk.

The SOEs themselves are not immune to undertaking politically sensitive activities. Although recent developments may represent a shift in strategy, the publicly owned maritime shipping monopoly Royal Arctic Line (the successor to the colonial trade monopoly) has over the past few years pursued a strategy of investing in infrastructure and processes that could assist in diversifying Kalaallit Nunaat's export markets and undermine the territory's dependence on Danish suppliers and wholesalers. Besides controversy over the financial costs and service disruptions related to this strategy, the very aim of globalizing Kalaallit Nunaat's imports and exports became a matter of political contention.

Indeed, the question of whether or how to decrease links with and dependence on Denmark is central to many political conflicts in today's Kalaallit Nunaat. Would making English (rather than Danish) the second language in education increase or decrease young people's future opportunities, increase or decrease the territory's potential to move toward independence? Danish is currently needed for Kalaallit students who wish to undertake higher-level educations in Denmark and for individuals who wish to work alongside Danes in high-level jobs in Kalaallit Nunaat itself. These arguments are simultaneously reasonable and self-fulfilling though, given that the present dominance of the Danish language is used to justify the need for continued focus on Danish-language education. The two parties that support close links with Denmark and rely significantly on votes from Danes living in Kalaallit Nunaat (Demokraatit and Samarbejdspartiet/Suleqatigiissitsisut) are unsurprisingly opposed to deprioritizing Danish, while the nationalist parties with their heartlands on the periphery (Siumut, Partii Naleraq, and Nunatta Qitornai) are unsurprisingly less attached to the Danish language. It is indicative of Kalaallit Nunaat's political complexities though that Inuit Ataqatigiit, the socialist party that has represented Siumut's primary electoral competitor for the past two decades, has found the language question difficult to grapple with, given that the party is, on the one hand, staunchly pro-independence and frequently preferred by cosmopolitan individuals with a global outlook and, on the other hand, dependent on voters who are more likely to live in Nuuk, more likely to be bilingual products of Danish-Kalaallit families, more likely to be better educated, and/or more likely to have strong personal connections with Denmark. Kalaallit Nunaat politics cannot be interpreted in terms of straightforward Kalaallit-Danish, left-right, or center-periphery divisions.

Foreign Relations

Kalaallit Nunaat's foreign relations remain under the authority of Denmark, yet as we have seen, many aspects of domestic policy are actually relevant to and perceived through the prism of the territory's relations with Denmark and other states. In addition, Kalaallit Nunaat is both undertaking increasing engagement with governance and business actors outside the Danish Realm and increasingly becoming a place of interest to governance and business actors outside the Danish Realm. As patterns of global trade and movement shift, as climate change harms some industries (such as subsistence hunting activities) while presenting new opportunities for others (such as container shipping), as the global balance of political and economic power tilts in new directions, Kalaallit Nunaat is coming under new pressures and being offered new possibilities.

The past few years have seen rising tensions in the three-way relationship between the governments of Kalaallit Nunaat, Denmark, and the USA over American military presence in the territory. The forced removal of the population of Pituffik to make way for the Thule Air Base has been used as an example of Denmark's and the USA's lack of interest in Kalaallit needs. In 2016, it came to light that climate change could lead to the leakage of nuclear waste from the abandoned American military facility Camp Century (Colgan et al. 2016). Subsequent debate between Kalaallit Nunaat, Denmark, and the USA over who was responsible for paying for the clean-up, coupled with the transfer of the service contract for Thule Air Base from a Kalaallit Nunaat SOE to a subsidiary of an American corporation, led to deepening resentment over the perceived lack of benefits that Kalaallit Nunaat obtains from the American presence and the Danish government's perceived inability to stand up for the interests of its autonomous territory.

The start of the Trump presidency in the USA in 2016 seemed to herald the continuation of the USA's gradual pivot away from the Arctic, yet as China has widened its sphere of projection of political and economic power beyond Asia and into Europe and the Arctic, Kalaallit Nunaat has once again come to be regarded as strategically important by international actors. The September 2018 decision by the Danish government to support investment in Kalaallit Nunaat's airport expansion and construction project (Olsen 2018), which had previously been roundly criticized by both Danish politicians and the Danish press, is widely regarded as a result of the Kalaallit Nunaat government's refusal to rule out the hiring of a Chinese contractor—despite veiled threats by both the governments of Denmark and the USA that they would not countenance Chinese involvement in strategic Kalaallit Nunaat infrastructure (Klarskov 2018). Tellingly, there has been a tendency for international commentators to portray this chain of events as a case of Kalaallit Nunaat rejecting a Chinese bid in favor of a Danish one (e.g. Gronholt-Pedersen 2018), though this is not—or at least, not yet—the case.

For its part, the Chinese government has maintained the position that it is not interested in interfering with Danish sovereignty over Kalaallit Nunaat. However, the lines between the economic and political are increasingly blurred in the context of China's ever-expanding Belt and Road Initiative, which is simultaneously a top-down central government strategy, a collection of bottom-up local strategies for harnessing Chinese investment, and a framework by which Chinese companies (in some cases SOEs) can expand their activities internationally. With the emergence of a Chinese 'Polar Silk Road' strategy (Eiterjord 2018), it may no longer be enough to say that Chinese interest in Kalaallit Nunaat is focused on business and scientific research, given that business and scientific research are themselves of strategic significance (Dubois and Gagaridis 2018). Furthermore, the Chinese government's willingness to entertain a mixed government and business delegation from Kalaallit Nunaat in 2017 suggests an enhanced openness in China to regarding Kalaallit Nunaat as a political actor in its own right, rather than simply as part of Denmark. The USA's recent—and somewhat ambiguous—expression of interest in investing in mixed-use civilian/military airport infrastructure in Kalaallit Nunaat similarly suggests that the USA is increasingly willing to engage with Kalaallit Nunaat per se.

It is sometimes too easy to discuss Kalaallit Nunaat's role in the international system in this manner—as a piece of land and marine territory that is the object of competition by large state actors. Yet it is the government of Kalaallit Nunaat which decided to engage in paradiplomacy by sending a delegation to China, which has sought to take a key role in the Inuit Circumpolar Council, and which has exploited the uncertain boundaries of its powers as set forth in the Self-Government Act. Pincus and Berbrick (2018) can write for an American readership about the dangers that Chinese involvement in Kalaallit Nunaat poses for American interests, and no less than the Prime Minister of Denmark himself can speak of the dangers that Chinese involvement poses to Denmark (ctd. in Klarskov 2018), but the interests, agency, and perspectives of the people of Kalaallit Nunaat are often lost in these discussions. When researchers such as Jakobsson (2018) warn that, in its relations with China, Kalaallit Nunaat risks becoming a 'vasal state', dominated by a foreign people and power, many Kalaallit will agree, but many other Kalaallit will respond that Kalaallit Nunaat *already* finds itself in this situation, dominated by Danes and Denmark. Future research calls for an understanding of Kalaallit values and principles and requires a recognition of Kalaallit understandings of self-interest.

CONCLUSION

Kalaallit Nunaat has long been portrayed as standing at the crossroads. Indeed, the territory's political, economic, and cultural development over the past seven decades has been a series of crossroads, and as Kalaallit Nunaat has changed, the world has changed around it. It is likely that the coming years will see greater engagement between Kalaallit Nunaat and international actors, but the recent airport investment by Denmark—paradoxically prompted by the threat of foreign engagement—may also herald a long-term development toward an improved relationship with Kalaallit Nunaat's former colonial power. Meanwhile the Kalaallit Nunaat state apparatus will continue to seek to serve the needs of a small but widely dispersed population, facing unusual challenges that have been conditioned by the territory's precise geographical, historical, and cultural context.

Kalaallit Nunaat hosts a highly developed system of Indigenous politics, embedded however in a Danish governmental and bureaucratic structure and held to Danish standards (Grydehøj 2018). There is no guarantee that the Kalaallit will continue to regard economic independence and political independence as fundamentally linked, and there is no guarantee that the current way of doing domestic politics and foreign relations will persist as Kalaallit Nunaat becomes increasingly engaged in a globalized world.

REFERENCES

Andersen, A., L. Jensen, and K. Hvenegård-Lassen. 2016. Qullissat: Historicising and Localising the Danish Scramble for the Arctic. In *Postcolonial Perspectives on the European High North*, ed. G. Huggan and L. Jensen, 93–116. London: Palgrave Macmillan.

Colgan, W., H. Machguth, M. MacFerrin, J.D. Colgan, D. As, and J.A. MacGregor. 2016. The Abandoned Ice Sheet Base at Camp Century, Greenland, in a Warming Climate. *Geophysical Research Letters* 43 (15): 8091–8096.

Committee for Greenlandic Mineral Resources to the Benefit of Society. 2014. *To the Benefit of Greenland*. Nuuk and Copenhagen: Ilisimatusarfik & University of Copenhagen.

Dubois, K., and A. Gagaridis. 2018. The Security Implications of China-Greenland Relations. *The Polar Connection*, July 10. http://polarconnection.org/security-china-greenland-relations/.

Eiterjord, T.A. 2018. The Growing Institutionalization of China's Polar Silk Road. *The Diplomat*, October 7. https://thediplomat.com/2018/10/the-growing-institutionalization-of-chinas-polar-silk-road/.

Gad, U.P. 2016. *National Identity Politics and Postcolonial Sovereignty Games: Greenland, Denmark, and the European Union*. Copenhagen: Museum Tusculanum.

Gronholt-Pedersen, J. 2018. Greenland Picks Denmark as Airport Project Partner Over Beijing. *Reuters*, September 10. https://www.reuters.com/article/us-china-silk-road-greenland/greenland-picks-denmark-as-airport-project-partner-over-beijing-idUSKCN1LQ2BX.

Grydehøj, A. 2014. Constructing a Centre on the Periphery: Urbanization and Urban Design in the Island City of Nuuk, Greenland. *Island Studies Journal* 9 (2): 205–222.
———. 2016. Navigating the Binaries of Island Independence and Dependence in Greenland: Decolonisation, Political Culture, and Strategic Services. *Political Geography* 55: 102–112.
———. 2018. Decolonising the Economy in Micropolities: Rents, Government Spending and Infrastructure Development in Kalaallit Nunaat (Greenland). *Small States & Territories* 1: 69–94.
Gulløv, H.C. 2008. The Nature of Contact Between Native Greenlanders and Norse. *Journal of the North Atlantic* 1 (1): 16–24.
Hendriksen, K. 2017. Landsplanlægning: en nødvendighed. *Sermitsiaq*, December 8, pp. 44–45.
Jakobsson, A.K. 2018. Kina har store ambitioner om strategiske investeringer i Grønland. Ser vi en vasalstat i vente? *Ræson*, September 15. https://www.raeson.dk/2018/ph-d-andre-ken-jakobsson-kina-har-store-ambitioner-om-strategiske-investeringer-i-groenland-ser-vi-en-vasalstat-i-vente/.
Klarskov, K. 2018. Løkke advarer Grønland mod at lade kinesere bygge lufthavne. *Politiken*, May 24. https://politiken.dk/indland/politik/art6535020/L%C3%B8kke-advarer-Gr%C3%B8nland-mod-at-lade-kinesere-bygge-lufthavne.
Lov om Grønlands Selvstyre. 2009, June 12. http://www.stm.dk/_p_5490.html.
Naatsorsueqqissaartarfik. 2016. *2016-imi innuttaasut.* Nuuk: Naatsorsueqqissaartarfik.
———. 2017. *Nationalregnskab.* Nuuk: Naatsorsueqqissaartarfik.
Olsen, L. 2018. Løkke vil bruge 700 millioner på lufthavne i Grønland: 'En historisk dag'. *DR*, September 10. https://www.dr.dk/nyheder/politik/loekke-vil-bruge-700-millioner-paa-lufthavne-i-groenland-en-historisk-dag.
Pincus, R., and W.A. Berbrick. 2018. Gray Zones in a Blue Arctic: Grappling with China's Growing Influence. *War on the Rocks*, October 24. https://warontherocks.com/2018/10/gray-zones-in-a-blue-arctic-grappling-with-chinas-growing-influence/?fbclid=IwAR3DD3wBQ32bmCgGQTH7_QGzM-qbATadaI7H2gTDtPaO4fZJ9WblW3P8DvY.
Rud, S. 2014. Governance and Tradition in Nineteenth-Century Greenland. *Interventions* 16 (4): 551–571.
Siorarsiorfik Nuuk City Development. 2017. *Siorarsiorfik Nuuk city development.* https://siorarsiorfik.sermersooq.gl/kl/.
Skydsbjerg, H., and W. Turnowsky. 2016. Namminiilivinnissamut angertut amerlanerussuteqarluartut. *Sermitsiaq*, December 1. http://sermitsiaq.ag/kl/namminiilivinnissamut-angertut-amerlanerussuteqarluartut.
Statsministeriet. n.d. Den grønlandske hjemmestyreordning. http://www.stm.dk/_p_5507.html.
Turnowsky, W. 2017. Ajornerulertoqassanngippat aatsaat namminersulivikkusupput. *Sermitsiaq*, April 1. http://sermitsiaq.ag/kl/ajornerulertoqassanngippat-aatsaat-namminersulivikkusupput.
———. 2018. Naalakkersuisoqatigiit isasoorput: Partii Naleraq Naalakkersuisunit tunuarpoq. *Sermitsiaq*, September 9. http://sermitsiaq.ag/kl/naalakkersuisoqatigiit-isasoorputpartii-naleraq-naalakkersuisunit-tunuarpoq.

CHAPTER 15

Arctic Policy of the United States: An Historical Survey

Stephen Haycox

Of the eight Arctic nations, the United States has been notably slow to develop clear and coherent policies for the region, both domestic and international. Perhaps because Alaska is not contiguous with the other states of the federal union, and the great majority of the country's citizens lack any Arctic experience or exposure, Alaska's Arctic has been rather out of sight and out of mind for American policy makers. But, new developments have prompted new attention to the region. Recently, global climate change has affected the Alaskan Arctic dramatically. Fierce storms caused by warmer weather cycles have eroded shoreland on which Inuit villages are located, which has generated the challenge of moving the villages. Shrinking annual polar ice has opened new access to the Northwest Passage. At the same time, reduced ice coverage of the sea has diminished hunting habitat for polar bears and haul-out environments for walrus, leading ecologists to ponder the future of these species. Melting permafrost has brought instability to some areas of the built environment, and has begun to release unprecedented levels of methane into the atmosphere. Having to confront these developments, American policy makers have directed new attention to the Arctic. Shifting decisions on off-shore oil and gas exploration and production, together with the pending sale of leases in the Arctic National Wildlife Refuge reflect new considerations in national energy policy. At the same time, growing awareness of Russian and other countries' interest in Arctic sovereignty and in pursuing new opportunities for economic development have raised significant questions regarding national security. These changed conditions have encouraged new consideration of the nation's Arctic policy and the need for greater administrative attention.

S. Haycox (✉)
University of Alaska Anchorage, Anchorage, AL, USA
e-mail: swhaycox@alaska.edu

© The Author(s) 2020
K. S. Coates, C. Holroyd (eds.), *The Palgrave Handbook of Arctic Policy and Politics*, https://doi.org/10.1007/978-3-030-20557-7_15

233

Alaska's remote location relative to the contiguous US states explains much of America's long lack of attention to its Arctic policy. That remoteness has been exacerbated by history, a history that has encouraged Americans to imagine that what happened on the far shores of the Arctic Ocean had little relevance to them because Alaska was isolated, a function of the psychology of non-contiguity.

Before Alaska was part of the United States, it was part of the Russian empire. But while in North America, the Russians did not penetrate the continent's Arctic region. The Russian maritime explorer Vitus Bering sailed through Bering Strait in 1728, but without realizing it was a strait. Russian enterprise on the continent took place far to the south. Two decades before the American acquisition of Alaska, in 1867, whalers, mostly American, began to frequent the Arctic Ocean north of Bering Strait in search of humpback whales, for their blubber, which was rendered into oil, and for their baleen, which was used in a variety of products. In 1848, the American captain Thomas Welcome Roys in his small ship, *Superior*, hunted 250 miles north of Bering Strait, obtaining a full complement of oil and baleen. When he reached Honolulu on his return from the north, the news of his success sparked an exploitation of the western Arctic whale fishery that lasted seven decades. 2700 voyages netted over 20,000 bowhead whales, leading nearly to their extinction; more than 150 whaleships were lost, mostly through being caught in ice and crushed (Bockstoce and Burns 1993, pp. 565–567).

Inuit people in what is now Alaska traded across Bering Strait and along the Arctic coast before contact. Scholar John Bockstoce has described the nature of that trade and how it was disrupted by European and American traders beginning in 1819 (Bockstoce 2009). The Inuit were accomplished traders with well-established transportation routes and localized monopolies of various trade articles. Not only did the arrival of European traders disrupt this trade, but the arrival of the whalers altered traditional Inuit lifeways. The whalers employed Inuit men as deckhands and hunters, and the women as seamstresses. The Inuit began increasingly to exchange their labor for goods, rather than using barter, as before. And as the men were occupied during the hunting season, the food supply in the villages was correspondingly reduced. Meantime, traders and whalers continued to barter tobacco, liquor, and other items for ivory and furs. Receiving more value in trade from the Europeans, the Inuit reduced their trade with Inland Inuit and Interior Athabaskans, many of whom then moved to the coast (Bockstoce 2009, 263ff, 324ff). In the 1880s, whalers established shore stations, the first, by the Pacific Steam Whaling Company in 1884; within a few years, there were 15 stations along the American Arctic coast. One of the most successful, at Point Barrow, was manned by Charles Brower, who lived there over half a century and became legendary as a businessman, and for his friendship with and advocacy of Alaska's Inuit people (Brower 1942). In 1890, a steam vessel pushed eastward to Herschel Island near the mouth of Canada's Mackenzie River, harvesting and processing the most valuable cargo of whale product recorded; many steam whalers followed. By the end of the first decade of the twentieth century, however, the demand for baleen (for corset stays, collar stiffeners, buggy whips, crinoline petticoats, and parasol ribs) had ebbed, and petroleum lubricants had replaced whale oil. Whale harvest in the American Arctic effectively disappeared, save continuing hunting by Inuit. In 1937, the United States signed the first International Whaling Agreement, and in 1946, the International Convention for the Regulation of Whaling which created the International Whaling Commission, intended to regulate commercial whaling. In 1977, Inuit established the Alaska Eskimo Whaling Commission which works with the IWC to establish annual quotas for the taking of whales by 11 Alaska Native villages for community subsistence.

Because the US Congress ceased treaty-making after the Civil War, there were no traditional indigenous land reservations in Alaska, and most villages were independent, relying on traditional subsistence harvest of regional resources for sustenance and stability. In the second decade following the American purchase of Alaska, both the federal government and private missionary societies began to construct schools in Inuit villages, and the US Revenue Cutter Service (forerunner of the US Coast Guard) began annual patrols to the Arctic to police whaling and attend to the welfare of the Natives. In 1890, crews of a merchant vessel and a Revenue Cutter ship constructed a school at Cape Prince of Wales (Darnell 1949). Other schools followed. Most of the Inuit villages had become highly dependent on the whale industry; its collapse after 1912 forced villagers to turn to the sale of land mammal pelts for livelihood. Traders who had worked in the shore stations and had stayed in Alaska helped facilitate the transition. Then, the influenza epidemic of 1918–19 took hundreds of lives across northern Alaska, 1500 within the Seward Peninsula alone. A fledgling federal agency, the US Bureau of Education, was given responsibility for providing services for Alaska's Native people, mostly education and medical attention. Later, in addition to education and medical services, the Bureau opened cooperative stores in several Inuit villages and maintained a marketing agency in Seattle for Native furs and arts and crafts.

Despite the annual voyages of the whaling fleet between 1884 and 1912, and the operation of the shore stations, Alaska's Arctic coast had not been fully mapped. In 1826, the British explorer John Franklin reached Prudhoe Bay from the east by land and Frederick Beechey in the *HMS Blossom* reached Point Barrow sailing from the west (Gough 1973). Later, Thomas Simpson traversed from the mouth of the Mackenzie River to Point Barrow in 1837. Little more was learned about Alaska's Arctic waters for nearly three-quarters of a century despite several voyages east from Bering Strait searching for the lost Franklin expedition. In 1906, John D. Rockefeller financed an Anglo-American Polar Expedition organized by a Dane, Enjar Mikkelson, and the American Ernest de Koven Leffingwell, to explore the Beaufort Sea and investigate a suspected Arctic land mass. The expedition did not complete its mission, but Leffingwell stayed in the Arctic from 1906 to 1914, spending six winters and nine summers there, camping on Flaxman Island at the mouth of the Canning River. He mapped the region including geologic prospects. With the help of Inuit villagers, he made 31 trips by dogsled and sledge. While he was engaged in this work, the US Navy explorer Robert Peary claimed to have reached the North Pole (Davies 1989). The consequent acclaim diverted attention from Leffingwell's remarkable contribution to Arctic geography (Collins 2017). Peary left a note at the end of his journey to the Pole claiming possession of "the entire region, and adjacent lands, for and in the name of the President of the United States of America" (Arctic Sovereignty 2014). The claim had no immediate consequence.

Two developments in the late nineteenth century brought Alaska and eventually the Arctic more within the purview of federal policy makers. The first was the rapid growth of near-monopoly corporate entities such as Carnegie Steel Company, Standard Oil Company and American Tobacco Company, and many others in the 1880s and 1890s. Critics interpreted as abuses of capitalism such corporate conglomerate practices as suppressing wages as prices rose, quashing union activity, maintaining unsafe working conditions in service to higher profit, the use of child labor, and the like. These generated a strong anti-trust movement in the United States which flourished after the turn of the century as the Progressive reform movement. This led to a shift from state to federal regulation, manifest in creation of the federal Interstate Commerce Commission in 1887, the Sherman Anti-Trust Act of 1890, leading ultimately to the Clayton Anti-Trust Act in 1914 and creation of the Federal Trade Commission in the same year (May 1989).

Following the 1898 Klondike and subsequent Alaska gold rushes, the largest mining enterprise in the United States, M. Guggenheim and Sons, headed first by Meyer and then his son Daniel Guggenheim, became interested in Alaska. They joined with America's most powerful banker, J.P. Morgan to form the Alaska Syndicate which developed the Kennecott copper mine, the Copper River and Western Railway, and had a controlling interest in the Alaska Steamship Company. The Syndicate planned construction of a railroad from tidewater to the Yukon River to develop mining prospects and a transportation network throughout the Alaska Yukon basin (Stearns 1936). This generated significant anti-trust sentiment in Alaska and in the federal government. One response was Congressional funding of a government-owned and operated railroad from tidewater at Seward to Fairbanks, the only federal railway in American states or territories. Congress committed as much as one-tenth of the total US annual budget to the project (Brehmer 2015).

President Theodore Roosevelt aggressively advanced progressive reform measures, supporting judicial action against trusts and vigorous anti-trust regulation (Miller 1992). He also actively advanced federal protection of the natural environment, setting aside large areas as national parks and nature reserves, establishing the US Forest Service and creating 150 national forests, 51 bird reserves and four game reserves (Brinkley 2009). He advocated passage through Congress of the 1906 Antiquities Act which authorizes the President to withdraw conservation lands over his signature alone, without corresponding action by Congress (Harmon et al. 2006). Learning from the Forest Service chief in Alaska of Guggenheim designs for Alaska, even before creation of the Alaska Syndicate, Roosevelt withdrew from mineral entry all coal lands in Alaska. Since the Syndicate planned to rely on Alaska coal to fuel their enterprises, Roosevelt's action had the effect of curtailing the Guggenheim expansion, as it was intended to do, both as an anti-trust initiative and a conservation measure. Congress would pass an Alaska Coal Lands Act in 1914 opening two areas for coal exploration and development, Bering River on the Gulf of Alaska coast, and Matanuska Valley, action coincident with its legislation authorizing construction of the Alaska Railroad. Roosevelt's action brought Alaska before federal conservation planners and monitors. Drawing on Leffingwell's surveys of the Arctic coast where he had found abundant petroleum seeps, in 1923 President Harding withdrew by executive order 23 million acres as Naval Petroleum Reserve No 4 (Rutledge 2006). Although oil discoveries in the contiguous states diverted attention away from Alaska, "Pet. 4" would be the beginning of Alaska's Arctic oil and gas regime.

The expansion of aviation to the Arctic in the 1920s would involve the United States and Alaska, but generated no significant Arctic policy initiatives or formulation. On May 9, 1926, American navy admiral Richard Byrd and his pilot Floyd Bennet made a historic first flight to the North Pole, from Spitzbergen. Two days later, Norwegian Arctic explorer Roald Amundson, American adventurer Lincoln Ellsworth, and Italian engineer and pilot Umberto Nobile departed Spitzbergen in the dirigible *Norge* for the North Pole, which they reached before flying on to Barrow and Teller, Alaska, on May 13. Two years later, the Australian explorer and American pilot Carl Ben Eielson flew from Point Barrow to Spitzbergen, the first cross polar flight (Reardon 2014). While these flights generated national interest in the heroism of the pilots and excitement in the scientific community over new information about the nature of the Arctic, they produced no reorientation of national policy regarding the Arctic.

World War II and the Cold War raised public awareness of Alaska in the United States. Following the Japanese attack on US bases at Pearl Harbor, Hawai'i, in December 1941, US military planners became aware, in late May 1942, of Japanese intentions to strike in the Aleutian Islands. On June 3 and 4, Japanese planes from a carrier task force bombed the fortified Aleutian town of Dutch Harbor, inflicting little substantial damage but killing 78 American service personnel. On June 6 and 7, Japanese forces invaded and occupied Kiska and Attu Islands in the Near Island group at the far west end of the Aleutians, taking captive 10 US Naval personnel on Kiska and 42 villagers and one non-Native school teacher on Attu, killing the radio operator husband of the school teacher. The captives were held for the duration of the war at Otaru City on Hokkaido where they suffered 40% mortality. In May 1943, a combined American and Canadian force fought a fierce battle at Massacre Bay on Attu, successfully retaking the islands. 549 Allied died, more than 1200 injured; the Japanese lost over 2351; only 28 Japanese were captured.

After the battle, war activity in Alaska subsided, military in the territory settling into a reconnaissance role. In addition to regular troops stationed across the territory, the US Army authorized an American US Army major, Marvin "Muktuk" Marston, to assemble a cadre of Native Alaskans as coastal scouts, organized as the Alaska Territorial Guard. Many were Inuit from the various Bering Sea and Arctic coastal villages. Established in June 1942, over 6000 villagers served under 21 staff officers, training, and watching coastal waters for signs of enemy activity. Even though authorized by the Alaska Defense Command, military personnel refused offers of service by the scouts other than watching the ocean. The Guard was disbanded in March 1947 (Marston 1969).

During the war, the United States transferred more than 8000 military aircraft to the Soviet Union under Congress's "lend-lease" plan to supply allies with war materiel (Dolitsky et al. 2016). The planes were flown by US pilots to Ladd Field near Fairbanks and to Nome and from those places they were flown by Russian pilots to European Russia for use there. The United States also shipped eight million tons of equipment by sea over the Great Circle route to Vladivostok. Over 500,000 tons were shipped through Bering Strait and along the Northern Sea route, as well (Motter 1952). But none of the World War II activities in Alaska motivated the United States to develop a comprehensive Arctic policy or strategy for applying it.

Issues of Arctic sovereignty for the United States arose right at the end of World War II. In September 1945, President Truman issued an executive order claiming for the US jurisdiction over all "natural resources of the subsoil and seabed" of America's continental shelf (Executive Order 9633 1945; McDorman 2012). This represented a change in the concept of high seas sovereignty, previously understood to include only the first three nautical miles seaward of the shoreline. The US State Department justified the order on the grounds of national security, asserting that to protect the country and conduct foreign relations, the United States had paramount rights in and power over the shelf, including full dominion over the natural resources, including oil. Added to the sovereignty claim, this was a significant departure from the centuries-old doctrine of freedom of the seas, which respected a 12-mile limit to national sovereignty in coastal waters. Truman's order was intended as much to clarify internally federal jurisdiction in relation to that of the several states as it was directed to the international community. The domestic aspect was soon tested before the US Supreme Court (*U.S. v. California* 1947). Internationally, the proclamation suggested that all coastal states have an exclusive right to resources within their continental shelf. In terms of the Arctic, Truman's residential proclamation set up a potential conflict with Canada which had in 1925 amended its Northwest Territories Act to require all foreign scientists and explorers in the Arctic to obtain government permits, an expression of Canadian sovereignty (McCormick 2014). Within a few years, the United Nations would convene discussions on a comprehensive agreement on the law of the sea.

In 1958, following a campaign characterized by vigorous citizen advocacy based mainly on the similarity of Alaskans and their institutions to those in the contiguous states, the US Congress granted statehood to Alaska, making it the 49th state to enter the union. Two provisions of the act, contradictory in nature, would establish the context for Alaska politics for more than a generation, with important ramifications for Arctic policy (Haycox 2016). One section authorized the new state to select 28% of the total land area, 104 million acres, for state title; the selections had to be made from land not already reserved by Congress or the President for specific uses such as national parks, forests, refuges, or other categories. Another section of the act, however, stipulated that the state and people of Alaska disclaimed any right or title to land that might be subject to Native title (72 U.S. Statute 1958). Unlike the contiguous states, there were no reservations for indigenes in Alaska. Traditional Indian reservations in those states were the result of treaties which extinguished Native title, called aboriginal title in law, outside the reservation boundaries. But, because Congress ceased making treaties with American indigenes after the Civil War, in 1871, there were no treaties with Alaska Native people, and therefore no reservations, as noted above (Cohen 1960). In 1941, the US Supreme Court confirmed the existence of aboriginal title, which is defined as Native title to land indigenes ever occupied or utilized, whether or not they continued to occupy or utilize it, unless the US Congress had formally extinguished that title, as it had done by treaty in all of the contiguous states (*United States v. Santa Fe* 1941). Most of Alaska, then, except lands already withdrawn by Congress or the President for specific uses (54 million acres as of statehood), was potentially subject to Native title, and therefore unavailable for selection by the state for its 104-million-acre entitlement. Conflict between the state and Native groups over which lands were in fact available for state selection would lead in 1971 to Congress's landmark Alaska Native Claims Settlement Act which titled 44 million acres of Alaska land to Alaska Natives (43 U.S.C. 1971; Mitchell, *Take My Land*, 1971).

Alaska achieved statehood in the midst of the Cold War. Alaska played a significant geopolitical role in the conflict, for it would be over Alaska that strategic air forces bound for the Soviet Union would last pass over friendly land; by the same token, Alaska would be the first enemy land Soviet air forces would pass if bound for the contiguous United States. Intercontinental ballistic missiles would supersede air delivery of nuclear weapons in the early 1960s and none were deployed in Alaska. However, before that time strategic air forces represented the primary offensive capability of the United States and its allies; air defense nuclear-armed Nike Hercules ground-to-air missiles, which were deployed in Alaska, were the first line of defense (Hollinger 2004). The US Army and US Air Force manned at Fairbanks and Anchorage nuclear-armed surface to air missiles intended to destroy invading Soviet bombers. Mindful of the potential needs of the US military in Alaska, the Alaska Statehood Act provided that the President could, on his signature, withdraw for purposes of national defense all land in Alaska north of a line paralleling the Porcupine, Kuskokwim, and Yukon Rivers, which included all of Arctic Alaska (U.S. Stat. 72, 1958). In addition to forward air bases in Interior and Northwest Alaska, the United States undertook construction of a Distant Early Warning line of interlinked radar stations from the Faroe Islands, Iceland, Greenland, northern Canada, and Alaska, including the Aleutian Islands, with accompanying communications networks (Coates et al. 2008). Twenty-two of the sixty-three DEW Line sites were in Alaska. While the United States did not have an inclusive Arctic policy after World War II, it had a fully developed air and ground military defense capability as part of its overall geopolitical relationship with its allies and with the Soviet Union.

The implicit notion in the 1945 Truman proclamation on sovereignty over the continental shelf prompted other countries to examine their policies toward shelf resources and many to assert similar claims, some imposing a 200-mile zone along their sea borders. This led to involvement of the United Nations, where Arctic nations met to discuss national policy differences. Their discussions produced in 1958 a UN Convention on the Continental Shelf. While the agreement represented an attempt to prevent conflict, it provided only an ambiguous definition of the rights of states. Nations were assured sovereignty over subsurface resources out to a depth of 200 meters, or to a point where the depth of the water allowed exploitation. With rapidly developing underwater technologies, that definition was not particularly effective (Myers and Barker 2013). Nonetheless, all the Arctic nations signed the convention except Iceland.

In August 1958 two US nuclear-powered submarines transited under Arctic sea ice to the North Pole. The following March one of the same vessels surfaced at the Pole. Their voyages exacerbated differing views of Arctic sovereignty advanced by the United States and Canada. These were the first operational nuclear submarines; they had been sent to the North Pole to demonstrate American security capability and preparedness, especially to the USSR which had recently successfully launched the Sputnik satellite (Byrne 2015). The exigencies of the Cold War militated against an open dispute between the United States and Canada over territoriality, but did not eliminate the claims.

The confusion and potential for conflict remaining after the 1945 agreement seemed an appropriate debate for United Nations action, and discussions began there in 1954. Building on the continental shelf convention, a 1958 UN Conference on the Law of the Sea (often referred to as UNCLOS I) generated four separate international treaties. These are sometimes gathered under the general title, Geneva Convention on the Continental Shelf; they dealt with territorial limits, what would be permitted on the continental shelf, cooperation on the high seas and fisheries (Morell 1992).

The United Nations continued to address issues involving ocean sovereign rights and international cooperation, staging a second conference in 1960 which produced no lasting results. However, conversations continued, eventually involving 160 countries. Officially, the third Conference on the Law of the Sea remained in session from 1973 to 1982 when participants reached a comprehensive agreement on a number of critical issues, discussed below.

In Alaska, though there was no effective US Arctic policy, there were significant developments in the region with the potential for major impacts on the

Arctic. The state had begun to select lands for which it desired state title soon after statehood became official January 3, 1959. The US Bureau of Land Management, which had jurisdiction over transfers of land title, shortly initiated confirmation of the state selections, and made the transfers. Very soon, however, Native groups rose to protest such transfers and where possible to prevent them. By the mid-1960s, the state was blanketed with proposed state selections and a myriad of Native protests and claims. Because of clouded land title, the conflict threatened further economic development in the state. Concerned about that issue, and sympathetic to Native claims and justice, the Secretary of the Interior, Stewart Udall, first postponed planned oil lease sales in northwest Alaska, and then, in 1966, enjoined the Bureau of Land Management from making further land transfers of federal land to state title until such time as Congress should craft a settlement of Native claims. At the time, the state had gained title to 12 million acres of its 104-million-acre entitlement (Mitchell 2001, p. 144). One area to which the state had gained title was a portion of the state's Arctic slope lands between Petroleum Reserve No. 4 on the west and the Arctic National Wildlife Range on the east, which included lands adjacent to Prudhoe Bay.

While discussions continued over the state's land selections and Alaska Native land claims, the state granted exploration leases on the Prudhoe Bay lands to oil companies, and in 1968 Richfield Oil confirmed discovery there of the largest oil deposit in North American history. The state soon sold additional leases, and a consortium of oil firms undertook plans for transporting the oil to market. One member of the consortium, Humble Oil Co., organized a test of a seaborne route through the Northwest Passage. The company acquired the largest and most powerful oil tanker yet constructed, the *SS Manhattan*, had her reinforced with ice-breaking capability, and in 1969 sent her on a voyage from the east coast through the Passage to Prudhoe Bay and back (Coen 2012). Unable to negotiate McClure Strait because of heavy ice, the ship, accompanied by two US and one Canadian icebreaker, sailed through Prince of Wales Strait and south of Banks Island. The ship collected a token barrel of oil at Prudhoe Bay before returning east. Assessing the voyage, oil company executives concluded that using tankers over the sea route would be more costly than a hot-oil pipeline across Alaska from the Arctic to a warm water port on the Gulf of Alaska.

The *Manhattan* voyage brought issues of territoriality to the fore for Canada, the United States and Inuit people in both countries. Canadian Prime Minister Pierre Trudeau was in the process of developing and articulating Canada's Arctic policy at the time of the tanker's sailing. Canada officially requested that the United States seek permission to use the Passage, which the US refused to do. Canada views the Passage as part of its internal waters for historic reasons; they lie behind Canada's claimed continental baseline (1986) and Canada asserts full sovereignty over them, including the right to deny transit to any ship. The United States regards the Passage as an international strait. The U.N. Conference on Law of the Sea declares that right-of-transit through an international strait is non-suspendable, even if a country's baseline is recognized as legitimate. The United States does not accept Canada's claim of a territorial baseline which includes the Passage or the archipelago of islands north of the Canadian mainland. The United States and Canada continue amicable relations despite the disagreement (Lassere 2011).

The Canadian government did not register the only protest to the transit of the *Manhattan*. As the ship transited Lancaster Sound, two Inuit hunters drove dogsleds ahead and into the ship's path. The ship stopped and the hunters demanded that the captain request permission to sail on. The captain complied, the hunters granted the request, and the ship continued on. A point had been made about contested sovereignty (Wright 2014). And soon afterward Canada announced a unilateral extension of its territorial seas to 12 miles from shore, and Parliament passed the Arctic Waters Pollution Protection Act providing for Canadian environmental regulation of its Arctic seas to a distance of 100 miles from the coast (Emmerson 2010).

It was not long afterward the United States began development of a coherent, wide-ranging set of policies toward the Arctic within the context of its position as an Arctic nation. Those policies continue to evolve today. 1971 can be seen in retrospect as a halcyon period of environmental awareness and action in the United States. In 1964, the US Congress passed the Wilderness Act which provided a legal definition of wilderness, and established the National Wilderness Preservation System, wilderness areas managed variously by the National Park Service, the US Forest Service, the US Fish and Wildlife Service, and the Bureau of Land Management. Wilderness, Congress determined, is "an area where the earth and its community are untrammeled by man." Today the wilderness system includes 757 areas in 44 states and Puerto Rico comprising 109.5 million acres, 5% of the land in the country. Half of that acreage, 54.5 million acres, are in Alaska.

In 1969, Congress passed the National Environmental Policy Act (NEPA) which established environmentally sensitive guidelines for federal agencies and a Council on Environmental Quality to advise the President on environmental policy. In December 1970, President Richard Nixon established the Environmental Protection Agency to consolidate many of the federal government's environmental responsibilities in a single office. The EPA would develop sweeping authority, based on NEPA, to monitor the environmental impact of alteration of the landscape, the built environment, and watercourses. Congress passed a Clean Air Act in 1970, also, and in 1972, a Clean Water Act, both administered by the EPA (Hays 1989). Canada's Arctic claim following the voyage of the *Manhattan* drew a multi-dimensional response from the United States. Initially, American spokespersons insisted that Canada's action amounted to a unilateral extension of jurisdiction on the high seas. At the same time, as noted, the United States worked at the United Nations with 160 other nations addressing the law of the sea. A third UN conference convened in 1973. Questions of territoriality received much attention, but so did environmental matters. The conference lasted until 1982, producing the first version of the current international agreement under which world oceans are monitored. The agreement became effective in 1994 when the 60th nation ratified it. To date, the United States has not ratified, citing military constraints it is unwilling to accept. But on most aspects of the agreement, the United States has been cooperative, agreeing to disagree without sabotaging the accord. On its own, in 1976 the US Congress enacted the Magnuson-Stevens Fisheries Conservation and Management Act which promotes commercial exploitation of US coastal fisheries. A principal aim of the act is the preservation of fish stocks in perpetuity. It is administered by the US National Marine Fisheries Service.

Environmental debate in Alaska impinged heavily on American national consciousness in the period. Because it was intended to be a land disposal act and settle all questions of land title, Congress included in the landmark Alaska Native Claims Settlement Act in 1971 provisions for reserving new federal conservation units within the state. Throughout the 1970s, debate raged in Congress and across the nation on what should be the balance in Alaska between environmental protection and economic development. With other Arctic regions, Alaska was and is economically dependent on natural resource extraction. As the capital for industrial development of resource extraction does not reside in the state, extraction projects are dependent on absentee corporate investment. State leaders and actual and potential corporate investors, particularly oil companies, vigorously opposed environmental land withdrawals and management regulations that threatened economic development. After a bitter struggle between the state and developers, and the national environmental lobby, in 1980 Congress passed the Alaska National Interest Lands Conservation Act (ANILCA). It withdrew 104 million acres in new national parks, national forests, national fish and wildlife refuges, designated nearly half of the acreage as wilderness as defined by the 1964 Wilderness Act. A compromise measure, ANILCA included numerous exceptions from wilderness policy for both potential economic development and traditional Native subsistence harvest (Nelson 2004). The act expanded the size of the Arctic National Wildlife Refuge, which lies adjacent to the Canadian border, to 16 million acres, but left unsettled whether to open then 1.5-million-acre coastal plain of the Refuge to oil drilling.

Reflecting continue environmental sensitivity in the United States, Congress passed the Arctic and Policy Act of 1984 which, while dealing with national defense, also addressed commercial fishing, and environmental and climate research. In 1989, the United States joined with the other Arctic nations in three years of talks on protecting the Arctic environment. Meetings were held in Rovaniemi, Finland, Yellowknife, Northwest Territories, Canada, and Kiruna, Sweden. In 1991, the group produced the Arctic Environmental Protection Strategy, a non-binding agreement signed by all eight nations at the ministerial level. It provided for preservation of environmental quality and natural resources, accommodating environmental protection principals to the needs and traditions of Arctic Native people, monitoring environmental conditions, and reducing and eventually eliminating pollution of the Arctic environment (Rothwell 1996). The agreement was historic on several counts. First, it was initiated by one of the Arctic nations, Finland, and proceeded without oversight by an external agency. Second, it brought to bear on the Arctic the raised environmental consciousness that had swept politics in the developed world after World War II. Third, it was the first Arctic international agreement that included the Arctic's indigenous people. Three indigenous organizations took part in the meetings as observers: the Sami Council, the Inuit Circumpolar Conference, and the Association of Indigenous Minorities of the North, Siberia and the Far East of the Russian Federation. Finally, subsequent meetings of the group, in Nuuk, Greenland in 1993 and Inuvik, Northwest Territories, Canada in 1996 generated a new, permanent Arctic working group in which all eight Arctic nations participate, the Arctic Council. The Council has become a major force in focusing international Arctic policy on specific issues and projects. As it developed, additional indigenous organizations have been welcomed as participants. A parallel organization of subnational entities was founded in 1991, the Northern Forum. The Governor of Alaska is the state's official representative on the Northern Forum. The organization's focus is on elements shared by the state and regional areas which are members. These include economies based on extraction of natural resources, lack of internal capital resources, limited infrastructure development, harsh climates and vulnerable ecosystems, diverse and relatively strong indigenous cultures, and sparse populations (Northern Forum 2018; Arctic Council 2018).

An impetus to the United States joining in the talks in Rovaniemi in 1989 was the environmental disaster in Prince William Sound, Alaska, in March of that year when the oil tanker *Exxon Valdez* went aground, spilling 11 million gallons or more of crude oil into the pristine waters of the Sound and onto 800 kilometers of beaches and shoreline (Haycox 2012). The causes of the accident, outlined by the National Transportation Safety Board as a finding from its investigation, were instructive for persons concerned about the maritime environment, especially in cold climates. Those included the failure to properly maneuver the vessel, possibly due to fatigue and excessive workload; failure of the master to provide a proper navigation watch, possibly due to impairment from alcohol; failure of Exxon Shipping Company to properly supervise the master and to provide a rested and sufficient crew for the vessel; failure of the US Coast Guard to provide an effective vessel traffic system; and lack of effective pilot and escort services (NTSB 1990). The latter two were of particular concern to anyone reflecting on maritime navigation and economic development in the Arctic. The sobering reality of the *Exxon Valdez* spill coincided with Prof. James Hansen's testimony on climate change before the US Congress the previous year (Hansen 1988). Hansen's testimony and projections have been seen as the beginning of the raising of climate change awareness in the literate world.

Continuing discussion of climate change and rising concern over the Arctic's environmental vulnerability, together with the formation of the Arctic Council, were considerations in the formulation during the Clinton administration for the first integrated and broad articulation of United States Arctic policy. The relaxing of Cold War tensions also played a role, as did discussions at the United Nations conference on the Law of the Sea. In June 1994, the White House articulated US policy in Presidential Decision Directive NSC 26. In a measure of the unfinished nature of its deliberations, the directive was not made public, and it addressed the Antarctic as well as the Arctic. Nonetheless, it can be seen as a significant step in the development of a US Arctic policy (Corgan 2014). The United States has, the directive asserted, six principal objectives in the Arctic: meeting post-Cold War national security and defense needs; protecting the Arctic environment and preserving its biological resources; assuring that natural resource management and economic development in the region are environmentally sustainable; strengthening institutions for international cooperation among the eight Arctic nations; involving the region's indigenous people in the decisions that affect them; and enhancing scientific monitoring and research into local, regional, and global environmental issues (Presidential Decision 2016). Though the end of the Cold War allowed a significant shift of emphasis in US Arctic policy, the directive stated, the country needed to "maintain its ability to protect against attack across the Arctic, to move ships and aircraft freely under the principles of customary law reflected in the 1982 Law of the Sea Convention, to control its borders, and carry out military exercises." The third directive objective alluded to the debate over the 1980 Alaska lands act (ANILCA) in its call for balancing resource management and economic development. The directive noted that the 1991 Arctic Environmental Protection Strategy called for cooperative monitoring of radioactive and chemical pollutants, and directed the US Office of Science and Technology to work with the Interagency Arctic Research Policy Committee (a subcommittee of the cabinet-level National Science and Technology Council) on monitoring and remediation strategies. Mindful of the *Exxon Valdez* disaster, the directive advised that the United States should work with the other Arctic nations on measures to "protect the marine environment from oil pollution and other adverse effects resulting from existing and planned land-based and offshore development activities and from potential increased use of the Arctic Ocean as a shipping corridor." The Department of the Interior, the Environmental Protection Agency (EPA), the National Oceanic and Atmospheric Administration (NOAA), the Coast Guard and other relevant agencies were counseled to review the adequacy of current US measures. Alluding to the coastal plain of the Arctic National Wildlife Refuge, the directive called for cooperative monitoring of the health of the Porcupine caribou herd which utilizes the Refuge during its annual migration. It also advised cooperation with Russia in protecting seal and walrus stocks. And the policy statement called on the US Interior Department, Environmental Protection Agency, National Oceanographic and Atmospheric Administration, and other agencies to work with the State of Alaska to ensure that development planning takes into account cyclical economic impacts, social impacts on indigenous people, and long-term environmental impacts. In all, it was a comprehensive and instructive articulation of an Arctic policy.

This first iteration of a full US Arctic policy was coincident with the 1994 Law of the Sea treaty, UNCLOS III, which covered many elements of management of the world's oceans, including territorial claims, navigation, archipelagic status and related transit, exclusive economic zones, continental shelf jurisdiction, deep seabed mining and exploitation of undersea resources, protection of the marine environment, scientific research, and the settling of disputes. The documents were very compatible except in the security aspects of territorial claims, to which the United States objected.

The Arctic Council became the major face of and conduit for US Arctic policy. Chairmanship rotates every two years. Since its inception, the United States has chaired the Council twice, 1998–2000 and 2015–2017. Responding to rising concern over global climate change, under the first US chairmanship, the Council launched an initial Arctic Climate Impact Assessment, working with the International Arctic Science Committee; they released their final report in 2004 (ACIA 2004). The assessment represented significant work on climate change by various scientific bodies, especially in the United States (Weller 1998). Human health was another emphasis of the first US chairmanship of the Council, resulting in an International Circumpolar Surveillance, a disease surveillance system led by the US Centers for Disease and Prevention (Parkinson et al. 2008). The United States committed the endeavors of numerous scientific groups and organizations to both initiatives of the Council.

The Clinton-era Arctic policy directive, NSC 26, remained in force until superseded by a new directive promulgated as President George W. Bush was leaving office in January 2009. Written in the context of the terrorist destruction of the World Trade Center towers in New York in 2001, National Security Directive 66/Homeland Security Presidential Directive 25 built on the earlier directive (NS Pres. Dir. 2001). But where NSC 26 had emphasized environmental protection, the Bush directive emphasized homeland security. As an Arctic Nation, the United States set forth a number of issues for attention. These included international governance in the Arctic, continental shelf and boundary concerns, promotion of international scientific cooperation, maritime transportation, energy, and environmental protection. Considered as priorities, the rearrangement of concerns and the highlighting of national security reflected changes in national perception of the Arctic. Not only did environmental protection now seem displaced, but maritime transportation and energy were added to the policy objectives. The directive took note of the impact of climate change and increased human activity in the Arctic. It acknowledged the ongoing work of the Arctic Council and urged continued cooperation with the eight Arctic nations. It noted both the fragility and richness of Arctic natural resources. Protection of the Arctic environment and conservation of its biological resources remained from NSC 26, as did the prescription that natural resource management and economic development be environmentally sustainable. So did the intention to involve the region's indigenous people in relevant decisions, and the commitment of the US scientific community to research and monitoring of the environment.

The Obama administration crafted no new formal Arctic policy directive, and the United States continued to operate under the guidance of NSPS 66/HSPD 25. During US chairmanship of the Arctic Council in 2015–2017, climate change drove much of the discussion and many of the discrete projects undertaken. The Council pursued three themes: Arctic Ocean safety, climate change impacts, and improving economic and living conditions. Foreign ministers from all eight nations attended the final plenary meeting of the Council in Fairbanks on May 11, 2017. There, they signed a legally binding agreement on scientific cooperation. Earlier, in 2011, ministers of all the nations signed an agreement on Arctic search and rescue protocols. Additionally, the Council made important progress on telecommunications infrastructure, reduction of black carbon and methane emissions, and strategic initiatives on dealing with invasive species.

In 2018, the US Congressional Research Service reviewed America's Arctic perspective. Record low extents of sea ice have focused scientific attention on links to climate change, the assessment noted, and projections of ice-free seasons in the Arctic within decades (O'Rourke 2018). Melting sea ice has the potential to affect weather in the United States, access to mineral and biological resources in the Arctic, the economies of the culture and peoples in the region, and natural security. It may lead to increased commercial shipping (Arctic Now 2018). While warming temperatures likely will permit increased exploration for oil, gas, and minerals, melting permafrost may pose challenges to onshore activities. Such increased exploration activity, as well as new tourism, will increase pollution. The report noted that clean-up technologies and strategies for oil spills in ice-covered waters have yet to be developed. Both indigenous communities and the various mammal and fish stocks in the Arctic will be affected by climate change. Finally, the report noted that the United States has but one functional icebreaker; the US Coast Guard has initiated construction of three heavy polar icebreakers.

Two of the more important elements of US Arctic policy are the Interagency Arctic Research Policy Committee (IARPC), already noted, and the Arctic Research Commission (USARC) (IARPC 2018; USARC 2018). These bodies coordinate research priorities and policy across various agencies, departments, and offices of the federal government. USARC is led by a seven-member board which includes academics, representatives of commercial activities, and an indigenous representative of the several indigenous groups in the state. Their meetings are held in Alaska. In addition to others, USARC consults with the Federal Subsistence Board which establishes subsistence harvest quotas on federal conservation land in Alaska after taking testimony from resource users, and with the state Board of Fish and Game.

All of the issues noted in the Congressional Research Service report are addressed in the US Arctic policy directive currently in force. The United States is likely to continue its new attentiveness to Arctic issues and policy, and to manifest through the Arctic its efforts in meeting the policy objectives. The United States seems now to be making substantial progress in articulating and pursuing a comprehensive, effective Arctic policy.

Notably absent from recent policy statements on Arctic policy are expressions of American exceptionalism, the notion that American culture is uniquely and positively committed to freedom, and acts distinctly as a force for good in the world. Arctic policy formulations are couched much more in the language of realpolitik, the notion that nations act in their own best interest ahead of humanitarian or inclusivist values. This may be because the United States has lagged behind other Arctic nations in articulating a coherent policy, and has not played a leadership role. Russia has been much more aggressive in developing economic and national security interests in the Arctic, and for the United States, Scandinavia and, for the most part, Canada have not been significant Arctic policy concerns of the United States. Only in scientific assessment and research, the United States has displayed important leadership, but has stressed international cooperation. As the country moves forward, this is not likely to change. Only a small portion of the Arctic lies under American sovereignty, and any long-term, comprehensive Arctic policies must necessarily rely on international cooperation.

References

ACIA. 2004. ACIA. Scientific Report, Introduction. http://www.acia.uaf.edu/acia_review/ACIA_Ch01_text_JAN04.pdf.

Arctic Council. 2018. The Arctic Council and the Northern Forum: A Northern Forum Position Paper. https://oaarchive.arctic-council.org/handle/11374/499.

Arctic Now. 2018, February. https://www.arctictoday.com/lng-tanker-completes-first-winter-northern-sea-route-crossing-without-escort/.

Arctic Sovereignty: A Short History. 2014. *Foreign Policy.* http://foreignpolicy.com/2014/05/07/arctic-sovereignty-a-short-history/.

Bockstoce, J.R. 2009. *Furs and Frontiers in the Far North: The Contest Among Natives and Foreign Nations for the Bering Strait Fur Trade.* New Haven: Yale. 9–10 and *passim.*

Bockstoce, J.R., and J.J. Burns. 1993. Commercial Whaling in the North Pacific Sector. *The Bowhead Whale.* Society for Marine Mammalogy.

Brehmer, E. 2015. Anchorage, Railroad Have Ties That Bind. *Alaska Journal of Commerce,* May 6.

Brinkley, D. 2009. *Wilderness Warrior: Theodore Roosevelt and the Crusade for America.* New York: Harper Collins.

Brower, C.D. 1942. *Fifty Years Below Zero: A Lifetime of Adventure in the North.* London: R. Hale.

Byrne, M. 2015. Happy Birthday to the Cold War's Most Eerie Technology, the 'Atom Sub:' The Ghosts That Haunt Earth's Last Remaining Hiding Places. *Motherboard.* https://motherboard.vice.com/en_us/article/9akgwe/happy-birthday-to-the-cold-wars-most-eerie-technology-the-nuclear-submarine.

Coates, K., P.W. Lakenbauer, W. Morrison, and G. Poelzer. 2008. *Arctic Frontier: Defending Canada in the Far North.* Toronto: Thomas Allen Publishers.

Coen, Ross. 2012. *Breaking Ice for Arctic Oil: The Epic Voyage of the S.S. Manhattan Through the Northwest Passage.* Fairbanks: University of Alaska Press.

Cohen, L.K. 1960. *The Legal Conscience: Selected Paper of Felix Cohen.* New Haven: Yale University Press.

Collins, J.R. 2017. *On the Arctic Frontier: Ernest Leffingwell's Polar Expeditions and Legacy.* Pullman: Washington State. http://foreignpolicy.com/2014/05/07/arctic-sovereignty-a-short-history/.

Corgan, M. 2014. The USA in the Arctic: Superpower or Spectator. In *Security and Sovereignty in the North Atlantic*, ed. L. Heinnin. London: Palgrave Macmillan.

Darnell, F. 1949. Education Among the Native Peoples of Alaska. *Polar Record* 19: 431–446.

Davies, T.D. 1989. *Robert Peary at the North Pole: A Report for the Foundation for the Art of Navigation.* Rockville: The Navigation Foundation.

Dolitsky, A., D. Hagedorn, J. Cloe, and V. Glasgow. 2016. *Pipeline to Russia: The Alaska Siberia Air Route in World War II.* Washington, DC: National Park Service.

Emmerson, C. 2010. *The Future History of the Arctic.* New York: Public Affairs.

Executive Order 9633. 1945, September 28. http://www.presidency.ucsb.edu/ws/?pid=12332.

Gough, B. 1973. *To the Pacific and the Arctic with Beechy: The Journal of Lt. George Peard of H.M.S. Blossom, 1825–28.* Cambridge/Washington, DC: Hakluyt Society/National Park Service.

Hansen. 1988. Projections. *Real Climate.* https://www.ntsb.gov/investigations/AccidentReports/Reports/MAR9004.pdf.

Harmon, D., F. McManamon, and D. Pitchaithley. 2006. *The Antiquities Act: A Century of American Archaeology, Historic Preservation and Nature Conservation.* Tucson: University of Arizona.

Haycox, S. 2012, June. 'Fetched up:' Unlearned Lessons from the *Exxon Valdez. Journal of American History* 99: 219–228.

———. 2016. *Battleground Alaska: Fighting Federal Sovereignty in America's Last Wilderness.* Lawrence: University Press of Kansas.

Hays, S. 1989. *Beauty, Health and Permanence: Environmental Politics in the United States, 1955–85.* Cambridge: Cambridge University Press.

Hollinger, K. 2004. *Nike Hercules Operations in Alaska, 1959–79.* Anchorage: Conservation Branch, Directorate of Public Works, U.S. Army Garrison Alaska. http://dnr.alaska.gov/Assets/uploads/DNRPublic/parks/oha/publications/nikehercops.pdf.

IARPC. 2018. https://www.nsf.gov/geo/opp/arctic/iarpc/start.jsp.

Lassere, F. 2011, November. The Geopolitics of Arctic Passages and Continental Shelves. *Public Sector Digest.* https://www.ggr.ulaval.ca/sites/default/files/documents/Lasserre/Publications/geopolitics.pdf.

Marston, M. 1969. *Men of the Tundra: Eskimos at War.* New York: October House.

May, J. 1989. Anti-trust in the Formative Era: Political and Economic Theory in Constitutional and Anti-trust Analysis, 1890–1918. *Ohio State Law Journal* 50: 258–395.

McCormick, T. 2014. Arctic Sovereignty: A Short History. *Foreign Policy*, May.

McDorman, T. 2012. *Salt Water Neighbors: The Law and Politics of the Canada-U.S. Ocean Relationship.* New York: Oxford University Press.

Miller, N. 1992. *Theodore Roosevelt: A Life.* New York: William Morrow.

Mitchell, D.C. 2001. *Take My Land, Take My Life: The Story of Congress's Historic Settlement of Alaska Native Land Claims, 1960–71.* Fairbanks: University of Alaska Press.

Morell, J. 1992. *The Law of the Sea: An Historical Analysis of the 1982 Treaty and Its Rejection by the United States.* Jefferson, NC: Mcfarland and Company.

Motter, T.H. 1952. *The Persian Corridor and Aid to Russia.* Washington, DC: Office of the Chief of Military History.

Myers, M., and J. Barker. 2013. *International Law and the Arctic.* Cambridge: Cambridge University Press.

National Security Presidential Directives. 2001. NSPD 66/hspd 25. https://fas.org/irp/offdocs/nspd/nspd-66.htm.

Nelson, D. 2004. *Northern Landscapes: The Struggle for Wilderness Alaska.* Washington, DC: Resources for the Future.

Northern Forum. 2018. https://www.northernforum.org/en/the-northern-forum/about-the-northern-forum.

NTSB. 1990. *Marine Accident Report: Grounding of the U.S. Tankship Exxon Valdez.* Report No. 90/04. Washington, DC: National Transportation Safety Board. 170 and *passim.* https://www.ntsb.gov/investigations/AccidentReports/Reports/MAR9004.pdf.

O'Rourke, R. 2018. Changes in the Arctic: Background and Issues for Congress. Congressional Record Service, January 4. https://fas.org/sgp/crs/misc/R41153.pdf.

Parkinson, A., M. Bruce, and T. Zulz. 2008. International Circumpolar Surveillance: An Arctic Network for the Surveillance of Infectious Disease. *Emerging Infectious Diseases* 14: 18–24.

Presidential Decision Directive. 2016. NSC 26. https://fas.org/irp/offdocs/pdd/pdd-26.pdf.

Reardon, J. 2014. *Alaska's First Bush Pilots, 1923–30: And the Winter in Siberia for Eielson.* Anchorage: Graphic Arts.

Rothwell, D. 1996. *The Polar Regions and the Development of International Law.* Cambridge: Cambridge University Press.

Rutledge, I. 2006. *Addicted to Oil: America's Relentless Drive for Energy Security.* London: I.b. Tauris.

Stearns, R. 1936. The Morgan-Guggenheim Syndicate and the Development of Alaska, 1906–1915. M.A. thesis, Columbia University.

U.S. v. California. 1947. 332 U.S. 19.

United States v. Santa Fe Railroad Co. 1941. 314 U.S. 339.

USARC. 2018. https://www.arctic.gov/about.html, U.S.C. 43 1601 (1971).

Weller, G. 1998. Regional Impact of Climate Change in the Arctic and Antarctic. *Annals of Glaciology* 27: 543–552.

Wright, S. 2014. *Our Ice Is Vanishing, sikuvut nunguliqtuq: A History of Inuit Newcomers and Climate Change.* Montreal: McGill Queens University Press.

CHAPTER 16

Iceland as an Arctic State

Valur Ingimundarson

The elevation of the Arctic in global politics as a result of climate change has moved the region from the margins to the center of Icelandic foreign policy. As an Arctic state, Iceland has, in the last decade, made a stakeholding claim in the region based on political and economic interests (Ingimundarson 2011a, 2012, 2015; Bailes and Heininen 2012; Bailes 2015; Bailes et al. 2014; Ingólfsdóttir 2016). It is motivated by a desire to take advantage of Iceland's geostrategic location, to be involved in Arctic governance, and to benefit from future natural resource extraction and transarctic shipping. At the same time, there is a growing awareness of the ecological vulnerability of the Arctic and of the need to enforce strict environmental rules when it comes to commercial activities in the region. Increased Western military presence in the North Atlantic because of Russian naval activities has also refocused attention on Iceland, which played an important military role during the Cold War. These developments have not only influenced Iceland's dual identity projection as a North Atlantic and Arctic state but also its bilateral relations with its Western partners and with non-Western countries, such as China and Russia, as well as its position within regional institutions, notably, the Arctic Council.

Iceland is grappling with the same questions as other Arctic states: how to reconcile sovereign state interests with intergovernmental cooperation in the Arctic, to deal with problems associated with jurisdictional and maritime boundaries in the region, and to regulate natural resource exploitation and shipping in accordance with international law. To ensure a full-fledged decision-making role in Arctic governance, Iceland has stressed the central role of the Arctic Council and firmly resisted the development of a hegemony exercised by

V. Ingimundarson (✉)
University of Iceland, Reykjavík, Iceland
e-mail: vi@hi.is

K. S. Coates, C. Holroyd (eds.), *The Palgrave Handbook of Arctic Policy and Politics*, https://doi.org/10.1007/978-3-030-20557-7_16

251

the five Arctic littoral states: Russia, Canada, Denmark on behalf of Greenland, Norway, and the United States. To buttress its position in the Arctic Council, it has even made the case for counting it among the Arctic "coastal states," which have traditionally been limited to the Arctic Five, making territorial claims in the Arctic Ocean.

In this chapter, I focus on the evolution of Iceland's Arctic policy since the 1990s. The emphasis is on three themes: Icelandic historical and cultural attitudes toward the Arctic; Iceland's economic and political interests in the region, and its role in Arctic security and geopolitics. Apart from seeing the region as a source of prestige associated with a membership in an elite forum of Arctic states, Icelandic policymakers have adapted the "idea of the Arctic" to different interests and circumstances based on political expediency, economic expectations, and cultural imagination. I show how the Arctic has been used by Icelandic political elites to promote a backward-looking narrative on a "glorified" Arctic past; how it has served the purpose of redrawing attention to Iceland's geostrategic position after the end of the Cold War; how it has functioned as a domestic "displacement mechanism" designed to offer forward-looking economic visions in response to the recent financial crisis, and how it has been adopted both to reinforce traditional Iceland's Western orientation and to explore non-Western possibilities.

The Historical Dimension: Framing an Icelandic Arctic Narrative

While subscribing to a northern identity, Iceland has, historically, only paid real attention to the Arctic when economic interests have been at stake. Its wealth is more associated with the North Atlantic and its rich fishing grounds. Apart from fish exports, the economy has also benefitted from tourism, which, in the last decade, has become Iceland's main source of income. With a population of only 350,000 people, Iceland, whose size is about 103,000 sq.km, has one of the highest per capita gross domestic product (GDP) in the world. A materialist approach—the projection of Iceland as a Western developed country—has always taken precedence over territorial aspirations in the Arctic or over the display of cultural affinity with Arctic indigenous communities. In this regard, there has been no gap between elite and popular perceptions of the Arctic.

To be sure, there has sometimes been a tension between a preponderant commitment to a Western nationalist trajectory and a Third World anti-colonial narrative in Iceland's foreign policy. From the 1950s to the 1970s, these two strands of nationalism merged in the "Cod Wars" against the British over the extension of Iceland's fishery limits (Ingimundarson 2011b). On the one hand, Iceland was strongly influenced by classical Western nationalism based on sovereignty concerns and the need to protect Iceland's borders—especially, the fishery grounds around the island. Yet, its policy was anti-traditional in the sense that it was rooted in a modernity discourse, which, was by definition ambivalent in its very origin. It represented an attempt to do away with any notions of national backwardness. On the other hand, Icelanders have sometimes mobilized around an anti-Western discourse based on Third World nationalism against colonialism and on the reification of historical traditions. This line of argument proved powerful and effective, even if it was problematic because Icelanders did not as a rich nation identify themselves with the developing world. But what made it discursively viable was that the Icelandic economic system was so dependent on fishing and was fighting Britain with its imperialist and colonial record.

With this background in mind, it was very hard for Icelandic political elites to come to terms with the 2008 banking collapse at the height of the global financial crisis, for it destabilized prevalent Western identities and self-perceptions. One manifestation of a thinly veiled inferiority complex was the reaction of Prime Minister Geir Haarde, when Iceland was forced to ask for an International Monetary Fund (IMF) bailout. He stressed that he had received a promise from the IMF that Iceland would not be treated like a "Third World" state (Gunnarsson 2009). Such attitudes also help explain why the Arctic has never occupied a reified place in Iceland's political and cultural imagination.

Thus, the prioritization of the Arctic in Iceland's foreign policy is a recent phenomenon. As a way of protecting its economic interests, Iceland came to an agreement with Norway, in 1980, on the boundaries issue of Jan Mayen in the Greenland Sea, and, in 2006, the two sides—together with the Faroe Islands/ Denmark also settled the northern continental shelf boundaries beyond 200 miles as part of their efforts to influence the recommendation of the UN Commission on Limits of the Continental Shelf. Yet, shying territorial demands in the North, the Icelandic government traditionally looked southward—to the Hatton Rockall area near the British Isles—which is also claimed by Britain, Ireland, and Denmark (on behalf of the Faroe Islands). Iceland did not even become a party to the 1920 Svalbard Treaty until 1994 as part of its efforts to bolster its position in a dispute with Norway over Icelandic fishing in the Barents Sea.

In view of this ambiguous affinity with the Arctic, it should not come as a surprise that Icelandic political elites were slow to identify with it when it re-emerged as a geopolitical space following the end of the Cold War. From the 1990s until the mid-2000s, Icelandic foreign policy was preoccupied with maintaining a defense relationship with the United States, which had been forged in 1951 and on cementing institutional ties with the European Union through its membership in the European Economic Area. This is not to say that Icelandic politicians completely sidelined Arctic developments. Iceland had been a founding member of the Arctic Council in 1996, and, as a sign of interest in the Arctic Council's environmental scientific work, Iceland agreed to host two working group secretariats, PAME (Protection of the Arctic Marine Environment) and CAFF (Conservation of Arctic Flora and Fauna). In addition, the Stefansson Arctic Institute (SAI) was established in 1998 under the auspices of the Icelandic Ministry for the Environment to promote multidisciplinary research and sustainable development in the Arctic. Ólafur Ragnar Grímsson, the President of Iceland, promoted the Arctic cause, advancing the Northern Forum Research Network and other cooperative Arctic proposals.

Yet, even if more attention was given to the Arctic when Iceland held the chairmanship of the Arctic Council from 2002 to 2004, the region was not prioritized. The agenda of the Arctic Council was, at that time, mostly confined to Arctic environmental protection, but socio-economic development was also emerging as a relevant topic. During its chairmanship, the Icelandic government stressed these issues and supported the rights of Arctic indigenous peoples (Jónsson 2003). Despite attempts to draw attention to the Arctic as a source of future riches, the importance of the Arctic in Iceland's foreign policy was dwarfed by a crisis in its relationship with the United States stemming from the end of the Cold War. The Icelandic government wanted to maintain what it termed a minimum territorial defense based on the 1951 US-Icelandic Defense Agreement. The Bush Administration, however, saw no military value in Iceland when its attention was focused on wars in Afghanistan and Iraq, claiming that Russia was not a hostile power anymore. After a protracted diplomatic dispute, lasting from 2003 to 2006, the United States decided, unilaterally, to close down the base in Iceland (Ingimundarson 2005; Bjarnason 2008). Iceland's agreement with NATO, in 2007, on a rotational Alliance military presence—through air policing arrangements by individual NATO countries three or four times a year for several weeks—was a partial compensation for the end of a permanent US presence. But it was clear that Iceland's military importance for the United States had evaporated after the end of the Cold War.

In response, the Icelandic government began, actively, to focus on the Arctic in its foreign policy, stressing the idea of the material potential offered to Iceland (Sverrisdóttir 2007). It was based on a new Arctic frontier narrative, mixing romanticized historical accounts of Arctic exploration with scientific discussion of contemporary Arctic shipping and the future possibilities offered by the shortening of transport routes between Asia and Europe and North America. Iceland was considered ideal for a transshipment port, which could equally serve as a hub for sea transport across the North Atlantic and the Arctic Ocean passage when it opened.

Thus, when the media spotlight was turned on the Arctic in response to the Russian North Pole flag-planting episode in 2007, Icelandic politicians were already reworking historical and geographical mythologies about Iceland's role in the Arctic. References to the term "Arctic Mediterranean"—coined a century ago by the explorer Vilhjálmur Stefánsson who was of Icelandic extraction—were used to evoke future material gains based on the prospective opening of new transarctic trade routes as a result of Arctic ice melting. Icelandic officials appropriated, reformulated, and repacked Stefánsson's early twentieth-century vision of all-year commercial sea routes around the Arctic, with ports, naval stations, and weather stations on strategically placed islands (Sverrisdóttir 2007).

Claiming that a military vacuum had opened up in the North Atlantic following the closure of the US base, the Icelandic government also stressed linkages between the Arctic question, on the one hand, and NATO's air policing policy and Russian strategic aviation, on the other. The Russian decision to resume Cold War style bomber flights in 2007 was meant to restore, symbolically, Russia's military prowess for domestic political consumption and to underscore its geostrategic interests in places such as the Arctic. Until September 2008, Russian bombers flew, on average, once a month near Iceland. Since the flights were not seen as posing a military threat, the Icelandic government made no direct attempt to securitize the Russian practices, even if it did not shun away from venting its diplomatic irritation at the Russians, who claimed that their actions did not violate international law. Nonetheless, Iceland's limited inclusion in NATO's air policing policy was justified on the basis of the far more extensive Baltic precedent. It was, however, not a response to increased threat levels.

The criticism of Russia quickly subsided when the economic crisis hit Iceland in the fall of 2008. After being turned down for emergency aid by its Western allies, Iceland turned to Russia for a 5.4 billion dollar loan (Interview 2009). Betraying a sense of desperation, Prime Minister Haarde put it this way: "We have not received the kind of support that we were requesting from our friends" and in such a situation "one has to look for new friends". The Russians were initially positive, with Prime Minister Vladimir Putin allegedly approving the deal. But a premature announcement by the Icelandic Central Bank raised suspicions in Moscow that Iceland was using the request to get Western assistance. When Iceland was subsequently forced to ask the IMF for a bailout, the Russian government saw no reason to bear the brunt of a rescue package. While the overture to Russia led to nothing, it raised questions of Cold War legacies, Arctic futures, and the instability of Iceland's foreign policy identities. Iceland faced Western criticisms for its eagerness to accept aid from Russia (Interview 2008). The loan was immediately put in an anti-Western context: that Russia wanted to use Iceland as part of a strategy to bolster its claim to Arctic oil and gas resources. That Russia wanted to expand its geostrategic influence in the North was likely, even if Iceland's Arctic role should not be exaggerated.

The debate also led to farfetched media speculations about Iceland's intention of offering Russia base rights as a sign of gratitude for a potential loan. What gave them added currency was a criticism voiced by Icelandic President, Ólafur Ragnar Grímsson, who sharply criticized Iceland's traditional allies, especially Britain but also the United States and the Nordic countries for not coming to Iceland's rescue. It would, of course, not have been the first time that Iceland—which topped the list of US per capita foreign aid from the late 1940s until the mid-1960s (Eisenhower Library 1965)—would use its strategic importance to play off East against the West. In the 1950s, NATO decided to provide Iceland—in a first for a member country—with economic aid to prevent it from accepting a huge loan offer from the Soviet Union. But Iceland was in no situation, in 2008, to play such political games for economic benefit. And there were never any plans to offer the Russians base rights in Iceland. To accept a loan that amounted to one-third of Iceland's GDP was certainly open to debate. But Iceland would not have jeopardized its Western political, trade, and cultural ties by joining a Russian alliance.

As part of the government's handling of the 2008 financial crisis, the Arctic region took on a new psychological dimension about future economic prospects. It served, in a functional sense, as an anti-dote to the traumatic effects of a crash. Compared to some other Arctic states, Iceland was not among those poised to gain much from natural resource extraction. Yet, at this time, the government was making preparations for oil exploration in its Exclusive Economic Zone (EEZ) near the island of Jan Mayen. This idea was also tied to other opportunities, such as making Iceland a service center in connection with oil and gas exploitation and tourism in Greenland and a transshipment hub. Moreover, the government began putting more emphasis on West Nordic cooperation or the relationship with Greenland and the Faroe Islands. Its purpose was not only to promote friendly, political, cultural, and trade relations. The move has also to be seen within the framework of natural resource politics—to provide services to what was dubbed the "energy triangle": the space, encompassing North-East Greenland, Jan Mayen, and the *Dreki* area off the north coast of Iceland.

Iceland's decision to expand trade and economic relations with China reflected the preponderant economic dimension in the government's post-crisis Arctic agenda. In 2013, Iceland became the first European state to conclude a Free Trade Agreement with China (Icelandic Ministry for Foreign Affairs 2013). A year earlier, China and Iceland had signed a memorandum on Arctic scientific cooperation (Embassy of Iceland 2012). As part of that cooperation, a Chinese-funded science project, centering on the construction of an Arctic Observatory in the northern part of Iceland, was launched in 2018. Apart from Iceland's strategic location, China's interest was probably first motivated by its aspiration for an Observer seat in the Arctic Council, which it was granted in 2013 (Jakobson and Peng 2012; Sun 2013; Hong 2014; Lund et al. 2015).

After recovering fully from the banking collapse, Iceland had less need for the instrumentalization of the Arctic as part of a crisis management agenda. True, some of the Arctic projects that were touted by politicians in the aftermath of economic downturn are still alive. A German engineering company, Bremenports has, for example, invested in preliminary research on the possibility of building a port in the northeast of Iceland based on the feasibility of a future transarctic route. Recently, it reached an agreement with Icelandic partners to form a company with an aim of establishing a port and industrial location for future Arctic shipping as part of the transpolar route. Yet, other plans have been shelved, notably the oil exploration project—based on a joint partnership between an Icelandic company, Eykon Energy, China's China National Offshore Oil Corporation (CNOOC) and the Norwegian state-owned company Petoro. In addition, ideas about making Iceland a service center in connection with oil and gas exploitation in Greenland have not materialized. And despite the symbolic importance of the Free Trade Agreement with China, it has not led to a major boost bilateral trade.

The Role of Iceland in Arctic Governance

A comprehensive Arctic policy, which was approved by the Icelandic parliament in 2011, continues to form the basis for government policy toward the region. It reveals a straddling line between the Arctic as a high-stakes resource base and a geopolitical arena, on the one hand, and as an ecological frontier to be regulated by an international regime, on the other. While Iceland's approach toward the "North" is, as noted, based on its strategic location and on an awareness of the region's economic potential and environmental vulnerability, it is also geared toward multilateralism and "identity politics" in relations with other Arctic states. It reflects a firm commitment to the United Nations Law of the Sea Convention (UNCLOS) and to the Arctic Council as being as the only legitimate institutional governing forum for the area. The Council should be provided with more political weight and normative regulatory instruments to deal with issues, such as Arctic shipping, natural resource extraction, and tourism. Iceland has also favored efforts to elevate the Arctic Council from a pure decision-shaping intergovernmental body to a decision-making one in some areas, such as search and rescue and anti-pollution measures. It fits with the portrayal of the eight Arctic Council states—together with Arctic indigenous peoples—as equal stakeholders, enjoying a privileged position in the region, even if it does not entail opposition to the involvement of non-Arctic actors in the Council's work. It remains to be seen, however, whether Iceland can make progress toward these goals during its second chairmanship of the Council from 2019 to 2021.

Iceland's Arctic policy reinforces the opposition to the development of an Arctic Five venue. The Ilullisaat meeting of the foreign ministers of the Arctic Five in 2008—and the follow-up meeting in Chelsea in 2010—was seen as an attempt to bypass the Arctic Council and to exclude the three other Arctic states, Finland, Sweden, and Iceland, from their deliberations. The Icelandic government rejects the legal reading of the term "coastal state" based on the delimitation of the Arctic continental shelf according to UNCLOS and on being restricted to the Arctic Ocean proper. Since Iceland's EEZ extends to the Greenland Sea in the Arctic—which is either seen as being part of the Arctic Ocean or bordering it—the case has been made that it is an Arctic coastal state, except when it came to the continental shelf itself. So far, the five littoral states have not been willing to accept Iceland's argument, but some of them recognize that the location of Iceland makes it difficult to exclude it from deliberations on key Arctic issues. The Arctic Five have felt the need to invite selected stakeholders outside its venue, such as Iceland, the European, China, Japan, and South Korea, to discuss initiatives, such as the unregulated high seas fishing in the central Arctic Ocean, which resulted in an agreement in 2017. Iceland was, thus, part of this deal, which was meant to enhance the legitimacy of the proposition to implement a fishing moratorium in the Arctic Ocean before a regulatory framework was in place. However, when it comes to the management of migratory and transboundary fish stock in the Arctic region, Iceland has been keen on fighting any attempts to establish an international fishery management organization, which could limit its own influence as a major fishing country.

Despite the emphasis on economic possibilities, worries about the potential effects of climate change on the marine environment in the "North" provide the subtext of Iceland's Arctic policy. Given its small size, Iceland will not be able to invest much in maritime surveillance and resource evaluation. But there is awareness in Iceland that oil and gas shipments do not only offer potential economic opportunities but also environmental risks, especially the danger of oil spills. Indeed, given the stakes here, it is quite possible that Iceland will, in the future, seek ways to securitize its fishing grounds by insisting on stricter rules around transport routes in international waters. Increased surveillance involvement of the International Maritime Organization (IMO) in the Arctic is seen as a step in the right direction. The Search and Rescue and anti-pollution intergovernmental agreements, which were negotiated in the last few years between the Arctic states under the rubric of the Arctic Council—as well as the adoption of a Polar Code through the IMO—were strongly supported by the Icelandic government. If a serious environmental accident occurs in Icelandic waters, one option is to attempt to enforce regulatory changes through unilateral means if multilateral mechanisms—such as UNCLOS—are not considered adequate. But, in general, the Icelandic emphasis is primarily on intergovernmental cooperation among the eight Arctic states within the institutional framework of the Arctic Council.

SECURITY POLICY AND GEOSTRATEGIC POSITION

Since the Ukrainian crisis, the United States and NATO refocused their attention on Iceland as part of efforts to boost Western military presence in the North Atlantic. It is motivated by a willingness to respond to increased Russian naval activities in the area by patrolling nuclear-powered submarines with missile capabilities (Olsen 2017). While the United States has resumed irregular rotational Cold War style US maritime and submarine patrols with long-range aircraft from Iceland (*Foreign Policy* 2017), it is unlikely to lead to the reopening of the military base. Yet, NATO's 2018 decision to reestablish an Atlantic Command is consistent with the new focus on the North Atlantic, reflecting a backward-looking trend in military thinking. Cold War concepts like "deterrence," "the GIUK gap," and "collective defense" have been resurrected for this purpose (Smith and Hendrix 2017). The remilitarization of the North Atlantic has not led to the abandonment of Arctic cooperation narratives, which continue to be expressed in a depoliticized language by the Arctic states, except for the United States under Donald Trump. There is, however, an unresolved tension between the portrayal of the North Atlantic as a potential conflict area and that of the Arctic as a peaceful region. Military activities are taking place in the Arctic as well as the North Atlantic, even if the former has mostly been spared the political fall-out of the Ukrainian crisis.

As for soft security risk factors in the Artic, the enormous Search and Rescue Region of Iceland (SRR)—which is 19 times the size of the country itself—presents challenges for the Icelandic maritime preparedness system. Due to the lack of infrastructure, rescue operations are extremely difficult north of Iceland and in the Greenland Sea. Aging cargo vessels and tanker fleets also pose risks. While large cruise vessels have not grounded near Iceland, there have been incidents with fishing vessels, tankers, and smaller passenger boats. Since Iceland does not have a military of its own, its preparedness system is exclusively run by civilian governmental institutions, primarily the Icelandic Coast Guard. In terms of operational capability, it is highly dependent on regional and international collaboration with neighboring countries and within multilateral forums. Historically, this has been the case for a long time. Thus, while the main aim of the US military base in Iceland was to maintain a military strategic posture in the North Atlantic during the Cold War and beyond, it also assisted Icelandic civilian institutions in responding to emergencies at sea. Following the closure of the US base, Iceland negotiated bilateral arrangements with Norway, Denmark, Britain, and Canada, with emphasis on soft security, including maritime collaboration. It has also taken part in the Arctic Coast Guard Forum, which includes Russia and which has taken on an increasingly important role in Arctic security.

The prospects of increased Arctic maritime access—and the opening of new sea routes—as a result of climate change have fueled discussions on Iceland's future role in a shifting geography. The Icelandic government has evaluated the feasibility of establishing an International Rescue and Response Center in Iceland to increase support capability with respect to rescue and response operations in the Arctic and to offer facilities and opportunities for joint Search and Rescue (SAR) training (Interviews 2016). The outcome of the project will not only be contingent on what Iceland is prepared to commit in terms of material resources but no less on the interest of other Arctic countries. Key questions about the purpose and functional role of such a rescue center, under whose ministerial and institutional control it should be placed, and whether it should be limited to Iceland's closest Western security partners or include others, such as Russia, as well remain unanswered. For these reasons, the project is still in its initial stages and no decisions have been made about its implementation.

The Icelandic government has primarily been looking at non-military factors in its assessment of risk scenarios in the Arctic. A parliamentary committee responsible for developing an Icelandic national security policy, which was approved in 2016, defined environmental threats, sea pollution, or accidents due to increased maritime traffic in the Arctic as core risks because of Iceland's dependence on fisheries (Icelandic National Security Policy 2016). Yet, increased military presence in the North Atlantic and Arctic is likely to affect Iceland's security policy in the near future. Currently, Iceland cooperates extensively with the Danish Navy through the Danish Joint Arctic Command (JACO) on maritime safety and surveillance around Iceland, Greenland, and the Faroe Islands. The Coast Guard has also concluded a bilateral agreement with Norway to facilitate information exchange. The Icelandic Coast Guard provides US military forces with logistics support when stationed in Iceland as part of military surveillance activities. If there will be an increased US presence in Iceland, it is possible that it would serve maritime and SAR purposes. Information exchange and cooperation on maritime security are also outlined in an MoU between Iceland and Canada. The Canadians have shown increased interest in an Arctic surveillance role north of Iceland. Iceland is usually not in direct contact with Russia on maritime security, with Norway serving as an interlocutor between the two countries. But no problems have emerged when it comes communications between the two sides on maritime cooperation (Interviews 2017).

Preparedness institutions in Iceland have traditionally wanted to "avoid geopolitics" and to facilitate transnational collaboration on Search and Rescue. The development of an Arctic "security community" is impossible in the absence of shared political security identities among the NATO Arctic states, on the one hand, and Russia, on the other. Yet, the taboo on discussing Arctic military security within the Arctic Council and the lack of a formal intergovernmental or institutional venue to discuss Arctic security clearly raises questions of long-term regional management. This affects Iceland's security as much as that of other Arctic states in a tension-ridden geopolitical climate.

CONCLUSION

The Icelandic government's Arctic agenda has been influenced by "the politics of transition"—the end of the Cold War, the closure of the US military base, the financial crisis, and renewed focus on the North Atlantic. As I have stressed here, the first steps toward making the Arctic a core issue in Icelandic foreign and security policy were taken after the US departure. And, while Iceland's foreign policy experiments—the brief encounter with Russia and the expansion of ties with China—have not led to any changes in its structural reliance on Europe and the United States, they were based on its geographic position and rooted in Arctic geopolitics.

The prioritization of the Arctic has also reflected a domestic political consensus. One reason was that the Arctic assumed a loose, multi-functional role, which was used by Icelandic political actors in various ways to define an "incomplete" region. When the financial crisis hit, for example, the Arctic was evoked—as part of a crisis response—as offering the prospects of economic relief within an unspecified time frame. Earlier attempts to portray, in nationalistic terms,

Iceland's place in the Arctic and future shipping potential by referring to backward-looking medieval accounts were also shelved during the financial crisis; after all, metaphors about modern-day Viking territorial conquerors had become unusable after being popularized to advertise the disastrous Icelandic banking expansion abroad.

While Iceland's full economic recovery has eased the pressure, in the domestic political domain, to use the Arctic as a prospective dividend in connection with the opening of new sea routes, the region is still projected in terms of material promise. Thus, the idea of Iceland as a future transshipment hub continues to be a part of governmental discourse, even if it is far less visible than it used to be. Initially, its referent point was the Northern Sea Route, but later, it shifted to the transpolar route, when it became the buzz word for Iceland's strategic harbor potential. This focus, however, has raised environmental questions. Various Icelandic governments have been careful to stress the need to take pre-emptive action to prevent oil spills and accidents in connection with the increased Arctic tourism (Icelandic Foreign Ministry 2013). As an indication of future clashes, however, the deep-water port project in the northeast of Iceland has been criticized by environmentalists on the grounds that more emphasis should be put on fighting climate change than short-term profit.

There are underlying Icelandic insecurities regarding Arctic governance and the fear of being excluded from decision-making in areas considered important for Iceland's economic security and interests. This attitude has clearly affected Iceland's policies with respect to the Arctic Five and ocean management in general. The Icelandic government's insistence on being recognized as a coastal state is primarily geared toward an international audience. It is seen as a means to strengthen Iceland's position in the Arctic Council, and in its dealings with other Arctic coastal states. Since Iceland is not challenging UNCLOS with its coastal state demand, it has not generated protests among the other Arctic stakeholding countries.

Support for the rights of Arctic indigenous peoples is emphasized in Iceland's Arctic policy. It reflects Iceland's own experience as a small state that received sovereignty from Denmark hundred years ago. The emphasis on West Nordic cooperation is part of this engagement with Greenland and the Faroe Islands. When it comes to the Arctic, however, this position is arguably driven as much by specific political and economic interests: on the one hand, any hegemonic aspirations on the part of the Arctic Five would be undercut by their exclusion of the Arctic indigenous people, such as the Greenlanders; on the other hand, Iceland is seen as standing to gain economically, in the future, from servicing future extraction of gas and oil and other natural resources in Greenland.

With its small population, Iceland is only capable of making minimal Arctic investments and has, so far, reaped no real economic gains from the region. While a tourist boom brought Iceland out of the financial crisis, it did not stem from governmental policies on the Arctic. The projects touted by politicians— the transshipment hub, port, and a search and rescue center—are still on the drawing boards. To be sure, the Icelandic government has toned down its economic rhetoric, refraining from making overtly exaggerated claims of Iceland's special position in the Arctic or of the impending opening of new sea-lanes. The failure of the oil exploration project with the Chinese and Norwegians in the Dragon Zone near Jan Mayen is a case in point.

After a post-Cold War hiatus, Iceland's geostrategic importance has been revived due to increased Russian naval activities in Northern waters. Russian strategic aviation or submarine operations are not seen as posing a territorial threat to Iceland. However, old metaphors describing Iceland as "the unsinkable aircraft carrier in the middle of the Atlantic" have been resurrected to make the case for its key role in the defense of the region. Suggestions by some Western security think tanks that it was time to reopen the US Naval Air Station in Iceland are premature (*Grapevine* 2017; *Navy Times* 2018). Increased US maritime patrolling activities from Iceland, even if they can be characterized as a creeping military presence, have so far not been resisted in the Icelandic domestic arena. At some point, questions can be raised about the difference between a temporary and a permanent US stay in Iceland, especially, if the rotational operations increase substantially in number. However, any hint of a return of US troops could reopen Cold War debates in Iceland. The current Icelandic government—a three-party coalition spanning the left-right political spectrum—was formed in 2018 to stabilize a political situation after a period of volatility and to manage sustained economic growth resulting from Iceland's recovery from the banking collapse. In the absence of a geopolitical emergency, an attempt to change Iceland's defense policy could pose a threat to the government, which is led by a party that opposes NATO (Ingimundarson 2018).

With respect to Arctic risk scenarios, the Icelandic preparedness system is highly dependent on regional and international collaboration, especially with neighboring countries but also within multilateral forums, such as the Arctic Council, the International Maritime Organization (IMO) and the Arctic Coast Guard Forum. It still lacks resources to respond to large scale incidents and its exclusively civilian nature has led to complications in interactions with militaries. These flaws point to the need for a more holistic approach with respect to prevention measures, operational capabilities in crisis situations, and the development of post-emergency regional planning. It would include a more formalized institutional cooperation between Iceland and Denmark on joint surveillance around Greenland and Iceland as well as logistics related to SAR, pollution prevention and civil protection; an expansion of the operational cooperation between the Icelandic and US coast guards in the fields of Search and Rescue; and the deepening of relations with Norway by going beyond information exchange and regional exercises. Iceland's cooperation with Russia could even be extended to the fields of transport and fishing vessels patrols. Such a regional approach could not only help identify potentials for joint action to mitigate sea-based risks in a vast oceanic domain but also serve the general purpose of improving safety and security in the Arctic. Yet, the implementation of such ideas is dependent on the continued will of the Arctic states to maintain political stability in the Arctic through multilateral cooperation—an increasingly difficult task in an age characterized by Great Power strategic competition.

REFERENCES

Bailes, Alyson, and Lassi Heininen. 2012. *Strategy Papers on the Arctic or High North: A Comparative Study and Analysis.* Reykjavík: Centre for Small States.

Bailes, Bailes. 2015. An Economic, Environmental, and Energy Security Framework for Small Entities in the Arctic: Challenges for Iceland, The Faroe Islands and Greenland. In *Polar Law and Resources,* ed. Natalia Loukacheva, 135–144. Copenhagen: Nordic Council of Ministers.

Bailes, Alyson, Margrét Cela, Kristinn Schram, and Katla Kjartansdóttir. 2014. Iceland: Small But Central. In *Perceptions and Strategies of Arcticness in the Sub-Arctic Europe,* ed. Andris Spruds and Toms Rostoks, 75–98. Riga: Latvian Institute of International Affairs.

Bjarnason, Gunnar Þór. 2008. *Óvænt áfall eða fyrirsjáanleg tímamót? Brottför Bandaríkjahers frá Íslandi. Aðdragandi og viðbrögð* [An Unexpected Shock or a Predictable Turning Point? The Withdrawal of US Troops from Iceland: Prehistory and Reaction]. Reykjavík: University of Iceland Press.

Hong, Nong. 2014. Emerging Interests of Non-Arctic Countries in the Arctic: A Chinese Perspective. *The Polar Journal* 4 (2): 271–286.

Iceland's PM Against Increased American Military Presence. *Grapevine,* December 7, 2017.

Ingimundarson, Valur. 2005, December. Confronting Strategic Irrelevance: The End of a U.S.-Icelandic Security Community? *RUSI Journal* 150 (4): 66–71.

———. 2011a. Territorial Discourses and Identity Politics: Iceland's Role in the Arctic. In *Arctic Security in an Age of Climate Change,* ed. James Kraska, 174–189. Cambridge: Cambridge University Press.

———. 2011b. *The Rebellious Ally: Iceland, the United States, and the Politics of Empire, 1945–2006,* 99–134. Dordrecht and St. Louis: Republic of Letters Publishing.

———. 2012. Territorial Nationalism and Arctic Geopolitics [with Klaus Dodds]. *The Polar Journal* 2 (1): 21–37.

———. 2015. Framing the National Interest: The Political Uses of the Arctic in Icelandic Foreign and Domestic Policies. *The Polar Journal* 5 (1): 81–100.

———. 2018, August. A Fleeting or Permanent Presence? The Revival of US Anti-Submarine Operation from Iceland. *RUSI Newsbrief* 37 (7): 1–4.

Ingólfsdóttir, Auður H. 2016. Climate Change and Security in the Arctic. A Feminist Analysis of Values and Norms Shaping Climate Policy in Iceland. Ph.D. thesis, University of Lapland.

Jakobson, Linda, and Jingchao Peng. 2012. *China's Arctic Aspiration.* SIPRI Policy Paper No. 34. Stockholm: Stockholm International Peace Research Institute.

Lund, Leif, Jan Yang, and Iselin Stensdal, eds. 2015. *Asian Countries and the Arctic Future.* Singapore: World Scientific.

Olsen, John Andreas, ed. 2017. *NATO and the Northern Flank: Revitalising Collective Defense.* RUSI Whitehall Paper. London: Royal United Services Institute (RUSI).

Restoration of US Air Base in Iceland Does Not Mean Troops Will Follow, Navy Says. *Navy Times,* January 10, 2018.

Sun, Kai. 2013, November. China and the Arctic: China's Interests and Participation in the Region. East Asia-Arctic Relations: Boundary, Security and International Relations, Paper 2. GIGI.

Archival Material

Þjóðskjalasafn Íslands (ÞÍ) [Icelandic National Archives], sendiráð Íslands í Washington [Icelandic Embassy in Washington, DC].

B-9, Memorandum, Hans G. Andersen to Icelandic Foreign Minister and State Secretary, January 22, 198.

B-9, Report, Skýrsla og sáttatillögur ríkisstjórna Íslands og Noregs frá sáttanefnd varðandi landsgrunnsvæðið milli Íslands og Jan Mayen [Report and Proposals to the Governments of Iceland and Norway from the Mediating Committee with Respect to the Territorial Zone Between Iceland and Jan Mayen], Washington, 1981.

Dwight D. Eisenhower Library, Abilene, Kansas.

A Report "Foreign Aid – Size and Composition," No Date [1965] Dennis Fitzgerald Papers, 1945–69, Box 44.

Interviews

Interview with an Icelandic Official, September 28, 2006.
Interview with a High-Level Russian Official, June 22, 2009.
An Interview with a High-Ranking Icelandic Official, November 15, 2008.
Interviews with Icelandic Officials, June 13, 2016.
Interviews with High-Ranking Icelandic Officials, May 11, 2017.

Web Sources

Address by Ambassador Benedikt Jónsson on behalf of Iceland's Chairmanship of the Arctic Council, at the International Round Table – Indigenous Peoples of the North and the Parliamentary System of the Russian Federation: Experience and Prospects, Moscow, March 12–13, 2003. http://www.utanrikisraduneyti.is/frettaefni/ymis-erindi/nr/224. Accessed August 11, 2018.

Opening Address by Valgerður Sverrisdóttir, Minister for Foreign Affairs, Conference Proceedings, "Breaking the Ice: Arctic Development and Maritime Transportation: Prospects of the Transarctic Route – Impact and Opportunities, Akureyri," March 27, 2007, 4–5. http://www.mfa.is/news-and-publications/nr/3586. Accessed August 11, 2018.

Icelandic Ministry for Foreign Affairs, Free Trade Agreement between Iceland and China. Fact Sheet, April 15, 2013. http://www.mfa.is/media/ftakina/China_fact_sheet_enska_15042013. Accessed August 11, 2018.

Embassy of Iceland in Beijing, "China and Iceland Sign Agreements on Geothermal and Geoscience Cooperation and in the Field of Polar Affairs," April 23, 2012. http://www.iceland.is/icelandabroad/cn/english/news-and-events/china-and-iceland-sign-agreements-on-geothermal. Accessed August 11, 2018.

Gunnarsson, Styrmir. *Umsátrið – fall Íslands og endurreisn* [The Siege: the Collapse of Iceland and Its Reconstruction]. Reykjavik: Veröld, 2009.

Icelandic Prime Minister Office, "The Prime Minister of Iceland and China Issue a Joint Statement," April 15, 2013. http://eng.forsaetisraduneyti.is/news-and-articles/nr/7555. Accessed August 11, 2018.

Icelandic Prime Minister Office, "Icelandic National Security Policy, 13 April 2016. https://www.stjornarradid.is/verkefni/almannaoryggi/thjodaroryggisrad/thjoda-roryggisstefna-fyrir-island/. Accessed January 1, 2019.

"China's Arctic Policy". Beijing: The State Council Information Office of the People's Republic of China, 2018. http://english.gov.cn/archive/white_paper/2018/01/26/content_281476026660336. Accessed August 11, 2018.

Smith Julianne and Jerry Hendrix, "Forgotten Waters: Minding the GIUK GAP". Washington, DC: Center for a New American Security, 2017. https://www.cnas.org/publications/reports/forgotten-waters. Accessed August 11, 2018.

CHAPTER 17

Svalbard: International Relations in an Exceptionally International Territory

Adam Grydehøj

INTRODUCTION

Svalbard is a land of superlatives. It hosts the world's northernmost towns and, by extension, its northernmost post offices, pubs, pizzerias, schools, supermarkets, museums, music festivals, movie theater, filling station, airport with regular commercial flights, hotels, barber shop, gift shops, clothing shops, and so on. The people who live in and visit Svalbard are acutely aware of this and choose to travel to Svalbard for an easily accessible taste of Arctic adventure. To be in Svalbard is to be at the top of the world.

That is one way of understanding Svalbard—to understand it in terms of the people to whom it holds personal meaning. Most people who engage with Svalbard understand it in this manner.

Yet there is another way of understanding Svalbard. This is to say that Svalbard is an archipelago with a landmass of $61{,}020$ km², lying between mainland Norway and the North Pole. Its territorial delimitations and its jurisdiction are enshrined and assigned in international law by the Svalbard Treaty (1920, §1), which recognizes:

> the full and absolute sovereignty of Norway over the Archipelago of Spitsbergen, comprising, with Bear Island or Beeren-Eiland, all the islands situated between 10° and 35° longitude East of Greenwich and between 74° and 81° latitude North, especially West Spitsbergen, North-East Land, Barents Island, Edge Island, Wiche Islands, Hope Island or Hopen-Eiland, and Prince Charles Foreland, together with all islands great or small and rocks appertaining thereto.

A. Grydehøj (✉)
University of Prince Edward Island, Charlottetown, PE, Canada
e-mail: agrydehoj@islanddynamics.org

K. S. Coates, C. Holroyd (eds.), *The Palgrave Handbook of Arctic Policy and Politics*, https://doi.org/10.1007/978-3-030-20557-7_17

From its early-seventeenth century origins in both international relations and international law, Svalbard has been a boxed off area on the map, a collection of islands contained within lines drawn across the sea. From the very start, Svalbard has been the object of attempted—and failed—possession by various states. The rationales behind this desire to possess Svalbard have often been opaque or, at the very least, unequal to efforts expended upon achieving this possession. By the same token, Svalbard itself—Svalbard as a place—has often been less important in the international discourse than has Svalbard as a principle.

It is this understanding of Svalbard that dominates the international discourse, including much of the scholarly discourse. Such an understanding at best grudgingly acknowledges the existence of those people who live in and visit Svalbard, focusing instead on the states that wish to somehow make Svalbard their own. The way in which one approaches the topic of Svalbard in Arctic policymaking depends on which perspective—the human or the IR—one takes or on how one manages to balance these two perspectives.

It is one thing to question why anyone really cares about a cluster of rocks in the Arctic, and it is another to question why the international discourse has come to conceptualize this particular cluster of rocks as a distinct territory, largely without reference to the thoughts, aspirations, and concerns of the people who actually live there. The present chapter will thus explore not only Svalbard's role in international relations but also how this role relates and sometimes fails to relate to the political, economic, cultural, and social lives of Svalbard's residents. The chapter takes the position that Svalbard's place in Arctic international relations cannot truly be understood without reference to Svalbard's people and vice versa.

Origins of the Svalbard Treaty

At the start of the twentieth century, a number of land areas in the Arctic held ambiguous territorial status. East Greenland, Jan Mayen, Franz Josef Land, Wrangel Island, and Svalbard were generally treated as *terra nullius*, that is, places that had never been subject to any state. Unlike the conquest and settlement of expanses of continental land across much of the Arctic, none of these regions were home to Indigenous populations at the time of discovery by peoples of European descent. Furthermore, these places are all islands or archipelagos, geographical categories that have been and continue to be exceptionally prone to territorial contestation (Baldacchino 2017). In the first half of the 1900s, however, all the outstanding Arctic terrestrial territorial disputes were unambiguously resolved—with the exception of the dispute over Svalbard, which was only resolved with considerable ambiguity. (Note that the ongoing dispute between Denmark and Canada over Hans Island only arose in the 1970s (Rudnicki 2016, p. 313), in the context of the determination of maritime boundaries under the United Nations Convention on the Law of the Sea (UNCLOS).)

The dispute over Svalbard's jurisdiction has roots in the early history of polar exploration. The Dutch explorer Willem Barentsz discovered Svalbard in 1596, on his final voyage in search of a Northeast Passage. Although it is possible that Svalbard had been sighted or visited by humans prior to this time, the evidence to support the various theories along these lines ranges from uncertain (visits by Pomor hunters from present-day Russia) to mere wishful thinking (discovery by Norsemen in 1194) (Arlov 2003, pp. 47–54). Efforts to exploit the whale, seal, and walrus resources in and around Svalbard began already early in the seventeenth century, with the Muscovy Company, based in England, taking the lead in exploration and ultimately hunting activities. By the 1612 hunting season, the Muscovy Company, which claimed a monopoly over hunting in the region through a royal warrant, faced competition from other companies from England, France, the Netherlands, and Spain. The Muscovy Company's efforts to protect its business through force of arms led the Netherlands to deploy the *Mare Liberum* principle, asserting that the seas around Svalbard belonged to no one in particular and were open for economic activity. Ultimately, the Netherlands' granting of a trade monopoly to the newly formed Noordsche Compagnie in 1614 led to a pragmatic agreement to divide hunting rights between the Dutch and the English operations (Rijkelijkhuizen 2009), with smaller-scale hunting operations from other countries continuing alongside. In subsequent years, the Kingdom of Denmark-Norway claimed the sole right to grant dispensation for hunting around Greenland and Svalbard, though it lacked the military capacity to disrupt the ongoing Dutch and English operations, which were ultimately complemented by strong German activity from the 1640s (Arlov 2003, pp. 65–74). Despite the prodigious quantity of marine mammals that were hunted around Svalbard over the course of the 1600s and 1700s (Rossi 2016), research by Arlov (2003, pp. 86–90) suggests that individual companies struggled to make a reliable profit from the industry, depending instead on state support.

The nineteenth century saw Svalbard gain prominence as a site for early polar tourism and as a stopping point for polar research expeditions. The early confusion and conflict concerning which state or states had the right to undertake activities in Svalbard had resulted in the cementing of Svalbard's *terra nullius* status in practice, a status that was increasingly taken for granted as whale populations plummeted and given that no permanent settlements were established in Svalbard for over three centuries following its discovery. In 1871, however, Sweden-Norway laid claim to Svalbard and considered establishing a colony there, with the proviso of permitting maritime activity by other states around the islands. This plan faltered in part due to protests from Russia and in part due to Norwegian concern that year-round presence in Svalbard would be impossible to maintain (Wråkberg 2002, p. 183). By the end of the nineteenth century, a flourishing of national sentiment in Norway gave the soon-to-be-independent Norwegian state grounds for wishing to claim Svalbard as its own, but by this point in time, rising Russian and German interest in the archipelago put paid to any plans for a straight-forward allocation of sovereignty (Arlov 2003, pp. 228–235).

Although the potential of Svalbard as a site for mining operations had long been noted, attempts at commercial mining first began at the turn of the twentieth century, leading to the establishment of Svalbard's first permanent and year-round settlement in 1905, with the subsequent few years seeing further mining towns being founded on the basis of American, British, Norwegian, and Russian capital. This was also the period in which Longyearbyen, currently the largest town in Svalbard, was established by the Boston-based Arctic Coal Company, which was later replaced by the Norway's Store Norske Spitsbergen Kulkompani (hereafter, Store Norske).

After Norway became independent from Sweden in 1905, the Norwegian state sought clarification on Svalbard's jurisdiction. A joint sovereignty solution was proposed by Norway, Sweden, and Russia in 1910 but met with opposition from the United States and Germany, and subsequent discussions in 1912 and 1914 likewise proved unsuccessful (Numminen 2011; Machowski 1995). The outbreak of World War I dashed any hopes for an immediate settlement, but the situation was different in the aftermath of the conflict. At the 1919 Paris Peace Conference, the Allied Supreme Council granted Norway sovereignty over Svalbard, though with provisions for international activity on the islands, resulting in the 1920 Treaty Concerning the Archipelago of Spitsbergen (hereafter, the Svalbard Treaty). The timing and details of this solution were consequences of a number of factors, including the desire of World War I's victors to reward Norway for its assistance; the United States' declining economic interest in Svalbard; and the post-war disempowerment of Germany and unrecognized status of the Bolshevik government in the Soviet Union, both countries that had interests in Svalbard and that might otherwise have driven hard bargains at the negotiating table (Arlov 2011, pp. 32–34). The Svalbard Treaty had 14 original signatory states, with the Soviet Union and Germany signing up in 1924 and 1925 respectively.

ECONOMIC ACTIVITY IN SVALBARD

In the 1600s and 1700s, the hunting of marine mammals around Svalbard was of strategic importance to various states, even if it was not necessarily profitable as a business (Arlov 2003, pp. 86–90). Similarly, as shall be discussed below, Svalbard's mining operations have never been commercially viable but have instead come to serve as means to geostrategic ends. Nevertheless, in the words of Rossi (2016, p. 116), "from the moment of its discovery in the modern age, the history of Spitsbergen became associated with the exploitation of natural resources." It is through natural resource exploitation—through extractive industries and, ultimately, scientific activity—that various states have justified and enacted their interest in Svalbard. The Svalbard Treaty was thus explicitly framed in terms of the capacity to undertake economic activity in Svalbard.

For our present purposes, the key provisions of the Svalbard Treaty (1920) are as follows:

- §1: Norway has "full and absolute sovereignty" over "all the islands between 10° and 35° longitude East of Greenwich and between 74° and 81° latitude North."
- §§2–3: Citizens of all signatory states may undertake economic activities "on a footing of absolute equality," subject to Norwegian legislation.
- §6: Those who occupied land prior to the treaty's signing would be granted title to this land.
- §7: Citizens of all signatory states may acquire and exercise the property ownership rights (including mineral rights) on terms of "complete equality."
- §8: "Taxes, dues, and fees levied shall be devoted exclusively" to Svalbard's administration.
- §9: Svalbard "may never be used for warlike purposes."

Although the Svalbard Treaty grants Norway "full and absolute sovereignty," this is "not in accordance with the ordinary and plain meaning of that phrase" (Rossi 2016, p. 110), and the treaty was not regarded as an unalloyed success even within the Norwegian political establishment (Arlov 2011, pp. 42–45).

The Svalbard Treaty necessitated regulation of the mining industry via the Mining Code of 1925, amended by royal decree in 1975. The Mining Code (§2) entitles treaty signatory states to search for, acquire, and exploit mineral deposits on an equal footing, yet not all signatory states possess equal mining rights in practice. In accordance with §6 of the Svalbard Treaty, which is essentially a "grandfather clause," foreign states could only claim land in Svalbard for a limited period following the treaty's ratification (Lüdecke 2011). The Norwegian state was intent from the start on overcoming its weak jurisdiction over Svalbard by purchasing mining rights from other states whenever they became available, and the Soviet Union began doing the same in the 1930s. As a result, ever since the late inter-war period, only two states have engaged in mining in Svalbard (Arlov 2011, pp. 40–41).

The Norwegian and Soviet success in purchasing mining rights is in part explained by the fact that Svalbard's coal mining operations have historically been unprofitable. Neither Norway's Store Norske nor the Soviet Union/Russia's Trust Arktikugol mining companies have ever been anything more than intermittently profitable, and both companies rely on state subsidies. The closure of the Russian town of Pyramiden in 1998 reduced Svalbard to three towns: the Norwegian towns of Longyearbyen (year-round population around 2170) and Ny-Ålesund (year-round population around 40) and the Russian town of Barentsburg (year-round population around 470), supplemented by a number of tiny research and weather stations.

The Svalbard Treaty does not specify how Norway should govern or administer the archipelago. Norwegian sovereignty over the islands is exercised primarily through the office of the Governor (*Sysselmann*), a figure who is appointed by the Norwegian Council of State and reports to the Norwegian Ministry of Justice. Up until the latter decades of the twentieth century, however, all of Svalbard's settlements were being run as company towns, meaning that day-to-day activities throughout the archipelago were administered in practice by the state-owned mining companies. As such, there was a practical division between Norway's centralized administration of Svalbard as a space of international relations activity and the localized administration of Svalbard's communities.

The 1970s' Cold War context saw the start of the very gradual liberalization of Longyearbyen, which led to the Governor taking over tasks from Store Norske and establishing "effective enforcement of Norwegian sovereignty, especially towards foreign agents on the archipelago" (Government of Norway 1999, §5.4.1). A 1979 royal decree states:

> [The Governor] shall seek to coordinate state activities on the archipelago. He must keep himself informed about any activities that may have significance for this work. He shall work for the good of Svalbard and, in this context, take those initiatives he considers necessary. (qtd. in Government of Norway 1999, §5.4.1)

The Governor thus gained responsibility for environmental protection, policing, notarial duties, tourism coordination, transport, public information, and contact with foreign-run settlements in Svalbard (Government of Norway 1999, §5.4).

In 1989, Store Norske further relinquished powers by splitting its provision of community services from its mining activities, with the result that the subsidiary Svalbard Samfunnsdrift company (a fully public corporation from 1993) took over the running of the town of Longyearbyen itself (Government of Norway 1999, §7.3.1). Store Norske transferred other functions to other new companies: Svalbard Næringsutvikling AS (commercial development), Spitsbergen Travel AS (tourism operations), and Svalbard Næringsbygg AS (commercial property). The gradual disempowerment of Store Norske was intended in part to enhance the power of the Governor and in part to foster diversification of Longyearbyen's economy, serving in the long run to transform this unprofitable mining town into more of a normal town, a place that would strengthen Norway's claims to jurisdiction by virtue of its existence under Norwegian administration. That is, it has been assumed that the Svalbard Treaty's granting of full and absolute sovereignty to Norway presupposes that Norway is active in and actively governs Svalbard.

ECONOMIC CHANGE AND CONTESTATION OF SOVEREIGNTY

The specific wording of the Svalbard Treaty has at times incentivized what might be regarded as counterintuitive strategic directions by the Norwegian state. From early on, Norway sought to institutionalize scientific research in the archipelago as a means of reinforcing its sovereignty. On 5 November 1962, an explosion at Ny-Ålesund killed 21 Norwegian miners and rendered further mining at the site impossible in practice. Norway decided to transform the town into a scientific research settlement, partially on foreign policy grounds. As Arlov (2011, p. 247; translation my own) argues:

> Sovereignty did not inherently require a specific level of presence, but it was clearly desirable to have Norwegian settlement and economic activity multiple places in the archipelago. This in part involved balancing the Soviet presence and in part involved demonstrating Norwegian interests.

Norway's efforts to "balance" Russian interests have continued to this day. Ny-Ålesund currently hosts 16 research stations from ten countries. By providing space and a framework for international research operations, Norway has been able to degrade Russia's status as a uniquely privileged actor in the archipelago. Furthermore, by formally involving actors from other states in a Norwegian-administered town, Norway is co-opting these other states into reinforcing its own sovereignty. By the same token though, other states have been able to use their presence in Ny-Ålesund as a means of gaining a voice in Arctic affairs (Roberts and Paglia 2016; Grydehøj 2014; Bailes 2011). Similarly, although the Svalbard Global Seed Vault, which opened outside Longyearbyen in 2008, may indeed be an important effort to preserve food crop genetic diversity, its placement in Svalbard combined with its reliance on contributions from numerous countries represents a method by which Norway can demonstrate international assent to its sovereignty. As Roberts and Paglia (2016, p. 2) argue, science and research have been used by both Norway and other states to "articulate their own narratives of belonging on Svalbard in particular and in the Arctic more generally." In line with the Svalbard Treaty, Norway has chosen to interpret research and education not just as activities but as *economic activities*, a determination that the Government of Norway (1999, §3.3) itself has stated is aimed at "ensuring the continuation of Norwegian settlements."

As Ny-Ålesund has transformed into an international research community, Longyearbyen has gradually become a tourism town, at latest count hosting seven hotels and guest houses, over a dozen bars and restaurants, and numerous shops aimed at the tourist market. Significantly, recent years have seen the sell-off even of the former service-oriented subsidiaries of the Store Norske coal company: Svalbardbutikken (the former company store and Longyearbyen's only department store and supermarket) is now part of the Coop Norge group, and the Spitsbergen Travel tourism company is now part of the Hurtigruten corporation. Store Norske has reduced its mining activities to very low levels.

Russia's own attempts at economic diversification in Svalbard have been hindered by Norwegian interpretations of the Svalbard Treaty. Given that none of Svalbard's mining operations have ever been profitable for a sustained period, mining in Svalbard has long been an excuse for—rather than a reason for—settlement. Norway has now found other excuses for being in Svalbard (research and tourism), but Russia remains trapped by the Svalbard Treaty's §§2–3, which predicate Russia's ability to be active in Svalbard upon its undertaking economic activities. It is thus that Russia, following Norway's lead, has encouraged tourism in Barentsburg. The Svalbard Treaty, however, does not guarantee the right to undertake economic activity in absolute terms but only "on a footing of absolute equality" with Norway, allowing Norway to deploy domestic legislation to limit kinds of activities in which Norway either does not intend to engage or does not intend to engage at a particular time or in a particular way. For example, a piece of Norwegian legislation, the Svalbard Environmental Protection Act, tightly constrains Russia's ability to open new mines connected to Barentsburg, yet this law was only passed after Norway had improved its own mining. Similarly, efforts by the Trust Arktikugol mining company to initiate tourist-oriented helicopter transport to Barentsburg from Longyearbyen (the site of Svalbard's sole commercial airport) were halted by the Norwegian courts, citing environmental protection (Åmund 2009).

Norway may have good reason to feel that its jurisdiction over Svalbard requires constant reaffirmation. For example, the Svalbard Treaty's §9 sets forth that the archipelago "may never be used for warlike purposes," yet Svalbard was the scene of fighting in World War II—fighting that was, in fact, prompted by the decision by Allied forces to evacuate Svalbard's predominantly Russian population and pre-emptively destroy the archipelago's settlements. In 2010, Russia argued that military use of photographs taken by the Norwegian-operated Svalbard Satellite Station were in contravention of the Svalbard Treaty (Numminen 2011), and as recently as 2017, Russia protested against the holding of a NATO meeting in Svalbard (Kovalev 2017). For its part, Norway argues that the provisions of the Svalbard Treaty's §9 do not equate to "demilitarization," and it has studiously established a Norwegian coastguard presence and regular military visits to Svalbard (Staalesen 2017).

This is reflective of Norway's wider approach to the Svalbard Treaty, which involves embracing the treaty's allocation of "full and absolute sovereignty" while interpreting the treaty's limiting clauses in the weakest possible sense. Thus, for example, Norway argues that the treaty's allocation of sovereignty allows it to claim an exclusive economic zone around Svalbard but that the treaty's allocation of rights to undertake economic activity on equal footing apply only to land and territorial sea to a distance of 12 nautical miles (around 22 km). Norway's establishment of a 200-nautical mile (around 370 km) Svalbard Fisheries Protection Zone in 1977 drew Soviet protests. Conflicting interpretations of the rights Norway derives from the treaty have occasioned sporadic international incidents, most dramatically in the "Elektron Incident" of 2005, involving the attempted seizure of a Russian trawler fishing within the protection zone (Åtland and Ven Bruusgaard 2009). Russia implicitly compromised on its claims while negotiating a maritime border with Norway in 2010 (Moe et al. 2011), yet there are continuing disputes over rights in the seas surrounding Svalbard, with the Norwegian Coast Guard seizing the Russian trawler Sapphire II on 28 September 2011 (Nilsen 2011).

Russia is currently attracting reproach for its wider foreign policy, but when it comes to Svalbard, Russia's de facto privileged status in the archipelago as well as its contested maritime borders with Norway grant it a key role in resisting Norwegian attempts to more completely and exclusively exercise its sovereignty. As Rossi (2016, p. 104) notes in the context of Norwegian continental shelf claims and its generous reading of the Svalbard Treaty, "Coordinated opposition to Norway extends beyond Russia's historical view, signaling that multi-party disputes are consolidating around the binary positions of Norway and other Arctic stakeholders," with the United States, Canada, and numerous European countries showing a willingness to tacitly or explicitly support Russian protests against Norwegian Svalbard policy. Russian pushback against Norway's reading of the Svalbard Treaty has provided precedent for claims by other interested state and non-state actors, as exemplified by the European Union's recent efforts to issue snow crab fishing licenses for waters around Svalbard (Bolongaro 2017).

Norway's attempts in the first decade of the twenty-first century to weaken Russia's status within Svalbard by encouraging other states to engage in the archipelago has thus had unintended consequences, with the encouragement of international scientific research and the liberalization of Longyearbyen granting many states a foothold in the archipelago and hindering Norway's central government from effectively implementing policy on the ground (Grydehøj 2014).

LOCAL COMMUNITIES AS TOOLS FOR FOREIGN POLICY

Norway has been intent on treating Svalbard as a space for geopolitical activity, with the Norwegian towns in Svalbard being used to support Norwegian sovereignty, rather than with Norwegian sovereignty being used to support the Norwegian towns. That is, the interests of the people who live in Longyearbyen and Ny-Ålesund (not to mention Barentsburg) are incidental to Norway's wider foreign policy goals.

Historically, Norway was able to exert complete control over the development of its settlements through the combined authority of the Governor and the Store Norske mining company. Longyearbyen's gradual liberalization has, however, made such control more difficult. The greatest indicator of this is the emergence of local democracy in Longyearbyen. During the Cold War, Norway's Svalbard policy had been dominated by national security concerns, with the provision of local decision-making power coming low on the list of priorities (Government of Norway 1999, §14.2.1). Nevertheless, the Local Svalbard Council (*Det stedlige svalbardråd*)—later renamed the Svalbard Council (*Svalbardrådet*)—was created in 1971, playing a purely advisory role to the Governor. A 1974–1975 Norwegian government white paper discussed and rejected the establishment of local democracy, while another white paper in 1985–1986 articulated "political, practical and economic obstacles to the development of local democracy following the mainland model" (Government of Norway 1999, §14.2.1). The Norwegian government believed that a strong Governor was essential for protecting Norwegian interests relative to the Svalbard Treaty (Government of Norway 2001, §2.1). A 1990 report by Geir Ulfstein argued that:

> Norway, by virtue of its sovereignty, in principle has freedom of action in respect of increased local democracy. However, Norway has both the right and the obligation to exercise sovereignty, so that fully autonomous settlements cannot be established without conflicting with the Treaty. A basic assumption is that Norway must both formally and effectively have control of any exercise of authority which impinges on the treaty rights of other states. The Svalbard Council can have limited decision-making authority in matters relating to the Norwegian population, but only the right to give an opinion in matters which concern other states or which are of special importance. (Government of Norway 1999, §14.2.3)

Despite these concerns, the rapid change that Longyearbyen underwent in the 1980s and 1990s suggested that the Governor's absolute authority was untenable in the long term. As a result, the Norwegian government began assessing how local democracy might be established without weakening Norway's position relative to the Svalbard Treaty. Finally, in 2002, a local government body— Longyearbyen Community Council (*Longyearbyen lokalstyre*)—was created, representing the advent of democratic society in Svalbard.

Longyearbyen Community Council has developed into a body capable of significant local governance activities. Even as the Governor seeks to govern Svalbard as a whole, policymaking in Longyearbyen itself is increasingly driven by local objectives and concerns. Now that the privatization and split up of Longyearbyen's former company town apparatus is complete, and now that Longyearbyen's economy is dominated by tourism and scientific research rather than mining, the Norwegian government possesses few tools for controlling Longyearbyen's future development and can at best nudge it in certain directions, for instance through subsidies. Crucially, the Svalbard Treaty's §§2–3, which grant nationals of all signatory states (in practice, of all states) the right to undertake economic activities in Svalbard, mean that the diverse private businesses that now fill Longyearbyen are capable of employing staff from around the world, without needing to worry about visa regulations, work permits, or immigration law that might apply in Norway proper. Furthermore, because the Svalbard Treaty's §8 states that "taxes, dues, and fees levied shall be devoted exclusively" to Svalbard's administration, Longyearbyen remains a low-tax zone, increasing its attractiveness to prospective Norwegian and foreign residents.

As Longyearbyen has slipped out of the Norwegian government's direct control, it has both grown in size and internationalized. As late as 2001, the Government of Norway (2001, §2.2.4.) asserted that Longyearbyen should ideally have a maximum of 1300 residents. As of 12 April 2018, Longyearbyen and Ny-Ålesund had 2214 residents combined, with all but a few dozen of these living in Longyearbyen. The population of Longyearbyen and Ny-Ålesund is very young: 53.6% are aged between 20 and 44, and just 1.7% are aged 67+

(relative to 33.6% and 14.8%, respectively, for Norway as a whole). Svalbard's Norwegian settlements also have a highly international population, with 29.8% having a registered country of residence outside of Norway (relative to 16.5% for Norway as a whole). In 2016, 46 nations were represented in the permanent populations of Longyearbyen and Ny-Ålesund, with the best-represented countries of origin of non-Norwegian residents being, in descending order: Thailand, Sweden, Russia, Philippines, Germany, Denmark, Ukraine, UK, Poland, and the United States (Statistics Norway 2016). Because Svalbard is not covered by Norwegian immigration law, even prolonged residency in Svalbard grants no right to residency in mainland Norway, meaning that many foreign nationals living in Svalbard truly are tied to the archipelago: It is very much a distinct territory. Longyearbyen is not just a Norwegian town like any other.

The youthfulness of Longyearbyen's population is linked both to the kinds of employment available, especially in the tourism industry, and to the potentially age-specific attraction of pursuing an Arctic adventure. Most residents are thus more-or-less temporary: In 2015, Longyearbyen had around a 20% population turnover rate, and the average period of residence in Longyearbyen and Ny-Ålesund was about seven years (Statistics Norway 2016). This high turnover rate is also linked to the fact that, even today, Longyearbyen is not set up as a cradle-to-grave community. While Longyearbyen has childcare and schooling provision, there is no long-term healthcare and social care provision, even for Norwegian citizens. The Norwegian government neither wants nor expects people to retire in Longyearbyen, neither wants nor expects the community to pass from one generation to another, with residents acquiring a strong sense of local inheritance and territoriality. A sense of community feeling and cultural heritage has nevertheless developed in Longyearbyen, even in the absence of population stability: When everyone is a temporary resident, all residents acquire an equal degree of community ownership (Grydehøj 2010).

This does not, however, mean that all residents acquire an equal say in the formal democratic system. Although nationals from any state may move to and work in Svalbard, eligibility to vote in Longyearbyen Community Council elections follows domestic Norwegian law. A person was entitled to vote in the 2015 local elections if he or she was (1) a citizen of Norway or another Nordic state who would have turned 16 by the close of the election year and who had been resident in Longyearbyen at least four weeks prior to election day; or (2) a person who had been registered as resident in Longyearbyen and/or in Norway for at least three years prior to election day *and* who would have turned 16 by the close of the election year and who had been resident in Longyearbyen at least four weeks prior to election day (Longyearbyen lokalstyre 2015). Nordic citizens thus gain voting rights in Longyearbyen with significantly greater ease than other residents, despite the treaty-conditioned internationalization of the local community.

The 2015 election saw 15 representatives elected to the Longyearbyen Community Council. All but one of these representatives were Norwegians, with the sole exception being Khaniitha Sinpru, a Thai national who had been living in Longyearbyen for 13 years (Mogård 2015). Despite English being universally spoken within Longyearbyen, and despite the large number of relatively short-term residents from outside the Nordic region (who are not expected to learn the Norwegian language), council business and political activity is carried out almost exclusively in Norwegian. In this regard, it is telling that, unlike in so many other highly remote small island communities, Longyearbyen's electoral politics are dominated by local branches of mainland-based parties. Electorally speaking, Longyearbyen remains an outpost of Norway.

Whether this represents a democratic problem depends on one's perspective. It is not, however, inconceivable that conditions and power structures within Longyearbyen will change over the coming decades. Bearing in mind that there are no borders within Svalbard and that Longyearbyen is a "Norwegian town" while Barentsburg is a "Russian town" only on account of the companies active in these towns and the nationalities of their residents, Norway may someday struggle to use the existence of Longyearbyen as proof of its absolute sovereignty over Svalbard.

It is a paradox of Norway's Svalbard policy that efforts to boost sovereignty claims have simultaneously led to the loss of Norwegian state control over Longyearbyen and to the reinforcement of Russian state control over Barentsburg. By seeking to stymie private enterprise and economic diversification in Barentsburg, Norway has inadvertently prevented the settlement from evolving out of company town status and Soviet-style centralized control. Barentsburg's 470 residents remain fully dependent on Trust Arktikugol for jobs and services, meaning that there is not just a lack of local democracy but also a lack of reporting requirements and formal clarity regarding decision-making processes (Gerlach and Kinossian 2016, p. 2). There are nevertheless a number of similarities between Barentsburg and Longyearbyen, in that the Russian town, like the Norwegian one, offers various employment advantages and high-quality infrastructure and services to its more-or-less transient workforce, at least relative to conditions in Russia itself (Gerlach and Kinossian 2016, p. 8). Barentsburg is still dominated by its unprofitable, subsidized mining industry, but despite Norwegian interference, the town has seen a rapid growth in tourist numbers as a side effect of the flourishing of tourism in Longyearbyen. Especially in light of Russia's increasingly assertive foreign policy in general, it thus seems likely that Barentsburg will remain active and will remain resolutely Russian for the foreseeable future.

EFFECTS OF THE SVALBARD TREATY

The Svalbard Treaty represents a common thread in the above discussions. It is the terms of this treaty that structure life in Svalbard, but it is also evident that the ambiguities and temporally specific wording of the treaty began causing problems fairly early on. Designed prior to the UNCLOS era and indeed prior to present-day understandings of state territoriality and to the end of the colonial era, the Svalbard Treaty at times seems hardly fit for purpose. Certainly, although the Svalbard Treaty has resulted in a degree of stability, it is telling that it has not been used as a model for resolving future territorial disputes—including the apparently similar situation regarding Antarctica.

Maintenance of the Svalbard Treaty has required constant attention from Norway on the one hand and from the rest of the world (most frequently, Russia) on the other. Ever since the start of the Cold War, Norway and Russia have each felt compelled to invest vast amounts of human, monetary, and political capital into maintaining their claims to Svalbard, and they have been joined in recent years by numerous other states, which are willing to go through the motions of treaty-supported economic activity in order to get their voices heard on wider Arctic policy issues. China in particular has been assiduous in using Svalbard as a platform for promoting its activities in the Arctic, with the Arctic Yellow River Station in Ny-Ålesund serving as a precursor for the expansion of Chinese activity across the Arctic, with increased investment in Arctic technologies and research stations planned in both Iceland and Greenland. This is an example of how Svalbard is providing various states with "strategic services" (such as capacity building in Arctic research and technologies, a stronger position in Arctic diplomacy, and grounds for protesting against attempts by Norway and other Arctic states to lay claim to large swathes of the ocean and seabed). Such strategic services are in a sense traded: Svalbard provides strategic services and receives in return subsidies, investment, and visitor expenditure (see also Grydehøj 2018, regarding strategic services and the economy of Kalaallit Nunaat/Greenland).

How we judge the Svalbard Treaty today depends on our precise objectives. While the treaty's antiquated legal language and its lack of reference to now-prevalent legal precepts have proven problematic for the two states that had the strongest historical links to Svalbard and the most interest in Svalbard back in 1920, they have resulted in a kind of productive ambiguity for everyone else. States such as China, Thailand, South Korea, and the Philippines (which had no pre-treaty links to the archipelago and which were never intended as Arctic stakeholders by those who designed the treaty) have no doubt benefited from the Svalbard Treaty's lack of clarity. Furthermore, the inadequacies of the Svalbard Treaty from the perspectives of Norway and Russia have been crucial for ensuring the creation and maintenance of permanent settlements and, ultimately, towns in the archipelago. Svalbard possesses community life today precisely because the Svalbard Treaty was unable to provide Norway with unambiguous and lasting sovereignty over the islands. A contrast is provided by the remote Arctic island of Jan Mayen, which Norway succeeds in maintaining jurisdiction over even in the absence of a permanent population.

CONCLUSION

Saville (2018) expresses how the mainland-oriented imaginary of Svalbard ends up eliding the roles of researcher, tourist, and explorer (see also Prince 2018, regarding the imagination of the Kerguelen Islands). If transience is regarded as characteristic of Svalbard's communities, then everyone—from the long-term resident to the weekend visitor, from the Thai restaurant worker to the Norwegian miner—has an equal stake in and share of the territory. Everyone participates in a continual ritual of identity formation, heritage creation, and valorization of territoriality that occurs within a societal framework supplied by and in the interests of the Norwegian state. If Svalbard is not just a place of wild, untamed nature but a place of specifically *Norwegian* wild, untamed nature (Roberts and Paglia 2016), then activities undertaken in Svalbard by non-Norwegians can be either integrated into the narrative of a Norwegian Svalbard or challenged as inappropriate to the environment, depending on what is most convenient for the Norwegian state.

Even though no state has straightforwardly profited from activities in Svalbard since at least the late 1800s, many states have sought to maintain a stake and have expressed an interest in the archipelago. This is perhaps in part due to a suspicion that if any one state ever did gain full control over Svalbard, it would have the ability to more successfully exploit the territory's resources, resulting in a relative weakening of all other states' positions. It is in this sense that Norway's efforts to place Svalbard at the core of its Arctic or High North policy have not been an unmitigated success (Grydehøj 2014). As centralized Norwegian control over Svalbard has lessened, the communities of Longyearbyen and Ny-Ålesund have begun providing other states with strategic services as well. That is, life in the Norwegian towns remains structured by Norway's interpretation of the Svalbard Treaty, but it is now to a large extent local business owners and community members of who are capable of creating "policy" in these northernmost of communities. Meanwhile, Norway has painted itself into a diplomatic corner through decades of reiteration that its own jurisdiction over Svalbard is rooted in the presence of Norwegian towns and settlements in the archipelago. It is interest in the strategic services being traded by the communities in Longyearbyen, Ny-Ålesund, and Barentsburg that keep local life in Svalbard running.

It is thus not a question of whether Svalbard should be understood through the locally grounded perspective of the people who live and visit there or through the more abstract perspective of international relations. Neither Svalbard's community life nor its importance within Arctic international relations can be understood independently, without reference to the other.

References

Åmund, B. 2009. Dom stanser turistplaner i Barentsburg. *Svalbardposten*, November 20. http://svalbardposten.no/nyheter/dom-stanser-turistplaner-i-barentsburg/19.1162.

Arlov, T.B. 2003. *Svalbards historie*. Trondheim: Tapir.

———. 2011. *Den rette mann: historien om Sysselmannen på Svalbard*. Trondheim: Tapir.

Åtland, K., and K. Ven Bruusgaard. 2009. When Security Speech Acts Misfire: Russia and the Elektron Incident. *Security Dialogue* 40 (3): 333–353.

Bailes, A. 2011. Spitsbergen in a Sea of Change. In *The Spitsbergen Treaty: Multilateral Governance in the Arctic*, ed. D. Wallis and S. Arnold, 34–37. Helsinki: Arctic Papers 1.

Baldacchino, G. 2017. *Solution Protocols to Festering Island Disputes: 'Win-Win' Solutions for the Diaoyu/Senkaku Islands*. Abingdon and New York: Routledge.

Bolongaro, K. 2017. Oil Lurks Beneath EU-Norway Snow Crab Clash. *Politico*, June 18. https://www.politico.eu/article/of-crustaceans-and-oil-the-case-of-the-snow-crab-on-svalbard/.

Gerlach, J., and N. Kinossian. 2016. Cultural Landscape of the Arctic: 'Recycling' of Soviet Imagery in the Russian Settlement of Barentsburg, Svalbard (Norway). *Polar Geography* 39 (1): 1–19.

Government of Norway. 1999. *Report No. 9 to the Storting (1999–2000): Svalbard, 1999*. Oslo: Government of Norway. http://www.regjeringen.no/en/dep/jd/Documents-and-publications/Reports-to-the-Storting-White-Papers/Reports-to-the-Storting/19992000/reportno-9-to-the-storting-.html?id=456868.

———. 2001. *Ot.prp. Nr. 58 (2000–2001): Lov om endringer til svalbardloven mm. (innføring av lokaldemokrati i Longyearbyen), 2001*. Oslo: Government of Norway. http://www.regjeringen.no/nb/dep/jd/dok/regpubl/otprp/20002001/otprp-nr-58-2000-2001-/2.html?id=164762.

Grydehøj, A. 2010. Uninherited Heritage: Tradition and Heritage Production in Shetland, Åland and Svalbard. *International Journal of Heritage Studies* 16 (1–2): 77–89.

———. 2014. Informal Diplomacy in Norway's Svalbard Policy: The Intersection of Local Community Development and Arctic International Relations. *Global Change, Peace & Security* 26 (1): 41–54.

———. 2018. Decolonising the Economy in Micropolities: Rents, Government Spending and Infrastructure Development in Kalaallit Nunaat (Greenland). *Small States & Territories* 1 (1): 69–94.

Kovalev, A. 2017. NATO Gets All Hot & Bothered for Norwegian Archipelago, Russia Says Stay Out. *Sputnik News*, April 20. https://sputniknews.com/politics/201704201052827663-norway-russia-nato-archipelago-dispute/.

Longyearbyen lokalstyre. 2015. Lokalstyrevalget. *Longyearbyen lokalstyre*, March 31. https://www.lokalstyre.no/lokalstyrevalget-2015.5730792-347088.html.

Lüdecke, C. 2011. Parallel Precedents for the Antarctic Treaty. In *Science Diplomacy: Antarctica, Science, and the Governance of International Spaces*, ed. P.A. Berkman, M.A. Lang, D.W.H. Walton, and O.R. Young, 253–263. Washington, DC: Smithsonian Institution & Scholarly Press.

Machowski, J. 1995. Scientific Activities on Spitsbergen in the Light of the International Legal Status of the Archipelago. *Polish Polar Research* 16 (1–2): 13–35.

Moe, A., D. Fjærtoft, and I. Øverland. 2011. Space and Timing: Why Was the Barents Sea Delimitation Dispute Resolved in 2010? *Polar Geography* 34 (3): 145–162.

Mogård, L.E. 2015. Historisk thai-kvinne på Svalbard. *NRK*, October 27. https://www.nrk.no/troms/historisk-thai-kvinne-pa-svalbard-1.12624614.

Nilsen, T. 2011. Russians Must Follow Norwegian Law. *Barents Observer*, October 14. http://www.barentsobserver.com/-russians-must-follow-norwegian-law.4972044.html.

Numminen, L. 2011. A History and Functioning of the Spitsbergen Treaty. In *The Spitsbergen Treaty: Multilateral Governance in the Arctic*, ed. D. Wallis and S. Arnold, 7–21. Helsinki: Arctic Papers 1.

Prince, S. 2018. Science and Culture in the Kerguelen Islands: A Relational Approach to the Spatial Formation of a Subantarctic Archipelago. *Island Studies Journal,* ahead of print.

Rijkelijkhuizen, M. 2009. Whales, Walruses, and Elephants: Artisans in Ivory, Baleen, and Other Skeletal Materials in Seventeenth-and Eighteenth-Century Amsterdam. *International Journal of Historical Archaeology* 13 (4): 409.

Roberts, P., and E. Paglia. 2016. Science as National Belonging: The Construction of Svalbard as a Norwegian Space. *Social Studies of Science* 46 (6): 894–911.

Rossi, C.R. 2016. 'A Unique International Problem': The Svalbard Treaty, Equal Enjoyment, and Terra Nullius: Lessons of Territorial Temptation from History. *Washington University Global Studies Law Review*, 93.

Rudnicki, J. 2016. The Hans Island Dispute and the Doctrine of Occupation. *Studia Iuridica* 68: 307–320.

Saville, S.M. 2018. Tourists and Researcher Identities: Critical Considerations of Collisions, Collaborations and Confluences in Svalbard. *Journal of Sustainable Tourism*, ahead of print.

Staalesen, A. 2017. Russian Svalbard Protest Totally Without Merit. *Barents Observer,* April 21. https://thebarentsobserver.com/en/arctic/2017/04/russian-svalbard-protest-totally-without-merit.

Statistics Norway. 2016. *This Is Svalbard 2016: What the Figures Say.* Oslo: Statistics Norway.

Svalbard Treaty. 1920 [1988]. *Treaty Between Norway, the United States of America, Denmark, France, Italy, Japan, the Netherlands, Great Britain and Ireland and the British Overseas Dominions and Sweden Concerning Spitsbergen, Signed in Paris 9th February 1920.* Oslo: Royal Ministry of Justice. http://app.uio.no/ub/ujur/oversatte-lover/data/lov-19250717-011-eng.pdf.

Wråkberg, U. 2002. The Politics of Naming: Contested Observations and the Shaping of Geographical Knowledge. In *Narrating the Arctic: A Cultural History of Nordic Scientific Practices*, ed. M. Bravo and S. Sörlin, 155–198. Canton, MA: Watson.

CHAPTER 18

Europe's North: The Arctic Policies of Sweden, Norway, and Finland

Ken S. Coates and Carin Holroyd

Northern Europe has given the world some of the most enduring Arctic images and personalities: the century's long activities of Sami reindeer herders of the north, the impressive Arctic explorations of Roald Amundsen and Friodjof Nansen, Allied supply runs along Norway's coast to Murmansk during World War II, the iconic Svalbard Global Seed Vault, the giant and generation-old Kiruna mine in northern Sweden, Rovaniemi's impressive Santa Claus Village, and the globally significant server farmers being developed around Luleå, Sweden. The region has established itself as a formidable presence in Arctic affairs, highlighted by the establishment of the Arctic Council's headquarters in Tromsø, Norway, and annual gatherings of northern leaders at Arctic Frontiers (Tromsø, Norway) and the High North Dialogue (Bodo, Norway).

The northern reaches of Scandinavia challenge many of the existing stereotypes of Arctic life. The Sami people have better social outcomes, higher educational attainment, and greater engagement in the northern economy than most other Indigenous peoples in the Far North. The regional infrastructure (roads, airfields, airports, electrical grids, and public facilities) is strong, comparable to national norms in each of the countries. Northern Scandinavia hosts an impressive set of universities, including University of Tromsø, Nord University, Umeå University, Luleå University of Technology, University of Oulu, and the University of Lapland. The economies of the North are relatively stable, buttressed by a high profile off-shore oil industry off Norway, Finland's robust forest sector, and Sweden's substantial industrial presence in

K. S. Coates
Johnson-Shoyama Graduate School of Public Policy, University of Saskatchewan, Saskatoon, SK, Canada

C. Holroyd (✉)
Department of Political Studies, University of Saskatchewan, Saskatoon, SK, Canada

© The Author(s) 2020
K. S. Coates, C. Holroyd (eds.), *The Palgrave Handbook of Arctic Policy and Politics*, https://doi.org/10.1007/978-3-030-20557-7_18

the region. Taken as a whole, northern Scandinavia is culturally strong, cosmopolitan, connected to the modern high technology economy, with a strong political presence at the national and international level.

NORTHERN NORWAY

The region includes three counties—Nordland, Tromsø, and Finnmark—with approximately one-third of the country's population. Despite its northern location—Tromsø is at 70° north, hundreds of kilometers north of Whitehorse, Yukon, and 5° further north than Fairbanks, Alaska. But with one of the most benign climates in the Arctic (comparable to Iceland and Anchorage, Alaska) to offset the long, dark winters, the region has considerable agricultural activity. The region has rich coastal fisheries, Nordland in particular, and a substantial administrative presence in the major cities, large-scale hydroelectric power production, and considerable mining activity. Troms County has over 150,000 people and is home to Tromsø, the self-styled "Capital of the Arctic." The northern interior, Finnmark County, has long and cold winters, more limited economic prospects and a sparse population.

Northern Norway played a pivotal role in the launch of the global Indigenous rights movement. In the 1970s, the Government of Norway proposed the development of the hydroelectric potential of the Alta River, brushing aside the realization that the Sami community of Maze would be flooded out and local reindeer herding and fishing seriously disrupted. Regional Sami, supported by other Sami and international supporters, pushed back against the plans in 1978, demanding that they be abandoned. The Sami blocked construction and took their protest to the Norwegian capital of Oslo. An escalation of the opposition in 1981, highlighted by long-lasting standoffs between Norwegian policy and Sami protesters and their supporters, slowed but did not stop the development of the Alta dam, which opened in 1987. The Sami emerged re-empowered from the process and pushed hard against the Government of Norway, leading in 1989 to the opening of the Sami Parliament in Karasjok and the formalization of the Sami's place in national political and legislative processes.

The North's political culture is dominated by debate about the oil industry, which is a source of Norway's impressive national economy and government revenue, and by the steady expansion of Sami assertiveness and political engagement. Environmental concerns rank high in regional affairs, focused on intense debates about mining projects and the protection of traditional reindeer herding activities. The Sami play a significant role in discussions on resource development but focus their efforts on asserting authority over traditional lands and protecting cultural and harvesting activities. There are significant concerns about long-term economic development and concerted efforts to diversify away from natural resource development, focused largely on the University of Tromsø and Nord University. Following on the national tradition of government-led and supported commercial development, northern Norway's businesses attract substantial financial and logistical support from the government, with much of it focused on developing value-added enterprises and connecting the region to the global innovation economy.

NORTHERN SWEDEN

As in Norway and Finland, Sweden embraces its northern character, as seen in its celebration of winter sports, extensive vacationing in the North, and an affinity for northern life. Norrland makes up close to 60% of Sweden's land mass, organized in 9 provinces with a total population of more than 1.7 million people (with 77,000 in Luleå and 122,000 in Umeå). Repeating a common northern pattern, people have been moving out of the rural and northern areas with comparatively few of the immigrants who are adding to Sweden's growth making their way north. The region relied on traditional resource development for generations, supported by rich mineral deposits, abundant forests, and considerable hydroelectric potential. The availability of resources supported a robust industrial economy, including pulp and paper and lumber production plus iron and steel plants.

Northern Sweden is leading the Circumpolar world in commercial innovation. Communities such as Luleå, Skellefteå, and Umeå originated with secondary resource and basic industrial production, running into difficult economic times as these sectors declined after the 1990s. Skellefteå, facing large-scale commercial flight, made a successful collaborative effort to attract international businesses to their community. Umeå built off the success of their university and established an international reputation for environmental innovations. In one of the more imaginative moves, Haparanda, a medium-sized city near the Finnish border, generated considerable interest when they convinced Ikea to open a store in the community. Luleå, a declining industrial and shipping city in the early twenty-first century, rebuilt its economy around the Luleå University of Technology and discovered important regional opportunities in the combination of abundant energy and a cold climate. Luleå attracted major investments from Google and Facebook, establishing the city as a center of the digital economy in Europe. The presence of two of the world's largest new economy firms made Luleå attractive to other digital firms and professionals, demonstrating the transformative potential of new economic sectors in the Circumpolar world.

Sweden's northern regions enjoy a high level of prosperity. The Sami in the area, like their counterparts in Norway and Finland, continue to assert their Indigenous rights and to advocate for access to reindeer herding lands. A Swedish Sami parliament was established in Kiruna in 1993, providing a platform for Sami political engagement. Like all of Sweden, the people in Norrland enjoy a generally high standard of living supported by national-quality roads, railways, airfields, and energy systems. The region has innovative firms and is highly regarded for recent developments in forestry and mining. In the latter case, the technological intensification of the famed Kiruna mine secured the support of unions in return for job protection.

Northern Finland

As is the case across northern Europe, Finland's population is concentrated in the south, with substantial decline in rural areas and small towns. The sparsely populated North—Lappi Province (Lapland), has only 3.4% of the country's population—is experiencing a substantial transformation. The region has demonstrated considerable resilience in the face of economic change. On the southern edge of northern Finland, the high-tech city of Oulu, home to Nokia, demonstrated the ability of the region to complete in the global economy. When Nokia experienced a sharp commercial decline, many of the professional engineers and technicians remained in the region and started new technology firms. The anticipated collapse of Oulu did not occur to the extent anticipated. Throughout the North, however, the traditional resource economy, mining, and logging, remains dominant. The city of Rovaniemi, population 64,000, is expanding from its forest foundation, in part by capitalizing on its location close to the Arctic Circle. The large Santa Claus Village attraction draws tens of thousands of international visitors annually, becoming a formidable part of the local economy. While the Village attracts considerable attention, the University of Lapland has perhaps had a greater impact on the regional economy, with the affiliated Arctic Centre raising the community's profile as a focal point for Circumpolar science and research.

The region is currently engaged in an intense debate about the construction of a joint 2.9 billion Euro Norway-Finland Arctic Railway, designed to connect the Arctic Coast at Kirkenes with Europe. The controversial project is heavily promoted for its economic benefits, including the encouragement of resource development along and near the route and the further expansion of Barents Sea oil and gas activity. Sami in the region are not supportive, fearing substantial disruptions of reindeer herding and traditional harvesting as the railway would cross right through the center of a major reindeer herding area. Governments have made public comments about engaging the Sami in the planning, but the project reflects a long-standing national government belief that investments in regional infrastructure will spark the development of previously orphaned natural resources and promote prosperity in the area. The railway initiative has attracted its sceptics, many claiming that the project's business case does not work, but the Government of Finland seemed determine to press ahead.

Northern Finland more closely approximates the socioeconomic conditions in northern Canada and Alaska than those in Norway and Sweden, albeit with a generally higher standard of living and better-quality northern infrastructure. Finland has a Sami Parliament, but the Sami have few protections and attract less political support than in Norway and Sweden. This is, in large measure, a function of the small size of the Sami population and their comparative isolation in Lapland. As across the region, northern Finland fits with the country's self-image for outdoor heartiness and attracts considerable interest from hunters, fishers, campers, and visitors from the South. The region lacks the higher-level prosperity of northern Sweden and northern Norway, and public investments, while strong in Circumpolar terms, lag well behind the other two northern European countries. The Finnish-Norwegian Arctic Railway represents a national effort to boost the northern economy, albeit in a fairly traditional, resource-focused manner. Oulu, in contrast, rivals Lulua and Tromsø in its commitment to the new economy and to the development of a more contemporary, technology-based order.

SAMI IN NORTHERN EUROPE

The Sami traditional territories cover a large portion of Northern Norway, Sweden, and Finland, but the 80,000–100,000 people are outnumbered by the many non-Indigenous people in the region. In general, the Sami dominate the rural and small-town areas, many of which are predominantly Sami, with significant traditional language use and large commitments to Indigenous cultures. The prominent use of Sami symbols—flag, songs, traditional clothing—have become ubiquitous throughout the region, markers of growing Sami assertiveness and self-determinations. On a national and international scale, the Sami are at the vanguard of the regional environmental movement and lead northern debates about mining, climate change, fishing, and infrastructure development. The national Sami groups have individually and collectively played a significant role in international Indigenous affairs, often with the active engagement of the governments of Norway and Sweden.

Across all three countries, land rights and reindeer herding requirements remain the focus for Sami political engagement. Because of their long-standing use of traditional lands and the unique character of reindeer herding with its need for large grazing areas, Sami seek to defend both individual and collective rights to pasture and transit lands. They become particularly active when facing the prospect of road, railway, or hydroelectric development, for each of these infrastructure projects can bisect important migratory pathways. The land debates remain formidable, with the national governments asserting control over the territories. Major debates have emerged about proposed mining projects, which governments have pushed aggressively to promote generational national prosperity.

Defining Indigenous rights in the region has proven difficult. The Government of Finland requires that Sami claimants prove historical ownership of their lands, a high bar that has resulted in the loss of Indigenous territories. The situation in Sweden is similar, making it difficult to secure recognition of Sami rights. Governments and corporate proponents are expected to consult Sami before proceeding with development, but the legal backing behind these requirements are comparatively slim. This stands in sharp contrast to the situation in Canada, where official "duty to consult and accommodate" requirements now ensure that First Nations and Inuit have a substantial role in resource development. As in the rest of the Circumpolar world, Indigenous people fight for their rights both because of the fundamental importance of being recognized as the traditional owners of their lands and because the rights can, if implemented, provide a means of protecting the most essential elements of Sami culture, including language, values, spirituality, custom, and governance.

The Nordic Sami Convention, signed in January 2017, stands as one of the most important statements of Indigenous aspirations and accomplishment in recent decades. The agreement concluded after 11 years of often difficult negotiations, and remains a work in progress in terms of implementation. The Convention covers the Sami in all three countries and is notable for its extranational priorities and consensus. It represents, in sum, a collective effort by Indigenous groups and three national governments to reconcile the aspirations and heritage of the Sami with the sovereignty and political authority of Norway, Sweden, and Finland (Carstens 2016). It is important to consider the preamble to this vital document[1]:

NORDIC SAAMI CONVENTION
The Governments of Finland, Norway, and Sweden, affirming

- that the Saami is the indigenous people of the three countries,
- that the Saami is one people residing across national borders,
- that the Saami people has its own culture, its own society, its own history, its own traditions, its own language, its own livelihoods, and its own visions of the future,
- that the three states have a national as well as an international responsibility to provide adequate conditions for the Saami culture and society,
- that the Saami people has the right of self-determination,
- that the Saami people's culture and society constitutes an enrichment to the countries' collected cultures and societies,
- that the Saami people has a particular need to develop its society across national borders,
- that lands and waters constitute the foundation for the Saami culture and that hence, the Saami must have access to such,
- and that, in determining the legal status of the Saami people, particular regard shall be paid to the fact that during the course of history the Saami have not been treated as a people of equal value, and have thus been subjected to injustice, that take as a basis for their deliberations that the Saami parliaments in the three states,
- want to build a better future for the life and culture of the Saami people,
- hold the vision that the national boundaries of the states shall not obstruct the community of the Saami people and Saami individuals,
- view a new Saami convention as a renewal and a development of Saami rights established through historical use of land that were codified in the Lapp Codicil of 1751,

---

[1] For the text of the Nordic Sami Convention, see https://www.sametinget.se/105173.

- emphasize the importance of respecting the right of self-determination that the Saami enjoy as a people,
- particularly emphasise that the Saami have rights to the land and water areas that constitutes the Saami people's historical homeland, as well as to natural resources in those,
- maintain that the traditional knowledge and traditional cultural expressions of the Saami people, integrated with the people's use of natural resources, constitutes a part of the Saami culture,
- hold that increased consideration shall be given to the role of Saami women as custodians of traditions in the Saami society, including when appointing representatives to public bodies,
- want that the Saami shall live as one people within the three states,
- emphasize the Saami people's aspiration, wish, and right to take responsibility for the development of its own future,
- and will assert the Saami people's rights and freedoms in accordance with international human rights law and other international law, that have elaborated this convention in close cooperation with representatives of the Saami, deeming it to be of particular importance that the Convention, before being ratified by the states, be approved by the three Saami parliaments and that commit themselves to secure the future of the Saami people in accordance with this convention, have agreed on the following Nordic Saami Convention.

The agreement, which remains to be implemented and is the subject of intense debate, has the potential to set Europe's North on a new course, one based on cooperation and engagement with the Sami and with an unprecedented recognition of the extra-national rights and needs of Indigenous peoples. While the Convention is an important achievement for the governments of Norway, Sweden, and Finland, it is even more a testament to the persistence and political skill of the Sami, who have capitalized on growing global interest in Indigenous affairs, as shown in the United Nations Declaration on the Rights of Indigenous Peoples of 2007.

ARCTIC STRATEGIES IN THE EUROPEAN NORTH

National governments in the Circumpolar World have strategies for the development of their northern regions. The Arctic Council played a significant role in sparking the development of these national plans for the Arctic. The Norwegians jumped in quickly in 2005, declaring the Arctic to be the country's highest foreign policy objective. Finland, which announced its priorities in 2010, emphasized Arctic cooperation, in part to overcome the fact that, like Sweden, Finland did not have an Arctic Ocean coastline. In some nations and subnational jurisdictions, these government policies merge into comprehensive planning documents that lay out strategies for northern socioeconomic and cultural development. The nations of the European North emphasize state planning and, even if they shifted to a somewhat more neoliberal economic model, they would continue to take a thoughtful, future-oriented approach to the development of the North. Each of the countries of the European North has a strategy for northern development that reveal both common patterns and national differences, in each instance holding to the pattern of being the most engaged and forward-looking countries of the Circumpolar World.

Norway's Northern Strategy

Norway has been among the most progressive and engaged of the Arctic nations, eager to take the lead on Arctic academic activities and on the management of the Arctic Council, which is headquartered in Tromsø. The country has 10% of its population within the Arctic Circle, which is the highest percentage in the world. In a 2014 report on northern priorities, the government announced a focus on international cooperation, business, infrastructure and knowledge development, and environmental protection and emergency preparedness (Government of Norway 2014). In its April 2017 Arctic Strategy, the government stated:

> The Arctic is important for Norway and for the world as a whole. The Government's vision is for the Arctic to be a peaceful, innovative and sustainable region where international cooperation and respect for the principles of international law are the norm. Foreign and domestic policy are intertwined in the region, and people's everyday lives are affected both by high politic and by day-to-day issues. (Government of Norway 2017)

The country has the most optimistic and aggressive Arctic policy environment in the world, reflecting the strength of the oil and gas industry, a vibrant seafood sector, a strong and assertive Sami population, and a robust science and technology ecosystem developing around the University of Tromsø. The government's Arctic Strategy is bold and forward-looking:

> The Government aims to make North Norway one of the most innovative and sustainable regions in the country. We will create economic growth and future-oriented jobs in the north in a way that takes account of environmental and social considerations. We will build local communities that can attract people of different ages and genders, and with different skills and expertise. Areas such as education, business development and infrastructure are vital in our efforts to build a sustainable region. In this strategy, the Government has sought to give greater consideration to the domestic aspects of Norway's Arctic policies. Well-functioning communities are built by the people who live and work there. In the development of North Norway, it is the region's own citizens, companies and politicians that have the most important role to play. (Government of Norway, p. 3)

The Norwegian current strategy is based on five core themes:

> International Cooperation: Norway plans to build on their high level of Circumpolar engagement and to continue their effort to develop a sustainable,

high quality economy across the region. They intend to encourage environmental and, importantly, social sustainability. The country favours an integrated approach to Arctic policy development, with a strong emphasis on international cooperation and "green growth." The government emphasizes international cooperation and intraregional commitments to sustainable development. It has made long-term commitments to cooperative organizations, including the Arctic Council (including active participation in the Emergency Prevention, Preparedness and Response Working Group), enhanced relations with Russia and the Barents Region, continued active membership in NATO, including stronger Norwegian armed forces, and cooperation with other northern nations and the European Union through such groups as the Nordic Council of Ministers, work on the UN Convention of the Law of the Sea and the Oslo Declaration on high seas fishing.

Business Development: Norway clearly believes the future of the Arctic lies in building beyond the current foundation of resource development. This included a greater emphasis on the "blue economy," or the strengthening of the ocean-based industries of the North, adding value to current resources and encouraging greater collaboration between scientists. The expansion of its Arctic economy, the government asserted, should be undertaken while supporting Sami aspirations and protecting the Arctic environment. The tourism industry is clearly viewed as a central element in North Norway's economic future, with a particular effort to attract travellers in the winter months.

Education and Research: The Government of Norway sees education and research as integral to the economic and social future of the Far North. There have been formidable and ongoing commitments to the improvement of educational quality and outcomes, North-centred research activities and specializations, and a tighter integration of academic research and northern business development. Concerns about gaps in the northern workforce require, the government asserted, a more focused effort to match education and training with the demand for skilled labour. Their commitment included plans for expanded ocean research capabilities, the inclusion of Sami initiatives in the skills training programs, and a greater level of cooperation with the Sami parliament (Samediggi).

Infrastructure: Norway has made extensive investments in northern infrastructure and is committed to maintaining the roads, bridges, airfields, and the electrical supply at a high and reliable level. Furthermore, they indicated additional investments (NOK 40 billion for the 2018–2029 period), improvements in planning and approval processes, and a particular emphasis on digital connectivity through the Internet, satellite systems, Arctic communications. Norway planned to integrate its infrastructure plans with those of Russia, Finland and Sweden and to continue with the Barents Euro-Arctic Transport Area. While expanding these investments, the Government promised to pay close attention to climate change and broader environmental considerations.

Environmental Protection and Responsiveness: Not surprisingly, the Government of Norway made strong commitments to environmental protection and the improvement of Arctic safety and emergency response. They committed to international strategies, continued support for Svalbard protections, and greater attention to the oversight of the Norwegian Sea and Barents Sea. In keeping with its multi-front devotion to international collaboration, the Government supported the work of the International Maritime Organization and the Arctic

Council, particularly in Arctic environmental oversight and pollution response. Substantial investments were planned in oil spill response and regional ocean management systems.

Rich in investment, bold in scope, and comprehensive in intent, Norway's Arctic policy makes a formidable commitment to maintaining the trajectory of the Arctic's most successful region. The Sami are less than confident, it seems, about the Arctic Strategy and are, in particular, concerned about plans for expanded mineral development. And the strategy contains an obvious conflict between pro-development strategies, commitments to environmental protection, and support for Indigenous rights. Norway clearly intends to build on its active engagement in Circumpolar affairs and to back up its Arctic policy statements with substantial and sustained investments. Perhaps most importantly, Norway's strategy does not take a "poverty approach" to the Arctic, but rather focuses on a positive, expansive, "new economy" plan that highlights advanced education and training and that gives high priority to international competitiveness and innovation.

Sweden's Northern Strategy

Sweden laid out a comprehensive Arctic strategy, including a first, extended statement in 2011. As the Government declared:

> The Arctic region is in a process of far-reaching change. Climate change is creating new challenges, but also opportunities, on which Sweden must take a position and exert an influence. New conditions are emerging for shipping, hunting, fishing, trade, and energy extraction, and alongside this, new needs are arising for an efficient infrastructure. New types of cross-border flows will develop. This will lead state and commercial actors to increase their presence, which will result in new relationships. Moreover, deeper Nordic and European cooperation means that Sweden is increasingly affected by other countries' policies and priorities in the Arctic. It is in Sweden's interests that new emerging activities are governed by common and robust regulatory frameworks and above all that they focus on environmental sustainability. (Government of Sweden 2011, p. 4)

Sweden's cautious and historically based policy statement emphasized its place between Norway and Finland and its lengthy cultural, commercial, research, and other ties across the region. It emphasized Arctic cooperation, with the Arctic Council, the European Union and the Sami, among others. The Sweden strategy had a lengthy list of priorities, from the climate and the environment to economic development, education (with an emphasis on human factors), and a complex set of issues relating to the Sami, including language preservation, and Indigenous industries.

The strategy called for enhanced efforts on climate change and the protection of biodiversity and continued commitments to northern economic development, "albeit with consideration for the environment and the traditional lifestyles of indigenous peoples" (Government of Sweden 2011, p. 6). Indeed, the country declared an intention "to bring the human dimension and the gender perspective to the fore in Arctic-related cooperation bodies." This statement included, earlier and more forcefully than other Arctic nations, a deep concern about climate change and the need for greatly expanded international cooperation, a position shared by Finland. Sweden also emphasized the global standing of its northern industries, including export-oriented sectors like mining and forest products and locally important fields like reindeer herding, hunting, and fishing. In particular, Sweden emphasized its northern-centric commercial capabilities in construction, shipping, tourism, and resource extraction.

In 2016, Sweden released an environmental a strategy for the Arctic. As the government declared, "The Arctic acts as the planet's refrigerator. Its enormous white expanses of ice and snow reflect large parts of the sun's rays back into space, thus stabilizing the Earth's climate. The area is home to millions of people, including Indigenous peoples, and ecosystems of great importance. The Arctic environment and its fate concerns us all, directly or indirectly" (Polar Connection 2016). The government called for greater protection against oil spills, conservation of vulnerable Arctic ecosystems, and controls on fishing.

Sweden has not released an extended Arctic strategy since 2011, but not because of a lack of interest in northern affairs. The country assumed a prominent role in the Arctic Council and, like Norway and Finland, made major commitments to regional integration and general Arctic cooperation. More than Finland and Norway, Sweden is alert to the fact that it is not an Arctic coastal state and therefore has a somewhat lesser role in the Arctic. The country's strategy for its northern regions and for the Arctic as whole focused on an attempt to balance economic development and environmental protection, ensure protection for Sami interests, and maintained a strong international role.

FINLAND'S NORTHERN STRATEGY

The Government of Finland laid out an aggressive strategy for Arctic development in 2013, with strategic updates produced in 2016 and 2017. The strategy begins with an admission of the barriers facing the region: "Finland is an active Arctic actor with the ability to reconcile the limitations imposed and business opportunities provided by the Arctic environment in a sustainable manner while drawing upon international cooperation" (Government of Finland 2013, p. 7). In adopting a strategy that the government declared to be "more wide-reaching in scope," the authorities also committed themselves to "intensified efforts" to develop and implement policies and programs to achieve the stated objectives. Unlike Norway (a non-European Union country), Finland planned to capitalize on EU funding and program support as appropriate. Specifically, the Government identified four priority areas:

- Finland as an Arctic Country
- The Development of Arctic Expertise
- Sustainable Development and Environmental Considerations
- International Cooperation

The country, furthermore, pledged itself to active engagement in international initiatives, tracing its engagement in Arctic cooperation to Ministerial-level meetings held in 1991 and re-enforced by the Northern Dimension Policy of the European Union, adapted six years later. Significantly, Finland declared that the whole country was engaged with the North:

> For the Finnish economy, the Arctic region represents a growth market close to home where Finland enjoys a natural edge to be active and succeed. This is an area where Finland's geographical, cultural and competence-based advantages come to the fore. However, success calls for long-term, visionary cooperation and close networking between the authorities and private companies both at the national and international levels. (Government of Finland 2013, p. 8)

Finland's 2013 strategic document drew attention to areas of expertise and continued Circumpolar leadership: Arctic shipping and winter navigation, sustainable mining advanced energy systems. It also called for greater cooperation with Russia and Norway relating to the Barents region, potential contributions to the opening of the North-East Passage, and the possible development of additional transportation routes through Lapland. It emphasized the need for attention to security considerations, search and rescue capacities, and other standard Arctic considerations.

Like Norway and Sweden, Finland signaled a concern about environmental protection, improved planning, and careful management and stewardship of the Arctic's resources. The Government also highlighted the country's innovative track-record in forestry management and harvesting. Finland's plans called for special attention to the needs of northern peoples, including the Sami, who enjoy constitutional protection in the country but whose language and culture required additional support. The country wished to capitalize on its advanced standing in digital media and telecommunications to improve services to the Far North, and offered additional commitments to the development of northern tourism, the protection of Sami culture and wider-ranging support for environmental protection. Importantly, the country called for a solution "conducive to a good quality of life and specifically tailored for northern conditions. One such solution is Arctic design, which refers to design that draws upon an understanding of the Arctic environment and circumstances, while giving due consideration to the peoples' adaptation to Arctic conditions" (Government of Finland 2013, p. 11).

Finland's Arctic plan called for substantial investments in research, education, and training, building on a widely celebrated educational system. The country emphasized the work of the Arctic Centre, University of Lapland in supporting the EU Arctic Information Centre. Much of the research and training effort was targeted at environmental considerations, emphasizing a pan-Arctic concern about climate change and greenhouse gases and the expansion of ecologically sensitive conservation efforts. The country committed itself, further, to active engagement in the Arctic Council, the Barents Euro-Arctic Council, the EU's Northern Dimension Policy and enhanced engagement with Russia. As the strategy indicates: International cooperation in the Arctic is an essential element of Finnish foreign policy. Increasing attention is being paid to Finland's Arctic role in the context of diplomacy and the efforts to build up the country brand by making use of the Team Finland approach, among others. Finland's objective is to bolster its position as an Arctic country and to reinforce international Arctic cooperation (Government of Finland 2013, p. 14).

Finland's Arctic Strategy was updated in 2016 and again in 2017, in both instances stating the goal of asserting Finland's leadership in Circumpolar affairs. The 2016 statement re-enforced the nation's commitment to research, increased emphasis on sustainable tourism and Arctic travel, and a steady improvement of northern infrastructure, including wireless services. The following year, Finland restated its priorities, focusing on the European Union and continued multilateral collaboration, greater investment in sustainable tourism and global marketing of the Arctic, additional infrastructure investments, and greater emphasis on the development of Arctic-related businesses, focusing on the marine industries. Finland prioritized the "broker function" that "facilitates matching the needs of international organizations and Finnish know-how" (Government of Finland 2017, p. 4).

Examining Arctic Strategies of the European North

In the end, the Arctic strategies of Norway, Sweden, and Finland have proven non-controversial and predictable. This is due, in part, to the broad commitment of all three nations to providing high quality public services and infrastructure to all citizens and all areas. Fields that are controversial in other Arctic nations—housing, education, water supplies, and the Internet—are integrated into broad national priorities. Other Arctic commitments, particularly to regional cooperation, reflect crucial regional priorities and vulnerabilities. Geographic location creates opportunities and challenges that have to be addressed through policy initiatives and cooperative ventures.

An interesting study by Vincent-Gregor Schulze compares the Arctic strategies of all Circumpolar nations. Schulze identifies areas of high, medium, and low national priority and includes fields of "no relevance" to Arctic policy in the country. The following two tables indicate how the Northern European countries compare with reference to their Arctic strategies (Tables 18.1 and 18.2).

Schulze's analysis is interesting. Norway is less emphatic about Indigenous issues than Sweden and Finland and stronger than the others on issues related to the military, oil and gas, fishing, shipping, and research. Compared to the other two countries, Finland is strong on mining, regional development and

Table 18.1   Arctic strategy priorities I

	Research	Environment	Tech and innovation	Education	Regional dev.	Int. law	Indigenous peoples
Finland	X	XXX	XXX	XXX	XXX	X	XXX
Norway	XXX	XX	XXX	XX	XX	XX	X
Sweden	X	XXX	XXX	XXX	X	XXX	XXX

X, low priority; XX, medium priority; XXX, high priority

Source: Vincent-Gregor Schulze, Arctic Strategies Round-up 2017. http://www.arctic-office.de/en/in-ocus/arctic-strategies-round-up/. October 2017

Table 18.2   Arctic strategy priorities II

	Infrastructure	Transport	Shipping	Search and rescue	Fish	Tourism	Oil and gas	Mining	Military
Finland	XXX	XXX	X	XX	X	XXX	X	XXX	X
Norway	XXX	XXX	XXX	XXX	XXX	XXX	XXX	X	XXX
Sweden	XX	XXX	X	XXX	XX	XXX	XX	XXX	X

X, low priority; XX, medium priority; XXX, high priority

Source: Vincent-Gregor Schulze, Arctic Strategies Round-up 2017. http://www.arctic-office.de/en/in-ocus/arctic-strategies-round-up/. October 2017

less engaged on shipping, fishing, oil and gas, the military, research and international law. Sweden, in comparison to Norway and Finland, is weaker on infrastructure, regional development and, like Finland, low on research, shipping and the military. The country has comparatively strong commitments to mining, international and Indigenous peoples. Areas of lesser priority, like infrastructure in Sweden, do not indicate neglect but rather the already substantial national commitments in the area. Some of these observations are logical. Norway is much more engaged in shipping, fishing, and oil and gas than the other two countries. All three are strongly committed to transportation, tourism, and technology and innovation.

Looked as a whole, the Arctic strategies of Finland, Sweden, and Norway reflect strong commitments to social well-being and equality of services and infrastructure, a desire to improve or regularize relations with Russia, concern for climate change, and a belief that the northern economies can be developed substantially. All three nations are deeply engaged internationally, and strongly favor participation in the Arctic Council and various regional forums. They support the enhancement of Sami culture and well-being but stop short of strong commitments to Indigenous rights and giving the Sami a stronger presence in national decision-making. The strategies reflect a mild, but not overwhelming, sense of urgency around climate change and a general confidence about economic, political, and social development. The three Northern European nations are, moreover, strong and even assertive about their role in the Circumpolar World. Without overstating their international significance, Norway, Sweden, and Finland make it clear they are aware of the need for Arctic leadership and, equally, Circumpolar responsibility.

NORTHERN EUROPE, CIRCUMPOLAR ENGAGEMENT, AND THE ARCTIC COUNCIL

The intensity of Circumpolar internationalism in recent years has largely obscured the quiescence of Far Northern collaboration in the proceeding decades. Early connections generally did not include the Indigenous peoples of the European North. The Arctic Winter Games, first held in Yellowknife, NWT, in 1970, initially attracted participants from the Yukon, Northwest Territories and Alaska, with Greenland joining subsequently. The Inuit in the Canadian North, Alaska and Greenland played a major role in building circumpolar connections beginning in the 1970s with the creation of the Inuit Circumpolar Council in 1977. Later attempts to build Circumpolar connections expanded to include the USSR/Russia, with relations with the European North developing more slowly.

The Sami were active in international Indigenous affairs, but they worked more broadly than the Circumpolar World. Building off the connections launched with the Alta Dam protests in Norway, the Sami subsequently played a significant role in the formation of the World Congress of Indigenous Peoples, participating in the exploratory meeting in 1974. Representatives from Norway, Sweden, and Finland attended the first formal meeting the following year. Arctic connections emerged over time, through engagement among academics, business people, and government officials. Because of the striking dissimilarities between the European North and Siberia, Arctic Canada, and most of Alaska, it was not immediately evident that there was common political cause among the various Arctic states.

Discussions continued culminating in the establishment of the Arctic Council following the 1996 Ottawa Declaration. Norway, Finland, and Sweden became active members, with each of the countries taking their turn as Chair of the Council. In their role as Chairs, each Circumpolar nation sets the agenda for their two-year leadership period, typically using the opportunity to introduce topics of national concern to international partners. In 2000–2002, Finland used its first time in the Chair to enhance the Council's international profile, build stronger ties with the European Union, focus on the Arctic environment, build interest in Arctic research, economic and social development, and promote intraregional cooperation. Norway's turn came in 2006–2009, and it joined with Denmark and Sweden to promote longer-term attention, spanning three Chair-ships, to climate change, integrated resource management, the International Polar Year, and Indigenous peoples and local living conditions: When Sweden took the chair in 2011–2013, the government prioritized work on the environment, including oil emissions, climate change,

Arctic resilience, biodiversity, and environmental protection, and focused on how to create a stronger Arctic Council. Finland moved back into the Chair in 2017–2019, encouraging Circumpolar discussion of environmental protection, connectivity, meteorological cooperation, and education. The Arctic Council provided the Northern European nations with a stable and effective forum for engaging on Circumpolar issues, although the institution operated within narrow, non-legislative parameters that attracted more diplomatic engagement than public interest.

The European nations grew quickly into their roles within the Arctic Council and Arctic diplomacy generally. Norway, in particular, continues to step forward, assuming substantial leadership roles in a variety of Arctic initiatives. By establishing the headquarters of the Arctic Council in Tromsø, Norway re-enforced its pivotal role in Arctic diplomatic and academic affairs. Likewise, Northern Europe is a major contributor to such important collaborations as the University of the Arctic, an international collaboration of dozens of institutions across the Circumpolar world and in southern centers. Indeed, over the past decade Norway, Sweden, and Finland have made substantial fiscal contributions and provided vital logistical support for Circumpolar initiatives.

The attention to Circumpolar collaboration has important intraregional dimensions as well. Cooperation between Norway and Finland, for example, on the railway to the Barents Seas seeks to capitalize on shared and overlapping interests. Regional collaboration in research and advanced education continues to expand, as do formal and informal collaboration between northern businesses, particularly but not exclusively in tourism. Arctic Airlink opened in 2015, connecting Oulu, Luleå, and Tromsø, a five-day-a-week service designed to shorten travel times between key northern cities and to build East-West connections, challenging the North-South connections between governments and businesses that harmed the development of Circumpolar connections. The service close in 2018 when the service provider, NextJet, declared bankruptcy and no replacement could be found. Government and academic collaborations have expanded, with Norway, Sweden, and Finland participating enthusiastically in an ongoing series of international conferences and meetings, with considerable emphasis placed on student exchange and collaborative Circumpolar research programs.

Policy Options and Priorities in the European North

Norway, Finland, and Sweden have embraced their role as Arctic nations and their places within the Circumpolar world. Compared to most other Circumpolar nations, the people and societies of the European North experience stronger economies, better government support, and better social and cultural outcomes. All three governments maintain excellent northern infrastructure and provide official services throughout the region. Authorities in the three countries take the North seriously and their northern strategies are replete with fulsome commitments to improving the quality of Arctic life. It is important that the governments not devote too much effort to elements with limited chance of success, such as building East-West connections across the Arctic.

On the Indigenous legal-political front, the Sami have substantial and well-supported parliaments in each country, which helped coordinate Indigenous action and perspectives on policy issues. While they have impressive institutions, they do not have the legal and constitutional recognition of land and harvesting rights that have been established under Canadian and American laws. They have, as a result, had mixed results delaying or preventing resource projects or securing substantial financial and employment returns from such undertakings. They have maintained a sustained and even ferocious defense of reindeer herding, which has re-enforced their land and resource rights and facilitated careful consideration of cross-border Sami rights. With government support and at Sami instigation, the Indigenous people of the European North have developed extensive language and cultural programs, post-secondary institutions, and major Sami research projects.

As the nations and peoples of the European North define a path forward on the policy front, it is important that they move beyond current programs and commitments and continue an undeclared effort to keep the region at the forefront of Arctic policy development and the promotion of the interests of the Circumpolar world. This said, Norway, Finland, and Sweden need to appreciate and respect their uniqueness. It is important that the countries understand that the differences between the European North and other jurisdictions are real and substantial. The path forward for the three countries shares some bold challenges with the rest of the Circumpolar world—cold, long winters, darkness, isolation, political marginalization, prominent indigenous populations, and resource-dependent economies. Overall, however, the policy framework for the European North will stand alone, overlapping and intersecting with the environment in other countries only occasionally and in limited measures. Among the many options, priorities and policy opportunities before the governments of the Circumpolar world, six stand out:

A North-Specific Climate and Energy Plan: Arctic regions are extremely vulnerable to climate change, but they are not the cause of the dramatic and potentially dangerous ecological transitions. Watching the North go through expensive gyrations to conform to southern and urban standards for energy use is value-signalling and symbolic, but it is not particularly realistic. Efforts should be made to cut energy use and, if possible, costs. But this need not be a regional obsession. Northern life is energy intensive. The distances are long, the winters cold, and the need for fossil fuels largely unavoidable in the short-term. Northern Europe has, overall, cheap electrical power and, off the coasts of Norway and Alaska, world-class oil and gas deposits. The region's focus should be on environmentally sensitive buildings and a continuation of the North's pattern of innovative urban design, focusing on northern conditions and realities. The rest of the world caused the global climatic crisis that threatens the ecological integrity of the

Arctic. It is up to the rest of the world to solve the challenge—and not to add to the North's burdens.

Defining and Operationalizing Sami/Indigenous Rights: The European North has made significant advances in representative institutions for Indigenous peoples, but they lag in many other respects. Norway and Sweden both recognize ILO Convention 169 and speak positively about the United Nations Declaration on the Rights of Indigenous Peoples; Finland is less engaged. Across the region, the legal protection of Indigenous land and harvesting rights leave much to be desired and the processes of consultation and engagement on development issues lag behind Canada, Greenland, and Alaska. The key lesson from other Arctic jurisdictions is that the empowerment of Indigenous peoples in the resource field and their active engagement in regulatory decisions and business operations produces better environmental and economic outcomes for the region.

Keeping and Sustaining Youth People in the North: The "brain drain" from North to South in Northern Europe is offset by a highly significant "brain gain" connected to the impressive northern universities and the new economy businesses operating in Oulu, Luleå, Tromsø, and Bobo. Sami and non-Sami alike worry about the steady migration of young adults to the South and the subsequent challenges for northern regions. Concerted efforts are being made, including the promotion of young entrepreneurs, extensive youth programming focused on culture and regional pride, and an expansion of northern academic and training programs. As other northern regions have discovered, the major issue is not the movement of young adults to southern latitudes but, instead, the question of whether or not they return to the Arctic to build a family, career, and community.

Capitalizing on the North's Economy of Scale and the Commercial Value of Remoteness: In many commercial areas, the North lacks standard economies of scale. Markets are small and distances to urban markets often substantial. Northern Europe has real advantages compared to other parts of the Arctic, including well-developed infrastructure (road, rail and air services), excellent energy systems, and very good training facilities. In several areas, including mining, forestry, oil and gas, and Arctic shipping, the region's businesses have world-leading capabilities. The small size of the northern population makes it comparatively easy to collaborate within the northern parts of Norway, Finland, and Sweden, and across national borders, following the pattern set by the Sami. Because Northern Europe is advanced in many technical and commercial respects, and because of the strength of the region's post-secondary institutions, Northern Europe has the potential to build beyond Tromsø's credible boast to be "the capital of the Arctic" to developing globally significant businesses connected to the challenges of working and living in remote regions.

Deciding the Future of Small Arctic Towns: Across the world, most rural areas and small towns are facing precipitous declines. In the age of rapid urbanization, the demographic and commercial vulnerabilities of small Arctic towns lay exposed. Because of Northern Europe's commitment to inter-regional equality and because government services and infrastructure are uniformly strong across the North, the small towns and rural areas are among the most stable and prosperous in the world. The Arctic regions are not immune to the pressures of demographic and geographic change, and outmigration remains a serious concern. Norway, Sweden, and Finland have succeeded more than other Circumpolar areas in stabilizing the population, supporting technological transitions and laying the foundation for new economy competitiveness. Not all the communities will thrive, particularly those suffering from the closure of a mine, industrial plant, pulp mill, or sawmill. The protections of the welfare state permit declining communities to either close or reinvent themselves, giving the region's communities a chance to find a new equilibrium. A thoughtful, staged approach to the redesign, expansion and, in some instances, the closure of small Arctic towns provides the region with an opportunity to adapt and respond to new realities.

Northern Europe and the Re-Definition of the Global North

The European North has adapted well to the realities, opportunities and challenges of the twenty-first century. A quarter of a century ago, the region did not have a strong presence on the Circumpolar stage. That is no longer the case. Arctic Europe is arguable the most dynamic, focused, and innovative part of the Far North, with impressive achievements in everything from Sami revitalization and high-technology resource development to northern entrepreneurship and advanced education. The region has an opportunity to redefine the global North, creating a more inclusive and approach definition of the Arctic.

At present, the definition of the Arctic includes some oddities. By several criteria—winter temperatures and general climate, the northward extension of agriculture—significant parts of Northern Europe do not seem particularly "Arctic." Bodo, Norway, has much more in common with Prince Rupert, British Columbia, than Nome, Alaska, or Iqaluit, Nunavut. Rovaniemi is closer in socioeconomic circumstances to Prince George, British Columbia than Fairbanks, Alaska. Tromsø, one of the most compelling cities in the Far North, does not share much in common with Cambridge Bay, Nunavut, or Barrow, Alaska. Whitehorse, Yukon is included in the Canadian definition of the Far North but it is Sub-Arctic in character; Churchill, Manitoba and Thunder Bay, Ontario, which are quite northern by standard definitions, are not considered to be truly Arctic.

There is a vast area—the global Sub-Arctic or the "North Below the North"—that lacks the profile of the Far North. The Sub-Arctic gets much less attention from governments and is often left out of discussions about the future of the region. By most definitions of the North and the Arctic, much of the European North is environmentally Sub-Arctic. There are important reasons for the expansion of the definition of the North. The Sub-Arctic has a substantially larger population than the Arctic. Much of the North's resource wealth is in the Sub-Arctic—from the oil sands in northern Alberta to the vast forests of Sub-Arctic Russia, the Ring of Fire mineral deposits in northern Ontario to the hydroelectric resources of northern Norway and northern Sweden and northern Quebec.

The Sub-Arctic lacks political focus and international standing. The Arctic Council is deliberately—and appropriately—focused on High Arctic issues, peoples, and priorities. But the Sub-Arctic regions typically struggle for attention within their respective nations, have few international connections, and limited global profile. The European North is the major exception, in large measure because of its connections to the Far North and the Arctic Council. Northern Norway, Finland, and Sweden are world leaders in advancing the cause, constructively and collaboratively, of Sub-Arctic Europe. The European North has an opportunity to extend that influence by drawing in other Sub-Arctic regions—the Russian and Canadian North are the best examples—and creating global interest in the Sub-Arctic that could, in time, parallel the emergence of the Arctic as a major force in international affairs over the past 30 years.

CONCLUSION

In many important ways, Europe's North has made vital strides both in drawing northern and Arctic regions into the mainstream of national affairs and engaging with the broader socioeconomic, cultural, and political currents of the Circumpolar world. Many of the far North's most impressive institutions—the Arctic Centre in Rovaniemi, the Svalbard scientific complex, the University of Tromsø—Norway's Arctic University, and network of major commercial operations, museums, and scientific operations—are in Norway, Sweden, or Finland. With comparatively minor exceptions, the northern regions of the three countries have good infrastructure and public services, sizeable investments in regional economic development and sustained government help in responding to the special needs and opportunities of the Arctic. Europe's North takes the Arctic seriously, and in all three countries northern, winter, and remote regions are internalized in the national consciousness.

In Norway, Finland, and Sweden, northern districts do not have particular difficulty attracting the attention of national governments. Sami affairs continue to require attention, and all three governments appear to pull back from prioritizing Indigenous harvesting, land, and resource rights. There is substantial public and private investment in the region and strong Indigenous and non-Indigenous commitment to the North. The challenges before the region and its residents—cold, remoteness, climate change, energy, and sustainable economic activity—are commonplace across the Arctic, with the substantial difference that the governments of Norway, Finland, and Sweden are actively involved in trying to address regional needs and aspirations in a timely fashion. Europe's North is leading the Arctic in broad socioeconomic outcomes; the region's leadership in accomplishment has, over the past two decades, been matched by the countries' desire to play a critical role in shaping the defining the future of the Arctic.

REFERENCES

Carstens, Margaret. 2016. Sami Land Rights: The Anaya Report and the Nordic Sami Convention. *Journal on Ethnopolitics and Minority Issues in Europe* 15 (1): 75–116.

Government of Finland. 2013. *Finland's Strategy for the Arctic Region 2013*. Helsinki: Prime Minister's Office Publications.

———. 2017. *Action Plan for the Update of the Arctic Strategy*. Helsinki: Prime Minister's Office Publications.

Government of Norway. 2014. *Norway's Arctic Policy*. Oslo: Government of Norway. https://www.regjeringen.no/en/documenter/nordkloden/id2076193/.

———. 2017. *Arctic Strategy*. Oslo: Government of Norway. https://www.regjeringen.no/en/dokumenter.arctic-strategy/id2550081/.

Government of Sweden. 2011. *Sweden's Strategy for the Arctic Region*. Stockholm: Government Offices of Sweden.

Nordic Sami Convention, see https://www.sametinget.se/105173.

Polar Connection. New Swedish Environmental Policy for the Arctic. *The Polar Connection*, 10 May 2016.

Schulze, Vincent-Gregor, and Arctic Strategies Round-up. 2017, October. http://www.arctic-office.de/en/in-ocus/arctic-strategies-round-up/.

PART IV

The Arctic and International Relations

CHAPTER 19

The Arctic in International Affairs

Heather Exner-Pirot

Introduction

It is only in the last 20 years that the Arctic has become a subject, and not merely an object, in international affairs. Prior to the fall of the USSR, the Arctic was, from a global geopolitical perspective, primarily seen as a military theater. Although its strategic location, nuclear deterrence-wise, means this view will continue to be valid, the region has evolved to have its own unique characteristics, and has carved out its own place and role in international affairs. Chief among these are the region's focus on marine and environmental issues, as well as its privileging of Indigenous, scientific, and other non-state actors. The way the Arctic has evolved as a political region has had a strong impact on states' behavior there, and it is overwhelmingly a cooperative place. This chapter seeks to assess the Arctic in international affairs and the factors that have contributed to its development as a political region.

Above all, the Arctic is a homeland for approximately 4 million, ethnically diverse people, about 10% of whom are Indigenous and about half of whom are Russian. Given this chapter's intellectual orientation in international relations, which privilege the state, local perspectives on the Arctic in international affairs will not be the focus. However, the impact of northern culture and Indigenous self-determination are important features in evaluating the Arctic as an international region and will be discussed.

The Arctic in Foreign Policy: Context and Trends

The novice reader of Arctic politics will likely have strong assumptions about the region and its role in international affairs: climate change is melting the ice,

H. Exner-Pirot (✉)
Observatoire de la politique et la securite de l'Arctique, CIRRICQ, Quebec, Canada
e-mail: heather.exnerpirot@usask.ca

K. S. Coates, C. Holroyd (eds.), *The Palgrave Handbook of Arctic Policy and Politics*, https://doi.org/10.1007/978-3-030-20557-7_19

resources and shipping routes are newly accessible, territory is unsettled, and Russia is resurgent. However, this is a relatively new narrative. The current era of Arctic foreign policy only began in August 2007, when a Russian explorer, acting independently as part of a transnational scientific mission, planted a titanium flag at the sea bed of the North Pole. Almost at the same time then Canadian Prime Minister Stephen Harper announced a series of Arctic sovereignty and defense investments (based on strategic campaign promises, rather than a military-identified need; see Flanagan 2013); and commodities prices, particularly oil and gas, boomed, making the Arctic's offshore oil deposits not only accessible but possibly profitable for the first time. Littoral Arctic states furthermore became more active in preparing their submissions for claiming extended continental shelf on the seabed of the Arctic Ocean, under the provisions set in the 1982 United Nations Convention on the Law of the Sea (UNCLOS). Some have deemed this era in Arctic foreign policy as an "Arctic race" or "Scramble", though this narrative has seemed to run its course given the compelling and continuing evidence of collaboration, rules-based governance, and evidence-based decision-making.

While it is difficult to predict what discourse or narrative will next influence Arctic foreign policy, it is instructive to delineate past ones (see Table 19.1). Human history in the Arctic region goes back millennia. However, if one adopts the Westphalian, state-based era as a starting point for evaluating foreign policy, then it is European expansionism and colonization that first brought the Arctic into traditional international affairs. This initial expansion

Table 19.1   Eras in Arctic foreign policy

Explorer (sixteenth century–1939)	Arctic as a region of discovery, exoticism, foreboding. Missionary activity. Colonization. Transnational trade based, e.g. on fur trade, whaling, fish, and mining.
World War II (1939–1945)	North American Arctic becomes a militarized space: Alaska highway, Aleutian islands; Winter War.
Cold War (1945–1987)	Arctic as a military theater; DEW line.
Indigenous and scientific collaboration (1965–1987)	More formal environmental, Indigenous and scientific linkages develop across borders and exclusive of states; devolution & self-determination; Polar Bear Agreement.
Glasnost and Perestroika (1987–1991)	Gorbachev promotes Arctic as a Zone of Peace; pan-Arctic collaboration intensifies; Rovaniemi Process.
Institutional Development—Post-Cold War (1991–2007)	Arctic Council established; low politics, soft security issues dominate; non-state and local actors have a high level of influence. Human rights and security a priority in the international system.
Arctic Race—Post, Post-Cold War (2007–2018)	Climate change intensifies, commodities boom, shipping, globalization, rise of Asian interest in region, Russian remilitarization, UNCLOS claims.
What's Next? (2019–?)	Policy shaping and policy making; exceptionalism; sustainable economic development; regional innovation. Post-Westphalian, post-sovereignty. The Arctic as a model of regional cooperation; the Arctic as a transnational space.

and state competition were driven by economics, including the fur trade, whaling, fishing, and mining; as well as by religious missions.

World War II and the Cold War more firmly entrenched the Arctic into states' foreign policies due to its strategic location, especially in the nuclear era. The Arctic was not an area merely to map but to occupy and control, and the area become progressively more militarized.

At the same time, the rise of self-determination movements, environmentalism, and the popularization of global travel in the post-World War II era precipitated non-state linkages across the region in Indigenous, scientific, and environmental communities. This has had continuing effects on the nature of Arctic international affairs. Mikhail Gorbachev famously called for the Arctic to become an international "Zone of Peace" in 1987, and when the Soviet Union collapsed in 1989, the Russian half of the Arctic suddenly opened up to international cooperation. An international region was born.

The immediate post-Cold War period saw the Arctic primarily—when it was seen at all—as a domain for scientific research, environmental protection, and the promotion of Indigenous peoples' rights, with the institutionally weak Arctic Council as the primary vehicle for international cooperation. While it is true that the Arctic states, and some non-Arctic states, increased their military and constabulary investments in the Arctic after 2007 (see Huebert et al. 2012), the cooperative nature of the Arctic region not only persisted after 2007 but has arguably been strengthened, propelled by the rise in significance of the region and the Arctic states' common interests in it. These interests are displayed in the series of national strategies for the region that each Arctic state issued between 2009 and 2013 (Heininen 2012) and include security and sovereignty; business and economic development; regional and sustainable development; environmental protection and climate change; safety and Search & Rescue; human dimension and Arctic peoples; research & knowledge; and international cooperation. These common interests are the key to understanding the Arctic region. It is a peaceful and cooperative region not because Arctic states are altruistic and benign north of the 66th parallel, but because they have experienced common challenges and interests since 1991 that require regional-level governance and cooperation.

THE ARCTIC IN GRAND STRATEGY

Environment

The value of the Arctic to global interests has been well delineated in the past decade, repeated faithfully in mainstream and academic articles to the point of becoming truisms. Foremost is the Arctic's role in global climate changes: warming is occurring more rapidly in the Arctic than in any other region, and the resultant melting of sea ice and glaciers, and their impact on sea level and ocean current function, is impacting global weather patterns. There is a widespread sense that keeping the Arctic cool and protected is essential to mitigating the adverse impacts of global climate change (of course, there is no way to disassociate or compartmentalize industrial activities in the South from the environmental repercussions in the North but that is a topic for another paper).

The Arctic has other environmental exceptionalities, as a home for unique species or as an important route or destination for migratory animals, for example; as well as vulnerabilities, such as acting as a sink for Persistent Organic Pollutants and mercury.

Transportation

More recently, the Arctic region has become to be understood as a potentially important shipping route. Until the 1980s, transits through either the Northwest Passage on the North American side, or the Northern Sea Route along Russia, were perilous and rare due to a preponderance of sea ice. (The Barents region benefits from a warm gulf stream that keeps much of the area, from southwestern Greenland to the Kola Peninsula, ice free year-round.) More recently, the sea ice extent has been reduced, in theory making Arctic shipping more accessible.

Much has been written about the savings that could be gained by using the shorter route the Arctic provides, versus the Panama or Suez Canals, between Asia and North America (up to 7500 km shorter) and Asia and Europe (up to 7635 km) respectively (see e.g. Stephens 2016). However, in practice there are real and imposed limits to Arctic shipping. There are high environmental and safety regulatory burdens placed on shippers, as enshrined in the 2017 International Maritime Organization (IMO) Polar Code (or *International Code for Ships Operating in Polar Waters*). There is limited infrastructure available, including ports, icebreaking services, and search and rescue capabilities. Perhaps most importantly, ice and weather conditions are unpredictable, and in some parts of the Arctic the melting of older sea ice is resulting in more heavily ice infested areas.

Because the northern routes are unpredictable, and their infrastructure is relatively rudimentary, they are unlikely to ever compete with the more established global transportation links in the South. However, there are strategic advantages to having alternate shipping routes available if others become unavailable due to geopolitical or environmental reasons, something that is not inconceivable. It is speculated that a transpolar route—coming across the North Pole and largely traversing high seas areas—will become available in a few decades, which would be another attractive option strategically.

Geopolitical Location

Traditionally, the Arctic's geopolitical importance has been vested in its strategic location. This remains true, even as other factors have gained in importance. It has long been recognized that the shortest route for a nuclear attack between Russia and the United States, the world's biggest nuclear powers, is across the Arctic. In addition, Russia's only year-round open water access to the Atlantic Ocean, and all the nuclear deterrence capabilities that access affords, is located on the Kola Peninsula, where its Northern Fleet is located. It is common to inventory Russian military assets and assert that it has heavily militarized its Arctic, but the strategic intent of the Kola Peninsula's assets is primarily global, not regional.

The United States also has a heavy presence in Alaska, but again it is primarily to serve global strategic interests, not regional ones. While all Arctic littoral states have invested in Arctic-specific assets in the past decade (see e.g. Regehr and Jackett 2017), and while the potential for regional tension has served as instigation for some of this investment, much of it is a logical response to the increased activity and accessibility of Arctic waters. Constabulary capabilities, including search, and rescue and law enforcement remain limited across most of the Arctic due to remoteness, vast distances, harsh climatic conditions, and lack of infrastructure.

Contested Territory

It is not possible to conceive of a rational need for states to exert offensive power outside their own Arctic sub-regions. All of the Arctic littoral states share a similar problem—that of too much Arctic territory, not that of not enough. None is even close to maximizing their exploitation of minerals and hydrocarbons in their own onshore and 200 nautical mile Exclusive Economic Zone areas, and indeed would benefit from the economies of scale more concentrated resource extraction could bring, for example, in transportation links and labor pools. The one area of significant, disputed territory is the extended continental shelf found along the Lomonosov Ridge in the Central Arctic Ocean, between Canada, Denmark, and Russia. It will be divided up according to the terms of UNCLOS and by the Commission on the Limits of the Continental Shelf, to which Denmark and Russia have already submitted their Arctic claims (Canada plans to submit its claim in 2019). Although the mainstream media often portrays the littoral Arctic states as competing for this sea bed, the reality is that they are all gaining disproportionately vis à vis non-Arctic states from the dividing up of the Arctic continental shelf—a relatively shallow ocean with unique geologic features that means up to 90% may be "claimable". They are more likely to seek complicity with each other in this endeavor than provoke conflict.

NON-ARCTIC STATE INTEREST IN THE ARCTIC

The Arctic as an international region was largely ignored by those outside of it (and routinely by those inside of it) until the mid-2000s. The common explanation, as articulated above, is that climate change and a commodities boom conspired to make the region much more politically interesting to a global audience. But this is not the whole picture.

Although the pronounced impacts of climate change in the Arctic region have attracted considerable recent international research attention, many states have had long-standing interest in the Arctic as a region of exploration and scientific research, such as through participation in the International Polar Years of 1882–83; 1932–33; and 2007–08, as well as the related International Geophysical Year of 1957–58. Svalbard island is a good example of international scientific interest in the region. According to the Svalbard Treaty of 1920, while Norway has sovereignty over Svalbard, it cannot be used for military purposes and citizens of any country can live and work there. Ten states have therefore been able to establish research stations at Ny-Ålesund, a research town on Svalbard's Spitsbergen Island, including the United Kingdom, Japan, Germany, France, South Korea, India, China, the Netherlands, and Italy. And a number of non-Arctic countries have invested in the considerable expense of an icebreaker for research in the Arctic (as well as Antarctic duties) including China, France, Germany, the United Kingdom, Japan, and South Korea. But it is not just the traditional global powers who are invested in Arctic research; in addition to the United Kingdom, Germany and Japan, both Poland and the Netherlands were present at the signing of the Ottawa Declaration, establishing the Arctic Council, in 1996, an indication of their active research history.

Yet from an international affairs perspective, it is the rise of Asian interest in Arctic affairs in the mid-2000s that has piqued the most interest, as it reflects two significant geopolitical trends: (1) that the Arctic is opening up for more significant human use, including shipping and resource exploitation; and (2) that Asia, led by China, is assuming more power and prominence in the international system.

Until 2007, the Arctic Council—the region's preeminent intergovernmental forum—had been a rather sleepy forum, off the radar of most foreign policy agendas. China's application for Observer status in the Arctic Council in 2008 triggered concern both within and outside the Arctic.

Arctic states and the forum's Indigenous representatives, known as Permanent Participants, were concerned that the introduction of global interests in resource development, shipping, and climate change science, might squeeze out regional ones such as Indigenous self-determination, sustainable development, and environmental protection. In addition, some states were alarmed by statements from some high-ranking Chinese officials that the Arctic should considered a common heritage of mankind, beyond the jurisdiction of any one state. From outside the Arctic, a number of other Asian states took notice of China's interest and submitted their own applications for Observer status in the Arctic Council; Japan, South Korea, and Singapore, all of whom have particular interest in shipping and ship-building, as well as India applied.

After several years of deliberations, all five Asian states, as well as Italy, were accepted as Observers in 2013. The Arctic Council reconciled its concerns by including a stipulation that Observers "recognize Arctic States' sovereignty, sovereign rights and jurisdiction in the Arctic" and "respect the values, interests, culture and traditions of Arctic indigenous peoples and other Arctic inhabitants" as well as a number of other provisions (Arctic Council 2018).

It would not be fair to say that the Arctic Council members were opposed specifically to non-European interest in the region and the forum. At the same time as China submitted its Observer application, so too did the European Union (EU). The latter has yet to be accepted as an Observer in the Arctic Council, reportedly stymied first by Canadian opposition due to its ban on seal products, an important cultural and economic product for Inuit[1]; and then by Russia, intent on limiting the influence of the powerful supranational organization and opposed to the sanctions the EU imposed following Russia's incursion into Crimea.

Beyond Observer status in the Arctic Council, which frankly is a role of limited influence, the Asian states have been involved in Arctic affairs through fora such as the International Maritime Organization, under whose auspices the Polar Code was developed; and in the negotiation of the *Agreement to Prevent Unregulated High Seas Fisheries in the Central Arctic Ocean*, concluded in 2017 and officially signed in October 2018. The latter is particularly interesting as a model of how Arctic and non-Arctic states can collaborate on Arctic issues in which the Asian states have a legitimate interest. It was concluded based on the precautionary principle and the best available evidence and practices, not on relative or absolute power.

THE EMERGENCE OF THE ARCTIC AS A REGION

Although the Arctic is often looked upon as a globally strategic battleground between East and West, its dynamic as a region has been a much better predictor of Arctic states' behavior there recently. Barry Buzan, Ole Waever and Jaap de Wilde's respective works (1998, 2003) on regional security complexes is instructive. They identify the importance of a regional, rather than solely global, level of analysis to understand international relations. This is manifested in the concept of Regional Security Complexes: "a set of units whose major processes of complex securitisation, desecuritisation, or both are so interlinked that their security problems cannot reasonably be analysed or resolved apart from one another" (2003, p. 491). Given the mobility patterns of people, weapons, disease and commerce, region-specific dynamics become essential.

While the Arctic can be identified and characterized as such a region, its processes of securitization are unique. While most—arguably all—other regions in the world have evolved either around traditional security interdependence or economic integration, or both, the Arctic has built itself around common environmental and marine challenges. Following from this a whole suite of characteristics distinguish the Arctic.

---

[1] Even though the European Union ban on seal products has made an exception for Inuit goods harvested through subsistence means, the ban has been blamed for gutting demand, and thus market and prices, for seal products.

Ocean Versus Land Based

Most regions are defined by their accessibility to each other, delineated from other regions by mountains or large bodies of water. That accessibility results in easier economic integration as well as susceptibility to encroachment and attack.

The peaceability of the Arctic can be linked to its ocean-based nature. Historically there has been little international conflict in Arctic territory states due to the great difficulty in either reaching it, taking it, or holding it (the violence of European colonization is, of course, another matter). As former Canadian Chief of Defence Staff Walter Natynczyk put it in 2009, "If someone were to invade the Canadian Arctic, my first task would be to rescue them" (as quoted in Deshayes 2009).

Rather, the oceans-based nature of the region has meant that environmental and marine issues are paramount. This can be seen in the organization of the Arctic Council, the region's preeminent intergovernmental forum. Its six Working Groups, four of which are holdovers from the 1991 Arctic Environmental Protection Strategy (AEPS), include (1) Arctic Contaminants Action Program; (2) Arctic Monitoring and Assessment Program; (3) Conservation of Arctic Flora and Fauna; (4) Emergency Prevention, Preparedness and Response; (5) Protection of the Arctic Marine Environment; and finally (6) the Sustainable Development Working Group. Its major reports have been on climate change (2004 Arctic Climate Impact Assessment), shipping (2009 Arctic Marine Shipping Assessment) and the cryosphere (2017 Snow, Water, Ice and Permafrost in the Arctic).

The three legally binding agreements the Arctic Council has generated have also focused on marine issues: the 2011 Agreement on Cooperation on Aeronautical and Maritime Search and Rescue in the Arctic; the 2013 Agreement on Cooperation on Marine Oil Pollution Preparedness and Response in the Arctic; and the 2017 Agreement on Enhancing International Arctic Scientific Cooperation, which while not marine or environmental explicitly, is meant to facilitate scientific cooperation primarily in those areas.

Other, non-Arctic Council, international collaboration is similarly organized around common environmental and marine challenges. Since the Arctic Ocean, like other oceans, is subject to UNCLOS, there is a large body of jurisprudence to regulate it already. The delineation of extended continental shelf in the Arctic basin is occurring under its provisions. Several international environmental agreements, such as the *Minamata Convention* on mercury or the *Stockholm Convention* on persistent organic pollutants, have been strongly influenced by Arctic efforts. Other, Arctic-specific, agreements have arisen in the past five years as well, including the 2017 International Maritime Organization's (IMO) *Polar Code* on shipping regulations; the establishment of an eight-state Arctic Coast Guard Forum in 2015; and the conclusion of an Arctic fisheries moratorium in 2017.

Environmental issues are extremely well suited for regional and collaborative interaction. Unlike traditional security which can incline toward zero sum agreements, environmental protection can accommodate relative as well as absolute gains; states generally benefit when their neighbors invest in environmental protection and marine regulations. And unlike development issues, states cannot tackle governance of the ocean independently.

Non-State Actor Inclusion

The Arctic is further distinct as a political region in that the role of non-state actors is highly influential. This is a result of a combination of factors: the regional political focus on the environment, which creates space for scientists and non-governmental organizations; the geography and remoteness of the Circumpolar North, which centers Indigenous nations as objects and subjects of policy; and the era in which the Arctic matured as a region—the 1990s—which lends it a post-materialist focus.

There are three main categories of non-state actors in the Arctic: (1) Indigenous peoples; (2) academics; and (3) non-governmental organizations, especially environmental ones.

Indigenous peoples are often represented in regional Arctic affairs by the Arctic Council's six Permanent Participants: Inuit Circumpolar Council, Arctic Athabaskan Council, Gwich'in Council International, Russian Association of Indigenous Peoples of the North, Saami Council, and Aleut International Association. Although their financial and human resource capacity is limited and challenges their ability to participate consistently in Arctic foreign policy decision-making venue, it is widely acknowledged that their influence in Arctic affairs, especially in the Arctic Council, is real, not token. Furthermore, there have been very few open disagreements between the Permanent Participants and the Arctic states; their relationship is not one marked by discord, as it frequently is on domestic issues.

The academic/scientific community has been a consistent participant in Arctic affairs since the nineteenth century. Their influence has led to popular imaginations of the Arctic as, for example, a home to exotic wildlife such as polar bears and narwhals; or as ground zero for global climate change. These characterizations have had real policy-setting influence, from environmental protection to pollution prevention to shipping regulation.

Scientific work in the Arctic has long been typified by strong international linkages, going back to the International Polar Year 1882–83. Transnational relations have developed and strengthened as a result of scientific networks and their role in knowledge dissemination. Many of these were formalized after the fall of the Soviet Union, including the International Arctic Science Committee, the International Arctic Social Science Association, and the University of the Arctic. The most recent agreement negotiated under the auspices of the Arctic Council is the 2017 *Agreement on Enhancing International Arctic Scientific Cooperation.*

Environmental NGOs have similarly played a prominent role in regional agenda-setting. A number of international and UN agencies, for instance, have observer roles in the Arctic Council, and the Arctic has often been used as a case for global environmental action, for example on mercury, Persistent Organic Pollutants (POPs) and climate change. But perhaps the biggest influence has come from large NGOs such as Greenpeace and WWF, who have been effective, for better and for worse, in shaping urban southerners' perceptions and concerns about the Arctic. WWF, which is an Observer in the Arctic Council, is usually considered a vocal but effective partner in working with a range of stakeholders to inform and advocate for environmental conservation policies. Greenpeace, which isn't an Observer, is better known for its high profile, populist campaigns, such as criticizing the seal hunt or protesting oil exploration in the Arctic.

In addition, and in keeping with a privileging of local engagement, several northern sub-national polities, for example, territories (Canada), states (United States), republics (Russia), autonomous constituents (Greenland/Denmark) and counties (Nordic states), are also involved and relatively influential in Arctic affairs. The main organization representing them is the Northern Forum, a pan-circumpolar NGO established in 1991; however, its mandate and influence is limited. Other sub-regional fora, including the Nordic Council, the Barents Euro Arctic Region (BEAR), and the Pacific Northwest Economic Regions Northern Caucus (PNWER) (Alaska, Yukon, and NWT), also exert some influence, though international Arctic affairs have generally been conducted with them as members of their respective state delegations rather than as independent actors.

ARCTIC EXCEPTIONALISM

Due to its oceans-based, environmentally focused, and locally privileging nature, the Arctic can be described as "exceptional" in that states' behavior is modified in ways that they are not in other regions (see Exner-Pirot and Murray 2017). Analysis by non-regional specialists will often assume a much more competitive dynamic in the Arctic, due to historic and current relations between Russia and the West, and the presence of resources in territories whose boundaries are unresolved. But Russia's interests and motivations are not the same in the Arctic as they are at a global level. Whereas Russia is a revisionist power in the international order, and seeks to upend the status quo there, in the Arctic region it has an interest in maintaining the current balance of power.

The manifestation of Arctic exceptionalism has been an implicit policy of compartmentalization: preventing spillover from broader geopolitical tensions from affecting the good relations in the region. The greatest test of this came following the Russian incursion into Crimea in 2014. Western states retaliated in a number of ways, including sanctions and the suspension of joint military cooperation. However, Arctic cooperation was only minimally affected, despite the fact that Canada was chairing the Arctic Council at the time and its

Conservative government, led by Stephen Harper, was amongst the most vocal critics of Russia's move (Exner-Pirot 2016). In the two years following the Crimean invasion, Russia was a partner in a plethora of multilateral, Arctic-specific initiatives and agreements, including, as briefly articulated above: (1) the establishment of the Arctic Coast Guard Forum—an informal, operationally driven forum to foster responsible maritime activity; (2) the *Declaration Concerning the Prevention of Unregulated High Seas Fishing in the Central Arctic Ocean*, signed by the five Arctic coastal states, agreeing to abstain from commercial fishing until or unless a regional fisheries management organization is in place to regulate it (and later expanded to include the European Union, Iceland, Japan, China, and South Korea, as articulated above); (3) Denmark and Russia submitted their claims to the Commission on the Limits of the Continental Shelf, thereby endorsing the process; and (4) the negotiation of the International Maritime Organization's *Polar Code* regulating the operation of ships in polar waters, an area where Russia arguably has the most vested interest.

Despite very real problems in Russia's relations with the West elsewhere, in the Arctic cooperation has continued unmitigated. Some have compared this compartmentalization to relations with the International Space Station or the Olympics (Exner-Pirot and Murray 2017).

CONCLUSION: THE ARCTIC AS A MODEL FOR INTERNATIONAL COOPERATION?

This chapter has identified the characteristics that make the Arctic unique in the region, and the influence they have had on making the Arctic a predominantly peaceful and cooperative region. But can the Arctic be held as a model for other regions?

As Brigham et al. (2016) describe, contemporary achievements of Arctic relations include the continued cooperation between Russia and the West; the inclusion of Indigenous peoples and other non-states actors in policy development; the application of the precautionary principle to environmental management; and the adoption of an evidence-based approach to decision-making (p. 15). However, the authors also assert that the Arctic Council "has not so much blazed a trail as invented and occupied a unique space in international relations" (p. 9), a description that can be fairly applied to the region as a whole. In addition to the impact of its oceans-based nature, the Arctic is a product of its time: as arguably the only international political region of significance to evolve after the Cold War, or even World War II, it has been much more welcoming of the role of non-state actors than other regions that developed politically during more traditional, state-based eras.

While the Arctic is unique as a region, one could foresee the rise of other, oceans-based, environmentally focused, and constitutionally post-materialist regions in the future. The melting and reduction of the polar ice cap in the

Arctic was a rare trigger for states to come together within a new, or newly important, political region. The dramatic effects of climate change and other geographic and environmental phenomena could conceivably lead to others. Should such regions arise, they would do well to emulate the Arctic in their governance and regional society.

REFERENCES

Arctic Council. 2018. *Observers*. Accessed September 2, 2018. www.arctic-council.org/index.php/en/about-us/arctic-council/observers.

Brigham, Lawson, Heather Exner-Pirot, Lassi Heininen, and Joel Plouffe. 2016. Introduction. In *Arctic Yearbook 2016*, ed. Lassi Heininen, Heather Exner-Pirot, and Joel Plouffe, 9–15. Akureyri: Northern Research Forum.

Buzan, Barry, and Ole Wæver. 2003. *Regions and Powers: The Structure of International Security*. Cambridge: Cambridge University Press.

Buzan, Barry, Ole Waever, and Jaap de Wilde. 1998. *Security: A New Framework for Analysis*. Boulder, CO: Lynne Reiner Publishers.

Deshayes, Pierre-Henry. 2009. "Arctic Threats and Challenges from Climate Change," *Agence France-Presse*, as Quoted in Michael Byers (Summer 2014). "Does Canada Need Submarines?". *Canadian Military Journal* 14 (3): 7–14.

Exner-Pirot, Heather. 2016. Canada's Arctic Council Chairmanship (2013–2015): A Post-Mortem. *Canadian Foreign Policy Journal* 22 (1): 84–96.

Exner-Pirot, Heather, and Robert Murray. 2017. Regional Order in the Arctic: Negotiated Exceptionalism. *Politik* 20 (3): 47–64.

Flanagan, Thomas. 2013, August 21. Arctic Symbolism, Harper Stagecraft. *National Post*. Accessed September 5, 2018. https://www.theglobeandmail.com/opinion/arctic-symbolism-harper-stagecraft/article13876049/.

Heininen, L. 2012. State of the Arctic Strategies and Policies – A Summary. In *Arctic Yearbook 2012*, 2–47. Akureyri: Northern Research Forum.

Huebert, R.N., H. Exner-Pirot, A. Lajeunesse, and J. Gulledge. 2012. *Climate Change & International Security: The Arctic as a Bellweather*. Arlington, VA: Center for Climate and Energy Solutions.

Regehr, Ernie, and Micelle Jackett. 2017, January. *Circumpolar Facilities of the Arctic Five*. The Simons Foundation. Accessed July 24, 2018. http://thesimonsfoundation.ca/sites/default/files/Circumpolar%20Military%20Facilities%20of%20the%20Arctic%20Five%20-%20updated%20January%202017_1.pdf.

Stephens, Hugh. 2016, May. The Opening of the Northern Sea Routes: The Implications for Global Shipping and for Canada's Relations with Asia. *Canadian Global Affairs Institute*. SPP Research Papers 9(19). Accessed July 24, 2018. https://www.policyschool.ca/wp-content/uploads/2016/06/northern-sea-routes-stephens.pdf.

CHAPTER 20

East Asia (Japan, South Korea and China) and the Arctic

Carin Holroyd

Changes in the Arctic, especially the melting of Arctic sea ice and the possibility of a summer ice-free Arctic with the potential for new shipping routes and enhanced resource extraction, have sparked global interest in the region. Outside the Arctic countries themselves, few nations have been paying more attention than the countries of East Asia. While Japan, South Korea and China have somewhat different motivations and objectives underlying their interests in the region, each nation is likely to be a player of some significance in the Arctic's future. However, the concerns that East Asia is trying to take over the Arctic are very much overblown.

All three East Asian countries are dependent on international trade, active in shipbuilding and/or shipping, and remain deeply concerned about their nation's long-term energy demand and supply. Each has both scientific and economic interests in the Arctic and to varying degrees refuses to be left out of whatever developments or opportunities do arise. This East Asian interest in the Arctic has generated mixed reactions. Appreciation for scientific contributions (Japan has been active in polar research for over half a century) is combined with the fear that Arctic residents and some of their governments will be left behind as the wealthy and densely populated East Asian countries assume larger roles in the Arctic region.

It is possible—even likely—that the year 2040 could see South Korean-made LNG carriers transporting Russian gas to China on the Polar Silk Route. Japan could be extracting methyl hydrates from under the Arctic ice, transforming the natural gas market. Uncertainty about these countries' interests in the region and worries about how much respect there will be, particularly from

C. Holroyd (✉)
Department of Political Studies, University of Saskatchewan, Saskatoon, SK, Canada

© The Author(s) 2020   319
K. S. Coates, C. Holroyd (eds.), *The Palgrave Handbook of Arctic Policy and Politics*, https://doi.org/10.1007/978-3-030-20557-7_20

China, for the sovereign rights and jurisdictions of the Arctic countries makes Arctic observers nervous.

The Arctic nations worry that the rest of the world wants to remake the Arctic into a truly global zone, not subject to regional or national control. Asian aspirations for engagement with the Arctic seem by some to foreshadow such an effort but that assumption is exaggerated. (There is an important caveat about China's global aspiration, for China is the new USA, insisting on being involved everywhere and backing their interventions with the economic might to force its way into regional affairs, as American used to do.) Far more benign than is generally assumed, Japan, South Korea and China want to understand the impact of the changes occurring in the Arctic, participate in the economic opportunities that may arise and contribute to the development and scientific understanding of the region.

SCIENTIFIC RESEARCH AND CLIMATE CHANGE CONCERNS

Japan, South Korea and China are all actively engaged in scientific research in the Arctic. Of the three countries, Japan's Arctic involvement has been the longest. It was one of the original 14 contracting parties to the 1920 Svalbard (originally Spitsbergen) Treaty, which recognized Norway's sovereignty over Svalbard and the Spitsbergen archipelago. Japan has been actively engaged in polar science for more than 50 years. The country launched the National Institute of Polar Research (NIPR) in 1973 and the NIPR established the Arctic Environment Research Centre (AERC) in 1990, building on 30 years of polar research. In 1991, AERC established an observation research center in Svalbard, initially in conjunction with the Norwegians (Ohnishi 2014, p. 23). Japan was also a founding non-Arctic member of the International Arctic Science Committee in 1992 (Tonami 2016, p. 48). Japan has three icebreakers but none are active in the Arctic. The *Shirase* is operated by the Japan Maritime Self Defense Force and is used as a supply vessel for Japanese Antarctic research. The *Soya* and the *Tesio* are operated by the Japanese Coast Guard as patrol boats in northern Japan. Japan was part of the 1993–1999 international research program to study the feasibility of what is now known as the Northern Sea Route (NSR) from the Russian Arctic to Bering Strait. Called the International Northern Sea Route Program, this Norwegian–Japanese–Russian collaboration involved 450 researchers from 14 countries and produced and released 167 technical reports (Fridtjof Nansen Institute).

South Korea has been actively engaged in polar research but for many decades focused national efforts on the Antarctic. In 1996, it launched joint Arctic research with Japan and moved to independent research in 2001, starting the Korea Arctic Scientific Committee and joining the International Arctic Science Committee. In 2002, Korea established the small and not permanently manned Dasan research station in Svalbard. Two years later the Korea Polar Research Institute was established in Incheon (Park 2014a). South Korea has been operating the icebreaker *Araon*, built by Hanjin Heavy Industries, in both the Arctic and Antarctic since 2010. In July 2009, the Northern Sea

Route was used by Korean cargo ships for the first time. In 2011, 34 Korean vessels traversed the passage (Park 2014).

China is a relative latecomer to the Arctic, although it signed the Svalbard Treaty in 1925. In the following decade, China embarked on a small number of polar activities, mainly in the Antarctic. It founded the Polar Research Institute of China (PRIC) and renamed the Office of the National Antarctic Expedition Committee as the Chinese Arctic and Antarctic Administration in 1989 (Lackenbauer et al. 2018, p. 55). From the late 1990s, China became more active in the Arctic as, indeed, it did all over the world. China joined the International Arctic Science Committee in 1996 and established the Arctic Yellow River research station on Svalbard in 2004. In 2013, PRIC, in conjunction with major Nordic research institutions focused on the Arctic, launched the China-Nordic Arctic Research Centre so there would be a common research platform for Chinese and Nordic scholars (Tonami 2016, p. 23).

In 1993, China bought an icebreaking cargo and supply ship made in the Ukraine, named it the *Xue Long* (Snow Dragon) and converted it to a research and resupply vessel. The *Xue Long* is outfitted with research labs and weather observation and navigation equipment. Its first Arctic expedition was in 1999. By the end of 2017 China and the *Xue Long* had completed eight Arctic expeditions. Jiangnan Shipyard Corporation began construction on a new icebreaker, the *Xue Long 2*, in December 2016. Designed by the China State Shipbuilding Corporation and Finland-based Aker Arctic Technology, the *Xue Long 2* will be able to cut through 1.5 meter thick ice. The ship became operational in 2019.

Chinese, South Korean and Japanese scientists are all involved in climate change research, including that with an Arctic focus. All three countries believe they have much expertise to contribute and share concerns about vulnerability to changing weather and sea level rise. Both China and Japan believe that changes in the Arctic are having a direct impact on their own weather. Research cited by the Japan Agency for Marine-Earth Science and Technology Research reveals Japan's colder winter temperatures are the result of retreating sea ice (Ohnishi in Hara and Coates 2014, p. 21). Chinese scientists believe that the changing Arctic is adversely affecting China's environment with implications for weather, natural disasters, agricultural production and even national security (Sun 2014; Zhang and Yang 2016, p. 224; Tonami 2016, p. 33). As one study reported, "Research shows that the frequent incidents of extreme weather in China are closely associated with Arctic warming. The Arctic sea ice anomaly of less ice-cover than previously is one of the main causes of China's climate disasters in recent years" (Zhang and Yang 2016, p. 224).

The Japanese government underscores that its main Arctic priority is understanding of the Arctic environment, protecting and ensuring its peaceful use. "According to the Japanese government, the country's primary aim of engagement in the Arctic has been and remains understanding and protecting the natural environment. As the negative impacts of climate change became more apparent, policies related to scientific research were given higher priority"

(Tonami 2014, p. 52). Research on the Arctic region is mentioned in Japan's Fifth Science and Technology Basic Plan that states "we are working on observation and research, including development of technology for Arctic observations, and on predicting the possibility of navigating the NSR. Furthermore, for adapting to the impacts of climate change, we are also pursuing R&D on technologies for predicting and assessing climate change impacts and for climate risk management. In addition, we aim to capture information about the global environment in the form of 'big data' and to develop the Global Environment Information Platform for meeting the economic and social challenges arising from climate change, as well as pursue research in cooperation with stakeholders both in and outside Japan" (Government of Japan 2016, p. 27).

China has been increasing both its science and social science research on the Arctic. There are Polar Research Institutes at a number of Chinese universities (e.g. Fudan, Wuhan, Ocean University of China, Shanghai Jiao Tong University), resulting in a rapidly increasing body of scholarship on Arctic governance, geopolitics, environmental protection and other Arctic topics. China is also seeing the Arctic as a good testing ground for "green" technologies and in other science and technology areas (Zhang and Yang 2016, p. 225). Despite the increase in Chinese research on the Arctic, it is important to note that China's polar research budget allocates four times as much toward research on the Antarctic as it does to research on the Arctic (Lackenbauer et al. 2018, p. 23). In research fields, curiosity-based studies continue to trump the geopolitical considerations of the Far North.

International Shipping

Most of the world, East Asia included, is paying close attention to the melting of Arctic sea ice that has been gradually rendering northern waters navigable in the summer months (July to October). The two main potential international shipping routes are the Northern Sea Route (NSR) above Russia and the Northwest Passage (NWP), along the northern coast of Alaska and through the Canadian Arctic archipelago. Both routes could cut transit time between East Asia and North America and Europe dramatically. The NWP would be a much faster route from China to the US eastern seaboard but is overall the less attractive of the two routes due to its geography and the lack of infrastructure in the Canadian Arctic compared to that in Russia. Going from Shanghai to Hamburg would be 6400 km (4000 miles) shorter using the Northern Sea Route than the usual route through the Malacca Straits and Suez Canal (*The Economist* 2015). As Nong Hong wrote, "Taking into account canal fees, fuel costs, and other variable that determine freight rates, these shortcuts could cut the cost of a single voyage by a large container ship by billions of dollars a year" (Hara and Coates 2015, p. 149). NSR cargo transit traffic began in 2009 and increased to 71 ships in 2013. (This represented the peak number of annual international transits to date as of June 2018 according to *High North News*.) These transits, reported by Russia, include those that cross that most difficult part of the NSR and not necessarily the entire length. Most trips were between Russian ports. Of the 71 ships in the region in 2013, for example, only 28 began or ended outside of Russia. Between 2011 and 2013, nine of the NSR international transit trips originated in China and ten originated in South Korea. For 17 ships, the final destination was China. For 13 vessels, the journeys ended in South Korea. Four trips ended in Japan. The largest number of international transits started in western Russia and ended in Europe.

The potential of the NSR and the Northwest Passage entices Japan, China and South Korea for the time and money they could save and the opportunity to avoid piracy or blockades in the Malacca Straits and the Gulf of Aden. However, so far the number of vessels transiting either is only a tiny fraction of the 18,000 that use the Suez Canal route annually. In addition, although there is no question that the Arctic ice is melting, this does not mean that the remaining ice does not pose a serious hazard; the potential for grounding or collisions remains. Japan investigated the feasibility of the Northern Sea Route for the Japanese shipping industry twice, the latest in 2002–2005 and "concluded that the feasibility of the NSR was limited and that there were too many uncertainties to generate any financial benefits in the near future" (Tonami 2016, p. 49). Frédéric Lasserre has conducted numerous surveys and interviews with Chinese shipping and forwarding companies, most recently in 2013 for the interviews and 2016 for the survey, and found "few industry representatives expressing any real interest for Arctic shipping" (Lackenbauer et al. 2018, pp. 81–82). As Lasserre explains,

> While several interviewees expressed a belief in the potential of Arctic shipping, none had yet undertaken an extensive cost/benefit or 'SWOT' analysis of that potential. Chinese companies cited various problems with Arctic operations, including the high investment necessary to buy ice-strengthened ships; market constraints surrounding schedules and ship sizes limiting economies of scale; an Arctic market too small to build a profitable route and, therefore, a longer return on investment on costly ice-strengthened ships; as well as physical risks and high insurance costs. (Lackenbauer et al. 2018, p. 82)

While the prospects for transit shipping remain at best uncertain, destination traffic (to and from communities, primarily for resource exploration) does appear likely to continue to increase. The Yamal LNG project, part of the Yamal megaproject to develop onshore and offshore oil and gas on Russia's Yamal peninsula, launched production in December 2017 (Tonami 2016, p. 23). This project that included an LNG factory and a port has the potential to boost Arctic shipping dramatically. Gas from oil fields on the eastern side of the Yamal peninsula is being extracted. A LNG factory and a port have been built and the LNG will be shipped out using the Northern Sea Route. Novatek, a Russian gas company, developed the project and now owns 50.1% of the firm. Total S.A., a French oil and gas company, owns 20%. China National Petroleum

Corporation purchased 20% in early 2014. The remaining 9.9% is owned by China's Silk Road Fund (Moe 2016, p. 116; Filimonova and Krivokhizh 2018). The project is seen as crucial for the development of the Northern Sea Route as its success depends on being able to ship the LNG year round to China via the NSR. This is almost three weeks faster than going the usual route through the Suez Canal. So, the Yamal project will generate substantial traffic for the Northern Sea Route and get Russian energy to China, which has committed to buying three million tons annually (Moe 2016, p. 116; Filimonova and Krivokhizh 2018).

China is heralding the Yamal as the first Arctic project within the newly announced Polar Silk Road, a series of Arctic plans that connect Russia to China's Belt and Road Initiative. The Polar Silk Road was first raised by President Xi in Moscow in 2017. "China's official interest in including the Arctic Ocean in the Belt and Road Initiative was first expressed in the 2017 publication "Vision for Maritime Cooperation Under the Belt and Road Initiative," a document jointly released by China's National Development and Reform Commission and the State Oceanic Administration on June 20, 2017." The vision that the publication laid out included a "Blue Economic Passage … leading up to Europe via the Arctic Ocean." This notion of connecting Europe and Asia through the melting Arctic was subsequently expanded and dubbed the "Polar Silk Road" in Beijing's 2018 "Arctic Policy Whitepaper" (Asia Pacific Foundation of Canada website).

Japanese companies are also important players in the Yamal project. They are involved in design, procurement and construction; "some Japanese companies have contracts to carry LNG from Yamal LNG trains to Asia and Europe" (Hammond 2017). According to Japan's first Arctic ambassador, Kazuko Shiraishi, with an eye on Yamal's potential to offer a new supply of energy "Japanese participation in Yamal is very important and as an Arctic project even more so. Japanese companies are very much invested in the outcome of the Yamal LNG project" (Hammond 2017). The prefectural government in Hokkaido, Japan's northernmost major island, has also been investigating how it might be able to capitalize on its location to be an entry point into the Northern Sea Route. Hokkaido's Tomakomai port located in the Tsugaru Strait, local officials hope, could become a northeast Asian terminal (Tonami 2016, p. 59).

Korea is also enthusiastic about the Northern Sea Route that could allow Korean goods to get to European markets more quickly and less expensively. Korea has world class shipbuilders (Hyundai Heavy Industries, Samsung Heavy Industries, Daewoo Shipbuilding and Marine Engineering). They hope to benefit from the increased need for icebreakers and ice class ships. Yamal has offered South Korean shipping a great start. Daewoo won the option to build 16 icebreaking LNG carriers (a contract worth $4.8 billion), in 2013. These carriers can cut through 1.5 meters of ice with a continuous speed of 5 knots and through 2.1 meters of ice independently but more slowly (Moe 2016, p. 116). The first of these, the *Christophe de Margerie*, was completed in 2016 and docked at Yamal's LNG terminal in March 2017. South Korea and Russia are also pursuing partnerships on port development projects and in January 2014, the two countries signed a memorandum of understanding on the development of five Russian ports.

All three East Asian countries are monitoring closely the potential of Arctic shipping routes, especially the Northern Sea Route, and investing when and how it makes the economic sense to do so. While the NSR offers the likelihood of destination shipping and the potential of some transit shipping, particularly if the Yamal project is successful, Japan, South Korea and China are primarily hedging their bets. Their investments in the Arctic pale in comparison to those the East Asian countries have made in ports and terminals around the world. As Lackenbauer et al. 2018, p. 90) explain,

> Few stories of China's growing Arctic interest address how small this interest actually is, relative to China's massive and ever-expanding shipping interests elsewhere in the world…While there may be Chinese interest in Arctic routes, these investments elsewhere help to keep it in perspective. China remains overwhelmingly wedded to the classical global sea routes through Malacca, Suez, and Panama.

RESOURCE DEVELOPMENT

Estimates are that the Arctic region could contain 13% of the world's undiscovered oil and 30% of its undiscovered natural gas. It is, therefore, unsurprising that the East Asian nations are monitoring developments and investment opportunities in this resource-rich and politically stable region. Japan and South Korea are very dependent on Middle Eastern oil and are always looking for opportunities to diversify their supplies. They are the first and second largest importers of liquefied natural gas; Japan is the third largest importer of oil while Korea is the sixth largest. China has its own oil but rapid economic growth has left it desperate for more. Projections are that by 2030 China will be importing 70% of the oil and 40% of the gas it needs (Zhang and Yang 2016, p. 224). East Asian interest in Arctic oil and gas (and minerals) is only a small part of East Asia's global search for long-term resource supplies. When compared to the region's activities in other parts of the world, including the sub-Arctic, Australia, Asia and Africa, it becomes clear that the Arctic is only a tiny part of the region's resource exploration and development efforts.

East Asia companies are investing and investigating a range of oil and gas investments. In 2011, the Korea Gas Corporation (KOGAS) bought a 20% share of the Umiak gas field in Canada's Mackenzie Valley in the Northwest Territories. Japan Petroleum Exploration Company (JAPEX) has undertaken exploration in Greenland. Japan Oil, Gas and Metals National Corporation (also a major shareholder in the JAPEX Greenland project) is active in testing projects for methane hydrates on Alaska's North Slope and in the Mackenzie Valley. Japanese companies are involved in the Sakhalin-2 oil and gas development on the Russia's Sakhalin Island. Subsidiaries of Mitsui and Mitsubishi are part of the Sakhalin Energy consortium in charge of the development. Japanese companies were involved in construction of the LNG plant. The Japan Bank for International Cooperation loaned $3.7 billion to Sakhalin Energy. Chinese state-owned enterprises have bought stakes in a number of Russian Arctic energy projects, including Yamal. (Russian law limits foreign ownership in oil and gas to minority status.) Since the Russian invasion of Crimea and resulting Western sanctions, deals between China and Russia have escalated. In 2014, the China National Petroleum Corporation (CNPC) and Rosneft, a Russian government-controlled integrated energy company, have agreed to explore fields in the Barents and Pechoras seas (Lackenbauer et al., p. 117). CNPC also purchased a 10% share in the Vankor oil field and 10% of an Eastern Siberian unit of Rosneft (Lackenbauer et al., p. 117). The Canadian government halted oil and gas exploration in Arctic waters in December 2016, subject to five-year reviews based on environmental assessments. So, for the foreseeable future, any offshore oil and gas development will be in the waters of other Arctic states.

Japan, South Korea and China are also interested in the potential of the melting Arctic ice to provide potential access to previously inaccessible minerals. The Arctic region is known to contain commercial deposits of coal, iron, uranium, nickel copper, tungsten, lead, rare earths, zinc, gold, silver and diamonds. South Korea and Japan are not well-endowed with resources of this variety and assumed magnitude so the potential of the region is naturally of interest. China's size and growth assume a continuing need for all resources. However, again, to date active Asian interest in mineral exploration and development is limited (Stensdal in Lunde et al. 2016, p. 161). China National Bluestart bought a quartzite mine in Finnmark. Japanese Sumitomo Metal Mining Company and Sumitomo Corp have been operating the Pogo gold mine in Alaska since 2009. Daewoo owns 1.7% of the Kiggavik uranium mine in Nunavut. China Non-Ferrous Metal Industry's Foreign Engineering and Construction Co. Ltd has a non-binding Memorandum of Understanding with Australia's Ironbark Zinc for the construction and financing, with an option to purchase a stake in and purchase of the output, of the Citronen zinc project in northern Greenland (Stensdal in Lunde, pp. 162–163). In January 2015, the Chinese company General Nice purchased the proposed Isua iron ore mine in northern Greenland. While there are a number of sub-Arctic mining projects with Chinese investment and other potential investments in the Arctic, overall as Norwegian researcher Iselin Stensdal notes, "All in all, the Asian active interest for the Arctic is minimal compared to the investments and active interests elsewhere. The Asian share of all mineral investments in the Arctic region is miniscule" (Stensdal in Lunde et al., p. 163). While all three countries are keen to obtain much needed resources, their investments remain strategic and economically based.

Other potential areas of economic interest include fishing and tourism. Climate change may bring more fish north and could make new fishing areas commercially viable that could be of interest to Japan, South Korea and China. Canada, the United States and Denmark currently have a moratorium on High

Arctic commercial fishing until there is a better scientific understanding of its potential impact (Lackenbauer et al., p. 122). East Asian interest in the north, particularly the northern lights, has been bringing tourists to Alaska, the Canadian territories (especially Yukon and the Northwest Territories) and to northern Russia, Sweden, Norway and Finland. Although accurate statistics on how many international visitors venture north are hard to track (because once they clear customs at an initial point of entry, it is hard to check where they go within a country), the number of East Asian Arctic visitors is climbing steadily. Chinese visitors, in particular, appear to represent a significant percentage of Arctic visitors (Meesak 2018).

GOVERNANCE

Japan, South Korea and China (along with India, Italy and Singapore) were all granted Arctic Council Observer status in May 2013. China had applied in 2006, South Korea in 2008 and Japan in 2009. These applications were clear indications that East Asian interest in the Arctic was serious. The action made some of the Arctic states, particularly Canada and Russia, nervous. The East Asian countries have articulated the reasons for their interest: the opening of Arctic passages and the implications for shipping, ship and offshore platform construction and port development; the potential accessibility of energy and natural resources; concerns about Arctic climate change and its implications for the East Asian countries and their desire to use their scientific and technological expertise to help the Arctic develop sustainably and peacefully. Japan's position, as researcher Aki Tonami describes it, is that "scientific research is what it does best as a technologically advanced nation. Japan also believes this is what the AC expects it to do. The natural environment of the Arctic is fragile and requires large-scale, costly research in order to under the possible repercussions of climate change" (Tonami 2016, p. 61).

All three countries want to be closely aware of the changes occurring in the Arctic so that they can understand the economic and climate implications for the world and their own countries. China, more so that Japan or South Korea, also wishes to be part of decisions that are made about or around the Arctic. As a global power with a significant part of the world's population and as a "near Arctic state" (as it has labeled itself), China believes it deserves a say in Arctic matters. It is important to note that this is China's view on all global decision-making and is not unique to the Arctic. Just as the Arctic is only a small piece of China's global search for resources, so too is it only a small part of China's efforts to carve out its place in international affairs.

South Korea was the first of the three countries, and in fact the first non-Arctic state, to release a national Arctic policy, which it did in 2013. Japan followed in 2015 and China in 2018. The South Korean policy outlined the country's long-term focus and established four strategic priorities for 2013–17. These priorities were strengthening international cooperation through the Arctic Council and other Arctic forums and institutions; enhancing scientific research activities, developing and promoting Arctic business through, for example, supporting the development of the Arctic sea routes, shipping and port industries, cooperating in fishery management; and establishing legal institutions and laws to support Arctic activities and create an Arctic information center (Park 2014; Tonami 2016, pp. 76–77). South Korea's Arctic policy sums up the country's two-part Arctic agenda nicely. It would like to help solve the Arctic's environmental problems while exploring commercial opportunities, "to contribute to the international community in terms of climate and environmental research, as well as to cooperate with the coastal states is a policy. Secondly, by means of developing the Arctic sea routes, energy and marine resources, create new industries for South Korea" (Gong 2016, p. 240).

Japan's comprehensive Arctic Policy was released in October 2015 by Japan's Arctic Ambassador Kazuko Shiraishi. The policy outlined seven areas of focus that included global environmental issues, science and technology research and development, rule of law and international cooperation, sea routes, natural resources, indigenous peoples and sustainable use of Arctic resources. This broad policy demonstrates Japan's deep and historic interest and commitment to the region. In the short to medium term, Japan's sense is that the commercial potential of the Arctic is limited but nonetheless important enough to be monitoring and considering. As one analyst wrote in advance of the Arctic Policy's release, "When the government formulates its Arctic policy, the data and knowledge obtained from scientific research should be strategically used for planning and promoting the long-term perspectives on the economic benefits that Japan can draw from the Arctic" (Ohnishi 2014, p. 29).

China's Arctic Policy was released in early 2018. Describing itself in the policy statement as a near-Arctic state (although it doesn't have any territory above the Arctic Circle nor an Arctic coastline), China's Arctic Policy outlines the country's policy goals, principles and positions on the Arctic and participating in Arctic affairs. The conclusion to the policy summarizes this white paper and China's views nicely. It reads

> The future of the Arctic concerns the interests of the Arctic States, the wellbeing of non-Arctic States and that of humanity as a whole. The governance of the Arctic requires the participation and contribution of all stakeholders. On the basis of the principles of 'respect, cooperation, win-win result and sustainability', China, as a responsible major country, is ready to cooperate with all relevant parties to seize the historic opportunity in the development of the Arctic, to address the challenges brought by the changes in the region, jointly understand, protect, develop and participate in the governance of the Arctic, and advance Arctic-related cooperation under the Belt and Road Initiative, so as to build a community with a shared future for mankind and contribute to peace, stability and sustainable development in the Arctic. (China's Arctic Policy 2018)

GOING FORWARD

Although Japan, South Korea and China are clearly not as financially invested in either Arctic natural resource development or shipping as many Arctic residents and politicians have worried, they are indeed interested in the economic potential the Arctic may offer, ready and well placed to contribute scientifically and poised to take advantage of the opportunities that arise. In Japan's case, as Tonami and Watters (Tonami and Watters 2012) describe it, "one can perhaps view the overarching ambition of Japan's Arctic Policy as planting a flag today, to be used tomorrow."

All the countries recognize that for each of the potential areas of economic success, there are numerous forces that could alter the trajectory. The price of oil could fall making the cost of extraction from Arctic fields too expensive. As one commentator pointed out,

> Even at $100 a barrel, many fields were marginal because the environment is so extreme. Gazprom and Statoil, the Russian and Norwegian firms developing one of the largest gasfields ever discovered (the Shtokman field in the Barents Sea), mothballed the project in 2012. The boss of Total, a French energy firm, called Arctic drilling too risky even before prices started to fall. With oil at $50 a barrel, few Arctic fields would be economic. (*The Economist* 2015)

Mining prices are also volatile and need to be high enough and sufficiently stable to make exploration and development viable. And the actual viability of commercial shipping through the Northern Sea Route clearly remains in doubt. In one year, the sea ice might melt considerably more than the next, as happened from 2013 to 2015. The other major challenge with the Northern Sea Route is punctuality. "Cutting a week or two off transit time is not the benefit it may seem if the vessel arrives a day late. In shipping, just-in-time arrival matters, not only the speed" (*The Economist* 2015). If, on the other hand, the sea ice melts reasonably completely, the need for icebreaking ships might decline, which could have an impact on South Korea's shipbuilding plans.

China's ambitions are more far reaching than those of Japan and South Korea. China has on occasion made the Arctic countries nervous with some of the comments about Arctic sovereignty. As Sun Kai states,

> it is not difficult to understand why some Arctic countries have mixed feelings toward China's growing engagement in the Arctic. If China is reluctant to recognize the sovereignty, sovereign rights and jurisdiction of Arctic states in the region, then what are its intentions? Some 'radical views' expressed in the Chinese media such as "no country has sovereignty in the Arctic" and "the Arctic is no country's backyard" provoked fears in Canada, which is especially sensitive about sovereignty issues in the region. (Sun 2014, p. 49)

High-level Chinese officials have therefore made it a point to frequently state that China recognizes the sovereignty and jurisdiction of the countries of the

Arctic region. Nonetheless China is concerned that the Arctic states will gang up and exclude others from the riches of the Arctic (Lackenbauer et al. 2018, p. 134). As David Wright, author of <u>The Dragon Eyes the Top of the World: Arctic Policy Debate and Discussion in China</u>, writes (as quoted in Lackenbauer et al. 2018, p. 135),

> China does not want to lose any ground in its campaign to become a major player in the world in general, and increasingly for Beijing that means being a player in the Arctic… Chinese diplomatic gesturing should not be confused for acquiescence or lack of resolve on China's part. Despite its status as a non-Arctic country, China seems bound and determined to have a voice, perhaps even a say-so, in Arctic affairs.

Although Lackenbauer et al. reject what they call Wright's (and Canadian political scientist Rob Huebert's) "threat narrative"—suggesting that Canadians should be quite concerned about East Asian, particularly Chinese, involvement in the Arctic—they agree that " …although there is little evidence that China's intentions in the Arctic are malignant, it will not tolerate being excluded from the Arctic conversation" (Lackenbauer, p. 154). It is not just Canada that gets nervous about China's intentions. Huang Nubo, a Chinese real estate developer and former official in the Communist Party's Propaganda department, wanted to buy property (over 100 square miles at a value of approximately US$200 billion) and create a luxury tourist resort and eco golf course in Iceland. Despite positive China-Iceland relations, local nervousness and suspicions of military and other ulterior motives led the sale to fall through. Huang subsequently try to buy property in northern Norway and then on Svalbard. Neither of these were successful (Higgins 2013; Lackenbauer et al. 2018, p. 114).

In summary, the East Asian countries are interested and engaged, as much as currently makes economic sense, in scientific research on the Arctic, in the future potential of Arctic shipping lanes, ships and infrastructure development, and in natural resource development. These investments, however, pale in size, volume and value to those the three countries have made in other parts of the world. Japan, South Korea and China all want to play a role in Arctic governance and development. China is the most vocal and definitive about its belief that it should be a part of all future discussions on Arctic development. While Canada and the other Arctic states feel some nervousness about China's intentions and even whether China recognizes the sovereignty of the Arctic states, others point out that by becoming an Observer to the Arctic Council, signatories agreed to "recognize Arctic states' sovereignty, sovereign rights and jurisdiction in the Arctic." China, along with South Korea and Japan, agreed to this recognition when it became an Arctic Council Observer in 2013.

While the Arctic states need not be worried about self-interested motives on the part of their East Asian counterparts, they might well be concerned that the careful watch that both the companies and the governments of Japan, South

Korea and China keep over Arctic developments could mean that East Asia is ahead of the game when it comes to capitalizing on those opportunities that do arise. The scenario outlined at the beginning—Korean-made ships transporting vast amounts of Russian LNG to China while Japanese investment in methane hydrate developments sees the opening up a new branch of energy—could indeed come to pass, ensuring that East Asia has a prominent role in the future of the Circumpolar World.

References

Asia Pacific Foundation of Canada Website, Sebastian Murdoch-Gibson Blog.

China's Arctic Policy. 2018, January. http://english.gov.cn/archive/white_paper/2018/01/26/content_281476026660336.htm.

Filimonova, Nadezhda, and Svetlana Krivokhizh. 2018. China's Stakes in the Russian Arctic. *The Diplomat*, January 18.

Fridtjof Nansens Institute. https://www.fni.no/projects/international-northern-sea-route-programme-insrop-article318-277.html.

Gong, Keyu. 2016. The Cooperation and Competition between China, Japan and South Korea in the Arctic. In *Asian Countries and the Arctic Future*, ed. Leiv Lunde, Jian Yang, and Iselin Stensdal. Singapore: World Scientific Publishing Co. Pte. Ltd.

Government of Japan. 2016, January. *Fifth Science and Technology Basic Plan*. Ministry of Economy Trade and Industry.

Hammond, Joseph. 2017. Interview with Japan's Arctic Ambassador: Kazuko Shiraishi on Japan's Approach to the Arctic Region. *The Diplomat*, March 8. https://thediplomat.com/2017/03/interview-with-japans-arctic-ambassador/.

Hara, Kimie, and Ken Coates, eds. 2015. *East Asia-Arctic Relations: Boundary, Security and International Politics*. Waterloo: The Centre for International Governance Innovation.

Higgins, Andrew. 2013. Teeing off at the Edge of the Arctic? A Chinese Plan Baffles Iceland. *The New York Times*, March 23. https://www.nytimes.com/2013/03/23/world/europe/iceland-baffled-by-chinese-plan-for-golf-resort.html.

Lackenbauer, Whitney P., Adam Lajeunesse, James Manicom, and Frédéric Lasserre. 2018. *China's Arctic Ambitions and What They Mean for Canada*. Calgary: University of Calgary Press.

Meesak, Daniel. 2018. Expect More Chinese Tourists in the Arctic Region. *Jing Travel*, February 1. https://jingtravel.com/expect-more-chinese-tourists-in-the-arctic/.

Moe, Arild. 2016. Chapter 6: "International Use of the Northern Sea Route – Tends and Prospects". In *Asian Countries and the Arctic Future*, ed. Leiv Lunde, Jian Yang, and Iselin Stensdal, 115. Singapore: World Scientific Publishing Co. Pte. Ltd.

Ohnishi, Fujio. 2014. The Process of Formulating Japan's Arctic Policy: From Involvement to Engagement. In *Chapter 2 in East Asia-Arctic Relations: Boundary, Security and International Politics*, ed. Kimie Hara and Ken Coates. Waterloo, ON: The Centre for International Governance Innovation.

———. 2016. Japan's Arctic Policy Development: From Engagement to Strategy. In *Asian Countries and the Arctic Future*, ed. Leiv Lunde, Jian Yang, and Iselin Stensdal. Singapore: World Scientific Publishing Co. Pte. Ltd.

Park, Young Kil. 2014a. Arctic Prospects and Challenges from a Korean Perspective. In *Chapter 4 in East Asia-Arctic Relations: Boundary, Security and International*

Politics, ed. Kimie Hara and Ken Coates. Waterloo, ON: The Centre for International Governance Innovation.

———. 2014b, July. South Korea's Interests in the Arctic. *Asia Policy* 18: 59–65.

Shipping Traffic on Northern Route Grows by 40 Percent. http://www.highnorth-news.com/shipping-traffic-on-northern-sea-route-grows-by-40-percent/.

Stensdal, Iselin. 2016. Arctic Mining: Asian Interests and Opportunities. In *Chapter 9 in Asian Countries and the Arctic Future,* ed. Leiv Lunde, Jian Yang, and Iselin Stensdal. Singapore: World Scientific Publishing Co. Pte.

Sun, Kai. 2014, July. Beyond the Dragon and the Panda: Understanding China's Engagement in the Arctic. *Asia Policy* 18: 49.

The Arctic: Not So Cool. *The Economist,* January 31, 2015.

Tonami, Aki. 2014, July. Future-Proofing Japan's Interests in the Arctic: Scientific Collaboration and a Search for Balance. *Asia Policy* 18: 52–58.

———. 2016. *Asian Foreign Policy in a Changing Arctic: The Diplomacy of Economy and Science at New Frontiers.* Basingstoke: Palgrave Macmillan.

Tonami, Aki, and Stewart Watters. 2012. Japan's Arctic Policy: The Sum of Many Parts. *Arctic Yearbook* www.arcticyearbook.com/images/Articles_2012/Tonami_and_Watters.pdf.

Zhang, Pei, and Jian Yang. 2016. Changes in the Arctic and China's Participation in Arctic Governance. In *Chapter 13 in Asian Countries and the Arctic Future,* ed. Leiv Lunde, Jian Yang, and Iselin Stendsdal. Singapore: World Scientific Publishing Co. Pte. Ltd.

CHAPTER 21

The History of USA-Russia Relations in the Bering Strait

Rebecca Pincus

The maritime border between Russia and the United States runs directly through the Bering Strait, a narrow and shallow channel that is a critical corridor for Arctic marine species and subsistence hunting. As human activity in the Arctic increases, the Bering Strait is emerging as a maritime chokepoint, where a hostile state or nonstate actor could block access to or from the western Arctic by controlling the strait. In addition, increased traffic will pose a complicated management problem as navigational hazards, fragile ecology, and a variety of human uses must be balanced—by the two neighboring superpowers. This chapter will trace the contours of US-Russian relations in the Bering region and provide an introduction to this unique geopolitical arena.

First, the chapter will review the basic contours of the challenge confronting policymakers on either side of the strait: the problems associated with the geography and climate; the ecological significance of the region and its importance to sustaining subsistence practices for local communities; and then the weighty history of political and military development—stretching back to the sale of Alaska to the United States by Russia in 1867. After setting the scene, the chapter will explore the focal legal agreements that support co-management, and introduce the key actors who interface across the Bering Strait. Finally, the chapter will argue that the United States and Russia share some important goals in the Bering region, and that this rapidly changing maritime border area should be a region of increasing engagement to ensure resilience in the face of massive change.

This chapter reflects the author's own views and not the official position of the US Naval War College or US Navy.

R. Pincus (✉)
U.S. Naval War College, Newport, RI, USA
e-mail: Rebecca.h.pincus@uscga.edu

© The Author(s) 2020
K. S. Coates, C. Holroyd (eds.), *The Palgrave Handbook of Arctic Policy and Politics*, https://doi.org/10.1007/978-3-030-20557-7_21

CHALLENGES OF GEOGRAPHY AND CLIMATE

Physical

At roughly 85 km wide and 55 m deep, Bering Strait is a narrow and shallow passage that is covered by sea ice for several months of the year (Fig. 21.1). The US-Russian border and the international date line divide the strait; in addition, the two Diomede Islands, Big and Little Diomede, straddle the boundary line. Little Diomede is on the eastern (United States) side, and Big Diomede is on the western (Russian) side, and a narrow channel separates the two (Fig. 21.2). The Bering Strait connects the Bering Sea and the northern Pacific Ocean to the south with the Chukchi Sea and the Arctic Ocean to the north.

The Bering Sea is one of the stormiest places on earth, and is legendary for its lethality to mariners. According to Coast Guard analysis, the waters off Alaska are by far the most hazardous location for vessel casualties in US waters, with 457 fishing vessels lost between 1992 and 2007 (USCG 2008). Conditions for mariners may become even worse in the future: studies indicate that winds and waves are increasing in the Bering Sea, and farther north in the Chukchi and Beaufort Seas as well, due to warming temperatures and retreating sea ice (Wang et al. 2015).

Fig. 21.1   Physical geography of Alaska and the Bering Sea

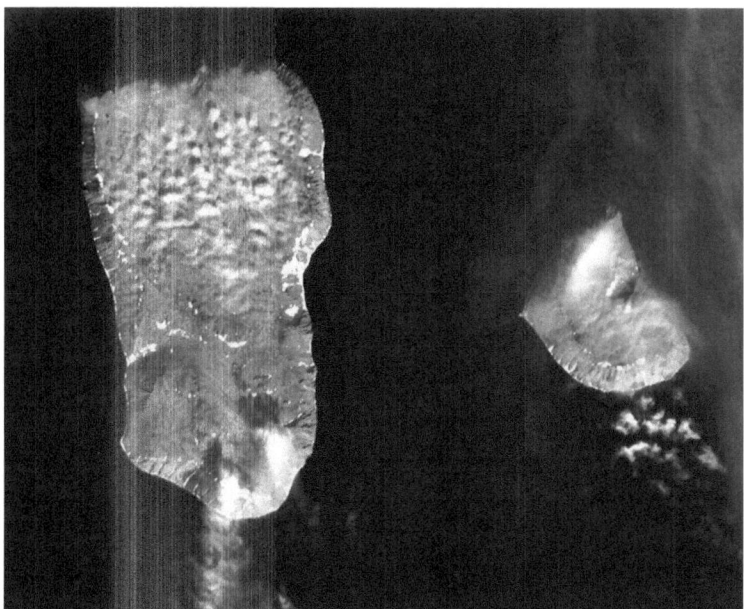

Fig. 21.2 Diomede Islands, Bering Strait. Source: NASA/GSFC/METI/ERSDAC/ JAROS, and United States/Japan ASTER Science Team

Part of the reason for the extreme storms and weather in the Bering Sea area is physical geography: the North American and Asian continents bend toward each other, and form a bowl via the string of volcanic islands in the Aleutian chain. As extra-tropical cyclones veer north through the northern Pacific Ocean, they drive into the mountains of the Aleutians and stall out, dumping precipitation and wind into the Bering Sea and Gulf of Alaska. The shallow waters in the Bering Strait, and northern Bering Sea, which also drive stormi-ness, are a product of the geologic history of the area. The Bering Strait was dry land—the Bering land bridge—between about 20,000 and 13,000 years ago and supported a shrub tundra environment similar to those found in Siberia and Alaska today (INSTAAR). As the glaciers retreated and sea level rose, the land bridge was once more submerged.

Human Geography

While archeologists are still refining our understanding of human migration between Asia and North America, it is clear from the record so far that humans have been in the area for millennia. Inuit or Eskimo peoples have lived in the Bering region for about 5000 years, although the archeological record includes evidence of premodern habitation dating back 10,000 years (Ahmasuk et al. 2007).

Today, there are three distinct linguistic and cultural groups in the Bering Strait region: the Inupiaq, Central Yupik, and Siberian Yupik. Siberian Yupik people live on St. Lawrence Island and in Chukotka, on the Russian side (Kawerak 2014). In the regional units on either side of the border, Indigenous peoples are minority populations: the Russian autonomous district of Chukotka, which has approximately 50,000 inhabitants, has the highest concentration of Indigenous peoples, with nearly 30% (Golunov 2016). In the United States, Alaska Native individuals are about 15% of the state's population, but comprise three-quarters of the 9000 inhabitants of the areas adjacent to the Bering Strait. Nome, with 3700 inhabitants, is the largest US settlement, and there are 15 year-round villages around Nome ranging from 161 to 798 inhabitants (Ahmasuk).

Ecology and Subsistence

Connector and Corridor

The Bering Strait is an important migration corridor for a huge variety of species; in addition, it is an important channel for water from the Pacific Ocean to flow in the Arctic basin. Water flowing northwards through the Bering Strait carries nutrients that upwell off the continental shelf in the northern Pacific, supporting rich primary and secondary productivity. Scientists point to this flux of nutrient-rich water through the Bering Strait as explanation for the incredibly high biological productivity of the Bering and Chukchi Seas, which, in spring and summer, "rival those of any location in the world ocean" and remain higher than any other part of the Arctic year-round (Cooper et al. 2006). Research suggests that northward flow through the Bering Strait acts as "a conduit" for warm water into the Arctic that may play an important role in overall Arctic warming (Woodgate et al. 2010).

The high primary productivity of the Bering Strait makes the region an "ecological hotspot of global proportions," according to Audubon Alaska, including birds, whales, seals, walrus, and fish. A huge variety of species travel to the Beaufort and Chukchi Sea area in the summer to take advantage of the explosion of productivity that occurs in the Arctic summer. "Because the Strait is the only passage between the Pacific and Arctic Oceans, all wildlife that live in the Chukchi and Beaufort Seas in the summer months funnel through the Bering Strait twice each year during spring and fall migration" (Audubon). In particular, "one of the largest marine migrations in the world" takes place each year as marine mammals and birds travel along the eastern side of the Bering Strait and along the Chukchi Sea coast (Pew). Martin Robards of the Wildlife Conservation Society describes this migration as "one of the great wonders of our planet," and observes that, "there is no getting around the sheer immensity of this movement of marine mammals back and forth through the international waters" between Chukotka and Alaska (Robards 2017).

Because of its importance to a wide variety of species, including fragile populations of migratory birds and marine mammals, which return to the same areas to feed and reproduce each year, and are therefore vulnerable if those areas are damaged, the Bering Strait region is a focus of advocacy and protection efforts for some of the most prominent environmental nongovernmental organizations (NGOs) in the world. In support of their advocacy efforts, leading NGOs have frequently served as intermediaries between Russian and American governments.

Subsistence

Given the remarkable ecological richness of the Bering Strait, it is also of great importance to the communities that live in the region and practice subsistence hunting and fishing. Traditional subsistence hunting of walrus, seals, and whales has provided food, shelter, and clothing for Bering region communities for thousands of years. Traditional hunting practices are based on respect for the animals and the broader ecosystem, and wastefulness, cruelty, or disrespect violates these norms. Specific values include practicing respect for individual animals and their communities; minimizing loss of dead or wounded animals; restricting harvest to what is needed; using all of the harvested animal and sharing the harvest; respecting other hunters; practicing stewardship of the land and ocean; and passing traditional knowledge and values to future generations (Gadamus and Raymond-Yakoubian 2015). Subsistence practices are shared across the Bering Strait. For example, the Kawerak nonprofit organization issued a poster explaining "Indigenous knowledge and use of ocean currents" in both English and Russian, which also defined important Inupiat and Chukchi words for currents.

In addition to the important cultural and community elements of subsistence practices, traditional foods are critical to supporting the food security of communities in the Bering region. In a region where wage-based employment is scarce, particularly once outside the major population centers of Barrow, Nome, and Kotzebue (about 5000, 4000, and 3000 residents, respectively), opportunities to earn cash are limited. In addition, since all goods need to be flown or barged in, prices for food and other goods are very high. As a result of these two factors, generally low cash income and high prices of food, traditional foods that are obtained outside of the cash economy are a critical element of the food security of the region. In 2000, Caulfield calculated that "the average rural Alaskan uses more wild meat and fish than the average American uses store-bought meats and fish (Caulfield 2000)." Similarly, in 2007, a comprehensive survey of subsistence harvest calculated that 12 communities on the US side of the Bering Strait collectively harvested over 750,000 lbs. of fish, and 3,062,395 lbs. of marine mammals per year (Ahmasuk). Of the total amount of food harvested through subsistence practices, marine mammals contributed 67.9%.

In the Bering Strait region, human and environmental health is inextricably linked. The reliance of the region's communities on locally harvested foods, in particular marine foods, makes environmental stewardship especially important for human health and community survival. The establishment of the Arctic Contaminants Action Program (ACAP) as one of the six permanent working groups of the Arctic Council is an indication of the scale of the threat that pollution poses in the Arctic region as a whole. In the Bering Strait region specifically, changing climatic conditions and rising maritime traffic pose significant additional threats.

Changing Environment

It is widely recognized that the Arctic is experiencing sudden and significant warming. In the Bering Strait region, warming has the potential to disrupt the ecological vitality and community integrity described above. Scientists characterize the changes taking place in the Bering Sea as a "major ecosystem shift" during which the reduction in sea ice, warmer water temperatures, and resulting shifts in productivity in upper and lower parts of the water column have led to changes in species distribution (National Oceanic and Atmospheric Administration, NOAA). For example, a 2008 survey in the Beaufort Sea, north of the Bering Strait, identified northern extensions to the ranges of four fish species, including varieties of cod, pollock, and sculpin, as well as increasing size and distribution of snow crab (Rand and Logerwell 2011). It is not yet clear how arrays of species will respond to changes in the environment, and it is therefore incumbent upon responsible authorities to continue to monitor changes in the Bering Strait region and develop plans for adaptation and enhanced resilience in the face of change.

Increasing Traffic

Increasing ship traffic in the Bering region adds to the challenge of managing for rapid environmental change. More ships in the area will increase underwater noise and baseline pollution due to regular operations. In addition, higher ship traffic raises the risk of episodic pollution associated with maritime incidents, as well as the likelihood of ship strikes harming both marine mammals and small vessels used by subsistence hunters. While vessel transits through the Bering Strait remain relatively few in number, reaching a high of 540 in 2015, the number of annual transits appears to be increasing (USCG 2016). Although projections of future traffic vary widely, the combination of increasing economic activity and reduced ice coverage has led to consensus that ship traffic will continue to grow in coming years.

The combination of change in both the environmental and human systems of the Bering Strait region heralds a new era of disruption. In earlier centuries, paradigmatic change brought transformation to the area in a series of successive waves. The first wave was comprised of commercial interests, in the form of fur traders and the colonial structures that were first established in the region in the nineteenth century. During the first penetration of the region by traders, including the Russian American Company, both Indigenous human populations and valuable animal species, including seals, otters, and some whales, were decimated. The second transformative wave followed the establishment of the Alaskan territory, during which later-stage industrial development generated environmental change through large mining operations and the refinement of industrial fishing techniques. During the war years, and the third phase of transformation, widespread infrastructure construction reengineered the landscape, and testing of nuclear and conventional arms increased pollution.

Through all of these changes, the Bering Strait region remained at the very fringes of two major states, which were both challenged by the distances and harsh climate to extend their power in the area effectively. The changes occurring today should be understood as comprising a fourth wave of transformation through the Bering Strait region, during which environmental change and advances in technology have the potential to significantly alter geopolitical realities.

POLITICAL AND MILITARY DEVELOPMENT

Sale of Alaska

On 30 March 1867, US Secretary of State William Seward and Russian Minister to the United States Edouard de Stoeckl signed the "Treaty concerning the Cession of the Russian Possessions in North America," which conveyed Russian-controlled territory in Alaska and the Aleutian Islands to the United States for $7.2 million. America's nineteenth-century westward expansion coincided with a period of Russian retrenchment, and the sale was seen as benefiting both parties.

A succession of extractive industries stripped resources from the rich territory: first the fur industry, then whaling, then salmon fishing and gold mining. The Klondike gold strike of 1896 drew an immense wave of immigration to Alaska, including prospectors and the industries serving them. Discoveries of gold in Seward and Nome sustained the rush, and the sudden growth finally drove a larger and more effective governmental presence.

Following World War I, the Alaskan territory saw a significant drop in population, as the wartime demands in the United States affected employment, and drops in copper and salmon exports also occurred. However, the advent of World War II marked a transformation across the Bering region, as a wave of militarization swept through Alaska and the Russian Far East. After decades of transient, avaricious interest by commercial actors, the Bering region would become the focus of two successive waves of militarization that would permanently reshape political, economic, and geographic realities on both sides of the border.

Militarization: World War II and Cold War

The advent of World War II, and then the Cold War, brought transformative change to Alaska and the Bering Strait. More than $1 billion of defense spending poured into Alaska between 1941 and 1945, remaking the territory and eventually contributing to Alaska's accession as the 49th US state in 1959 (Hummel 2005).

The Japanese invasion of the Aleutian Islands in June, 1942, opened a dramatic chapter in World War II and dramatically spurred militarization in the region. By August 1943, when the campaign ended, the US military presence in Alaska had expanded exponentially. New bases and airfields were constructed at Port Heiden, Adak, Umak, Cold Bay, and Amchitka. Ships and submarines were sent in to help oust the Japanese forces. Hundreds of thousands of troops eventually were involved in the campaign, which is remembered today for the disastrous effect of bad weather on operations across land, air, and sea.

In addition to serving as a theater of combat against Japan, the Bering Strait region also was a key location for the uneasy alliance between Washington and Moscow. Through the Lend-Lease program, the United States transferred supplies and military equipment to the Soviet Union to support its fight against the Nazi regime. Airplanes were transferred by the shortest route between the United States and USSR: the ALSIB, or Alaska-Siberian air route, which opened in August 1942 (Dolitsky 2016). By April 1943, 142 aircraft per month were transferred via the ALSIB route, although weather, cold, and logistics challenges hampered the process. Large shipments of supplies went via the Pacific route into Vladivostok, and studies examined the feasibility of shipping across the Bering Sea as well. The success of the northern Pacific supply route, which was virtually unthreatened by enemy attack, stood in contrast to the heavy losses inflicted upon Allied shipments through the North Atlantic into Murmansk and Archangelsk, and demonstrated that taking the kitchen door between Alaska and Siberia, although complicated by bad weather, was far safer than risking German interdiction.

However, growing hostility between Moscow and Washington led to an abrupt reversal in relations. In 1948, all of the Soviet Union east of the Urals was closed to Westerners, and the formerly cooperative relationship across the Bering Strait and northern Pacific turned hostile. Instead of supply missions, US planes and submarines quietly gathered intelligence, and Soviet reports even suggested the CIA was running Eskimo spies in Chukotka (Stephan 1994).

In the region of the Bering Strait, the Cold War saw the growth of submarine operations under the polar ice. In World War II, submarines had occasionally used the icepack as a cover, but without the technology to operate safely in this hazardous environment, they had not ventured deep into the Arctic. However, during the rapid advances in military science and technology that characterized the early Cold War, US submarines began developing the knowledge and tools to permit under-ice operations of increasing sophistication. Soviet submarine operations similarly advanced into Arctic areas. The concept of the under-ice region as an "Arctic sanctuary" for submarines armed with nuclear missiles drove development of technology and operational concepts by both American and Soviet navies (Michishita et al. 2016). When the first American submarine successfully crossed the Arctic Ocean, the USS NAUTILUS in 1958, the transit was characterized as "America's answer to Sputnik," demonstrating the importance of strategic Arctic submarine operations at the time (Leary 1999). With the dawn of the era of ballistic missile submarines, the Bering Strait was increasingly identified as a key maritime chokepoint permitting access to the Arctic basin.

The Ice Curtain

The separation of families and communities across the Bering Strait that began in 1948 continued through the Cold War, and was a politically sensitive issue in Alaska. In 1948, 26 men from Little Diomede who had rowed over to Big Diomede for a family visit were held by Soviet soldiers for 52 days, under hardship conditions that resulted in one man's death (Griffith 1975). After this experience, the previously friendly and fairly easy cross-strait relations plunged into a deep chill that persisted until the 1980s. The forced relocation of villages off of Big Diomede in the 1950s by Soviet authorities further separated families and friends.

Motivated citizens began campaigning for a relaxation of border controls in the 1980s. In the waning days of the Cold War, citizen-driven diplomacy in Arctic Alaska made a significant contribution to improving relations between the United States and USSR (Ramseur 2017). The proximity of Alaska and Chukotka facilitated a series of public-relations coups that helped drive political momentum toward more open relations.

In August 1987, US swimmer Lynne Cox made global news for her unprecedented 2.7 mile swim across the Bering Strait, between the islands of Big and Little Diomede. Wearing just a Speedo swimsuit, and beset by fog and strong currents, she made the swim in 44-degree weather and barely avoided hypothermia. The Soviet authorities, which had refused to grant permission until hours beforehand, laid on a welcoming party with TV cameras, a buffet, and a warming tent with sports medicine doctors (Roberts 1987).

In 1988, the Russia newspaper *Pravda* blared, "Alaskan Eskimos Visit Chukotka Capital." Along with Alaska Natives, the 87-person delegation of visitors from Alaska included Governor Cowper and Senator Frank Murkowski (father of current Alaska Senator Lisa Murkowski). *Pravda* noted, "This is probably the first delegation in the entire postwar history of Soviet-U.S. relations to take the shortest route to our country's territory: the flight took less than 30 minutes" (Foreign Broadcast Information Service (FBIS) 1988).

The remarkable exploits of activists utilized the proximity of the United States and Russia across the Bering Strait to viscerally drive home their arguments about closer relations across the border. In the Cold War context, the Bering Strait served as a tangible reminder of the possibilities for local communication between the two superpowers.

Legal Agreements and Key Areas of Cooperation

During the Cold War, while activists tried to force openness across the Bering Strait, the US and Soviet governments did take tentative steps toward developing a coordinated governance regime in the region. An early priority was fisheries management, since the rich Bering Strait fishery was plagued by spats over access, quotas, and accusations of illegal fishing. Environmental issues were also a relatively safe area for growing cooperation, and in 1972 Moscow and

Washington signed an Agreement on Cooperation in the Field of Environmental Protection. Other Arctic states facilitated cooperation between the Soviets and Americans, in part by encouraging environmental measures like the 1973 Agreement on the Conservation of Polar Bears, which was signed by the five coastal Arctic states. Following the end of the Soviet Union, cooperative agreements and projects flourished in the Bering region. Many of these avenues of cooperation are still active today, although in one key area—the maritime border itself—there is not yet a comprehensive agreement. The following section will review major areas of cooperation and the important agreements that underpin them.

Maritime Border

In 1990, the US Senate voted in favor of the "Agreement with the USSR on the Maritime Boundary," which had been signed in Washington DC, 1 June 1990. In his transmitting letter, President Bush noted, "I believe the agreement to be fully in the United States' interest." He added that the agreement "removes a significant potential source of dispute between the United States and the Soviet Union." Secretary Baker led negotiations on the agreement. The Agreement states: "From the initial point, 65 deg 30' N., 168 deg 58' 37" W., the maritime boundary extends north along the 168 deg 58' 37" W. meridian through the Bering Strait and Chukchi Sea into the Arctic Ocean as far as permitted under international law." The Agreement laid out the coordinates and track of the boundary line, and also included provisions pertaining to international maritime law. For example, the Agreement stipulated that both states would cede claims to "special areas" within their respective exclusive economic zones (EEZs) that fell on the opposite side of the boundary line.

Negotiators intended the agreement to settle a long-running dispute stemming from uncertainty over the type of line used to demarcate the US-Russian border in the 1867 Convention that concluded the sale of Alaska to the United States. The ratification of the Law of the Sea Convention (UNCLOS) and growing implementation of 200-nm exclusive economic zones (EEZs) in the 1970s drew attention to the lack of agreement between the two parties. In 1977, the United States and the USSR exchanged diplomatic notes, and talks commenced in the early 1980s (Konyshev and Sergunin 2014). Since neither party had preserved the maps used during the original negotiation, there was no authoritative way to settle the disagreement over which type of line was used in 1867, rhomb or geodetic, and which type of map, Mercator or conical. Each party based their interpretation on the line that gave their side the best and largest claim, with the result being that the US and Soviet claims overlapped to the tune of 15,000–18,000 square nm (Kaczynski 2007). Nine years of negotiations led to the final agreement.

However, the 1990 agreement was broadly criticized in Russia for failing to defend Russian fishing interests in the disputed area. The Russian Duma has never ratified the agreement, and it remains a sensitive topic.

Fisheries

Russian hesitation over the maritime border in the Bering stems in part from the outstanding biologic resources in the basin, which is one of the world's richest fishing grounds. Russian and American vessels work the area, along with foreign fleets from Asia and beyond. Serious problems with fisheries management served to galvanize efforts toward cooperation by Russian and American authorities. Experts point to evidence of significant overfishing on the Russian side of the boundary, including by members of organized criminal networks: the 'fish, crab, and caviar mafias,' which have impacted regional ecology and marine productivity (Konyshev). In 2014, the environmental NGO World Wildlife Fund (WWF) reported that illegal crab harvests in the Bering Sea were several times higher than the legal limit, and warned that the sustainability of several stocks was in jeopardy (WWF 2014). Pressure on the high-value Bering fishery may increase competition between fishing interests and fisheries regulators on either side of the boundary.

On the key issue of fishing in the Bering, the US Coast Guard works with the National Oceanic and Atmospheric Administration (NOAA), and the State Department, to enforce and execute agreements relating to fisheries management.

US-Soviet fisheries discussions were active during the depths of the Cold War. In 1975, the two states signed an agreement on fishing for king and tanner crab; and the following year, the two powers concluded an agreement on fisheries. In 1988, the United States and the USSR signed a comprehensive fisheries agreement, which was succeeded by a bilateral agreement between the United States and the Russian Federation. The agreement established the Intergovernmental Consultative Committee, which meets annually to discuss cooperation on fisheries science, joint ventures, and IUU fishing, as well as to allocate surplus resources and consult on issues of joint concern.

While the Russian Coast Guard/Border Guard conducts at-sea enforcement of fisheries, the Russian Federal Fisheries Agency manages policy questions and participates in the ICC. The agreement, which has been regularly renewed, will next expire 31 December 2018. In 2015, the US-Russia ICC signed an agreement on cooperation in enforcement against illegal, unregulated, and unreported (IUU) fishing in the North Pacific area. The 2016 meeting also emphasized the importance of law enforcement cooperation to prevent IUU fishing.

Fisheries policy is an important issue in the Bering Sea, and is of high political and economic importance in the region, which gives fisheries issues great salience in coast guard relations across the maritime border. On the Russian side, key federal authorities include the Russian Coast Guard, which is an element of the Border Guard Service of the Federal Security Service (successor to the KGB). There are two Border Guard districts adjacent to Alaska: Kamchatka and Chukotka. The US Coast Guard District 17 and the Kamchatka Border Guard Directorate have longstanding agreements that facilitate cooperation in fisheries enforcement. The district commanders meet regularly and US Coast Guard vessels occasionally make port calls. In 2013, USCG's District 17 and the Kamchatka Border Guard Directorate signed a joint agreement to coordinate maritime security and fisheries law enforcement in the Bering region (USCG 2013). The agreement was signed in Anchorage, during a biannual meeting that alternates between Alaska and Russia. In 2011, the US Coast Guard also hosted the meeting, which included a visit by the Russian cutter VOROVSKY, during which the crew conducted joint exercises and activities with the USCGC BERTHOLF and its crew (USCG 2011).

Maritime Safety and Security

In 1989, just two months after the Exxon Valdez oil spill in Prince William Sound, Alaska, the United States and the Soviet Union signed a treaty "Concerning cooperation in combatting pollution in the Bering and Chukchi Seas in emergency situations" which provided that the two countries "undertake to render assistance to each other in combatting pollution incidents which may affect the areas of responsibility of the Parties, regardless of where such incidents may occur."

Another key Russian actor is the State Marine Pollution Control Salvage and Rescue Administration (SMPCSRA), which is part of the Federal Agency for Maritime and River Transport. The SMPCSRA is the lead actor in managing search and rescue (SAR) and marine pollution incidents, and works with the US Coast Guard in those circumstances, relying on agreements that support communications and response. The US Coast Guard and the SMPCSRA participated in a bilateral Russian-led communications exercise in September 2012 (Duignan 2013). The Russian Ministry of Emergency Situations, or EMERCOM, also provides response assets to SMPCSRA in case of an emergency.

At the level of maritime policy, the US Coast Guard works with the Russian Ministry of Transport, specifically the Department of State Policy in Marine and River Transport, which has regulatory authority over areas like navigation safety and standards. In 2017, the United States and Russia jointly submitted a proposal establishing two-way routing measures and precautionary areas in the Bering Strait to the International Maritime Organization (IMO), a specialized agency of the United Nations that sets global shipping rules.

Park Diplomacy and Environmental Protection

Environmental protection is another significant track for cooperation between Russia and the United States in the Bering Strait region. In June 1990, Soviet leader Mikhail Gorbachev came to Washington for a heavily-scrutinized summit with US President George H.W. Bush. While most of the attention focused on arms-control agreements and talks over the future of Germany and the

Baltics, the two leaders also signed an agreement to establish an international peace park straddling the Bering Strait, the Beringian Heritage International Park (Gorman 1990). While legislation to formally establish the park never made it through the US Congress (it was introduced by Senator Pell in the form of S.2088, the Beringian Heritage International Park Act of 1991), in 2012 Secretary of State Hillary Clinton and Foreign Minister Sergey Lavrov signed a joint statement on developing the Beringia park. In her remarks, Secretary Clinton noted that the joint statement "signals our desire to collaborate more closely in the region where our countries are only miles apart" (State Dept. 2012). The National Park Service has administered the Shared Beringian Heritage Program since 1991, which "works to improve local, national, and international understanding" and "cultural vitality of Native peoples" of the region (NPS 2013).

In 1994, US and Russian authorities signed a treaty on "Cooperation in the Field of Protection of the Environment and Natural Resources." The responsible agencies are the Fish and Wildlife Service in the United States and the Ministry of Natural Resources and Environment in Russia. On environmental issues, the US Fish and Wildlife Service works closely with the Russian Ministry of Natural Resources and Environment to protect and manage species in the Bering region. In particular, the US-Russia Marine Mammal Working Group advances protection for marine mammals including whales, walrus, and otters. Working Group meetings occur every few years through the Wildlife Without Borders program, and rotate between Russia and the United States.

A special program deals specifically with polar bears. In 1973, the United States, Soviet Union, Canada, Denmark, and Norway all signed an "Agreement on the Conservation of Polar Bears," however, in 2000, the US and Russia signed another bilateral treaty specifically addressing their shared polar bear populations. The "Agreement between the Government of the U.S.A. and the Government of the Russian Federation on the Conservation and Management of the Alaska-Chukotka Polar Bear Population" established a US-Russia Polar Bear Commission that is responsible for management decisions, and created a Scientific Working Group to advise the Commission. The Commission meets annually and rotates between Russia and the United States.

The cooperative measures on environmental protection and management are also supported by formal cooperative structures that facilitate scientific information gathering and data sharing. For example, the Russian-American Long-term Census of the Arctic (RUSALCA, which is "mermaid" in Russian) grew out of a memorandum of understanding between NOAA and the Russian Academy of Sciences. RUSALCA led to joint scientific expeditions in the Bering and Arctic regions annually between 2004 and 2015, as well as the installation of a chain of scientific moorings across the Bering Strait, which sampled fluxes of heat, salt, and nutrients between the Bering Sea and the Arctic Ocean.

ROOM FOR COOPERATION

Since 2014, there has been very little contact between the US and Russian governments. In March of that year, in response to Russian actions in Ukraine, President Obama took a series of punitive steps, including the imposition of targeted sanctions through Executive Order 13660. The initial round of sanctions has since been expanded through additional orders. Among other targets, the sanctions specifically included the Russian Arctic oil sector. The Arctic sanctions led to embarrassment for then-Secretary of State Rex Tillerson, who had, as CEO of Exxon Mobil, concluded agreements with Igor Sechin, the head of Rosneft, for joint ventures in the Kara Sea that violated the sanctions regime. Exxon was fined $2 million by the Treasury Department in 2017 and in 2018 pulled out of the project, taking a $200 million loss (Rappeport 2017). The specific inclusion of Arctic projects in the sanctions regime, along with the highly-visible flap over Secretary Tillerson and Exxon's deals with Rosneft, threw an unusually strong spotlight on US-Russian relations in the Arctic region.

In the Bering Strait, however, the shared maritime border means that some interaction is still necessary. The government-government contact across the Bering Strait in recent years has been confined to clearly mandated missions. Areas of shared interest are clear: fisheries concerns dominate, and closely related concerns relating to maritime safety and environmental stewardship are also of evident importance.

Serghei Golunov (2016) describes the geographical proximity of the United States and Russia across the Bering Strait as "highly relative" and highly variable to changing perceptions and circumstances. While political factors and economic conditions affect perceptions of proximity, Golunov also argues that the difficulties inherent in US-Russian proximity—including the maritime border, and the climate and infrastructure hurdles—require an extra step. In order to be workable and exploited fully, US-Russian proximity requires "the willingness and opportunities of certain actors to cooperate."

The history of Russia-US relations across the shared maritime border in the Bering Strait reflects the variability Golunov describes. In addition, formal government-government relations have frequently been much different from informal connections between communities in the region. However, proximity to the border (as well as distance from the respective capitols and centers of government authority) has also served to facilitate grassroots activism aimed at altering government relations.

The special aspects of a shared maritime border lend another dimension to the complex and shifting tenor of Russia-US relations in the Bering Strait region. Without fixed border installations, and systems of infrastructure linking the two countries like road and rail connections, the border is less tangible and seemingly permeable. As one local inhabitant described it, "We do not know exactly where the line is, where the border is. It is sort of tempting; do you understand?" (Griffith 1975). The absence of a physical border, and the flow of marine life all around and through the Bering Strait, lends an element of surrealism only heightened by the extreme political swings: from post-revolutionary suspicion to the desperate partnership in war; then again to Cold War acrimony, followed by the 1990s thaw in relations—followed once more by rising tension in recent years.

Throughout the decades of drastic change, maritime authorities have been at the front lines of official contact and cooperation. The coast/border guards at sea, and the constellation of fisheries and environmental scientists and managers beyond, have sustained US-Russian relations in the Bering Strait region through political turmoil. As both countries prepare for a new wave of upheaval linked to both a changing climate and increasing human activity, these front-line authorities should be given the tools and latitude to advance cooperation and stewardship across the region. Although the relationship between Russia and the United States is at a modern low point, the shared interests present in the Bering Strait region—including sustainable and controlled fisheries, environmental stewardship, and safety of life at sea—are enduring. Any diminishment in cooperation in these areas will only harm US interests, now and in the future. Conversely, growing cooperation may offer room for positive improvements in influencing Russian positions and behavior.

In the circumpolar perspective, the cross-border cooperation that seems so astonishing between Russia and the US is less remarkable. The strength of cultural ties between communities around the Arctic has supported enduring cooperative relationships at the local level, which often cross international borders. For example, the subnational governments of Nunavut (Canada) and Greenland (Denmark) have had a long and robust relationship centered on shared heritage and common interests across their shared maritime border. Close cooperative ties also exist across land borders: the Nordic Sámi Convention is a governance initiative representing the interests of Sámi people in Norway, Sweden, and Finland, where national borders cut through the traditional areas inhabited by the Sámi people. However, the US-Russian example in the Bering region is an important marker for cross-border cooperation in the Arctic. As human activity increases in the Bering Strait, the stakes for cooperation will grow even higher. The Bering Strait is a fulcrum point where national and local levels of interests, diplomacy, and security interact and collide. As traffic increases, both US and Russian authorities in the region will face increasing pressure to hold these complex tensions in balance.

References

Ahmasuk, A., E. Trigg, J. Magdanz, and B. Robbins. 2007. Bering Strait Region Local and Traditional Knowledge Pilot Project; A Comprehensive Subsistence Use Study of the Bering Strait Region, North Pacific Research Board, Kawerak, Inc.

Audubon Alaska. "Bering Sea." http://ak.audubon.org/conservation/bering-sea.

Caulfield, R. 2000. Food Security in Arctic Alaska: A Preliminary Assessment. Recherche du Groupe d'études inuit et circumpolaires (GÉTIC) de l'Université Laval.

Cooper, L.W., et al. 2006. The potential for using Little Diomede Island as a platform for observing environmental conditions in Bering Strait. *Arctic* 59 (2): 129–141.

Dolitsky, A.B., ed. 2016. *Pipeline to Russia*. Alaska: National Park Service.

Duignan, K. 2013. Partnerships in the Arctic. Proceedings. 58.

Foreign Broadcast Information Service (FBIS-SOV-88-120). Pravda, "Alaskan Eskimos visit Chukotka Capital." 15 June 1988.

Gadamus, L., and J. Raymond-Yakoubian. 2015. A Bering Strait Indigenous Framework for Resource Management: Respectful Seal and Walrus Hunting. *Arctic Anthropology* 52 (2), 87–101. Also See Report from ICC-Alaska, "Bering Strait Regional Food Security Workshop." (2014).

Golunov, S. 2016. The Russian-U.S. Borderland: Opportunities and Barriers, Desires and Fears. *Eurasia Border Review* 7 (1): 31–50.

Gorman, Steven J. 1990. International Park Accord Signed. UPI, June 1. https://www.upi.com/Archives/1990/06/01/International-park-accord-signed/6019644212800/.

Griffith, W. 1975. Detente on the Rocks. *The New York Times*, August 17.

Hummel, L.J. 2005. The U.S. Military as Geographical Agent: The Case of Cold War Alaska. *Geographical Review* 95 (1): 47–72.

Kaczynski, V. 2007. US-Russian Bering Sea Marine Border Dispute: Conflict Over Strategic Assets, Fisheries and Energy Resources. *Russian Analytical Digest* 20: 2–5.

Kawerak, Inc. and Oceana. 2014. *Bering Strait: Marine Life and Subsistence Use Data Synthesis*. https://oceana.org/publications/reports/the-bering-strait-marine-life-and-subsistencedata-synthesis.

Konyshev, V., and A. Sergunin. 2014. Russia's Policies on the Territorial Disputes in the Arctic. *Journal of International Relations and Foreign Policy* 2 (1): 55–83.

Leary, W.M. 1999. *Under Ice: Waldo Lyon and the Development of the Arctic Submarine*, 131. College Station: Texas A&M University Press.

Michishita, N., P.M. Schwartz, and D.F. Winkler. 2016. Lessons of the Cold War in the Pacific. Wilson Center and Sasakawa Peace Foundation, pp. 5–6.

National Park Service. 2013. *National Park Service Programs*, pp. 63–63.

Pew Trusts. A Look at Important Marine Areas in the U.S. Beaufort and Chukchi Seas. http://www.pewtrusts.org/en/research-and-analysis/issue-briefs/2016/05/a-look-at-important-marine-areas-in-the-us-beaufort-and-chukchi-seas.

Ramseur, D. 2017. *Melting the Ice Curtain*. Fairbanks: University of Alaska Press.

Rand, K.M., and E.A. Logerwell. 2011. The First Demersal Travel Survey of Benthic Fish and Invertebrates in the Beaufort Sea Since the late 1970s. *Polar Biology* 34 (4): 475–488.

Rappeport, A. 2017. Exxon Mobil Fined for Violated Sanctions on Russia. *The New York Times, Politics*, July 20.

Roberts, R. 1987. Orange County Woman Swims Bering Strait. *Los Angeles Times*, August 8.

Robards, M. 2017. Rush Hour in the Bering Strait. *Scientific American, Observations*. https://blogs.scientificamerican.com/observations/rush-hour-in-the-bering-strait/.

Stephan, J.J. 1994. *The Russian Far East*. Stanford, CA: Stanford University Press.

University of Colorado Boulder Institute of Arctic and Alpine Research (INSTAAR). Postglacial Flooding of the Bering Land Bridge. http://instaar.colorado.edu/groups/QGISL/bering_land_bridge/.

U.S. Coast Guard. 2016. Preliminary Findings: Port Access Route Study: In the Chukchi Sea, Bering Strait, and Bering Sea. 17th Coast Guard District. 13.

U.S. Coast Guard News. 2011. Coast Guard Hosts Meetings with Russia's Northeast Border Directorate in Kodiak, Anchorage. http://coastguardnews.com/coast-guard-host-meetings-with-russias-northeast-border-directorate-in-kodiak-anchorage/2011/04/15/.

———. 2013. Coast Guard, Kamchatka Border Guard Sign Joint Agency Protocol Document in Anchorage, Alaska. http://coastguardnews.com/coast-guard-kamchatka-border-guard-sign-joint-agency-protocol-document-in-anchorage-alaska/2013/04/18/.

U.S. Coast Guard Office of Investigations and Analysis. 2008. *Analysis of Fishing Vessel Casualties*, p. 6.

U.S. Department of State. 2012. Signing Ceremony with Russian Foreign Minister Sergey Lavrov. 8 September 2012. https://2009-2017.state.gov/secretary/20092013clinton/rm/2012/09/197518.htm.

Wang, X.L., Y. Feng, V.R. Swail, and A. Cox. 2015. Historical Changes in the Beaufort-Chukchi-Bering Seas Surface Winds and Waves, 1971–2013. *Journal of Climate* 28 (19): 7457–7469.

Woodgate, R.A., T. Weingartner, and R. Lindsay. 2010. The 2007 Bering Strait Oceanic Heat Flux and Anomalous Arctic Sea-Ice Retreat. *Geophysical Research Letters* 37 (L01602). https://doi.org/10.1029/2009GL041621.

WWF. 2014. Illegal Russian Crab: An Investigation of Trade Flow. https://www.worldwildlife.org/publications/illegal-russian-crab-an-investigation-of-trade-flow.

CHAPTER 22

Canada and Russia in an Evolving Circumpolar Arctic

Ron R. Wallace

INTRODUCTION

Circumpolar interests have traditionally concerned eight nations: Canada, Denmark, Finland, Iceland, Norway, Russia, Sweden, and the USA. Notwithstanding these direct geographic and territorial interests other nations, such as the UK and, increasingly, China, have expressed commercial and security interests in the circumpolar region. Rapid climatic changes in the circumpolar Arctic have engendered unprecedented scrutiny related to evolving commercial interests and have led to associated challenges for northern governance, the environment, legal rights, and geopolitics.

As noted by the Jasper Innovation Forum:

> At present, circumpolar governance entails the multi-layered application of national and international laws and various non-binding international treaties and agreements. International cooperation occurs through mutually agreed upon governance frameworks such as the United Nations Convention on the Law of the Sea (UNCLOS), the Arctic Council and the United Nations Framework Convention on Climate Change (UNFCCC). Other governance mechanisms are regionally specific, such as the work of the Arctic Council and the Barents Euro-Arctic Council or functionally specific, such as the guidelines for shipping developed under the auspices of the International Maritime Organization (IMO). Governance is further complicated by the fact that the Indigenous peoples of several northern nations have a variety of legally recognized rights and treaty agreements over large tracts of their traditional lands. (JIF 2011)

R. R. Wallace (✉)
Canadian Global Affairs Institute, Calgary, AB, Canada

© The Author(s) 2020
K. S. Coates, C. Holroyd (eds.), *The Palgrave Handbook of Arctic Policy and Politics*, https://doi.org/10.1007/978-3-030-20557-7_22

The population that centers along the rim of the Arctic Ocean are largely centered along the northern European coastlines of Russia, Norway, Sweden, and Finland with the Canadian Arctic territories being the least populated. These geographic realities have significant consequences for military and commercial outcomes throughout the Arctic, not the least of which are the development of potential trade and shipping routes through the Arctic region. Russia has capitalized on these geographic advantages to incrementally develop the Northern Sea Route into a potential strategic commercial Arctic trade route.

A Russian Renaissance

Nobody listened to us. Listen now.
V. Putin, Annual State of the Nation Address. Moscow, March 1, 2018

The Russian Federation emerged from the ashes of the former Soviet Union after having experienced a near-total economic and military collapse. With a gross domestic product (GDP) roughly similar to that of Australia, the Russian Federation has achieved a remarkable comeback, one that has endured the weight of escalating western economic sanctions. With an economy heavily dependent on hydrocarbon sales and exports and with real incomes dropping Russia has pursued a form of state capitalism led by oligarchs and political insiders that seemingly ignores western economic principles while it attempts aggressively to re-assert itself on the geopolitical stage.

In what has been described elsewhere as "aggressive isolationism" (Holmes and Krastev 2015) the rising presence of Russia in the world of international affairs, from its bolstered military establishment (with defense expenditures roughly a tenth that of the USA) through to the highly successful electronic hacking of western political establishments, Russia has demonstrated that it is willing to use force to achieve its political ambitions. Recall that since 2000 Russia has deployed armed forces in Chechnya, the Caucasus border regions, Georgia, the Ukraine Donbass Region and, not least, Syria. The Russians have also pioneered a Eurasian Economic Union between Belarus, Kazakhstan, Armenia, and Kyrgyzstan and developed a significant economic rapprochement with China.

Relations between Russia and the West have been characterized by rising diplomatic and political tensions accompanied by escalating reciprocal diplomatic expulsions. One commentator concluded: "Either Russia is soberly deciding to trade wealth for prestige, or Putin is distracting from the poor economy with 'wins' abroad" (Hopper 2017). After the humiliating economic and political collapse of the Union of Soviet Socialist Republics (USSR), Russia has rekindled a mantra of a remerging country that is "great again" with an elected Duma, and a geopolitically consequential military led by a popular President who appears to understand that economic development is essential for successful political and diplomatic strategies: "We need to make a decisive breakthrough in the prosperity of our citizens—falling behind is the main threat, that's our enemy" (Tanas and Biryukov 2018).

Clearly, no commentary on Russia would be complete without a careful examination of the rise in power and policies of Russian President Vladimir Putin. There are insightful studies documenting the rise of Putin and his former associates in the Federal Security Service (FSB) (The Russian FSB [ФСБ] is concerned with Russian security. Following in the footsteps of the notorious Komitet Gosudarstvennoy Bezopasnosti [KGB], the FSB deals with internal affairs inside the country carrying out counterintelligence, internal and border security, counterterrorism, and surveillance, including the investigation of serious crimes. On 25 July 1998, B. Yeltsin appointed V. Putin as Director of the Federal Security Service [FSB]. Clearly, no commentary on Russia would be complete without a careful examination of the rise in power and policies of Russian President Vladimir Putin. There are insightful studies documenting the rise of Putin and his former associates in the FSB (2003), including the work by Shevtsova (2003) in which she presciently noted: "Russia continues to matter…. To secure Russia's integration into the international community is becoming one of the most ambitious challenges for the west in the twenty-first century."

Felshtinsky and Pribylovsky (2008) also chronicled Putin's astonishing rise to power. Many Russian journalists, telecasters, bankers, and former FSB agents were soon to experience forcefully, some lethally, the overwhelming shift of power to Putin, who during his time as Prime Minister, asserted to an audience of FSB agents in Moscow: "We are in power again, this time forever." Such attitudes, combined with Putin's profound mistrust of liberal democracies and of any hint of Russian citizen activism, have had consequences for the entire globe.

In March 2014, Vladimir Putin made an emotional address to members of the Russian State Duma in the Grand Kremlin Palace to announce that Russia had reclaimed Crimea. Putin extolled the restoration of Russia following the humiliating collapse of the former Soviet Union, which he considered to have been the greatest geopolitical catastrophe of the twentieth century. He has denounced the alleged allied global domination of the West, led by one superpower: "They cheated us again and again, made decisions behind our back, presenting us with completed facts—that's the way it was with the expansion of NATO in the east, with the deployment of military infrastructure at our borders. They always told us the same thing: Well, this doesn't involve you" (Putin 2014). Putin has embarked upon a sharp turn in Russian international diplomacy, one based on re-establishing its military capabilities: "Russia's growing military might is a reliable guarantee of peace on our planet because it ensures the strategic balance in the world" (Holmes and Krastev 2015).

Many journalists describe how Putin regularly cites a long list of grievances that started with the transfer of Crimea to the Ukrainian Republic in 1954, through to the 1999 war in Kosovo and the conflict that led to the assassination of Col. el-Qaddafi in 2011. These events and observations are significant because they provide insight into the mindset of Putin and, importantly, how he has tapped into the deepest feelings of patriotism and resentment of the Russian people, policies that have worked to keep him in power since 2012. (During his first term as Prime Minister, Vladimir Vladimirovich Putin served as Acting President of Russia due to the resignation of President Boris Yeltsin.

Putin has retained power longer than many western Presidents and Prime Ministers. At age 65 he won re-election for a fourth term in 2018, extending his term of office as President until 2024. Born on 7 October 1952, he is a former intelligence officer rising to serve as President of Russia since 2012. He previously held the position of President from 2000 until 2008. He was Prime Minister of Russia from 1999 until the beginning of his first presidency in 2000, and again between presidencies from 2008 until 2012 while serving as Chairman of the United Russia Party.) With stellar approval ratings, significant electoral success since the 2014 annexation of Crimea and in spite of growing political discontent resulting from anemic economic growth, Putin has issued ever more stern warnings to the West. Nonetheless, as threatening as these actions and words may sound, as Holmes and Krastev (2015) cautioned: "In reality, Russia's policies have almost nothing to do with Russia's traditional imperialism or expansionism, nor is cultural conservatism such a decisive factor as some commentators allege. Putin does not dream of conquering Warsaw or re-occupying Riga. On the contrary, his policies are an expression of aggressive isolationism. They embody his defensive reaction to the threat posed not so much by NATO as by global economic interdependency" (Holmes and Krastev 2015).

Lindley-French (2019) described three elements of Russian strategy that provide what he termed as an:

> all-important strategic rationale for Russia's military modernization: intent, opportunity and capability. The intent of Moscow's complex coercive strategy is driven by a world-view that combines a particular view of Russian history with the Kremlin's political culture, which is little different from that of Russia prior to the October 1917 Bolshevik Revolution. For Russia, the end of the Cold War was a humiliating defeat which saw power in Europe move decisively away from Moscow to Berlin and Brussels. For Moscow, the loss of all-important prestige was compounded by NATO and EU enlargement as proof of an insidious West's designs to destroy what Russians see as the legitimate legacy of the Great Patriotic War and with it, Russian influence in Europe.

In particular, Lindley-French (2019) cites the 2014 Association Agreement between Ukraine and the European Union as one that increased the Kremlin's paranoia:

> The traditional Russian reliance on force as a key component of its influence reinforced the Putin regime's tendency to imagine (and to some extent manufacture for domestic consumption) a new threat to Russia from the West. The increasingly securitized Russian state thus has come to see the threat of force as a key and again legitimate component of Russian defence, albeit more hammer and nail than hammer and sickle. Hard though it is for many Western observers to admit, it is also not difficult to see how Russia, with its particular history, and Putin's Kremlin with its particular world-view, has come again to this viewpoint. The West's mistake would be to believe that such a world-view is not actually believed at the pinnacle of power in Russia. It is.

For these, and many other reasons, many observers consider that Russian global commercial, strategic, and military policies are closely aligned globally and that such approaches include the Arctic.

RUSSIAN AND CANADIAN OBJECTIVES IN THE CIRCUMPOLAR ARCTIC

Events since 2000 have demonstrated a fierce Russian political resolve to secure its economic independence from the West en route to re-establishing itself on the world geopolitical stage. Russia traditionally has prioritized secure borders that are safe from any perceived encroachment. What does this have to do with the circumpolar region in general and the Canadian Arctic in particular? Here, it is argued that Canada needs to seriously address these emerging realities.

Russia has implemented development policies that embrace the economic significance of its offshore Polar Region. Moreover, there is every indication that Russia is prepared to assert and defend its ownership of those resources, as evidenced by the largest Russian military buildup in the Polar region since 1991. Significantly, Canada regards the Arctic in fundamentally different terms. As Bercuson (2018) noted: "there is virtually no chance that the (Northwest) passage will be used for regular freight traffic for many years due to the unpredictability of ice conditions there in the summer, let alone the winter. No company will issue insurance for passage in those waters until there is a high predictability of sea/ice conditions from season to season, which is certainly not the case now."

In particular, Canada has increasingly embraced risk-averse resource development policies that are more concerned with Arctic offshore conservation than economics. As Dr. Rob Huebert (2017) recently opined:

> On December 20, 2017, Canadian Prime Minister Justin Trudeau and U.S. President Barack Obama announced that their countries were banning oil and gas development in northern waters—the U.S. indefinitely and Canada for a five-year period. One month later, the Russian President Vladimir Putin announced the opening of several major oil and gas pipelines that will bring large amounts of oil and gas from their northern fields in Yamal into production. One of those pipelines will double the capacity of the Nord Stream that connects Russian Arctic gas production to Germany.

Wallace (2018) noted how Canada ignored the results of the 2016 US election when it raced to embrace the policies of the outgoing Obama administration by agreeing to jointly "launch actions ensuring a strong, sustainable and viable Arctic economy and ecosystem, with low-impact shipping, science-based management of marine resources, and free from the future risks of offshore oil and gas activity."

In so doing, Canada demonstrated that it was more concerned with "risks" of northern development than the benefits that might accrue from economic advancement in the north. Northwest Territories Premier Bob McLeod was aghast at the lack of governmental consultation that preceded the unilateral announcement from Ottawa. McLeod noted that the announced unilateral drilling ban negated key benefits of the NWT's 2014 Devolution Agreement that had allocated province-like powers to the GNWT and provided for co-management of the offshore along with resource revenue sharing. In an announcement termed a "red alert," delivered one year after the unilateral declaration of a five-year moratorium on Arctic offshore oil and gas development, the Premier described the attitude of southern Canadians as one that regarded the Territory as a "large park" and "but one example of our economic self-determination being thwarted by Ottawa" (Forrest 2018).

There is perhaps no better illustration of the relentless focus by central Canada to diminish the economic decision-making of devolved northern governments in favor of the ideological and political forces associated with the climate change agenda. Nonetheless, it was a short-lived triumph for North American climate diplomacy. In April 2017, Trump's "America First Offshore Energy Executive Order" explicitly reversed the Obama administration's ban on Arctic leases, a policy reversal that immediately placed Canada's northern development policies at odds with both the Trump administration—and with Russia. The US Department of the Interior subsequently announced plans to offer offshore leases for Arctic oil and gas exploration with access to previously inaccessible acreages and overturned the prior indefinite drilling bans in much of the Arctic Ocean announced by the Obama administration. Several aspects of the latter US policy reversals remain controversial in Canada (Weber 2018a).

In 2012, just days before the Presidential elections Putin appeared to extend a challenge to Canada to establish a "joint scientific council" to assess issues associated with Arctic sovereignty—an apparent initiative that was soon consumed and lost in the whirlwind of the Russian pre- and post-election process (Wallace 2012). Meanwhile, the Yamal Nenets herders, faced with the consequences of an enormous liquefied natural gas (LNG) development on their traditional lands, were attempting to gain the attention of decision-makers in Moscow just at a time when Canada was to assume chairmanship of the Arctic Council (Wallace 2013; Wallace and Dean 2013).

By comparison, Russia's determination to integrate deeply its Arctic resources into the economic fabric of not just the European Union but of Asia is demonstrated by the completion of the Bovanenko-Ukhta natural gas pipeline that feeds directly into the Nord Stream energy system, a development that has significantly increased German reliance on Russian Arctic-based resources.

With expanded access to northern sea lanes in a warming Arctic, Russia is historically, geographically and militarily positioned to create and control any new Arctic trading routes. Hence, the Russian icebreaker fleet is virtually unchallenged as a geopolitical, commercial instrument for Arctic commerce, through Arctic routes which, as early as 2011, Putin opined could rival the Suez Canal as the primary link from Europe to Asia, as it would: "rival traditional trade lanes in service fees, security and quality" (Bryanski 2011).

Another example is the Russian Arctic Yamal LNG project, a liquefied natural gas plant located at the northeast of the Russian Yamal Peninsula. In October 2010, the Russian government chose Novatek to initiate a US $27 billion Arctic pilot project (which by December 2014 reportedly required a 150 billion rouble subsidy from the Russian government). Construction of the port began in 2013 with commercial operations opened on 8 December 2017, at an event attended by President Vladimir Putin in the presence of Saudi Arabia's energy minister Khalid al-Falih (Foy 2018).

In addition to the LNG plant, the project includes production from the huge Yuzhno-Tambeyskoye gas field along with a power plant, 180 km rail line, seaport, and airport. The prime export market for the LNG, to be shipped through the Russian Northeastern Passage, is China. (Daewoo Shipbuilding & Marine Engineering has been contracted to build up to 16 Arc7 double acting ice-class gas tankers to be chartered and operated by Sovcomflot.[1] Designed in Finland by Aker Arctic Technology Inc. and constructed at the Daewoo Shipbuilding & Marine Engineering [DSME] shipyard in South Korea, each icebreaking LNG tanker is designed to operate year-round in ice of up to 2.5 m.) A partner, Total S.A., has subsequently announced that 15 LNG icebreakers will be commissioned between December 2016 and 2019 (Roston 2018).

Gosnell (2018) cautioned that, while it should not be considered as portending the arrival of a serious competitor to the Suez Canal, the August 2017 transit of the Russian icebreaking (LNG) carrier Christophe de Margerie (300 m long and with a capacity of 172,600 m^3 to sail in temperatures of -52°C and in ice thickness in excess of 2.0 m with an open water speed of 19.5 knots) is designed to carry resources from Yamal and Murmansk to Asia, Europe, and India. The vessel achieved a remarkable record-setting transit of the Northern Sea Route (NSR) (Fig. 22.1):

> The ship transited the 2,193 nautical mile NSR in just six days, twelve hours, and fifteen minutes. It completed the entire journey from Hammerfest, Norway, to Boryeong, South Korea, in nineteen days—nearly thirty percent faster than the traditional Suez Canal route. During the transit, the vessel averaged just over fourteen knots, remarkable given that part of the transit was through ice fields that were 1.2 meters thick. Sovcomflot's unique LNG carrier sets new record with Northern Sea Route transit in just 6.5 days.

The implications of the revolutionary *Christophe de Margerie* are significant in that it has demonstrated operational capabilities that:

> brings forth tremendous operational capabilities to the Arctic. The class was designed to service the Yamal LNG project, with anticipated year-round navigation through the Arctic, in accordance with a twenty year contract signed by Yamal Trade and Fluxys LNG for transshipment of up to eight million tons of

[1] As described in World Maritime News, "Russia, 2013: Sovcomflot, NOVATEK and VEB to Cooperate in Yamal LNG Project," June 21, 2013.

Fig. 22.1   LNG tanker *Christophe de Margerie* at the dock in Yamal. Photo: Dimitriy Monakov. From: Gosnell, R. (2018)

> LNG per year, supporting deliveries from the Yamal Peninsula to Asian markets. Indeed, the Christophe de Margerie arrived at the Yamal berth in November to load the inaugural liquified natural gas cargo destined for China. The second ship in the new class, the Boris Vilkitsky, berthed alongside the Christophe de Margerie for the inaugural loading. (Gosnell 2018)

Significantly, the Yamal LNG plant could generate 16.5 million tons of lique-fied natural gas per year when fully operational. With the first train opera-tional in late 2017, the plant is projected to reach full capacity by 2021 with another LNG Plant (Arctic LNG) also proposed near the Gydan Peninsula on the Ob river estuary. Notably, while Russia's Novatek owns 50.1% of the company, Total S.A. and China National Petroleum Company (CNPC) each own 20%, with China's Silk Road Fund having signed an agreement to pur-chase 9.9% stake. Total and Novatek announced in May 2018 that binding documents for the Arctic 2 Project, were jointly signed in 2018 by Presidents Macron and Putin at the St. Petersburg International Economic Forum. The transaction to be closed no later than March 31, 2019, has a value of $25.5 billion (US). The Arctic LNG 2 project envisages construction of three LNG trains to annually produce 6.6 million tons from a gravity-based struc-ture (GBS) offshore platform to be constructed in Murmansk and positioned in the Ob Bay. Noting the huge LNG and liquids resource potential of the region that led themselves to "scalable LNG projects," NOVATEK Chairman Leonid Mikhelson remarked: "The entry of such a professional partner to

Arctic 2 already at an early stage confirms the outstanding economic attractiveness and huge perspectives of LNG projects on the *Yamal and Gydan peninsulas*" (Kobzeva and Golubkova 2015).

Furthermore, early in 2018 the *Eduard Toll* became the first LNG tanker to complete the Northern Sea Route in winter traveling from South Korea to the Yamal LNG plant at Sabetta to deliver its cargo to France, a feat that eliminated 3000 nm from the traditional sea routes through Suez. Then in July 2018, two LNG cargoes were delivered from Yamal to the Chinese port of Jiangsu Rudong, without the use of icebreaker support, to complete a delivery in 19 days as compared with traditional Suez routes that take up to 35 days.

The potential economic and strategic significance of a developed Northern Sea Route, one extending from Murmansk to the Alaskan Bering Strait, has long been an ambition of Russia. Beginning with the *Yermak* the world's first icebreaker commissioned in 1898, Russia began a long tradition of icebreaking. Beginning with the nuclear-powered icebreaker *N.S. Lenin*, once a point of supreme pride of the Soviet Union (now decommissioned), Russia has maintained an unmatched polar capability in commercial icebreaking operations. At a time when Canada's heavy icebreaker fleet is rapidly aging and US operational capability has declined (with but one operational heavy icebreaker the *Polar Star* and one medium icebreaker [the *Healy*]), Russia has maintained its capability as the foremost military and shipping power in the circumpolar region (Fig. 22.2) with one of the world's largest icebreakers and 11 more under construction. (In May 2018 Canada and the USA announced the Critical Infrastructure Protection and Border Security [CIPBS] Agreement with

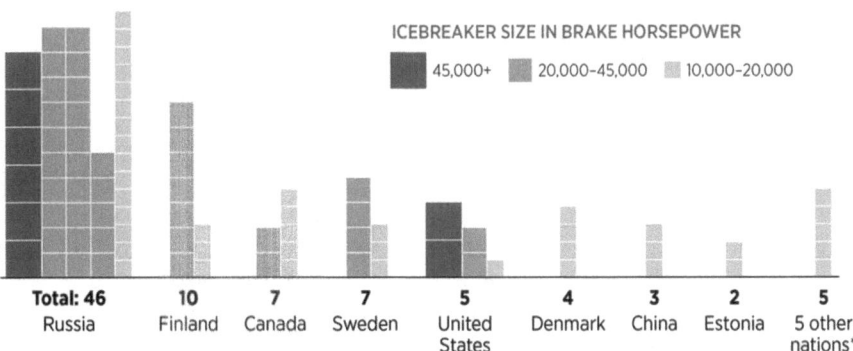

* Norway, Germany, Latvia, Japan, and South Korea.

Fig. 22.2   International icebreaking capabilities (BHP). Source: Ronald O'Rourke, "Coast Guard Polar Icebreaker Program: Background and Issues for Congress," Congressional Research Service *Report* RL34391, July 9, 2018, p. 10, https://fas.org/sgp/crs/weapons/RL34391.pdf (accessed August 1, 2018). Note: List includes both government-owned and privately-owned icebreakers. List excludes icebreakers for southern hemisphere countries Chile, Australia, South Africa, and Argentina

Canada's National Research Council [NRC] to facilitate collaborative testing and evaluation in icebreaking ship technologies. The US Coast Guard [currently overseen by the US Department of Homeland Security] and the US Navy have not commissioned a polar icebreaker for decades).

Moreover, the Russian effort to strengthen its Arctic presence includes a substantial modernizing of its icebreaker fleet. The new $1.74 billion Russian nuclear-powered, double-hulled icebreaker *Arktika* (Arctic) launched from St. Petersburg in June 2016 is, at 567 feet and 33,500 tons currently the world's largest, designed to escort oil and gas shippers from the Yamal Peninsula and Gdansk oil fields to Asia-Pacific regional markets.

Among a total operational fleet of approximately 40 icebreakers, 10 are nuclear-powered (9 icebreakers and 1 icebreaking container ship). Russia's nuclear-powered icebreaking fleet is operated by Rosatomflot (Murmansk). The *NS Sibir* initiated the first two tourist cruises to the North Pole in 1989 and 1990, while in 1991 and 1992 similar tourist trips were undertaken by *NS Sovyetski Soyuz*. The *NS Yamal* completed three expeditions there in 1993 while the nuclear-powered icebreaker *NS 50 Lyet Pobyedi* ("50 Years of Victory") embarking from Murmansk on June 24, 2008, on its maiden voyage traveled to the North Pole, completing a total of three expeditions that same year. These commercial tours have continued to the present day.

In a parallel development, Russia has also recently unveiled a floating nuclear power plant, a development that Massachusetts Institute of Technology (MIT) engineering professionals have described as being "light years ahead of us." The platform is capable of generating 70 MW of electricity with an operational lifetime before refueling of 12 years. The *Akademik Lomonosov* is to be towed into position to a remote Siberian port near Pevek, a feat that represents a new chapter in Russian northern development.

In 2014, Russia became the first nation to ship oil drilled from an *offshore* Arctic platform as Russian President Putin viewed the event via a video link when oil was loaded onto a tanker from the Prirazlomnoye drilling platform. While this indeed was a first for offshore oil produced from a platform, some authors commented incorrectly that the Russian achievement marked "the first time that oil had been extracted, and shipped from, above the Arctic Circle." As an aside, it would be more correct to assign that honor to Canada, not Russia. In 1985, Panarctic Oils became the first Arctic commercial oil producer, albeit on an experimental scale. Shipments from the Bent Horn oil field on Cameron Island, NU to Montreal began with a single 100,000 barrel (16,000 m³) tanker load of oil shipped via the *MV Arctic*, which carried two shipments per year until cessation of operations in 1996. A total of 2.8 million barrels was produced by the time that the field was abandoned in 1997. Bent Horn was the most northerly producing oil field in the world with the unrefined oil of such high quality that it was used to fuel electrical power generators at Resolute Bay and a lead-zinc mine on Little Cornwallis Island. [Panarctic was a consortium established between as many as 37 private companies and Canada to explore for oil and gas in the Canadian Arctic Islands. Panarctic drilled 150 wells with the most northerly well located approximately 80°45′ N on Ellesmere Island and the most southerly well was at 72°40′ N on Prince of Wales Island. Of these wells 38 were drilled offshore from floating ice platforms in water depths of up to 550 m. A large amount (17.5 trillion ft³) of natural gas reserves was discovered over this period.]

China has also paid growing attention to the potential for northern passage, characterizing Arctic sea routes in a January 2018 policy document, as the "Polar Silk Road." China, earlier having secured observer status on the Arctic Council, recently announced the construction of the new icebreaker *Xuelong 2* (Snow Dragon) designed jointly by the China State Shipbuilding Corporation and Finland-based Aker Arctic Technology. Scheduled for delivery in 2019 the *Xuelong 2* will join China's sole icebreaker, the *Xuelong*, for polar scientific missions. The *Xuelong*, China's first icebreaker, was built in Ukraine and entered service in 1994. Many writers have chronicled the developing Chinese presence in, and their attentions paid to, the Arctic as China develops its position as a "near Arctic state."

In consideration of such events, Gosnell (2018) noted that:

While the adage "High North, low tension" has long characterized the Arctic, there are justifiable concerns that the region will instead be colored by the evolving economic, environmental, military, and geopolitical global environment. Russia has enacted a robust Arctic policy that reflects both Putin's revanchist policies as well as the country's significant Arctic interests. With about half of the Arctic's population and coastline, Russia sees the region as increasingly vital to its interests. The Arctic Zone yields about 10 percent of Russia's GDP and accounts for 20 percent of its exports. The country is pursuing the construction of new icebreakers and maintains approximately 40 icebreakers, seven of which are nuclear. While these numbers dramatically dwarf other icebreaker fleets, it must also be noted that Russia requires a robust icebreaker fleet to escort commercial vessels along the Northern Sea Route (NSR)—a maritime route established in 1936. The NSR is becoming increasingly important to the Russian economy, though passage is still limited during much of the year.

In addition to Russia's commercial interests in the Arctic, Gosnell (2018) commented:

Congruent with its Arctic economic interests—and its geographic limitations for fleet basing—Russia's Northern Fleet is its largest, with reportedly forty-one submarines and thirty-seven surface ships. The reopening of Soviet-era Arctic bases, establishment of an Arctic Command, and conduct of regional military exercises are prompting concerns over Arctic militarization. Indeed, the 2015 Maritime Doctrine of the Russian Federation devotes an entire chapter to "The Arctic Regional Priority Zone." In June, the Northern Fleet completed a no-notice snap exercise that involved 36 ships, submarines, and support vessels—its' largest in 10 years. The Arctic forces further played a role in Russia's Vostok 2018 exercise, with Northern Fleet and Russian Marine units conducting a mock amphibious assault on the coast of the Chukchi Sea. The Northern Fleet warships sailed more than 4,000 nautical miles along the NSR to participate in the largest military drills—totaling nearly 300,000 personnel—that Russia has held since the end of the Cold War.

While the Arctic may yet be considered a low risk environment for military conflict, North Atlantic Treaty Organization (NATO) allies are increasingly cognizant of the fact that such activities by Russia will, at least, compel heightened cooperation among northern states: *The Arctic remains challenging to operate in due to the unpredictable weather and hostile environment—regardless of ice coverage. Likewise, military operations in the region will remain limited, as Arctic operations are challenging and costly. Following a congressionally mandated review, the U.S. Government Accountability Office recently reported that the Navy's June 2018 report on capabilities in the Arctic aligned with Department of Defense assessments that "the Arctic is at low risk for conflict." Yet the overall increase in activity will require greater maritime domain awareness, search and rescue capacity, and security presence. It will also require greater cooperation amongst Arctic stakeholders to protect the natural resources and environment, as well as sovereignty of the Arctic States* (Gosnell 2018).

CANADIAN ARCTIC PERSPECTIVES

Meanwhile, significant challenges have occurred as a result of Canadian northern policies. Collins (2019) described the problems, and resulting opportunities, that have beset Canadian shipbuilding efforts:

> Deep personnel cuts between 1989 and 1997 had erased much of the Department of National Defence's (DND) institutional memory on shipbuilding. Additional challenges came in the form of skyrocketing global shipbuilding material and labour costs of 200 per cent to 300 per cent, and inadequate shipyard infrastructure. Because of these reasons the first attempt at getting AOR replacements led to noncompliant bids. In fact, when the JSS was cancelled in August 2008, the Harper government cut the Canadian Coast Guard's (CCG) $750 million Mid-Shore Patrol Vessel for largely the same reasons. The two JSS bidders meanwhile lost an estimated $20 million to $30 million on bid preparation.
>
> The JSS failure necessitated a rethink on the federal government's approach to domestic shipbuilding. Both DND and the CCG knew as far back as the early 2000s that a minimum of 30 ships was needed to replace both services' aging fleets over the coming decades. This presented an opportunity. An interdepartmental National Shipbuilding Procurement Office struck in 2008–2009 in the wake of the JSS cancellation recommended moving beyond the boom-and-bust history of Canadian shipbuilding to a continuous-build strategy that would help avoid the inevitable economic impact of closed shipyards and lost shipbuilding skills of the country's previous project-by-project efforts. The result is the National Shipbuilding Procurement Strategy (NSPS), launched in June 2010.

As a consequence of these developments, the importance of a viable and capable icebreaking fleet so vital to Canadian offshore Arctic territorial claims appears to have degraded or, at least, become less focused due to program cancellations, changing budget commitments and unanticipated attritions. While Canadian icebreakers have successfully operated throughout the Northwest Passage for decades this capability is rapidly eroding. As Spears (2018) noted:

> A number of sources have raised concerns about Canada's impeding "icebreaker gap", which was first addressed three years ago in a Canadian Sailings article entitled Canada's Icebreaker Gap. Little, if anything, has been done to alleviate this problem during the past years, which has now assumed critical proportions.... Canada presently has 15 icebreakers in operation along with two air cushion vehicles that are utilized for icebreaking and flood control along the St. Lawrence River. CCG currently has two (2) Heavy Ice Breakers (HI), four (4) Medium Icebreakers (MI), and nine (9) Multi-Task Light Icebreakers in its inventory. CCG deploys these vessels in Canada's Arctic waters during the late-June to mid-November period (the Arctic season), and South of 60° latitude from the December to May period (the Southern season).

Ominously, in 2015, the Shipping Federation of Canada also warned:

> The past two years have demonstrated the limits of CCG's icebreaking fleet as it dealt with icebreaking in the Arctic, the Great Lakes, the St. Lawrence River and Eastern Canada. Already operating with a limited and aging number of assets over a very large geographical area, the conditions demonstrated the breaking point for the system and the need for more icebreakers as soon as possible to meet adequate levels of service and safety.

Similarly, the Emerson Report (2016) graphically described Canada's Coast Guard fleet "as the oldest in the world" noting that: *At that rate, the median age of the fleet will not decrease. Other strategies, such as outsourcing or leasing, are not part of the strategy and thus cannot be deployed to meet short-term requirements.*

In January 2018 Canada announced the start of negotiations with Quebec's Davie Shipyard to lease icebreakers as "Project Resolute," designed as a "P3 Project" to provide Canada with four existing foreign flag icebreakers to be modified in Davie's facilities. Under "Project Resolute" Davie offered to convert the *MV Aiviq*, a heavy icebreaker built in 2012 for Shell's Alaska drilling program along with three Norwegian-built medium icebreakers to provide the Coast Guard with interim capacity. Ken Hansen, a retired Canadian navy commander commented (Berthiaume 2018): *When government can't or won't put money in to replace equipment, you end up in situations like this... This is crisis planning, when government resorts to things like special contracts to Davie. Unusual purchases and repairs are a sign of illness in the system.*

Prior to the June 2018 Advanced Contract Award Notice (ACAN), in March 2018 the Canadian Coast Guard announced new administrative powers to call on the private sector for short-term help with duties such as ice-clearing in the St. Lawrence Seaway and the Great Lakes, a vital corridor through which much of Canada's foreign shipping flows. Regrettably, longer-term plans to replace the aging Canadian icebreaker fleet appeared then to be in chaotic flux, one that reflected an aging coast guard icebreaking fleet that increasingly experienced material losses in operational days due to mechanical breakdowns. Then, in October 2018 the Canadian Coast Guard announced that three previously-used "interim icebreakers" had been purchased from European sources for use over the next 15 to 20 years as part of a $610 million, sole-source agreement with Quebec-based Davie Shipbuilding. The purchase of the three ships eliminated a previously predicted Canadian trade surplus for August and replaced it with a deficit as reported by Statistics Canada: *Most of this revision was due to the import of three high value ships, which were reported after the publication of August data,* "the agency said of the transaction, which on its own added $598 million to the monthly import number."

As for the Canadian Navy, it is currently procuring vessels under the Arctic Offshore Patrol Ship (AOPS) program (*HMCS Harry DeWolf* is currently under construction at the Irving Shipbuilding in Halifax) in advance of the commencement of work on new frigates. Not designed as icebreakers, the frigates will have slightly less ice capability than the coast guard's medium icebreakers, designed for operations in summer ice up to a meter in thickness. The procurement has been further complicated by questions related to Chinese-made equipment largely related to American concerns regarding Chinese telecom firm Huawei and other concerns in light of the, as yet unresolved, case of Qing Quentin Huang of Lloyd's Registry who was charged in 2013 with an attempt to pass on design information about Canada's proposed Arctic ships to the Chinese government.

RUSSIA AND THE CANADIAN POLAR INTEREST

The marked "divergence" in consequential northern resource policies between Canada and Russia was highlighted by Huebert (2017) who noted that: "as Canada and the United States decide not to develop their Arctic offshore oil and gas, Russia is moving forward with growing intensity to develop its resources. This not only highlights the differences that exist in oil and gas development regimes, but also in the thinking of the leadership of all three countries."

Russia is accelerating its efforts to reopen abandoned former Soviet military, air and radar bases throughout Siberian Arctic lands and islands while also building new operational bases. These activities have drawn the attention of UK lawmakers and policy experts: "A recurring theme in Russian military strategy is the ability to combine various tools simultaneously, to give a fully integrated, comprehensive approach" (Russia 2016). While MPs accepted there was a "divergence of views on Russia's motivation" they cautioned: "It is difficult to conclude that this build-up of military strength is proportionate to an exclusively defensive outlook" (Nicholls 2018).

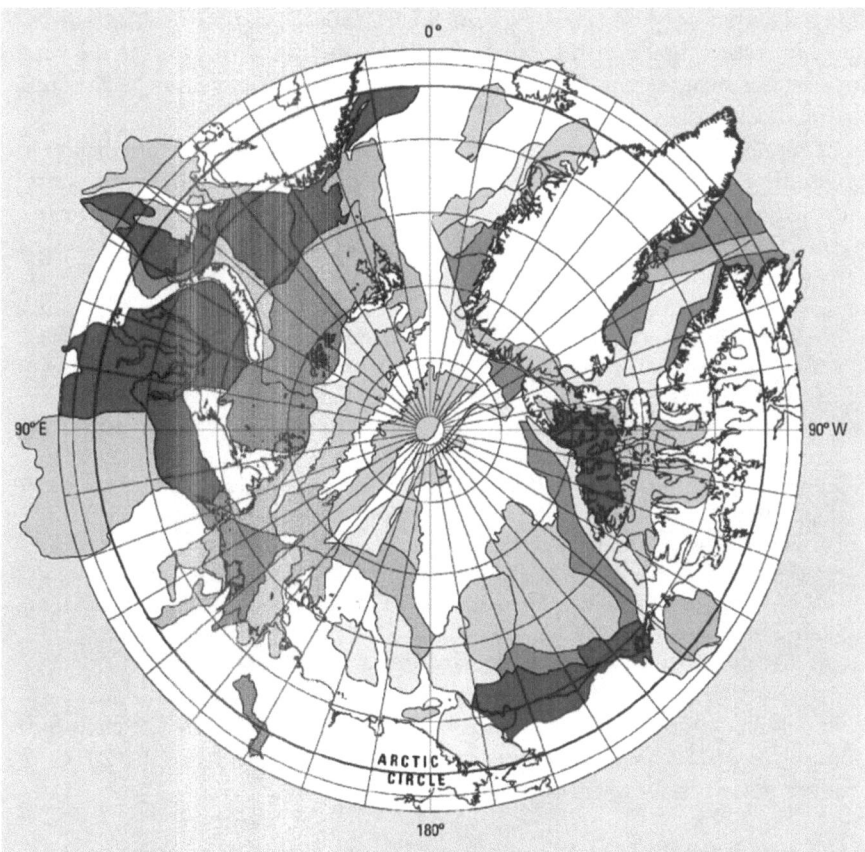

Fig. 22.3 Probability of the presence of undiscovered Arctic oil/gas fields with significant amount of recoverable resources (>50 million barrels of oil equivalent) with darker hues designating higher probabilities. Courtesy of the US Geological Survey. Urban, O. 2015. Future of Arctic Oil reserves. http://large.stanford.edu/courses/2015/ph240/urban2/

The US Geological Survey has concluded that the Arctic Basin may contain oil and gas reserves equivalent to 412 billion barrels of oil (constituting approximately 22% of global undiscovered reserves) (Fig. 22.3).

Protecting their existing and potential future development claims to these resources is obviously of interest to the Russian Federation. Along with its surface icebreaking and subsurface nuclear fleets, Russia is reopening 6 Arctic military bases, 16 deep water ports and 13 airbases to be equipped with advanced S-400 long-range surface to air rockets. By comparison, the USA has no major military bases north of the Arctic Circle. Osborn cited comments from Moscow Defense Brief Editor Mikhail Barabanov: "The modernization of Arctic forces and of Arctic military infrastructure is taking place at an unprecedented pace not seen even in Soviet times," further noting that two Russian Arctic brigades had been established, something not witnessed even in Soviet times with plans well advanced to form a third brigade for special Arctic coastal defense divisions (Osborn 2017).

At his 2017 confirmation hearings in Washington, US Defence Secretary James Mattis characterized Moscow's Arctic initiatives as "aggressive steps" and pledged to give priority to the development of new US strategies. In a recent 2018 CGAI Policy Paper, the authors observed that the North American defense environment is undergoing a significant transformation:

> occasioned by dramatic changes in the geostrategic/political landscape and the development of new generations of weapon systems....The air threat has now returned with the resumption of long-range Russian flights across the Arctic and down the Atlantic and Pacific coasts. This threat, however, is of a different character because of the development of a new generation of Russian long-range air-launched cruise missiles, as well as sea-launched cruise missiles, which have direct implications for NORAD's capacity to deter, detect and defend, as well as for its current area of operations and mission suites....Finally, the consequences of an attack against North America, alongside potential catastrophic natural disasters relative to the role of military forces in support of civil authorities, raises issues for both Canada and the U.S., and thus NORAD, regarding the most efficient and effective means to respond. (Charron and Fergusson 2018)

A parallel 2018 UK Defence Sub-Committee report (UK Defence Sub-Committee 2018), one that followed closely the conclusions of the 2016 UK Defence and Security review, warned:

> Our view is that the UK and its allies should be extremely wary of Russia's intentions in the region. It is difficult to credit that the scale and range of military capabilities being deployed by Russia in the Arctic fulfil solely defensive purposes. Russia has shown itself to be ready to exploit regional military advantage for political gain. While the Arctic remains a region of low tension, this could change quickly.

The 2016 report had previously warned:

> Some 30% of Russia's territory, land and sea, is within the Arctic. It has recently increased its military presence in the region. We assess that, for a variety of reasons, both economic and military, Russia will protect its Arctic assets and influence strongly. It has developed a new Arctic Command *and increases exercise activity levels.* Russian military activity in Arctic territory has so far been both legal and reasonable. However, future tensions over Arctic resources and freedom of navigation in newly opened seas routes could create tension in the region.

Significantly, the accelerating Russian presence and military capability in the Arctic have occasioned not just re-evaluations, but a reinvigoration, of Arctic defense postures among NATO circumpolar allies including the UK. Defence Secretary Gavin Wilson commented that a new Defence Arctic Strategy was crafted to "to allow Britain to effectively monitor Russian submarine activity and ensure that the Armed Forces are 'well placed' to respond to any threats."

Goldstein (2018) observed that the West "got a fresh jolt from Moscow" with the announcement by Russian defense minister Sergey Shoygu of a military exercise dubbed "Vostok-18" [East-18]. The exercise, which perhaps provide insights into the prevailing mentality in the Kremlin, was to be carried out by Russian defense forces on a scale not seen since the early 1980s with more than one thousand aircraft, almost 300,000 soldiers and nearly all Russian military installations in the Central and Eastern military regions, including also the Northern and Pacific fleets. China also provided 3200 Chinese soldiers along with 30 aircraft. Significantly, announcements clarified that Vostok-18 was not a "joint exercise" but rather constituted Chinese participation in a large-scale Russian exercise. The event was marked by comments that the relationship between Moscow and Beijing had reached new levels of cooperation, scope and intensity through the participation of the Chinese with Russian units. However, Goldstein cautioned that:

> the location of the exercise no doubt reflects the Kremlin's desire to cool down tensions in the European theater. At a time of emerging fissures within the Trans-Atlantic Community, such an enormous exercise close to NATO countries would be excessively provocative and counter to Russia's interests. That the Kremlin understands this is no doubt a good thing for European security. The other important point that has not registered in most Western analyses is the confluence of the September Eastern Economic Forum in Vladivostok and the Vostok-18 exercise. It's easy to forget that six months ago, it looked more than a little likely that a massive war would engulf Northeast Asia. The exercise was most likely put together as a show of force meant to favorably impact diplomacy and the related "correlation of forces" in and around the Korean Peninsula.

Russian–Chinese strategic cooperation has undoubtedly reached a new stage, one that is exemplified by new bilateral relationships that include the new concept of a "Polar Silk Road"—one that appears to be related to the larger Chinese Belt and Road initiative. China's announcement of a new nuclear icebreaker, likely to be constructed with Russian technical assistance, is an unambiguous acknowledgment of a serious commitment to the Russian Northern Sea Route (NSR) that, along with increasing Chinese investment and strategic, commercial attentions, is essential to the realization of an operational, maritime commercial pathway along the Russian northern coastline.

Meanwhile, Hage (2018) noted that the USA and Canada continue to engage in a largely silent duel over the status of Canadian claims to the Northwest Passage a debate that appears to ignore developing Russian advances in the Arctic. Combined with potential Chinese interest to access the Canadian Passage as a potential shortcut for Pacific to Atlantic commercial commerce, these facts present new challenges for Canadian Arctic interests: "With the rising Russian military threat in the Arctic, melting Arctic ice and the possibilities of possibilities of Chinese and other cargo ships using the Passage to shorten the route between Asia and Europe—along with cruise ships bringing tourists on Arctic adventures—the water's status continues to be a question."

Canada's hard-won, well-established, historic claim to its northern regions, including the Northwest Passage, should represent more than a geographic claim to Arctic sovereignty. It should represent a commitment by and for all Canadians, most certainly the Inuit, that assumes that Canada will responsibly discharge its obligations for northern economic development, security, and environmental protection. As John Higginbotham of the Center for International Governance Innovation observed (Weber, June 2018b):

In Norway, the North is the first thing they think about in the morning and the last thing they think of at night. It's probably the same for (Russian President Vladimir) Putin. In Canadian consciousness, it's not a consistent national priority. There's no one worrying about long-term economic development in the Arctic and making the kind of investments we need.

The current Canadian government, approaching the end of its mandate, continues to "consult" on a formal Arctic policy having, by contrast rapidly implemented without any significant prior northern consultations or studies on the socio-economic consequences for northerners, a five-year moratorium on Arctic offshore energy development. Reflecting an obvious frustration with traditional centralized Canadian Arctic policy development and implementation, Northwest Territories Premier Bob McLeod commented: "We're strongly suggesting it would be much better to have an alliance between the Arctic territories, Greenland, Iceland, the Faroe Islands—all of those jurisdictions have similar issues."

Conclusion

Following the collapse of the Soviet Union, Canada and the Russian Federation have pursued divergent strategies for economic development and security in the circumpolar Arctic with very different northern strategic outcomes. Material differences in geography, population distributions, and histories have unquestionably influenced Canadian and Russian northern development. However, changing geopolitical strategies and attitudes toward northern economic imperatives have emerged as a significant force in shaping future political and economic outcomes in the circumpolar Arctic.

With the potential for increased maritime access to northern latitudes resulting from a warming Arctic, Canada's Northwest Passage and most particularly Russia's Northern Sea Route have attracted global commercial and strategic military attentions. As a result of policies that reflect environmental, investment, and market priorities, after decades of Arctic offshore exploration and development Canada has, at least temporarily, chosen to defer opportunities for offshore hydrocarbon exploration. By comparison, Russia has actively pursued the commercial development of its northeast trade route along with hydrocarbon exploration, development, and transportation. Overall, Canadian policies appear to have drifted toward declining strategic icebreaking capabilities accompanied by a lack of pipeline or shipping access to "stranded" northern Canadian hydrocarbon resources.

In order to secure access to Asian and European markets, Russia has aggressively pursued completion of Arctic gas pipelines and major transhipment facilities for natural gas and LNG. These developments of global economic and strategic significance have been accompanied by expansive Russian military deployments throughout the Arctic. Accordingly, Russia has increasingly assumed a position of strategic advantage throughout the entire circumpolar region. There are significant international implications resulting from an increasingly militarized, destabilized, circumpolar region.

Hence, Canada's challenges in the Arctic cannot be reduced to simplistic concepts of "sovereignty." As Charron and Fergusson observed:

> Referencing "Arctic" and "sovereignty" in the same sentence is generally a recipe for alarmist and precipitous action. It is usually translated into a demand for a more military presence which, while a ready answer for the Canadian government, ignores the fact that sovereignty issues today are settled in the courtrooms. There are no de jure or de facto threats to Canadian Arctic sovereignty. If Russia is a real threat, it is to Canada and its allies as a whole. Indeed, the Arctic is one issue area in which Russian co-operation has been tremendously helpful...For now, however, Canadians should replace Arctic sovereignty with homeland defence and devote attention to issues which relate to how the federal government exercises its sovereign authority over the people who live in its Arctic territory and how it will work with allies now and in the future to defend Canada. (Charron and Fergusson 2018)

Unquestionably, it is important for Canada to develop policies that address rapid climatic change in the Arctic. However, there is also a parallel need to pursue beneficial economic and resource development policies for Canada's northern peoples that accommodate informed principles for enhanced economic development and northern conservation. The compelling realities of an accelerating Russian presence and capability in the Arctic have occasioned a re-evaluation and reinvigoration of Arctic defense postures among circumpolar allies, one that reflects an evolving security environment throughout the region. The strategic importance, and growing significance, of Russia's expanding Arctic military capabilities cannot be ignored. Indeed, it should be a matter of growing significance to Canada and its northern NATO allies. Nonetheless, the consequences of current Canadian Arctic policies should also be of concern to northerners, Canadian economic policy makers and defense strategists. Issues include potential civilian air or sea incidents, involving elements of Search and Rescue (SAR) and related air-sea rescue operations that have to be initiated from distant southern bases, a flagging capability for Arctic icebreaking assistance and a policy shift away from offshore resource development in favor of measures associated exclusively with Arctic conservation.

The West increasingly is being forced to re-examine the heightened degree of Russian Arctic military activities to determine if indeed they are designed solely for defensive purposes. As these geopolitical events are of direct consequence for Canada, a careful reconsideration of its economic priorities for the Arctic, along with heavy icebreaking and military capabilities, is overdue. As such, issues of "sovereignty" should not be used to compensate for Canada's ongoing neglect of its northern heritage and circumpolar responsibilities. Such unfocussed, singular attentions to sovereignty have tended to hamper discussions of the need for economic development, self-determination and political autonomy for Canadian northerners. By comparison, Russia's determination to integrate deeply its Arctic resources into the economic fabric of not just the European Union but of Asia may serve as a useful demonstration for Canada as it develops future policies for northern economic development needed to bolster and realize its sovereign Arctic claims (Wright 2018).

Lindley-French (2019) described three elements of Russian strategy that provide an all-important strategic rationale for Russia's military modernization: intent, opportunity and capability:

> The intent of Moscow's complex coercive strategy is driven by a world-view that combines a particular view of Russian history with the Kremlin's political culture, which is little different from that of Russia prior to the October 1917 Bolshevik Revolution. For Russia, the end of the Cold War was a humiliating defeat which saw power in Europe move decisively away from Moscow to Berlin and Brussels. For Moscow, the loss of all-important prestige was compounded by NATO and EU enlargement as proof of an insidious West's designs to destroy what Russians see as the legitimate legacy of the Great Patriotic War and with it, Russian influence in Europe.

This article is based on, and adapted from, one published by the Canadian Global Affairs Institute Calgary, Alberta "The Arctic is Warming and Turning Red: Implications for Canada and Russia in an Evolving Polar Region." January 2019, 29 pp, ISBN: 978-1-77397-056-1. https://www.cgai.ca/the_arctic_is_warming_and_turning_red_implications_for_canada_and_russiainan_evolving_polar_region.

Acknowledgments The author thanks Dr. David Bercuson, Mr. Kelly Ogle, and Mr. Adam Frost of the University of Calgary Canadian Global Affairs Institute for supporting the research and independent peer review done in the preparation of this chapter and for the editorial support of Dr. Ken Coates and Dr. Carin Holroyd. The author accepts responsibility for any errors or omissions that may have occurred.

REFERENCES

Bercuson, D. 2018. Canada's Sovereignty: The Threats of a New Era. *The Global Exchange* XVI (III): 7–10.

Berthiaume, L. 2018, March 2. *Private Companies Set to Help Canadian Coast Guard's Aging Fleet with Icebreaking*. Toronto, ON: Canadian Press.

Bryanski, G. 2011. Russia's Putin Says Arctic Trade Route to Rival Suez. *Reuters – Daimler*, September 22. https://www.reuters.com/article/us-russia-arctic-idUSTRE78L5TC20110922.

Charron, A., and J. Fergusson. 2018. From NORAD to NOR{A}D: The Future Evolution of North American Defence Co-Operation. Canadian Global Affairs Institute Policy Paper. University of Manitoba, May 2018.

Collins, J.F. 2019. *Overcoming 'Boom and Bust'? Analyzing National Shipbuilding Plans in Canada and Australia*. Canadian Global Affairs Institute. 19 pp. ISBN: 978-1-77397-058-5.

Emerson, D. 2016, February. *Pathways Connecting Canada's Transportation System to the World*. Canada Transportation Act Review. Ottawa. 83 pp.

Felshtinsky, Y., and V. Pribylovsky. 2008. *The Corporation*. New York and London: Encounter Books. 537 pp.

Forrest, M. 2018, December 13. NWT Not Interested in Joining Provinces in Carbon Tax Fight. Territory Forges Ahead with Own Plan. *National Post*, p. A7.

Foy, H. 2018. Russia Ships First Gas From $27bn Arctic Project. *Financial Times*, December 8, 2017.

Goldstein, L. 2018. What Russia's Vostok-18 Exercise with China Means. Moscow Knows What It Is Doing and Washington Should Take Note. *The National Interest*, September 5. https://nationalinterest.org/feature/what-russias-vostok-18-exercise-china-means-30577.

Gosnell, R. 2018. The Complexities of Arctic Maritime Traffic. *The Arctic Institute – Center for Circumpolar Security Studies* 2019. 19 pp.

Hage, R. 2018, September. *Rights of Passage: It's Time the U.S. Recognizes Canada's Arctic Claim*. Ottawa, ON: Canadian Global Affairs Institute.

Holmes, S., and I. Krastev. 2015. Russia's Aggressive Isolationism. *The American Interest* X (3): 13–18.

Hopper, T. 2017. Why Russia Runs the World. *National Post* A2, January 17.

Huebert, R. 2017. Russia and North America Diverge on Arctic Resources. 2017. *Arctic Deeply*. https://www.newsdeeply.com/arctic/community/2017/02/03/russia-and-north-america-diverge-on-arctic-resources.

JIF. 2011. Global North 2050. Summary Report Jasper Innovation Forum 2011: The Global North 2050. Jasper, Alberta, Canada. November 22 to 25, 2011. Alberta Innovates – Technology Futures April 16, 2012. 93 pp.

Kobzeva, O., and K. Golubkova. 2015. Russia's Sberbank Says to Decide on Yamal LNG Financing Terms by Month-End. *Reuters*, September 8.

Lindley-French, J. 2019, January. Complex Strategic Coercion and Russian Military Modernization. *Canadian Global Affairs Institute*, 5 pp.

Nicholls, N. 2018. Defence Cuts, Skill Fade and Lack of 'Intellectual Investment' in Arctic Threatens Britain's Northern Flank. *The Telegraph*, August 15.

Osborn, A. 2017. Putin's Russia in Biggest Arctic Military Push Since the Fall of the Soviet Union. *Reuters*, February 1.

Putin, V. 2014, March 8. en.kremlin.ru/events/president/news/20603.

Roston, E. 2018. Russia Is Building $320 Million Icebreakers to Carve New Arctic Routes: The 1,000-Foot-Long Vessels for Hauling Liquefied Natural Gas Can Cut Through Ice up to 7 Feet Thick. *Bloomberg*, July 9.

Russia: Implications for UK Defence and Security: Government Response to the Committee's First Report of Session 2016–17 and the Findings Set Out in the Committee's Report (HC 107). Published 5 July 2016. https://publications.parliament.uk/pa/cm201617/cmselect/cmdfence/107/107.pdf.

Shevtsova, L. 2003. *Putin's Russia*. Washington, DC: Carnegie Endowment for International Peace. 457 pp.

Spears, J. 2018. Closing Canada's Icebreaker Gap. *Canadian Sailings*, March 18.

Tanas, O., and A. Biryukov. 2018. Nobody Listened to Us. Listen Now: Putin Warns the U.S. with Nuclear Weapons Display. *Bloomberg News. National Post*, March 1.

UK Defence Sub-Committee Report. 2018. On Thin Ice: UK Defence in the Arctic. https://publications.parliament.uk/pa/cm201617/cmselect/cmdfence/668/66804.htm.

Wallace, R. 2012. Putin Proffers and Arctic Gauntlet – How Should Canada Respond? CDFAI. *Calgary*, March 2012.

———. 2013. The Case for RAIPON: Implications for Canada and the Arctic Council. CDFAI. *Calgary*, February 2013. 14 pp.

Wallace, R.R. 2018. The Tortuous Path to NEB 'Modernization'. *The Energy Regulation Quarterly Journal* 6 (2): 23–39.

Wallace, R., and R. Dean. 2013. *A Circumpolar Convergence: Canada, Russia, the Arctic Council and RAIPON*. Toronto, ON: Munk School of Global Affairs and the Walter & Duncan Gordon Foundation.

Weber, B. 2018a, February 1. *Canada Far Behind in Arctic Economic Development: Think Tank*. Canadian Press (Citing Higginbotham Study).

———. 2018b, June 19. Canadian Press: Canadians Sign Letter Opposing U.S. Arctic Drilling in Wildlife Sanctuary. https://www.ctvnews.ca/canada/canadians-sign-letter-opposing-u-s-arctic-drilling-in-wildlife-sanctuary-1.3979769.

Wright, D.C. 2018, September. *The Dragon and Great Power Rivalry at the Top of the World*. Canadian Global Affairs Institute. https://www.cgai.ca/the_dragon_and_great_power_rivalry_at_the_top_of_the_world.

PART V

Arctic Legal and Institutional Systems

CHAPTER 23

The United Nations Convention on the Law of the Sea and the Arctic Ocean

Nigel Bankes and Maria Madalena das Neves

INTRODUCTION

The United Nations Convention on the Law of the Sea (LOSC 1982) was adopted in 1982 and entered into force in 1994. It currently has 168 parties. According to the Preamble, the Parties recognized the "desirability of establishing through this Convention, with due regard for the sovereignty of all States, a legal order for the seas and oceans which will facilitate international communication, and will promote the peaceful uses of the seas and oceans, the equitable and efficient utilization of their resources, the conservation of their living resources, and the study, protection and preservation of the marine environment." Often referred to as "a constitution for the oceans" (Gavouneli 2007, 1; Churchill 2015, 44), the 17 Parts of the Convention deal with everything from establishing different maritime zones (Parts I–VI) and rules for the high seas and deep seabed area (Parts VII and XI), to provisions dealing with the protection of the marine environment (Part XII), marine scientific research (Part XIII), and dispute resolution (Part XV). The LOSC is a framework Convention in the sense that it contemplates that it will be further elaborated through bilateral, regional, and multilateral instruments.

N. Bankes (✉)
University of Calgary, Calgary, AB, Canada

University of Tromsi - Norway's Arctic University, Tromsø, Norway
e-mail: ndbankes@ucalgary.ca

M. M. das Neves
University of Tromsi - Norway's Arctic University, Tromsø, Norway
e-mail: maria.m.neves@uit.no

© The Author(s) 2020
K. S. Coates, C. Holroyd (eds.), *The Palgrave Handbook of Arctic Policy and Politics*, https://doi.org/10.1007/978-3-030-20557-7_23

There is only one Arctic-specific provision in the LOSC and even this does not reference the Arctic by name. This is the well-known Article 234 dealing with ice-covered areas (discussed further below).

This chapter begins with an analysis of the legal status of the LOSC in the Arctic and a brief review of its content emphasizing the different maritime zones established by the LOSC. It then examines the rules established by the LOSC for the delimitation of overlapping maritime claims as well as the rules for extended continental shelf claims by coastal States. The chapter then discusses two more specific issues: first, the LOSC provisions with respect to the protection of the marine environment with more detailed examination of Article 234 and the related question as to the legal relationship between national measures under Article 234 and the Polar Code recently adopted by the International Maritime Organization (IMO); and second, the LOSC provisions with respect to high seas fisheries and a more detailed examination of the recently concluded Agreement to Prevent Unregulated High Seas Fisheries in the Central Arctic Ocean.

The principal importance of the LOSC for this chapter is that it provides the framework rulebook for the use of the Arctic Ocean including the maritime entitlements of the coastal states, navigation and shipping, fisheries, resource development, and environmental protection. Insofar as the Arctic Ocean is an ocean surrounded by land territories it is hard to overestimate the significance of the Convention for the coastal states of the Arctic but also for all other countries and users of the oceans.

Legal Status of the LOSC in the Arctic

All Arctic States are party to the LOSC except the United States. As a non-party the United States is not bound by the LOSC as a treaty. However, commentators generally agree that most of the provisions of the Convention (beyond the detailed provisions dealing with the deep seabed) represent customary international law and thus bind the United States as such (e.g. Gavouneli 2007, 4; Churchill 2015, 37 [with some qualifications]). Thus, while references to the LOSC are references to the Convention, general references to "the law of the sea" are neutral as to the source of obligation and thus embrace both the Convention (as applicable) as well as customary international law and general principles of international law.

In addition to the Convention there are two so-called implementing agreements. These are the Agreement Relating to the Implementation of Part XI of the Convention (the deep seabed regime) and the Agreement for the Implementation of the Provisions of the United Nations Convention on the Law of the Sea of 10 December 1982 relating to the Conservation and Management of Straddling Fish Stocks and Highly Migratory Fish Stocks (Churchill 2015, 42–43). Negotiations have started with respect to a third possible such agreement dealing with the conservation of marine biological diversity in areas beyond national jurisdiction (ABNJ).

The fact that large parts of the Arctic Ocean may be covered with sea ice in different forms for the entire year, or significant parts of the year, does not mean that the LOSC (or the law of the sea more generally) is inapplicable. The five Arctic coastal states (Canada, Denmark (in respect of Greenland) Norway, Russia, and the United States) reaffirmed this a decade ago when they adopted the Ilullisat Declaration in 2008 (Ilullisat Declaration 2008). The five States recalled "that an extensive international legal framework applies to the Arctic Ocean" and concluded that "the law of the sea provides for important rights and obligations concerning the delineation of the outer limits of the continental shelf, the protection of the marine environment, including ice-covered areas, freedom of navigation, marine scientific research, and other uses of the sea. We remain committed to this legal framework and to the orderly settlement of any possible overlapping claims."

Maritime Zones

The LOSC recognizes that ocean space may be divided into a number of maritime zones principally comprising the internal waters of a coastal State, a 12 nm territorial sea, and a 200 nm exclusive economic zone (EEZ). Beyond the EEZ there are the high seas. The seabed is subject to the continental shelf entitlement of the coastal state and beyond those collective entitlements there is the deep seabed (the Area), the common heritage of humankind.

The internal waters of a coastal State are those waters that lie to the landwards of the baseline from which the territorial sea is measured (Article 8). The sovereignty of the coastal State extends to both its internal waters and its territorial sea (Article 2.1). The baseline for measuring the territorial sea is normally the low-water line along the coast (Article 5) although the coastal State may employ straight baselines (Article 7.1) "where the coast is deeply indented and cut into, or if there is a fringe of islands along the coast in its immediate vicinity." For example, Norway drew straight baselines along its Arctic coast early in the twentieth century. The International Court of Justice upheld the validity of that approach in the Anglo-Norwegian Fisheries Case (1951). A more recent example is that of Canada that purported to draw straight baselines around its Arctic Archipelago in 1985 (Schönfeldt 2017, 551). The validity of these baselines is contested by both the United States and the European Union (Schönfeldt 2017, 559). Within the territorial sea the coastal State's sovereignty is subject to the right of innocent passage of all vessels (Article 17).

Beyond the territorial sea the claims of the coastal State are more limited and functional in nature—but still extensive (Gavouneli 2007). Thus, within the EEZ, the coastal State has sovereign rights (Article 56) "for the purpose of exploring and exploiting, conserving and managing the natural resources, whether living or non-living, of the waters superjacent to the seabed and of the seabed and its subsoil, and with regard to other activities for the economic exploitation and exploration of the zone, such as the production of energy from the water, currents and winds." It also has jurisdiction (Article 56) "with regard to: (i) the establishment and use of artificial islands, installations, and structures; (ii) marine scientific research; (iii) the protection and preservation of the marine environment." The maximum breadth of the EEZ is 200 nm measured from the same territorial sea baseline described above (Article 57). The coastal State has a duty to promote the objective of optimum utilization of the living resources of the EEZ. Where it is unable to harvest the entire total allowable catch (TAC) it is to afford other States access to the surplus (Article 62). This duty is modified somewhat in the case of a coastal State (Article 71) "whose economy is overwhelmingly dependent on the exploitation of the living resources of its EEZ."

Other States have rights within a coastal State's EEZ and in particular they continue to be able to exercise most traditional high seas freedoms (Articles 58 and 87) such as navigation and overflight; but all rights to exploit resources within the EEZ are the exclusive prerogative of the coastal State. The interaction between the rights of the coastal state and traditional maritime freedoms may cause some controversy. For example, in the Arctic Sunrise Arbitration a Netherlands flagged Greenpeace vessel, the *Arctic Sunrise*, placed protesters on board an offshore oil platform, the *Prirazlomnaya*, operating in the Pechora Sea in the south east part of the Barents Sea within Russia's EEZ. Russian authorities subsequently boarded, seized, and detained the *Arctic Sunrise* and its crew. The arbitral tribunal concluded that while Russia as the coastal state had a legitimate interest in protecting its sovereign rights within its EEZ, it must do so using measures that were reasonable, necessary, and proportionate. Furthermore, the coastal state must "tolerate some level of nuisance through civilian protest" and thus give "due regard … to rights of other States, including the right to allow vessels flying their flag to protest" (Arctic Sunrise Award 2015, 326–328). In so concluding, the tribunal recognized that "Protest at sea is an internationally lawful use of the sea related to freedom of navigation" (Arctic Sunrise Award 2015, 227). Accordingly, the tribunal concluded that Russia was in breach of the obligations it owed under the LOSC to the Netherlands as the flag state of the *Arctic Sunrise*.

The continental shelf of a coastal State "comprises the seabed and subsoil of the submarine areas that extend … throughout the natural prolongation of its land territory to the outer edge of the continental margin" or out to 200 nm from the baseline (Article 76). In other words, a coastal State will always (subject to overlapping entitlements with an opposite State) have a continental shelf of 200 nm but may also have an extended shelf where the continental margin extends beyond that out to 350 nm (or 100 nm from the 2500 m isobath) provided that certain technical conditions can be established. A coastal State making a claim to an extended shelf must provide information with respect to its claim along with supporting technical data to the Commission on the Limits of the Continental Shelf (CLCS). The CLCS examines the claim and supporting data and then makes (Article 76(8)) "recommendations to coastal States" with respect to establishing the outer limits of their shelves. It is then up to the coastal State to establish the limits of the shelf "on the basis of these recommendations" which limits "shall be final and binding" (Elferink 2001; Macnab 2004; McDorman 2013). States may submit full or partial claims (i.e. with respect to only part of their coasts). All Arctic coastal states (with the exception of the United States) have submitted full or partial claims. Canada recently (May 2019) filed its claim with respect to the Arctic Ocean part of its coast.

The Commission faces a considerable backlog in its work. Norway is the only Arctic State to have received the Commission's advice. In the case of Russia (the first coastal State to file) the Commission has asked Russia for additional supporting information, which Russia submitted in August 2015. A final recommendation of the CLCS on Russia's submission is still pending. Although frequently portrayed in the media as a competing "scramble for resources" as between Arctic coastal states, these States have actually demonstrated a remarkable degree of cooperation in collecting the necessary scientific data to support their extended shelf claims (Bankes and Koivurova 2015, 245–247). Once having delineated their extended shelves it will still be necessary for Arctic coastal states to reach delimitation agreements where those claims to extended shelves overlap. In other words, the processes of delineation of an extended shelf and delimitation of that extended shelf are two different processes. The Commission's only responsibility is to provide advice on delineation.

The coastal State has exclusive sovereign rights for the purpose of exploring and exploiting (Article 77) the "mineral and other non-living resources of the seabed and subsoil together with living organisms belonging to sedentary species ...." The continental shelf does not affect the status of the superjacent waters (Article 78) and thus, much as with the EEZ, other States may continue to exercise traditional high seas freedoms. In this case the freedoms would include the freedom of fishing (for all species except sedentary species) in relation to the extended shelf (i.e. the area beyond 200 nm).

It should be apparent from this account that all coastal states, including Arctic coastal States, were major beneficiaries of the 1982 LOSC. They made significant gains in terms of coastal state powers and jurisdiction by comparison with the last major treaty settlements of 1958 pertaining to ocean space (the Geneva Conventions).

The "high seas" in a formal sense (Article 86) refers to all of those waters that are not included within the EEZ, territorial sea or internal waters of a coastal state. The high seas are "open to all States" and within the high seas all States may exercise the "freedom of the high seas" that freedom includes (Article 87):

1. freedom of navigation
2. freedom of overflight
3. freedom to lay submarine cables and pipelines ...
4. freedom to construct artificial islands and other installations permitted under international law ...
5. freedom of fishing ...
6. freedom of scientific research ...

In the case of the Arctic, this means that there is an area (the central Arctic Ocean) in which the flag vessels of all States (including non-Arctic coastal States and land-locked States) may exercise these high seas freedoms. Furthermore, it will be recalled that those high seas freedoms that do not conflict with the EEZ rights of a coastal State may also be exercised in the 188 nm zone between a coastal State's territorial sea and the outer limits of its EEZ. In that limited sense the EEZ has some of the characteristics of the high seas (Gavouneli 2007).

These high seas freedoms are not absolute. Article 87.1 of the LOSC indicates that the freedoms may only be exercised "under the conditions laid down by the Convention and by other rules of international law." Article 87.2 further specifies that

> These freedoms shall be exercised by all States with due regard for the interests of other States in their exercise of the freedom of the high seas, and also with due regard for the rights under this Convention with respect to activities in the Area

While these generic references to "other rules of international law" and "due regard" may seem too general to have any real bite, recent arbitral awards have confirmed that these types of open-textured and ambulatory provisions may serve to incorporate a broad range of norms. For example, in the Chagos Award the arbitral tribunal concluded that these references served to oblige the United Kingdom to recognize certain commitments made to Mauritius in respect of the Chagos Islands at the time that Mauritius achieved independence. The United Kingdom breached these commitments when it unilaterally declared a marine protected area the whole of the EEZ pertaining to the Chagos Islands (Chagos Arbitration Award 2015).

Finally, there is the Area. The Area means (Article 1.1(1)) "the sea-bed and ocean floor and subsoil thereof, beyond the limits of national jurisdiction." It follows from this that the Area begins at the end of the shelves or the extended shelves of the coastal States. Since most extended shelves of most Arctic States have yet to be delineated it is as yet unclear whether any part of the seabed of the Arctic Ocean falls within the Area. The "Area and its resources are the common heritage of mankind." The detailed rules with respect to the Area are established by Part XI of the LOSC (as modified by the Part XI Implementing Agreement).

While these are the main marine spatial zones recognized by the Convention, the Convention also recognizes some specialized zones or areas such as archipelagic waters (Part IV) and enclosed or semi-enclosed seas (Part IX); and there is, in addition, a regime for straits used for international navigation between one part of the high seas (or an EEZ) and another part of the high seas (or an EEZ). There are many details to the straits regime (Part III) but the basic concept is that all ships and aircraft enjoy an unimpeded right of transit passage through such a strait (Rothwell 2015).

DELIMITATION RULES

It is necessary for adjacent and opposite coastal States to delimit their international maritime claims and boundaries where those claims overlap (Evans 2015). In the case of territorial sea claims Article 15 instructs that, failing agreement, delimitation shall be on the basis of equidistance except where required otherwise "by reason of historic title or other special circumstances." An Arctic example of a territorial sea delimitation is that between Russia and Norway in respect of the Varangerfjord effected by treaty in 1957 on the basis of a median (i.e. equidistance) line (Schönfeldt 2017, 699). This line was later extended out to a 12 nm territorial sea (and even some distance beyond) by a subsequent treaty in 2007 (Schönfeldt 2017, 700).

The Convention offers less definitive guidance with respect to the delimitation of the EEZ and the continental shelf. In identical language in Articles 74.1 (EEZ) and 83.1 (the shelf), the Convention stipulates that "The delimitation of the continental shelf [EEZ] between States with opposite or adjacent coasts shall be effected by agreement on the basis of international law, ...in order to achieve an equitable solution." Where agreement cannot be reached the States concerned should avail themselves of Part XV of the Convention dealing with the settlement of disputes. "Pending agreement the States concerned, in a spirit of understanding and cooperation, shall make every effort to enter into provisional arrangements of a practical nature and, during this transitional period, not to jeopardize or hamper the reaching of the final agreement" (LOSC, Articles 74.3 and 83.3).

Although the guidance offered by the Convention is limited, successive decisions of the International Court of Justice (ICJ), the International Tribunal for the Law of the Sea (ITLOS) and various arbitral awards over the last decade have largely settled on an agreed delimitation methodology designed to achieve "an equitable solution." The process begins by identifying the relevant coasts that give rise to the overlapping entitlement (Evans 2015, 267). Next, the tribunal constructs a provisional equidistance line. It then determines whether there are reasons to adjust that line. Relevant reasons might include the configuration of the coasts (concave or convex), proportionality (i.e. relative lengths of the opposing coasts), the presence of islands, and in exceptional cases the distribution of resources (Tanaka 2015, 209–224). Finally, the tribunal is to consider whether the line leads to an inequitable result (and if so adjust the line accordingly).

Another feature of modern delimitation practice is for adjacent and opposite state to agree on a single maritime boundary rather than separate boundaries for the EEZ and the continental shelf (although the continental shelf boundary between adjacent states may extend beyond 200 nm).

State practice in the Arctic offers examples of delimitation by agreement, by adjudication and by voluntary conciliation and perhaps an example of a provisional arrangement (Bankes 2016; Bankes and Koivurova 2015).

An early Arctic example of a continental shelf delimitation in the Arctic is the 1973 Agreement between Denmark and Canada between Greenland and Canada based upon a median line (Schönfeldt 2017, 733). A more recent example is the 2010 Treaty between Norway and the Russian Federation concerning Maritime Delimitation and Cooperation in the Barents Sea and the Arctic Ocean 2010 (Schönfeldt 2017, 702). In this case the delimitation line begins where the 2007 Varangerfjord Agreement ended and extends northward

(Moe et al. 2011). The delimitation line is not a median line. Henriksen and Ulfstein (2011, 7) observe that the "The treaty's approximate equal division of the disputed area ... raises the question of whether the agreed boundary is best described as a modified median line (as argued by Norway) or a modified sector line (as argued by Russia)." In addition to the delimitation, the Treaty includes two annexes, one dealing with fisheries matters (continuing two earlier treaties of 1975 and 1976), and a second dealing with transboundary hydrocarbon deposits.

An example of a delimitation by adjudication in the Arctic is that effected by the ICJ between Norway and Denmark with respect to the maritime boundary between Greenland and the Norwegian island of Jan Mayen (ICJ 1993).

An example of a delimitation facilitated through voluntary conciliation is the 1981 agreement between Iceland and Norway with respect to the maritime areas between Iceland and Jan Mayen. This agreement affords Iceland a full 200 nm EEZ and also establishes a joint development zone on terms that are very favorable to Iceland. (Richardson 1988).

Provisional arrangements are appropriate where adjacent or opposite States have yet to agree upon a delimitation. While Arctic States have made considerable progress in resolving their maritime boundary claims (Bankes and Koivurova 2015) some boundary issues remain unresolved, notably the Beaufort Sea boundary between Canada and the United States (Baker and Byers 2012). In this context it is interesting to examine a joint announcement by the United States and Canada in December 2016 in which the two governments reaffirmed their commitment to ensure that commercial activities would only occur within the Arctic "if the highest safety and environmental standards are met." More concretely, the announcement went on to say that the United States was "designating the vast majority of US waters in the Chukchi and Beaufort Seas as indefinitely off limits to offshore oil and gas leasing." Canada in turn made a similar commitment that Canadian Arctic waters would be off limits to "future oil and gas licensing" subject to review every five years "through a climate and marine science-based life-cycle assessment." Although Prime Minister Trudeau was criticized by northern leaders for making this announcement without consultation with northerners, and although President Obama's commitment was soon abrogated by President Trump (Executive Order 2017), it is the unusual footnote to this part of the announcement that it is of particular interest. In that footnote the two governments indicated that they were making these commitments in light of their obligations "under international law to protect and preserve the marine environment, these steps also support the goals of various international frameworks and commitments concerning pollution." The footnote goes on to say that "with respect to areas of the Beaufort Sea where the U.S.-Canada maritime boundary has not yet been agreed, these practical arrangements are without prejudice to either side's position and demonstrate self-restraint, taking into account the principle of making every effort not to jeopardize or hamper reaching a final maritime boundary agreement." Much of this language is drawn from the provisional arrangements provisions of Articles 74(3) and 83(3) of the LOSC. Given that the United States is not a party to the LOSC, the express adoption of LOSC text seems particularly significant.

Part XII of the LOSC deals with the protection and preservation of the marine environment. Article 234 dealing with ice-covered areas falls within Part XII and hence it is important to have some appreciation of the entire Part in order to understand the significance of Article 234. Part XII imposes obligations on all states to protect and preserve the marine environment (Article 192) and to prevent, reduce, and control pollution of the marine environment from any source (Article 194) including land-based sources (Article 207), seabed activities (Article 208), dumping (Article 210), and from vessels (Article 211). With respect to pollution from vessels, the general principle is that measures to prevent reduce and control pollution should be taken through "the competent international organization" (in practice the International Maritime Organization (IMO)) rather than unilaterally. Exceptionally (Article 211(6)), coastal States may propose special rules to the IMO for their EEZs on the basis of oceanographical or technological reasons, but such measures can only be implemented with the approval of the IMO.

With this background in mind the chapter now turns to examine the "Arctic exception" of LOSC (Article 234) in more detail as well as the relationship between Article 234 and the recently adopted Polar Code.

ARTICLE 234 OF THE LOSC

Background and Negotiation of Article 234

Article 234 is the only provision of the LOSC specifically tailor-made for the Arctic. Although the provision itself contains no express reference to the Arctic (and indeed the provision applies to ice-covered waters in general), the history of its negotiation clearly shows that it was drafted with the Arctic in mind. In effect, Article 234 reflects the shared and competing interests of Arctic States in relation to navigation in Arctic waters, in particular the interests of Canada, Russia (at the time the Union of Soviet Socialist Republics), and the United States. The text was first negotiated between those three States and then submitted for negotiations with the other States involved in UNCLOS III.

The initiative to include an "Arctic exception" in the LOSC came from Canada, which sought acceptance by the international community of its Arctic Waters Pollution Prevention Act (AWPPA 1985), enacted as a response to the voyage of the SS Manhattan in 1969 through the Northwest Passage (NWP) (McRae 1987; Mestral 2015). It essentially created a special regime for vessels navigating in Canadian Arctic waters and reflected Canada's growing concerns with increasing navigation in the Arctic and the concomitant risk of vessel-source pollution that could have disastrous impacts on the fragile Arctic marine environment. Enactment of the AWPPA did raise protests from a number of countries, notably from the United States, which insisted on the status of the

NWP as a strait used for international navigation subject to the right of innocent passage (which would later become the right of transit passage in the LOSC), and argued that the AWPPA was in breach of international law (Roach and Smith 2012; Bartenstein 2011).

These different views on the AWPPA and on navigation rights in the NWP were also relevant for navigation through the Russian Northern Sea Route (NSR). The negotiations were held during the Cold War and both the United States and Russia had strategic interests in navigation in Arctic waters. Finally, the background also included the three major maritime casualties that occurred immediately prior to and during the negotiation of UNCLOS III: the *Torrey Canyon* in 1967, the *Sea Star* in 1972, and the *Amoco Cadiz* in 1978. These major oil spills assisted in raising consensus as to the need to protect the Arctic marine environment and to afford extended powers to coastal States in relation to ice-covered areas.

The Scope of Article 234 of the LOSC

As previously stated, Article 234 is included in Section 8, Ice-Covered Areas, of Part XII of the LOSC that deals with the protection and preservation of the marine environment. Article 234 expands a coastal State's prescriptive and enforcement jurisdiction over vessel-source pollution, as it allows a coastal State to adopt and enforce regulations for the protection of the marine environment that are more stringent than generally accepted international rules and standards (GAIRS). This right is, however, subject to several conditions. Article 234 states the following:

> Coastal States have the right to adopt and enforce *non-discriminatory* laws and regulations for the prevention, reduction and control of marine pollution from vessels in *ice-covered areas within the limits of the exclusive economic zone, where particularly severe climatic conditions and the presence of ice covering such areas for most of the year* create obstructions or exceptional hazards to navigation, and pollution of the marine environment *could cause major harm to or irreversible disturbance of the ecological balance*. Such laws and regulations *shall have due regard to navigation* and the protection and preservation of the marine environment *based on the best available scientific evidence*. (Emphasis added)

The final text of Article 234 has been characterized as "probably the most ambiguous, if not controversial, clause in the entire treaty" (Lamson 1987). The text prompts a number questions: (i) When can an area be construed as being ice-covered for most of the year? (ii) At what point are climatic conditions particularly severe? (iii) What does the expression "within the limits of the exclusive economic zone" entail? (iv) What can be construed as exceptional hazards to navigation? (v) When is the threshold for major harm to the marine environment met? (vi) What is the content of the due regard obligation? (vii) What can be construed as best available scientific evidence? Establishing the normative content of this provision thus requires resorting to the interpretation rules contained in the Vienna Convention on the Law of Treaties (VCLT 1969), namely, Articles 31 and 32. Several authors have sought to clarify the exact content of each of the conditions laid in Article 234 of the LOSC (see, e.g. Bartenstein 2011; Luttmann 2015). Still, the interpretation of Article 234 remains somewhat controversial.

Article 234 of the LOSC and the IMO's Polar Code

A final and more recent issue is that of the relationship between Article 234 and the mandatory International Code for Ships Operating in Polar Waters (Polar Code) adopted by the International Maritime Organization (IMO) in 2015 through amendments to the International Convention on the Safety of Life at Sea (SOLAS 1974) and to the International Convention for the Prevention of Pollution from Ships (MARPOL 1972). While it is beyond the scope of this chapter to engage in an in-depth discussion of this issue, it is nonetheless appropriate to outline some brief remarks on the topic (for more extensive coverage see McDorman 2015) both because it is important in its own right but also because the relationship between Article 234 and the Code illustrates more broadly the interplay between the LOSC and on the ongoing work of the IMO in formulating general standards and thereby effectively elaborating on some of the more open-textured provisions of the LOSC.

Article 234 and the Polar Code have different objectives. Whereas Article 234 grants coastal States unilateral prescriptive and enforcement jurisdiction concerning the protection of the marine environment from vessel-source pollution, subject to the conditions therein outlined, the Polar Code enshrines a body of general rules and standards that should be applied in a uniform manner by States. The adoption of unilateral rules under Article 234 that diverge from the Polar Code could undermine the Code's relevance. Conversely, if it is deemed that the Polar Code takes precedence over Article 234 it could have the effect of rendering this provision useless. This issue is one that particularly interests Canada and Russia since both rely extensively on Article 234 as the legal basis for their prescriptive and enforcement jurisdiction in respect of navigation both in the NWP and the NSR. Moreover, Canada and Russia also favor Article 234 since it does not require either State to seek the prior approval of the IMO of their national legislation (Bognar 2016). Both Canada and Russia sought to safeguard their national legislation adopted under Article 234 in the context of the IMO and the Polar Code negotiations, though without success.

The question of which instrument takes precedence can only be resolved by the interpretation of both treaties in light of any relevant relationship clauses (i.e. a clause identifying how treaty A is to relate to earlier and/or later treaties dealing with the same subject matter) in the treaties and in light of the rules of the VCLT, specifically Article 30. The Polar Code, in what seems to be a conscious decision to depart from IMO's general practice of including a relationship clause deferring to the LOSC does not contain a relationship clause. But other provisions are also relevant to this question including Articles 297 and 311 of the LOSC, Article 9(2) of the MARPOL, Chapter XIV Regulation 2.5 of the SOLAS and Article 30 of the VCLT (dealing with the specific issue of successive treaties) A number of authors have examined this issue more closely. McDorman seems to suggest that Article 234 of the LOSC would prevail (McDorman 2015). Jensen similarly argues that the Polar Code does not affect the rights of coastal States under Article 234 of the LOSC (Jensen 2016). Conversely, Roach and Smith, as well as Fauchald, have outlined contrary views, that is, that the Polar Code would have priority over Article 234 of the LOSC (Roach and Smith 2012, 494; Fauchald 2011, 83). This debate is likely to continue.

THE AGREEMENT TO PREVENT UNREGULATED HIGH SEAS FISHERIES IN THE CENTRAL ARCTIC OCEAN

Background

While there is a long history of fishing activities in the Arctic (especially in the Barents and Bering Seas) to date there are still no commercial fishing activities taking place in the high seas portion of the Central Arctic Ocean (CAO). This does not mean that there is a legal vacuum concerning fisheries in the high seas of CAO (Molenaar 2016). Article 87 of the LOSC affirms the freedom of fishing in the high seas (which includes the high seas portions of the CAO), subject only to the conditions relating to the conservation and management of living resources of the high seas, included in Articles 116–120 of the LOSC. Article 118 of the LOSC, as well as Article 5 of the Fish Stocks Agreement (one of the LOSC implementing agreements referenced above dealing with straddling fish stocks and highly migratory fish stocks) (UNFSA 1995), require all States to cooperate in the conservation and management of living resources in the high seas. Such cooperation can occur via the conclusion of agreements and establishment of subregional or regional fisheries organizations. Furthermore, Article 197 of the LOSC prescribes that States are to cooperate globally or regionally for the protection and preservation of the marine environment, which includes living marine resources (Chagos Arbitration Award 2015), while taking into account characteristic regional features. Moreover, for those that consider the Arctic Ocean to be a semi-enclosed sea (e.g. Pharand 1992; Corell 2007), which is arguable, Article 123 of the LOSC would also serve as a basis for cooperation between the Arctic States and other interested States on conservation and management of fisheries resources. Finally, Articles 237 and 311 of the LOSC inform that the States can conclude other agreements as long as they do not conflict with the general principles and objectives of the LOSC.

Relying on these provisions of the LOSC the Arctic States have recently taken the initiative to assume a stewardship role and broker an agreement seeking the sustainable management of fisheries in the Central Arctic Ocean thereby further implementing these more general provisions. On November 2017, the five Arctic coastal States (Canada, Denmark, Norway, Russia, and the United States) together with China, Iceland, Japan, South Korea, and the European Union (EU), concluded negotiations on an Agreement to Prevent Unregulated High Seas Fisheries in the Central Arctic Ocean (CAOF Agreement).

This Agreement has already been dubbed by some as "historic" and "unprecedented." Taylor, for example, noted that "[t]his will be the only ocean in the world that humankind have agreed to not fish in until we have a scientific understanding of what's there and the management regime under which to operate" (Taylor as quoted by Sevunts 2017). This statement may go too far with its reference to "humankind" given the principle that a treaty does not bind a non-party (VCLT, Article 34) without its consent but, it does adequately portray the novelty of the Agreement and the enthusiasm with which it has been greeted. If signed/ratified by all parties and it enters into force, this Agreement will have implications for a large share of the world's fishing fleets, and certainly represent an important step toward the conservation and sustainable management of fish stocks of the Central Arctic Ocean. Moreover, the CAOF Agreement unequivocally manifests the acknowledgment by all negotiating parties, and more importantly by the United States, of the causal relationship between climate change, diminishing ice in the Arctic Ocean, the potential for future commercial fisheries in the Central Arctic Ocean, and the fragility of the Arctic ecosystem. The novel aspect of the CAOF Agreement lies in its preemptive nature and true embodiment of the precautionary approach. That is, the Agreement seeks to regulate commercial fishing activities in the Central Arctic Ocean before they actually take place, and to eventually authorize fishing activities only on the basis of scientific information. To the authors' knowledge this is the first time within international fisheries law that an agreement tries to regulate an activity a priori instead of a posteriori as a reaction to existing problems.

Key Features of the CAOF Agreement

The CAOF Agreement is a relatively straightforward document comprising a preamble and 15 provisions. What follows is a brief overview of its key aspects. The CAOF Agreement applies to fishing activities in the high seas portion of the Central Arctic Ocean, and aims to prevent unregulated fishing in that area through the application of precautionary conservation and management measures (Article 2). While the Agreement implies an indefinite abstention from commercial fishing activities, it does not preclude the possibility for such fisheries pursuant to conservation and management measures of existing or future regional or subregional fisheries management organizations or arrangements (such as the North East Atlantic Fisheries Commission and the Joint Norwegian Fisheries Commission); or pursuant to interim conservation and management measures established by the Parties (Article 3(1)). The CAOF Agreement also does not preclude exploratory fisheries, that is, fisheries conducted for the "purpose of assessing the sustainability and feasibility of future commercial fisheries by contributing to scientific data relating to such fisheries" (Article 3(3)). That is, as long as exploratory fisheries are conducted in accordance with the conservation and management measures that may be established by the Parties under Article 5(1)(d), and that they do not undermine the purpose of the Agreement. In this respect, the CAOF Agreement puts into place the basis for implementing the obligations to cooperate and to adopt conservation and management measures prescribed by the LOSC in Articles 116–120, 123, and 197. The CAOF Agreement also places great emphasis on the need for the Parties to make decisions based on the best scientific evidence, and on the need to cooperate in scientific activities contributing toward increasing the knowledge on the Central Arctic Ocean's ecosystem and status of living marine resources (Article 4). Such cooperation is to occur through a Joint Program of Scientific Research and Monitoring. This is in line with Article 119 of the LOSC, which requires States to adopt the best scientific evidence available when taking measures relating to the conservation of the living marine resources of the high seas, and to share available scientific information. The emphasis that the CAOF Agreement places on best scientific evidence, the precautionary approach, the value of protecting the marine ecosystems, and cooperation is also in line with the UNFSA. Additionally, the Agreement also refers to the interests of Arctic residents and indigenous peoples, and indicates the desire to promote indigenous and local knowledge of Arctic Ocean living marine resources and ecosystem (paragraphs 10–12 of the Preamble and Article 5(1) (b) and Article 5(2)).While the LOSC does not expressly mention indigenous peoples, it does nonetheless aim to contribute to the maintenance of peace, justice, and progress for all peoples of the world (LOSC preamble). Here the CAOF Agreement takes a step further by specifically addressing the interests and possibility of participation of indigenous peoples. Still, it remains to be seen how indigenous peoples and Arctic communities will be allowed to participate in the context of the CAOF Agreement.

Finally, the Agreement addresses the possible unwillingness of third States to abstain from fishing in the Central Arctic Ocean. Article 8(1) of the CAOF Agreement requires the Parties to encourage non-parties to respect the purposes of the Agreement and to take measures consistent with it. This aspiration is, at the same time, strengthened by Article 8(2) that compels the Parties to take measures against vessels from non-parties that undermine the effective implementation of the agreement. Such measures need nonetheless to be consistent with international law, for example, the observance of Article 87(1)(e) of the LOSC on freedom of high seas fisheries. In order to try to attain greater acceptance among other States, Article 10 also foresees the possibility of accession by other States with a real interest. However, this may only occur upon invitation of the Parties. It is also not clear what can be construed as "real interest."

This brief outline of the key features of the Agreement demonstrates that the Agreement establishes the bases on which to further implement the general obligations laid out in the LOSC. At the same time, it shows that there are still a number of details to be worked out within the first years following the entry into force of the Agreement.

Conclusions

This chapter has examined the legal status of the LOSC in the Arctic and how the LOSC addresses maritime delimitation, environmental protection, and fisheries in the Arctic. The chapter makes it clear that, contrary to some depictions, the Arctic Ocean is not a frontier area void of legal rules where States race to capture all available resources. The LOSC is a universal and global convention that applies to the Arctic Ocean just as it applies to all marine areas in the world. As a framework convention, the LOSC identifies the essential rights and obligations of States regarding maritime claims in the Arctic, and all other uses and activities in the Arctic Ocean. Still, because the LOSC is a framework convention it requires further implementation by additional agreements and instruments whether developed by global institutions such as the IMO (e.g. the Polar Code) or regional arrangements (e.g. the CAOF Agreement), or even bilateral arrangements (e.g. bilateral delimitation agreements).

References

Agreement to Prevent Unregulated High Seas Fisheries in the Central Arctic Ocean (CAOF Agreement). https://eur-lex.europa.eu/legal-content/en/TXT/?uri=CELEX:52018PC0454.

Arctic Sunrise Award. 2015. Award on the Merits. Registry: Permanent Court of Arbitration. https://pcacases.com/web/sendAttach/1438.

AWPPA. 1985. Arctic Waters Pollution Prevention Act, Revised Statues of Canada, chapter A-12.

Baker, J., and M. Byers. 2012. Crossed Lines: The Curious Case of the Beaufort Sea Maritime Boundary Dispute. *Ocean Development and International Law* 43: 70–95.

Bankes, N. 2016. The Regime for Transboundary Hydrocarbon Deposits in the Maritime Delimitation Treaties and Other Related Agreements of Arctic States. *Ocean Development and International Law* 47 (2): 141–164.

Bankes, N., and T. Koivurova. 2015. Legal Systems. In *Arctic Human Development Report*, 2nd ed., 221–252. http://norden.diva-portal.org/smash/record.jsf?pid=diva2%3A788965&dswid=-3505.

Bartenstein, K. 2011. The "Arctic Exception" in the Law of the Sea Convention: A Contribution to Safer Navigation in the Northwest Passage? *Ocean Development & International Law* 42 (1–2): 25–27.

Bognar, D. 2016. Russian Proposals on the Polar Code: Contributing to Common Rules or Furthering State Interests? *Arctic Review on Law and Politics* 7 (2): 111–135.

Chagos Arbitration Award. 2015. Merits. Registry: Permanent Court of Arbitration. https://pca-cpa.org/en/cases/11/.

Churchill, R. 2015. The 1982 United Nations Convention on the Law of the Sea. In *The Oxford Handbook of the Law of the Sea*, ed. D.R. Rothwell, A.G.O. Elferink, K.N. Scott, and T. Stephens. Oxford: Oxford University Press.

Corell, H. 2007. Reflections on the Possibilities and Limitations of a Binding Legal Regime. *Environmental Policy and Law* 37: 322.

Elferink, A.G.O. 2001. The Outer Continental Shelf in the Arctic: The Application of Article 76 to the LOS Convention in a Regional Context. In *The Law of the Sea and Polar Maritime Delimitation and Jurisdiction*, ed. A.G.O. Elferink and D.R. Rothwell. The Hague: Martinus Nijhoff.

Evans, M.D. 2015. Maritime Boundary Delimitation. In *The Oxford Handbook of the Law of the Sea*, ed. D.R. Rothwell, A.G.O. Elferink, K.N. Scott, and T.Stephens. Oxford. Oxford University Press.

Executive Order. 2017. Implementing an America-First Offshore Energy Strategy, Executive Order 13795, April 28, 2017. https://www.federalregister.gov/documents/2017/05/03/2017-09087/implementing-an-america-first-offshore-energy-strategy.

Fauchald, O.K. 2011. Regulatory Frameworks for Maritime Transport in the Arctic: Will a Polar Code Contribute to Resolve Conflicting Interests? In *Maritime Transport in the High North*, ed. J. Grue and R. Gabrielsen. Oslo: Norwegian Academy of Science and Letters and Norwegian Academy of Technological Studies.

Gavouneli, M. 2007. *Functional Jurisdiction in the Law of the Sea*. Leiden/Boston: Martinus Nijhoff.

Henriksen, T., and G. Ulfstein. 2011. Maritime Delimitation in the Arctic: The Barents Sea Treaty. *Ocean Development and International Law* 42: 1–21.

ICJ. 1993. Maritime Delimitation in the Area between Greenland and Jan Mayen (Denmark v. Norway). *ICJ Reports, 38*.

Ilullisat Declaration. 2008. The Ilullisat Declaration. Arctic Ocean Conference, Ilullisat Greenland, 27–29 May 2008. http://www.oceanlaw.org/downloads/arctic/Ilulissat_Declaration.pdf.

Jensen, Ø. 2016. The International Code for Ships Operating in Polar Waters: Finalization, Adoption and Law of the Sea Implications. *Arctic Review on Law and Politics* 7 (1): 75–77.

Lamson, C. 1987. Arctic Shipping, Marine Safety and Environmental Protection. *Marine Policy* 11: 3.

LOSC. 1982. United Nations Convention in the Law of the Sea, December 10, 1982, 1833 *U.N.T.S.* 397.

Luttmann, P. 2015. Ice-Covered Areas under the Law of the Sea Convention: How Extensive are Canada's Coastal State Powers in the Arctic? *Ocean Yearbook Online* 29 (1): 85–124.

Macnab, R. 2004. The Outer Limit of the Continental Shelf in the Arctic Ocean. In *Legal and Scientific Aspects of Continental Shelf Limits*, ed. M.H. Nordquist, J.N. Moore, and T.H. Heidar, 304–305. Leiden: Martinus Nijhoff.

MARPOL. 1972. International Convention for the Prevention of Pollution from Ships, 1859 *U.N.T.S.* 2.

McDorman, T.L. 2013. The International Legal Regime of the Continental Shelf with Special Reference to the Polar Regions. In *Polar Law Textbook II*, ed. N. Loukacheva, 77–93. Copenhagen: Nordic Council of Ministers.

———. 2015. A Note on the Potential Conflicting Treaty Rights and Obligations between the IMO's Polar Code and Article 234 of the Law of the Sea Convention. In *International Law and Politics of the Arctic Ocean: Essays in Honor of Donat Pharand*, ed. Suzanne Lalonde and Ted L. McDorman, 141–159. Brill Nijhoff.

McRae, D.M. 1987. The Negotiation of Article 234. In *Politics of the Northwest Passage*, ed. F. Griffiths, 98–114. Kingston: McGill-Queen's University Press.

Mestral, A. 2015. Article 234 of the United Nations Convention on the Law of the Sea: Its Origins and Its Future. In *International Law and Politics of the Arctic Ocean: Essays in Honor of Donat Pharand*, ed. Suzanne Lalonde and Ted L. McDorman, 111–112. Brill.

Moe, A., D. Fjaertoft, and I. Øverland. 2011. Space and Timing: Why was the Barents Sea Delimitation Dispute Resolved in 2010? *Polar Geography* 43 (3): 145–162.

Molenaar, E.J. 2016. International Regulation of Central Arctic Ocean Fisheries. In *Challenges of the Changing Arctic: Continental Shelf, Navigation, and Fisheries*, Centre for Oceans Law and Policy Series, vol. 19, ed. Myron H. Nordquist, John Norton Moore, and Ronán Long, 429–463. Brill Nijhoff.

Pharand, D. 1992. The Case for an Arctic Region Council and a Treaty Proposal. *Revue générale de droit* 23: 163–195.

Polar Code and SOLAS amendments adopted at the 94th session of IMO's Maritime Safety Committee (MSC), in November 2014; Environmental Provisions and MARPOL amendments adopted at the 68th session of the Marine Environment Protection Committee (MEPC) in May 2015. Polar Code entered into force on 1 January 2017.

Richardson, E.L. 1988. Jan Mayen in Perspective. *American Journal of International Law* 82: 443–458.

Roach, A.J., and R.W. Smith. 2012. *United States Responses to Excessive Maritime Claims*. Brill Nijhoff.

Rothwell, D.R. 2015. International Straits. In *The Oxford Handbook of the Law of the Sea*, ed. D.R. Rothwell, A.G.O. Elferink, K.N. Scott, and T. Stephens. Oxford: Oxford University Press.

Schönfeldt, K., ed. 2017. *The Arctic in International Law and Policy*, Series on Documents in International Law. Oxford and Portland: Hart Publishing.

SOLAS. 1974. International Convention on the Safety of Life at Sea, 1184 *U.N.T.S.* 277.

Tanaka, Y. 2015. *The International Law of the Sea*. Cambridge: Cambridge University Press.

Taylor, T. 2017. As Quoted in Levon Sevunts, Arctic Nations and Fishing Powers Sign 'Historic' Agreement on Fishery. *The Barents Observer*, December 1. https://thebarentsobserver.com/en/ecology/2017/12/arctic-nations-and-fishing-powers-sign-historic-agreement-fishery.

UNFSA. 1995. United Nations Agreement for the Implementation of the Provisions of the United Nations Convention on the Law of the Sea of 10 December 1982 Relating to the Conservation and Management of Straddling Fish Stocks and Highly Migratory Fish Stocks, December 4, 1995, 2167 *U.N.T.S.* 88.

US\Canada. 2016. United States-Canada Joint Arctic Leaders' Statement, December 20. https://pm.gc.ca/eng/news/2016/12/20/united-states-canada-joint-arctic-leaders-statement.

VCLT. 1969. Vienna Convention on the Law of Treaties, 1155 *U.N.T.S.* 397.

CHAPTER 24

Arctic Policy Developments and Marine Transportation

Lawson W. Brigham

INTRODUCTION

Extraordinary changes have come to the maritime Arctic early in the twenty-first century. The expanded development of Arctic natural resources is linking the region to global markets and increasing the need for a host of policy measures to address the safety and effectiveness of Arctic marine transportation systems. Hydrocarbon developments in coastal Norway and in the Russian Arctic have stimulated increases in Arctic marine traffic, and Russia's Northern Sea Route (NSR) has witnessed a resurgence of marine traffic in summer consisting principally of liquefied natural gas (LNG) carriers, tankers, and bulk carriers (The Moscow Times 2013). Advanced research icebreakers continue to explore all regions of the central Arctic Ocean during summer in support of science and the gathering of data for delimitation of the outer continental shelf by the Arctic Ocean coastal states. Large cruise ships and specialized expeditionary vessels have ventured into most Arctic waters on summer voyages of "discovery." Marine access is also changing in unprecedented ways as Arctic sea ice undergoes a profound retreat and transformation in extent, thickness, and character influenced by global and regional anthropogenic warming. Longer seasons of Arctic navigation are becoming much more plausible. Rapid changes are transforming the entire maritime Arctic that is significantly more accessible in all seasons.

The research and writing of this chapter were supported by US National Science Foundation grant award number 1263678 to the University of Alaska Fairbanks.

L. W. Brigham (✉)
University of Alaska Fairbanks, Fairbanks, AK, USA
e-mail: lwbrigham@alaska.edu

© The Author(s) 2020
K. S. Coates, C. Holroyd (eds.), *The Palgrave Handbook of Arctic Policy and Politics*, https://doi.org/10.1007/978-3-030-20557-7_24

The central challenge for the Arctic states and the global maritime community is how to implement effective protections for Arctic people, the marine environment, and the safety of shipboard crews during an era of expanding marine use. This new era has evolved rapidly with few international shipping regulations and rules that have binding or mandatory Arctic-specific provisions. And the lack of marine infrastructure such as adequate charting, marine observations, ports, salvage, and emergency response capacity in most Arctic regions (except for areas along the Norwegian and Icelandic coasts and in northwest Russia) remains a fundamental, serious limitation to significant increases in Arctic marine traffic (AMSA 2009). Fortunately during the past 15 years the Arctic states at the Arctic Council have focused significant cooperative efforts and attention on marine safety and environmental protection issues; key progress has also been made on response issues. The International Maritime Organization (IMO) has developed and negotiated a new *International Code for Ships Operating in Polar Waters* (the "Polar Code") which fully came into force on 1 July 2018. The Arctic states must continue to identify their common interests and develop unified positions at IMO and other international maritime bodies. The real keys for further advancing Arctic marine safety and environmental protection will be the engagement of non-Arctic states and the marine industry in the processes at IMO (and other international bodies), and the degree to which the Arctic states proactively enforce the Polar Code and communicate with the global maritime community on the critical need to implement the Code and refine its rules in the years ahead. This chapter explores recent and ongoing maritime policy initiatives to strengthen marine safety, environmental protection, and response measures in the Arctic. Figure 24.1 illustrates various marine routes in the Arctic Ocean including Russia's NSR and the Northwest Passage through the Canadian Arctic.

Factors Shaping Arctic Marine Navigation

The maritime Arctic is being connected to the global economy because of the region's abundant natural wealth. Although Arctic sea ice retreat provides greater marine access (and longer seasons of navigation), the leading driver of today's Arctic marine traffic is the development of natural resources influenced by global commodity prices and in the long-term, scarcer resources around the globe (AMSA 2009; Brigham 2011). The Arctic Council's Arctic Marine Shipping Assessment (AMSA) conducted 2005–2009 used a scenarios creation process to identify the main uncertainties and factors shaping the future of Arctic navigation. Among the most influential driving forces of some 120 factors were global oil prices, new Arctic natural resource discoveries, the marine economic implications of seasonal Arctic marine operations, global trade dynamics and world trade patterns, climate change severity, a major Arctic marine disaster, transit fees on Arctic waterways, the safety of other global maritime routes, global (IMO) agreements on Arctic ship construction rules

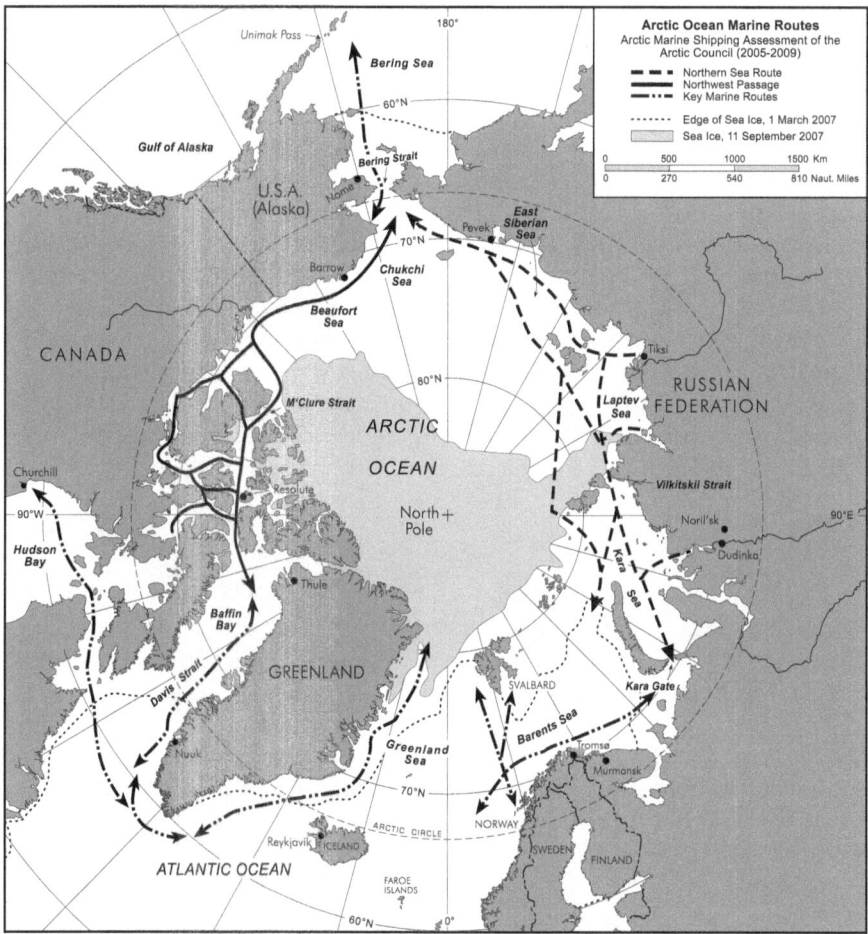

Fig. 24.1 The Arctic Ocean and marine routes (Source: L.W. Brigham, University of Alaska Fairbanks)

and standards, the legal stability and overall governance of Arctic marine use, and the entry of non-Arctic flag state ships into the maritime Arctic (AMSA 2009).

Of importance to the AMSA scenarios effort was the identification of two primary drivers as the axes of uncertainty in the scenarios matrix that was used to develop four plausible futures of Arctic marine navigation (to 2020 and 2050). Among the many uncertainties and drivers, degree of plausibility, relevance to Arctic maritime affairs, and being at the right threshold of influence were three criteria which resulted in the selection of two primary factors: *resources and trade*—the demand for Arctic natural resources influenced by the uncertainty of global commodity markets and market developments—and *governance of Arctic marine activity*—the degree of stability of rules and standards for marine use both within the Arctic and internationally (AMSA 2009). Again, climate change and Arctic sea ice retreat are fully considered by the AMSA scenarios as key to improving marine access, and these changes were understood to continue through the century. However, throughout the conduct of AMSA, global economic factors driving Arctic natural resource developments consistently loomed large as the major determinants of future Arctic navigation (Brigham 2010; Mikkola and Kapyla 2013). A primary example today is the growth in numbers of large tankers and LNG carriers out of the Yamal Peninsula along Russia's NSR (Staalesen 2017; Brigham 2013). The fact that large bulk carriers, tankers, and LNG carriers will be sailing sooner in Arctic waters in greater numbers requires complex regulatory and policy measures, as well as much greater cooperation between the Arctic states and the global shipping enterprise. Such polar voyages also demand that Arctic marine infrastructure improvements will have to be made much earlier than anticipated to keep pace with the rapid increase in use of Arctic coastal waterways and to provide robust systems for safe navigation.

Arctic Marine Accessibility and Policy Constraints

A critical point from the perspectives of marine users and policies focused on marine safety and environmental protection is that the Arctic Ocean remains fully or partially ice-covered for much of the winter, spring, and autumn. It is not an ice-free environment (there is certainly less coverage and the ice is thinner) that is to be regulated, but one with sea ice present that may be more mobile. Therefore, future ships navigating in Arctic waters will most likely be required to be built with some level of polar or ice-class capability so that they can safely and efficiently sail during potentially extended seasons of navigation. Global climate models do project continued Arctic sea ice reductions with plausible ice-free conditions for a summer time period by mid-century or earlier. Such a period would mark the disappearance of old or multi-year sea ice leaving the Arctic Ocean covered by only seasonal, first year ice which is more navigable by ship. Research has focused on how changes to Arctic marine access can be evaluated by using the global climate model sea ice simulations and a range of polar class ship types (Stephenson et al. 2013). Higher class ships (Polar Class 3) are found to gain significant marine access nearly year-round for much of the Arctic Ocean (Stephenson et al. 2013). Changing sea ice conditions by mid-century may also allow lower polar class vessels (Polar Class 6), and perhaps even non-ice strengthened (open water) ships, to cross the Arctic Ocean in September (Smith and Stephenson 2013). However, none of these results indicate regular trade route are possible, just that certain type ships may or may not have marine access for select times of the year given a range of climatic projections. This research does provide important new information about what may be plausible (and technically possible) seasons of Arctic navigation. The type of cargoes and the economics of global shipping, along with governance and environmental factors, will determine which Arctic routes might be viable (Brigham 2011; Carmel 2013).

Changing sea ice conditions and the resulting increases in Arctic marine accessibility can shape as well as constrain a broad array of policy initiatives such as marine infrastructure siting, usable technology and investment protocols; emergency response strategies; marine safety systems options (e.g., marine routing in straits); environmental protection measures for sensitive marine areas; and operational decision-making (e.g., establishing the length of the ice navigation season for commercial ships with or without icebreaker escort). New marine policies by the Arctic states are needed to address multiple use challenges between indigenous and commercial users in newly open coastal waters and sensitive natural areas. Increases in the lengths of the ice navigation seasons will require risk analyses to identify the potential locations of emergency response equipment (for search and rescue, and environmental response) and marine salvage to respond timely to Arctic marine incidents. Specific regulations may be required for application to very large ships sailing in Arctic waters focusing on maximum draft (and size) limitations, technical challenges, and operational constraints. Further policy development is needed to identify the constraints, risks, and challenges for the cruise ship industry operating in expanded seasons of operation in Arctic waters.

THE ARCTIC COUNCIL AND ARCTIC MARINE SHIPPING ASSESSMENT

The Arctic Council, an intergovernmental forum, has been a proactive international body focusing on the challenges of Arctic marine safety and environmental protection and shaping maritime policy. Established by the Ottawa Declaration in 1996, the Council focuses on sustainable development and environmental protection in the Arctic (Ottawa Declaration 1996). Key to the functioning of the Council is that six indigenous Arctic people groups (named the Permanent Participants) sit with the eight Arctic state delegations in "active participation" and "full consultation" in all Council activities (Ottawa Declaration 1996). Scientific and policy assessments, and special reports, are developed in six Arctic Council Working Groups, the technical expertise of the Council: Arctic Contaminants Action Program (ACAP); Arctic Monitoring and Assessment Programme (AMAP); Conservation of Arctic Flora and Fauna (CAFF); Emergency Prevention, Preparedness and Response (EPPR); Protection of the Arctic Marine Environment (PAME); and the Sustainable Development Working Group (SDWG). Recent work has included cross-cutting projects and activities among the groups; for example, AMAP, CAFF, and SDWG have participated with PAME in an Ecosystem Approach expert group, and EPPR has worked closely with PAME on the implementation of key recommendations from AMSA. Engagement and input of ideas and issues from non-Arctic state observers, other Council observers, and outside experts are primarily handled through the working groups which are led by Arctic state delegations (subject matter government experts) with Permanent Participant representation.

The most relevant and visible Arctic Council policy document on marine safety and environmental protection issues is the Arctic Marine Shipping Assessment conducted by PAME for the Arctic Ministers during 2004–2009 and, importantly, an outgrowth of the Council's Arctic Climate Impact Assessment which gained global attention when released in 2004. More than 200 experts led by Canada, Finland, and the United States focused the assessment on marine safety and environmental protection issues, consistent with the Council's mandate; 13 major workshops were held on key topics such as scenarios, human dimensions, environmental impacts, and infrastructure, and 14 AMSA town-hall meetings were held in Arctic communities to gain insights into the concerns and shared interests of indigenous residents. Ninety-six findings are presented in the *Arctic Marine Shipping Assessment 2009 Report* and a select list of key findings is presented in Table 24.1. The entire body of work in AMSA can be viewed in three related ways:

1. As a *baseline assessment* and snapshot of Arctic marine use early in the twenty-first century (developed from data collected by the Arctic states on ship/vessel type, marine use, season of operation, and region of operation);
2. As a *strategic guide* to a host of states, Arctic residents, users, stakeholders, and actors involved in current and future marine operations; and
3. As a *policy framework* document of the Arctic Council and the Arctic states focused on protecting Arctic people and the environment.

Certainly the key aspect of the *AMSA 2009 Report* is that the 17 recommendations were negotiated by the Arctic states and consensus reached so that final report could be approved by the Arctic Ministers at the April 2009 Arctic Council Ministerial meeting in Tromsø, Norway. The AMSA report is a maritime policy document approved at the highest levels of the Arctic states. The work of AMSA continues to this day as follow-up status reports have been requested by the Arctic Ministers and the Senior Arctic Officials. Four status reports on the implementation of the *AMSA 2009 Report* recommendations have been issued by the Arctic Council: in May 2011 (Nuuk, Greenland), May 2013 (Kiruna, Sweden), April 2015 (Iqaluit, Canada), and May 2017 (Fairbanks, USA). In this way AMSA is a "living" document and a robust process with a worthy, long-term goal of implementing all 17 recommendations, each an integral part of a whole maritime policy strategy.

AMSA's 17 recommendations as approved in 2009 focus on three, interrelated themes: (I) Enhancing Arctic Marine Safety; (II) Protecting Arctic People and the Environment; and (III) Building the Arctic Marine Infrastructure. Table 24.2 indicates the specific recommendations and actions required under each of these three broad themes. All of the recommendations require increased international cooperation, among the Arctic states, among the maritime nations at IMO (and other international bodies), and in the development of new public-private partnerships. There is little doubt the most significant recom-

Table 24.1   Select findings of the Arctic Council's Arctic Marine Shipping Assessment (AMSA 2009)

- *Arctic Sea Ice*—Global climate model simulations indicate a continuing retreat of Arctic sea ice through the twenty-first century; however, all simulations indicate an Arctic sea ice cover in winter.
- *Governing Legal Regime*—The Law of the Sea, as reflected in the 1962 *United Nations Convention on the Law of the Sea* (UNCLOS), sets out the legal framework for the regulation of (Arctic) shipping according to maritime zones of jurisdiction.
- *Key Drivers of Arctic Shipping*—Natural resource development and regional trade are the key drivers of increased Arctic marine activity. Global commodities prices for oil, gas, hard minerals, coal, and so on are driving the exploration for Arctic natural wealth.
- *Destinational Shipping*—Most Arctic shipping today is destinational (vice trans-Arctic), moving goods into the Arctic for community resupply or moving natural resources out the Arctic to world markets. Nearly all marine tourist voyages are destinational as well. Regions of high concentration of shipping occur along the coasts of northwest Russia, and in ice-free water offshore Norway, Greenland, Iceland, and the Bering Sea.
- *Impacts of Arctic Shipping on Arctic Communities*—Marine shipping is one of many factors affecting Arctic communities, directly and indirectly. The variety of shipping activities and the range of social, cultural, and economic conditions in Arctic communities mean that shipping can have many effects, both positive and negative.
- *Most Significant Environmental Threat*—Release of oil in the Arctic marine environment, either through accidental release or illegal discharge, is the most significant threat from shipping activity.
- *Special Areas*—There are certain areas of the Arctic region that are of heightened ecological significance, many of which will be at risk from current and/or increased shipping.
- *Charting and Marine Observations*—Significant portions of the primary Arctic shipping routes do not have adequate hydrographic data, and therefore charts, to support safe navigation. The operational network of meteorological and oceanographic observations in the Arctic, essential for accurate weather and wave forecasting for safe navigation, is extremely sparse.
- *Marine Infrastructure Deficit*—A lack of major ports and other maritime infrastructure, except for those along the Norwegian coast and the coast of northwest Russia, is a significant factor (limitation) in evolving and future Arctic marine operations.
- *Uncertainties of Arctic Navigation*—A large number of uncertainties define the future of Arctic shipping activity, including the legal and governance situation, degree of Arctic state cooperation, climate change variability, radical changes in global trade, insurance industry roles, an Arctic maritime disaster, new resource discoveries, oil prices and other commodity pricing, multiple use conflict (Indigenous and commercial), and future marine technologies.
- *Central Arctic Ocean*—Increased traffic in the central Arctic Ocean is a reality (in summer)—for scientific exploration and tourism.
- *Ice Navigator Expertise*—Safe navigation in ice-covered waters depends much on the experience, knowledge, and skill of the ice navigator. Currently, most ice navigator training programs are ad hoc and there are no uniform international training standards.

mendation in Theme I was the establishment of mandatory IMO standards and requirements for ships operating in Arctic waters, and the augmentation of *existing* IMO ship safety and pollution prevention conventions with Arctic-specific requirements. Another recommendation notes the importance of strengthening passenger ship safety in Arctic waters. Theme II has a key recommendation for the need to conduct comprehensive surveys of indige-

Table 24.2 The Arctic Marine Shipping Assessment Recommendations by Theme—a framework policy for the Arctic Council (AMSA 2009)

I. *Enhancing Arctic Marine Safety:*
A. Linking with International Organizations
B. IMO Measures for Arctic Shipping
C. Uniformity of Arctic Shipping Governance
D. Strengthening Passenger Ship Safety in Arctic waters
E. Arctic Search and Rescue (SAR) Instrument

II. *Protecting Arctic People and the Environment:*
A. Survey of Arctic Indigenous Marine Use
B. Engagement with Arctic Communities
C. Areas of Heightened Ecological and Cultural Significance
D. Specially Designated Arctic Marine Areas
E. Protection from Invasive Species
F. Oil Spill Prevention
G. Addressing Impacts on Marine Mammals
H. Reducing Air Emissions

III. *Building the Arctic Marine Infrastructure:*
A. Addressing the Infrastructure Deficit
B. Arctic Marine Traffic System
C. Circumpolar Environmental Response Capacity
D. Investing in Hydrographic, Meteorological, and Oceanographic Data

nous marine use. These are very necessary if integrated, multiple use management principles, or marine spatial planning concepts, are to be applied to Arctic areas. Also, there are calls for identifying areas of heightened ecological and cultural areas, and exploring the need for specially designated Arctic marine areas (e.g., IMO Special Areas or Particularly Sensitive Sea Areas) (IMO 2012). The elements of the third recommendation theme on marine infrastructure were believed by the AMSA team to be of critical importance. Most the Arctic marine environment is poorly charted and requires increased hydrographic surveying to support safe Arctic navigation. The region is in need of many key investments for improved communications, an effective monitoring and tracking system, more observed environmental information (weather, climate, sea ice, and more), and environmental response capacity. The infrastructure initiatives are all complex projects and long-term, and each will require significant funding.

Although AMSA was focused appropriately on Arctic marine safety and environmental protection, the assessment did provide an overview of some of the issues and challenges of trans-Arctic navigation (AMSA 2009). As noted earlier, the AMSA scenarios creation effort indicated the primary driver of marine traffic would be Arctic natural resource development; regional traffic levels would be influenced by onshore and offshore development, and the shipping of resources out of the Arctic to global markets would be primarily on destinational voyages. How trans-Arctic routes might develop will depend on the continuing presence of sea ice; the seasonality and reliability of Arctic navigation routes will be key factors in trying to integrate Arctic routes into most global marine operations. Any integration effort involving Arctic ships (Polar Class vessels) will contend with many uncertainties and potentially high operating costs. Although many new icebreaking carriers (e.g., the LNG icebreaking carriers operating to the port of Sabetta on the Yamal Peninsula) are designed to operate independently in ice, in some regions along the NSR, escort by icebreaker and mandatory pilotage will pose significant economic issues relevant to the viability of commercial voyages. Potentially long voyages in ice beyond the summer season (presenting risks for ships and cargo), the lack of marine infrastructure as a safety net, and schedule disruptions will be key factors for the marine insurance industry in establishing Arctic rates. While the conduct of trans-Arctic navigation is technically possible today with advanced icebreakers and Polar Class carriers, the operational, economic, and environmental challenges for routine voyages are not yet fully understood. Implementation and strict enforcement of the IMO Polar Code will be the key policy strategy for ships on whether they are on future destinational or trans-Arctic voyages.

ARCTIC STATE AGREEMENTS AS POLICY RESPONSES

Since the release of AMSA in April 2009 two key recommendations have been acted on by the Arctic states using the Arctic Council process (with Permanent Participant and observer involvement) to negotiate binding agreements. A treaty on Arctic search and rescue (SAR), the *Agreement on Cooperation on Aeronautical and Maritime Search and Rescue in the Arctic*, was signed by the Arctic Ministers of the eight Arctic states during the Arctic Council Ministerial meeting in Nuuk, Greenland on 12 May 2011 (Arctic SAR Agreement 2011). It is a binding or mandatory agreement to strengthen SAR cooperation and coordination in the Arctic, and establishes areas of SAR responsibility for each of the Arctic states. These areas of responsibility, noted in the agreement, do not prejudice any other boundaries between the states or their sovereignty. The agreement also fosters the conduct of joint Arctic SAR exercises and training, lists information on the Arctic states' rescue coordination centers, and addresses the issue of requests to enter the territory of a Party for SAR operations. The Arctic SAR agreement entered into force on 19 January 2013 after it had been ratified by each of the eight (Arctic) signatory states. Figure 24.2 illustrates the SAR boundaries and areas of responsibility of each Arctic state under the terms of the agreement.

A second agreement negotiated under the auspices of the Arctic Council is the *Agreement on Cooperation on Marine Oil Pollution Preparedness and Response in the Arctic* signed by the Arctic Ministers in Kiruna, Sweden on 15 May 2013 and ratified by the Arctic states on 25 March 2016 (Arctic Oil Pollution Agreement 2013). This treaty focuses on Arctic oil spills and addresses a range of practical issues: requiring a national 24-hour system for response, facilitation of cross-border transfer of resources, notification of the

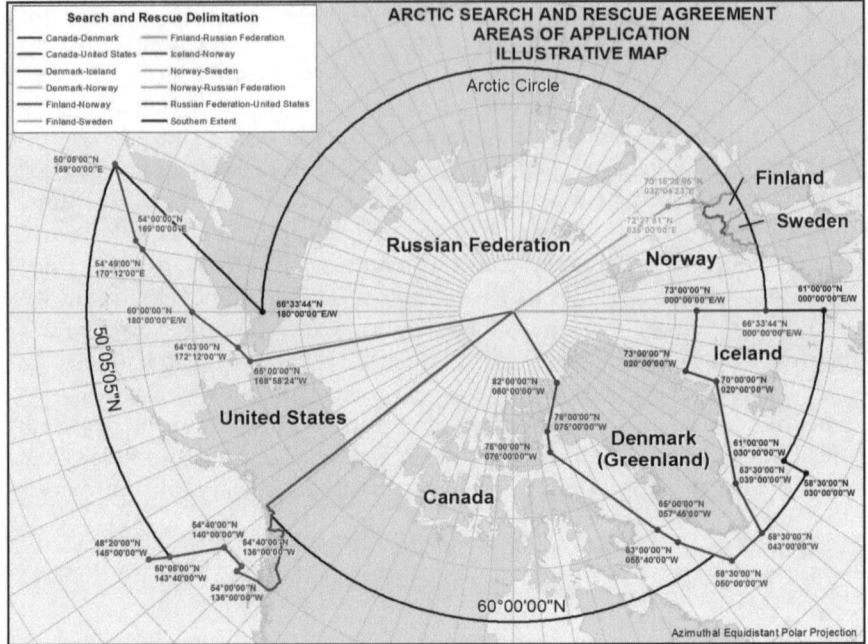

Fig. 24.2 The areas of application and responsibility of the Arctic states under the Arctic Search and Rescue Agreement of 2011 (Source: Illustrative Map provided by the U.S. Department of State)

Parties, monitoring of spills, conduct of exercises and training, joint reviews of responses to Arctic spills, and a set of operational guidelines in an appendix. Both treaties are in their implementation phases and the Arctic Council and maritime community will be able to follow the progress of the Arctic states in developing their close cooperation in the practical aspects of Arctic emergency response.

The Arctic Coast Guard Forum, established in 2015 among the Coast Guards of the Arctic states, held its first live SAR exercise near Reykjavik, Iceland in September 2017; a second exercise hosted by Finland was held in April 2019 in the northern Baltic Sea. The Russian Federation and the United States submitted a proposal to the IMO in November 2017 for a system of two-way marine routes for vessels to follow in the Bering Strait and Bering Sea. The proposal was approved by IMO in May 2018 for six two-way routes and six precautionary areas, the first IMO-sanctioned ship routing measures in polar waters. The new rules went into effect on 1 December 2018 and are voluntary for all international and domestic vessels sailing in these waters. These safety measures do not limit subsistence hunting or commercial fishing.

Non-Arctic State Observers to the Arctic Council

As of May 2017 there are 13 non-Arctic state observers in the Arctic Council: China, France, Germany, India, Italy, Japan, Republic of Korea, the Netherlands, Poland, Singapore, Spain, Switzerland, and the United Kingdom. A key challenge for the Arctic states and these national observers is how to facilitate non-Arctic state contributions into the work of the Arctic Council. How can experts from the non-Arctic states bring meaningful and useful concepts and information to the Council's working groups, especially on global maritime affairs? From symbolic and diplomatic perspectives, it is important for these observer states to be present at the Ministerial and Senior Arctic Officials meetings of the Council. While their roles are limited and constrained at these high-level meetings, it is important for the Arctic community, and for the observer's diplomats, that they witness the dialogue and broad range of Arctic issues being addressed by the Council. It is also critical that the observers see firsthand the role of the Permanent Participants in the Council's deliberations and how indigenous issues are woven into the Council's deliberations. The Senior Arctic Officials have adopted an observer manual to provide guidance to the working groups and other Council bodies on the roles to be played by the observers and meeting logistics (Kiruna Declaration 2013). Of key importance "observers may, at the discretion of the Chair, make statements, present written statements, submit relevant documents and provide views on the issues under discussion" (Arctic Council 2013). Thus, the Arctic Council is facilitating and encouraging the observers to make contributions primarily at the working group or subsidiary body level.

For Arctic marine safety and environmental protection measures and most maritime policy issues, EPPR and PAME are the most appropriate Council working groups for engagement by the non-Arctic state observers. Their maritime ministries, coast guards, and response organizations have technical and scientific expertise that can be invaluable in the deliberations and review of joint PAME-EPPR special reports, and development of guidelines and strategies. PAME is focused on continued implementation of AMSA's 17 recommendations, revising the Council's *Arctic Marine Strategic Plans*, and forming expert groups to continue work on an ecosystems approach to management and Arctic marine protected areas. The non-Arctic state observers have a broad selection of maritime themes in which to contribute and to observe Arctic marine policy developments in PAME, and participate in the formulation of response strategies in EPPR. Arctic marine policies will be much stronger and gain wider acceptance with broad input from non-Arctic states from the global maritime community.

International Maritime Organizations and IMO Polar Code

All of the Arctic states and most of the non-Arctic state observers to the Arctic Council (total of 21 states) are members of IMO, the International Hydrographic Organization (IHO), the World Meteorological Organization (WMO), and the International Association of Marine Aids to Navigation and

Lighthouse Authorities (IALA). Perhaps this is not surprising since these states all have a rich maritime heritage and an active involvement in global maritime operations and cooperation. Importantly, each of these international bodies has a key role to play in shaping marine policies for the future of the "new" maritime Arctic; each has specific polar initiatives underway where member states can contribute their expertise and voice their concerns, hopefully in more unified approaches in Arctic affairs.

The IMO is central to any discussion of Arctic maritime affairs, especially marine safety and environmental protection issues. The complex process that developed a mandatory IMO *International Code for Ships Operating in Polar Waters* provided a unified approach to Arctic marine safety and environmental protection with leadership by the Arctic states and broad support and contributions of the global maritime community. The work on the Polar Code has a lengthy history dating to the early 1990s; an IMO Outside Working Group of technical experts met from 1993 to 1997 and drafted an early version of a "polar code" for the IMO (Brigham 2000). The Polar Code was never intended to duplicate or replace existing IMO standards for safety, pollution prevention and training; the Polar Code is a set of *amendments* to three existing IMO instruments: the International Convention for the Safety of Life at Sea (SOLAS), the International Convention for the Prevention of Pollution from Ships (MARPOL), and the International Convention on Standards of Training, Certification and Watchkeeping for Mariners (STCW). Early measures focused on polar ship construction standards, polar marine safety equipment, and ice navigator standards for training and experience. These elements were included in IMO's voluntary *Guidelines for Ships Operating in Polar Waters* (IMO 2009). Continued work on the Polar Code focused on defining the risks for various class ships operating in ice-covered and ice-free polar waters, identifying hazards, and then relating how the marine hazards can be adequately mitigated to lower (and acceptable) levels. A second challenge was how to include select environmental protection measures in a Polar Code as Arctic-specific annexes to major IMO conventions such as MARPOL. The final mandatory IMO Polar Code now in force includes the following major elements for commercial carriers and passenger ships 500 tons and higher (IMO 2016):

- Polar ship structural and equipment standards (Ice classes: PC1 to PC7).
- Marine safety and lifesaving equipment.
- Training and experience of polar mariners.
- A *Polar Ship Certificate* issued by the flag state (in one of three classes: A, B, and C).
- A *Polar Water Operations Manual* that is ship specific.
- Polar environmental rules in the form of amendments to MARPOL annexes: Annex I and no discharge of oil and oily mixtures; Annex II and no discharge of noxious liquid substances; Annex IV with specific sewage regulations; and Annex V with specific food waste and garbage regulations.

Each of the above elements has ramifications for the global maritime industry operating in the Arctic. However, none of these should come as a surprise and several are focused on establishing uniform and non-discriminatory IMO rules and regulations for polar waters. The Polar Code is applicable in all Antarctic waters of 60° south. This boundary around the Antarctic continent corresponds to the northern boundary of the Antarctic Treaty. The Polar Code boundary in the Arctic includes adjustments for the warmer waters in the North Atlantic (and the higher latitude of the maximum extent of Arctic sea ice). In the Bering Sea the Polar Code boundary is set at 60° north. The boundary moves slightly south to include all of Greenland and then runs northeast along the east Greenland coast and north of Iceland until it intersects with the Russian Arctic coast in the Barents Sea. All of Iceland, Norway, and the Kola Peninsula in northwest Russia are not included within the Polar Code area since these regions are ice-free throughout the year (IMO 2016).

The International Hydrographic Organization (IHO), established in 1921, is a key maritime organization—an intergovernmental consultative body—that supports safety of navigation and the protection of the marine environment. It coordinates the activities of the national hydrographic offices, sets standards to foster worldwide uniformity in nautical charts, and supports development of new techniques for conducting and exploiting hydrographic surveys, tasks critical to Arctic maritime affairs. Since its inception IHO has established 15 regional hydrographic commissions. The 16th commission, the Arctic Regional Hydrographic Commission (ARHC), was established in October 2010 by the five Arctic Ocean coastal states, Canada, Denmark, Norway, Russia, and the United States; Finland and Iceland are now observers to ARHC. The Arctic Ocean coastal states recognized the need for such a body in an era of increases in Arctic traffic with little availability of reliable navigation and environmental data; ARHC noted that today less than 10% of Arctic waters are charted to modern international navigation standards (IHO 2010). The establishment of ARHC is an important contribution to improving Arctic marine infrastructure, and its commitment to cooperate with the marine transportation community and other intergovernmental bodies bodes well for a sharing of critical navigation information related to evolving Arctic safety and protection measures. IHO member states can contribute to the work of ARHC and foster cooperation of ARHC with their national hydrographic offices. The IHO, ARHC, and its member states could explore with the global maritime industry the potential for public-private partnerships in surveying and mapping the extensive uncharted waters of the Arctic marine environment.

As a specialized agency of the United Nations (e.g., IMO), the World Meteorological Organization (WMO) is a global body focusing on weather, climate, and hydrology. WMO has promoted the establishment of worldwide networks for a broad range of meteorological, climatological, hydrological, and geophysical observations. WMO fosters the standardization of data and facilitates the global free exchange of information and observations. Increasingly engaged in climate change issues, WMO is a leading organization for global monitoring, protecting the environment, and developing adequate monitoring/observing systems. WMO, in concert with IMO and IHO, established five new WMO METAREAs (IMO NAVAREAs) covering the Arctic. The new areas became operational in June 2011 with Canada, Norway, and Russia taking responsibility for providing services (IMO 2011). WMO has also linked with the International Ice Charting Working Group (IICWG), a forum of the national ice services, to develop and implement policies and procedures for sea ice mapping, ice forecasts, and ice edge information (IICWG 2007).

The development of future Arctic observing systems is another area where the membership of WMO, IMO, IHO, and IICWG might seek to develop public-private funding mechanisms and partnerships so that a comprehensive net of observations can support safe Arctic navigation. The involvement of Arctic marine industries in such an initiative—for example, commercial shipping, cruise ship tourism, and offshore hydrocarbon exploration and development—is essential as these are key providers of regional data as well as significant marine users.

The International Association of Marine Aids to Navigation and Lighthouse Authorities (IALA) is a non-governmental organization and international technical association that fosters the harmonization and development of marine aids to navigation (IALA 2006). Members include National Authorities responsible for marine aids to navigation, associate members (other service or scientific agencies), and importantly, industrial members (manufacturers, distributors, and technical service providers). IALA has recently developed a Northern (Arctic) Strategy in support of the design and operation of Arctic aids to navigation as well as support infrastructure such as vessel monitoring systems and remote communications. IALA is also addressing the overall information needs to enable safe Arctic navigation and the technical challenges of virtual aids to navigation. Two strengths of IALA are apparent: it continues to focus on Arctic navigation infrastructure issues and it has members with technical expertise from the marine industry. All maritime states should proactively support the work of IALA as an integral component to the establishment of safe and efficient Arctic navigation systems for individual ships and vessel traffic. IALA's promoting close, international cooperation between national agencies and the maritime industry is a key strategy to using the latest technologies and advancing best practices for newly deployed Arctic navigation networks.

CONCLUSIONS AND THE FUTURE POLICY WORK

Significant progress has been made during the past decade in developing Arctic policies related to marine transportation issues that have addressed critical challenges in marine safety, environmental protection, and maritime response. Close Arctic state cooperation in the Arctic Council and global maritime cooperation, particularly at IMO with adoption of the Polar Code, have been positive responses to increase in marine operations and shipping in the Arctic Ocean. Two binding agreements of the Arctic states on Arctic SAR and Arctic oil pollution provide new governance mechanisms on practical maritime issues. A number of key issues remain that will require innovative policy measures and continued, robust international maritime cooperation:

- Full implementation and enforcement of the Polar Code by the flag states and Arctic states; strict enforcement of the Polar Code by the Arctic states using port state control measures.
- Restrictions on heavy fuel oil in Arctic waters.
- Development of a marine emissions control zone for the Arctic Ocean.
- Strengthening of passenger ship safety in Arctic waters with enhanced national guidelines and expanded sharing of best practices.
- Addressing the uniformity of Arctic marine shipping regulatory regimes and further protection measures for the central Arctic Ocean (beyond coastal state jurisdiction).
- Continued development of an effective monitoring and surveillance system (an Arctic domain awareness system) using IMO mandatory Automatic Identification System (AIS) transponders, and sharing the data among the maritime agencies of the Arctic states for safety and enforcement.
- Development of creative public-private partnerships and strategies for investments in marine infrastructure such as charting, ports, ocean observing systems, communications, aids to navigation, and response capacity.

Two of the clear benefits of closer international cooperation in Arctic marine transportation are the creation of effective marine policies and the fostering of regional stability. Close cooperation between Arctic and non-Arctic states on the practical aspects of Arctic marine safety and environmental protection sets the stage for development of uniform rules and regulations (at IMO), and builds lasting relationships with the maritime community who will operate in a future maritime Arctic. Addressing together the many environmental security challenges of Arctic navigation can foster an era of unprecedented cooperation among the maritime states, the people who live in the Arctic, and the global maritime enterprise.

REFERENCES

Agreement on Cooperation on Aeronautical and Maritime Search and Rescue in the Arctic (Arctic SAR Agreement). 2011. Nuuk, Greenland, 12 May 2011.

Agreement on Cooperation on Marine Oil Pollution Preparedness and Response in the Arctic (Arctic Oil Pollution Agreement). 2013. Kiruna, Sweden, 15 May 2013.

AMSA. 2009. *Arctic Marine Shipping Assessment 2009 Report (AMSA)*. Arctic Council.

Arctic Council. 2013. Arctic Council Observer Manual for Subsidiary Bodies. Kiruna, Sweden, May 2013.

Brigham, L. 2000. The Emerging International Polar Code: Bi-polar Relevance? In *Protecting the Polar Marine Environment*, ed. D. Vidas, 242–262. Cambridge, UK: Cambridge University Press.

———. 2010. Think Again: The Arctic. *Foreign Policy*, September/October.

———. 2011. Marine Protection in the Arctic Cannot Wait. *Nature* 478: 157, October 13.

———. 2013. Arctic Marine Transport Driven by Natural Resource Development. *Baltic Rim Economies Quarterly Review* 2: 13–14.

Carmel, S. 2013. The Cold, Hard Realities of Arctic Shipping. *U.S. Naval Institute Proceedings* 139/7/1,325.

International Association of Marine Aids to Navigation and Lighthouse Authorities (IALA). 2006. IALA Constitution, 23 May 2006.

International Hydrographic Organization (IHO). 2010. Statement of the Arctic Regional Hydrographic Commission (ARHCC1-07D), October 2010.

International Ice Charting Working Group (IICWG). 2007. Charter and Terms of Reference, October 2007.

International Maritime Organization (IMO). 2009. *Guidelines for Ships Operating in Polar Waters.* Adopted by the IMO General Assembly on 2 December 2009, Resolution A. 1014(26).

International Maritime Organization. 2011. *Expansion of World-Wide Navigational Warning System into Arctic Waters by IMO, WMO and IHO Chiefs.* IMO Briefing Paper, 8 March 2011.

———. 2012. List of Special Ares under MARPOL and Particularly Sensitive Sea Areas, London, MEPC.1/Circ. 778/Rev.1, 16 November 2012.

———. 2016. International Code for Ships Operating in Polar Waters. Consolidated Text of the Polar Code.

Kiruna Declaration. 2013. Arctic Council Secretariat, Kiruna, Sweden, 15 May 2013.

Mikkola, H., and J. Kapyla. 2013. *Arctic Economic Potential.* The Finnish Institute of international Affairs Briefing Paper 127, April 2013.

Ottawa Declaration. 1996. Declaration on the Establishment of the Arctic Council, 19 September 1996.

Smith, L., and S. Stephenson. 2013. New Trans-Arctic Shipping Routes Navigable by Midcentury. *Proceedings of the National Academy of Sciences Plus.* https://doi.org/10.1073/pnas.1214212110.

Staalesen, A. 2017. New Era Starts on the Northern Sea Route. *The Independent Barents Observer*, December 8.

Stephenson, S., L. Smith, L. Brigham, and J. Agnew. 2013. Projected 21st-Century Changes to Arctic Marine Access. *Climatic Change.* https://doi.org/10.1007/s10584-012-0685-0.

The Moscow Times. 2013. Northern Sea Route Slated for Massive Growth, June 4.

CHAPTER 25

Emergence of a New Ocean: How to React to the Massive Change?

Timo Koivurova, Pirjo Kleemola-Juntunen,
and Stefan Kirchner

Introduction

Before the states commenced negotiating what became the 1982 UN Convention on the Law of the Sea (UNCLOS), there was not much discussion of how laws of the sea applied in the Arctic Ocean, given that the normal uses of the sea could not be practiced there (but cf. Kikkert and Lackenbauer 2014). From the beginning of the twentieth century, there had been academic discussion as to whether the Arctic Ocean sea ice cover could be claimed as land (ice-is-land theory) or via sectoral claims (similar to those that took place in the other Pole, Antarctica). Yet, as *Erik Franckx* has shown, most active in this respect were scholars, not the states and their civil servants (this discussion took place in particular in Canada and the Soviet Union, given that it is these two states that have the longest coastlines to the Arctic Ocean, see Franckx 1993).

Even if UNCLOS did address the Arctic Ocean with one article (Art. 234), it is fair to say that there was limited discussion over what rules govern the very limited uses of the Arctic Ocean and adjacent seas (UNCLOS III 1982, 41–43). This was due to the fact that it was largely an ice-barren ocean, where only limited amount of human activities took place, which were dependent on a delicate ecological balance.

All this is now changing, and at an accelerating pace. Climate change has progressed more intensely in the Arctic, which has caused and is causing

T. Koivurova (✉) • P. Kleemola-Juntunen • S. Kirchner
University of Lapland, Rovaniemi, Finland
e-mail: timo.koivurova@ulapland.fi; pirjo.kleemola-juntunen@ulapland.fi; stefan.kirchner@ulapland.fi

© The Author(s) 2020
K. S. Coates, C. Holroyd (eds.), *The Palgrave Handbook of Arctic Policy and Politics*, https://doi.org/10.1007/978-3-030-20557-7_25

concrete and dramatic impacts on the extent and breadth of the sea ice. The Arctic Ocean has already lost approximately 40–50% of its sea ice volume in about 40 years. The first ice-free summer season is projected to occur sometime between 2030 and 2040 (Onarheim et al. 2018). Today, we are literally witnessing a full-scale transformation of the Arctic marine areas, their ecosystems and the amount of human activity taking place in these marine areas. It is hence no wonder that there is ever more discussion on how UNCLOS and various maritime treaties apply in the Arctic waters and how they should be adjusted to accommodate the unique Arctic maritime conditions.

In this article, we will refer to the Arctic Ocean and the associated seas as the Arctic marine areas. We will adopt the widely used definition that is used by the Arctic Monitoring and Assessment Programme (AMAP)—one of the working groups of the Arctic Council. It uses the working definition of marine areas north of the Arctic Circle (66°32′N), and north of 62°N in Asia and 60°N in North America (as modified to include the marine areas north of the Aleutian chain, Hudson Bay and parts of the North Atlantic Ocean, including the Labrador Sea). It also seems widely accepted that there are only five coastal states to the Arctic Ocean, namely, Canada, Denmark (through Greenland), Norway, the Russian Federation and the United States.[1]

Before examining how the Arctic marine areas are currently regulated, it is important to examine who is legally competent to regulate the uses of the Oceans and where such competences apply. With this background, it is possible to study how the Arctic states have countered the vast challenges posed by climate change and other drivers behind the emergence of what is, for many purposes, a new Ocean. Finally, before drawing conclusions, we will also examine how to address some of the more difficult issues that are emerging in this melting Ocean and adjacent seas.

MARITIME ZONES IN THE ARCTIC: WHICH POLICY ENTITY IS COMPETENT AND WHERE?

From World War II onward, coastal states have gradually asserted more powers over their adjacent sea areas. This phenomenon, referred to as "creeping jurisdiction" (Ball 1996), has taken place all over the Planet, including the Arctic marine areas. Because of this growth in claims to jurisdiction, coastal states gained an increasing amount of rights over the coastal areas near their land. Most notably, the territorial sea has expanded from a long-established customary law maximum of three nautical miles to the current maximum (under both customary international law and Article 3 UNCLOS) of 12 nautical miles, and new maritime zones have gradually become accepted, including exclusive economic zones (EEZ) and the continental shelf. In addition, many coastal states have taken liberal use of the straight baselines method to separate their internal waters and territorial sea (Haacke 2016). This means that larger marine areas

[1] Even though Iceland has occasionally tried to contest this.

are within the full sovereignty of coastal states as internal waters, where other state's vessels do not enjoy innocent passage rights as they do in the territorial sea. Since maritime zones are measured from these straight baselines,[2] the outer limits of the territorial sea, EEZ and in some cases the continental shelf are pushed further out to the sea. Since the sovereignty and sovereign rights of coastal states have grown larger, these zones overlap more and more between neighboring states and hence there has been a need to resolve maritime boundary disputes. It has also meant that there is less space for the high seas and the deep seabed, both of which are areas beyond national jurisdiction. This development highlights the gains of the coastal states in the second half of the twentieth century, in particular through UNCLOS.

All this has taken place also in the Arctic marine areas. The Arctic coastal states have enlarged their maritime zones, claiming larger territorial seas and exclusive economic zones, as well as continental shelves, in line with the aforementioned development of customary and codified law of the sea, with the result that neighboring state's now claim overlapping maritime zones (Economist 2014). Gradually, one by one, Arctic marine states resolved these boundaries via treaties, sometimes with the help of arbitration, conciliation or even the International Court of Justice. Most boundaries have, however, simply been negotiated amicably between the Arctic states, such as the long maritime boundary between Norway and Russia in the Barents Sea. Canada and the United States, however, have had difficulties in negotiating their maritime boundary in the Beaufort Sea (Byers 2014, 56–92). Coastal state proposals to extend continental shelf limits have exposed new areas of overlapping territory that will require negotiation or litigation to delimit boundaries. This will likely be a long process, as the Commission on the Limits of Continental Shelf (CLCS) still needs to make recommendations to many Arctic Ocean littoral states and thereafter the states themselves are the only competent ones to negotiate the boundary between them. Since, for instance, Canada is yet to make its submission to the CLCS as regards the Central Arctic Ocean, we can expect that the process will take a long time, as the CLCS has a long queue of submissions to process. Already in the 2008 Ilulissat Declaration, the Arctic Ocean coastal states held that

> By virtue of their sovereignty, sovereign rights and jurisdiction in large areas of the Arctic Ocean the five coastal states are in a unique position to address these possibilities and challenges. In this regard, we recall that an extensive international legal framework applies to the Arctic Ocean as discussed between our representatives at the meeting in Oslo on 15 and 16 October 2007 at the level of senior officials. **Notably, the law of the sea provides for important rights and obligations concerning the delineation of the outer limits of the continental**

---

[2] UNCLOS recognizes four types of baselines for drawing maritime zones: straight, normal, archipelagic, and closing lines across river mouths and bays, see Articles 3, 33, 47, 57, and 76.

shelf, the protection of the marine environment, including ice-covered areas, freedom of navigation, marine scientific research, and other uses of the sea. We remain committed to this legal framework **and to the orderly settlement of any possible overlapping claims**. (Ilulissat Declaration)[3]

So far, they have proceeded very cooperatively in addressing these possible overlapping continental shelf entitlements, though there remain differences regarding the legal status of the Lomonosov Ridge currently before the CLCS (Russian Federation 2015).

As the Arctic marine states have advanced their maritime sovereign rights further, the areas for high seas and deep seabed have diminished. There is a 2.8 million km² high seas area in the central Arctic Ocean, and smaller pockets of high seas in the Barents Sea, Bering Sea, and Northeast Atlantic. In these areas, all the traditional high seas freedoms are, in principle, available for all vessels of the world. Presently, about 40% of these high seas areas are open waters during the summertime.[4] It is likely that there will remain only two pockets of deep seabed, part of the common heritage of humankind, in the Arctic Ocean after the coastal state's submissions to the CLCS are processed and the coastal states have negotiated the boundaries between each other. In addition, those remaining areas are likely located in places where it is technically difficult (and therefore most likely not yet profitable) to even explore minerals under the Part XI and the regulations of the International Seabed Authority (ISBA).

REGULATORY ENVIRONMENT

As is clear from the above, UNCLOS is a type of "constitution for the oceans" (Koh 1982) and it is the cornerstone legal framework for regulating uses of Arctic marine areas. The United States, although not a party to UNCLOS, accepts most of the rules that have been codified within UNCLOS as amounting to norms of customary international law; it is therefore useful to refer to UNCLOS as the overarching framework.[5] There are also many other treaties that govern the uses of also the Arctic marine areas, including international environmental treaties and legally binding norms which have been created under the auspices of the International Maritime Organization (IMO).

United Nations Convention on the Law of the Sea and the Arctic

As mentioned above, there is only one article in the whole convention specifically tailored to take Arctic conditions into account, Article 234. Article 234, on ice-covered areas, was negotiated mainly between the two Cold War

---

[3] Emphasis added.

[4] Data from National Snow and Ice Data Center, available at: https://nsidc.org/cryosphere/sotc/sea_ice.html.

[5] Yet, it is also correct to point out that the states used the term "law of the sea" in Ilulissat Declaration, simply because the United States is not a party to this Convention.

superpowers: the United States and Soviet Union, but also importantly with Canada. It was, in effect, Canada which catalyzed the development for negotiations on this article by enacting domestic legislation which contravened of the law of the sea since it enabled marine environmental protection measures to be adopted and enforced outside of the territorial sea against all vessels on a non-discriminatory basis (Arctic Waters Pollution Prevention Act 1970). Yet, Canada was able to convince the two superpowers to endorse Article 234 of UNCLOS, and since it became an article in UNCLOS, it also binds all contracting states to the UNCLOS, and its contents have, arguably, become a norm of customary international law. Article 234 reads as follows:

Ice-covered areas
Coastal States have the right to adopt and enforce non-discriminatory laws and regulations for the prevention, reduction and control of marine pollution from vessels in ice-covered areas within the limits of the exclusive economic zone, where particularly severe climatic conditions and the presence of ice covering such areas for most of the year create obstructions or exceptional hazards to navigation, and pollution of the marine environment could cause major harm to or irreversible disturbance of the ecological balance. Such laws and regulations shall have due regard to navigation and the protection and preservation of the marine environment based on the best available scientific evidence. (Art. 234 UNCLOS)

This norm thus provides coastal states with legal powers to enact and enforce domestic legislation up until the limit of the exclusive economic zone to prevent, reduce, and control marine pollution from vessels in ice-covered areas—competencies that coastal states would otherwise not have in their EEZs. Such laws need to be non-discriminatory and have due regard to navigational interests. While it is unclear if these powers will continue to apply if warming trends accelerate and the EEZ parts of the Arctic marine area are no longer covered by ice "most of the year," it appears that states may continue applying Article 234 to their Arctic marine area regardless of the ice cover (Dremliuga 2017).

Some scholars have also suggested that Arctic Ocean coastal states could make use of Articles 122 and 123 of UNCLOS, as for them the Arctic Ocean could qualify as a semi-enclosed sea and hence would entail, arguably, legal obligations for the coastal states (Scovazzi 2009). According to Article 122:

For the purposes of this Convention, "enclosed or semi-enclosed sea" means a gulf, basin or sea surrounded by two or more States and connected to another sea or the ocean by a narrow outlet or consisting entirely or primarily of the territorial seas and exclusive economic zones of two or more coastal States.

This provision identifies two types of sea areas to be within its scope: either those that are covered primarily by territorial seas and EEZs of coastal States or those that are connected to other sea areas only by a narrow strait. Since the terms used in Article 122 are vague, it is difficult to provide a clear-cut answer as to whether the Arctic Ocean is an enclosed or semi-enclosed sea in the meaning of Article 122. As regards the first type of sea area, it is important to note that a large part of the Arctic Ocean consists of high seas and thereby would not convincingly satisfy the requirement of "primarily." As regards the second type of sea area, in comparison to the seas that are clearly enclosed or semi-enclosed—such as the Baltic or Mediterranean Seas—the Arctic Ocean opens relatively broadly to the North-East Atlantic.If the Arctic Ocean could be considered a semi-enclosed sea based on Article 122, the coastal states would be under the obligations laid down in Article 123:

> States bordering an enclosed or semi-enclosed sea should cooperate with each other in the exercise of their rights and in the performance of their duties under this Convention. To this end they shall endeavour, directly or through an appropriate regional organization: (a) to coordinate the management, conservation, exploration and exploitation of the living resources of the sea; (b) to coordinate the implementation of their rights and duties with respect to the protection and preservation of the marine environment; (c) to coordinate their scientific research policies and undertake where appropriate joint programmes of scientific research in the area; (d) to invite, as appropriate, other interested States or international organizations to cooperate with them in furtherance of the provisions of this article. (Art. 123 UNCLOS)

According to the phrasing of this provision, it seems more adequate to interpret Article 123 as encouraging regional sea cooperation over marine environmental protection, management of living resources, and marine scientific research rather than imposing on coastal States a legally binding obligation to do so. In international treaty practice, "should" is normally used to denote non-legally binding guidance rather than a legal obligation (for which "shall" or "must" are used). Moreover, the use of "shall" in the second sentence is significantly qualified by the term "endeavour." It seems, hence, a better argument that Article 123 merely contains a weak obligation to cooperate, but it does urge the coastal States—perhaps together with other States and international organizations—to engage in regional cooperation over the policy areas enumerated in the provision.

If the coastal States were to regard the Arctic Ocean as an enclosed or semi-enclosed sea in the meaning of Article 122, and if they were to be prepared to commence negotiations over how to implement cooperation in the fields mentioned in Article 123, they would also need to define the relationship between this initiative and the Arctic Council, given that the Council's work so far also extends to marine environmental protection and scientific research in the Arctic Ocean. Yet so far they have not invoked Articles 122 and 123 as the basis of their marine cooperation. Not even the Ilulissat Declaration, which identified possible areas of cooperation between the five Arctic Ocean coastal states, referred to these provisions.

New Regulatory Measures to Regulate the Arctic Marine Environment

Arctic states have responded to the dramatically changing marine area in a variety of ways. First, they have taken regulatory action under the Arctic Council, the predominant intergovernmental forum dedicated to environmental protection and sustainability in the Arctic marine area. The Council has used both soft- and hard-law measures to advance marine governance of Arctic waters. Second, Arctic states have acted on a sectoral basis outside the Council to introduce stricter shipping and fisheries regulations.

Ocean-Related Efforts of the Arctic Council

The Arctic Council has had the marine agenda from the very beginning, first as the Arctic Environmental Protection Strategy (AEPS) of 1991, and then integrated with several other working groups into its current structure in 1996. The Protection of the Arctic Marine Environment (PAME) was a working group already during the AEPS and it continued its functioning under the umbrella of the Arctic Council as the main working group dedicated to the Arctic marine affairs. The Conservation of Arctic Flora and Fauna (CAFF), Emergency Prevention, Preparedness and Response (EPPR), and Arctic Monitoring and Assessment Program (AMAP) working groups have also conducted important maritime activities under the Arctic Council.

There is a vast amount of marine relevant policy activities undertaken in PAME—and in other working groups—over the years (Koivurova and VanderZwaag 2007). Examples of recent soft-law activities include the 2009 Arctic Marine Shipping Assessment (AMSA) Recommendations (PAME 2009). The Arctic Offshore Oil and Gas Guidelines, were first adopted in 1997 and have already been revised twice (most recently as the AOOGG 2009), contain a non-binding set of suggested best practices for oil and gas extraction designed to advise industry officials and government regulators. PAME also has a long-standing strategy to work for ecosystem-based management and marine protected areas for the Arctic marine area, the most recent one adopted up until 2025 (AMSP 2015). Some additional activities will be studied below.

Treaties Negotiated Through Arctic Council Task Forces

The Arctic Council has also recently recognized three legally binding agreements that are independent of the Arctic Council but negotiated through Arctic Council task forces: the Arctic Search and Rescue Agreement, the Agreement on Cooperation on Marine Oil Pollution Preparedness and Response in the Arctic, and the Agreement on Enhancing International Arctic Scientific Cooperation. Each agreement will be discussed below.

At the Arctic Council Ministerial Meeting in Nuuk on 12 May 2011, the eight Arctic countries concluded the Agreement on Cooperation on Aeronautical and Maritime Search and Rescue in the Arctic (Arctic SAR Agreement/Arctic SAR 2011). The Agreement is the first legally binding treaty relating particularly to the Arctic negotiated under the auspices the

Arctic Council. The objective of the Agreement is to strengthen aeronautical and maritime search and rescue cooperation and coordination in the Arctic (Art. 2 Arctic SAR 2011). The Arctic SAR Agreement contains 20 Articles, an Annex delimiting the area of each State's search and rescue jurisdiction and three Appendices, which define competent authorities, search and rescue agencies, and rescue coordination centers of each Party. The agreement provides delimitation of the air and, in particular, sea rescue regions between the parties up to the North Pole. Thereby, the Arctic SAR Agreement covers the whole Arctic Ocean and many other sub-Arctic marine areas including the Bering Sea, Irminger Sea and Labrador Sea. Finland, Norway, Russia and Sweden apply the agreement in the regions north from the Arctic Circle. The agreement contains provisions on the competent authorities, as well as arrangements for cooperation regarding alerting, the conduct of operations and the exchange of information. The authorities responsible for air and sea rescue operations with their powers and existing resources ensure the fulfillment of the obligations of the parties to the agreement.

The Arctic SAR Agreement is mainly based on previous international agreements, takes into account established practices and is applied in compliance with the international aeronautical and maritime search and rescue manual. The existing international conventions to which the Agreement refers to are the 1979 International Convention on Maritime Search and Rescue (SAR Convention 1979) and the 1944 Convention on International Civil Aviation (Chicago Convention 1944). The Agreement relies on these Conventions for terms and definitions as well as the scope and the enunciated measures. The provisions of the Agreement are in line with the provisions and obligations of these two broader universal Conventions, and exceed them by detailing how the parties carry out their SAR Convention cooperation obligations regarding sharing information and experience and the carrying out of joint research and training activities (Arts. 9 and 10 Arctic SAR 2011).

The Agreement also implements the obligations set out in Art. 98 (2) UNCLOS, which provides that, where needed, neighboring states shall cooperate through regional agreements to promote and maintain adequate and effective search and rescue services. While the Agreement does not establish its own institutional arrangements like a Secretariat, Committees, or Working Groups, the parties will meet regularly "in order to consider and resolve issues concerning practical cooperation" (Art. 10 Arctic SAR 2011). To accomplish this, the EPPR established a SAR Expert Group to facilitate the exchange of best practices.

The second treaty negotiated under the auspices the Arctic Council is the Agreement on Cooperation on Marine Oil Pollution Preparedness and Response in the Arctic (MOPPRA 2013) signed at the Kiruna Ministerial Meeting in May 2013. The objective of the Agreement is "to strengthen cooperation, coordination, and mutual assistance among the Parties on oil pollution preparedness and response in the Arctic in order to protect the marine environment from pollution by oil" and in doing so to increase collective capacity in spill response operations (Tanaka 2015, 322).

The Agreement builds on the 1990 International Convention on Oil Pollution Preparedness, Response and Co-operation (OPRC 1990) to which the Eight Arctic States are all parties, and applies the general principle of "polluter pays" (Sands et al. 2018, 642). MOPPRA treaty provisions are in line with the content and wording of the OPRC. The added value of MOPPRA is that within its framework, the parties will create a narrower network of Arctic operators for OPRC cooperation. There was previously no legally binding, specific multilateral marine oil pollution response instrument for the Arctic, where spills of any significant magnitude may exceed any one Arctic State's ability to address it alone (Byers 2014, 212–213). The Agreement includes demarcation lines, requirements for monitoring, cooperation and exchange of information, joint exercises and training, joint reviews of any oil pollution incident response and for reimbursement for the costs of providing assistance in certain circumstances (Arts. 7–13 MOPPRA 2013). The parties also agreed to meet on a regular basis to review MOPPRA's practical implementation (Art. 14 MOPPRA 2013).

In addition, some states in the region are parties to the Agreement concerning Cooperation in Taking Measures against Pollution of the Sea by Oil or other Harmful Substances (Copenhagen Agreement) which went into effect in 1971 and were updated by a 1993 superseding agreement that came into effect in 1998 (Copenhagen Agreement 1993). This agreement imposed monitoring, investigation, reporting, preparedness, assistance, information exchange and reimbursement obligations on the parties similar to what was later included in MOPPRA.

The third legally binding agreement negotiated through the framework of the Arctic Council is the Agreement on Enhancing International Arctic Scientific Cooperation signed on 11 May 2017 at the 10th Ministerial Meeting in Fairbanks, Alaska, and entered into force on 23 May 2018 (ASC 2017). The aim and purpose of the Agreement is to improve practical research collaboration, facilitating permitting procedures for the mobility of researchers, samples, and research equipment across borders, as many areas of research require large infrastructures including extensive datasets and exploration vessels for which individual research institutions do not have the resources or capacity. The Agreement is of a general nature and does not prejudice the sovereignty and sovereign rights of the parties in their maritime zones granted by the LOSC relating to the access to research areas or alter the rights and obligations of any party under the Part XIII of the LOSC. In addition to the framework provided by international law in general, the Agreement is to be implemented in accordance with applicable national laws, regulations, procedures and policies of the parties concerned (Art. 10 ASC 2017).

Scientific Research Catalyzed by Arctic Council Task Forces

In addition to the normative legal activities catalyzed by the Arctic Council's working groups, scientific reporting about environmental problems in the Arctic marine areas has long been a core aspect of the work of the Arctic Council. For instance, the 2013 Arctic Biodiversity Assessment (ABA) gave an alarm on the threats to the Arctic marine ecosystems and biodiversity: "There is increasing concern that the global demand for seafood outside the Arctic combined with increasing accessibility of Arctic seas as a result of sea ice loss creates the potential for increased risks to poorly known fish and crustacean stocks" (CAFF 2013, 14). In general, "habitat loss and degradation pose the main threats to biodiversity. The relative well-being of many Arctic ecosystems today is largely the fortuitous result of a lack of intensive human encroachment," which are now being affected by increasing human activities (CAFF 2013, 8). ABA also states that climate change is "by far the most serious threat to biodiversity and exacerbates all other threats" in the Arctic, and in particular the marine Arctic (CAFF 2013, 9). Of key concern is the rapid loss of multi-year ice in the central Arctic basins and changes in sea ice dynamics on the extensive Arctic shelves, which affect the biodiversity and productivity of marine ecosystems. Additionally, The AMAP Working Group adopted the Arctic Ocean Acidification Overview Report in 2013. It found that the Arctic Ocean is rapidly accumulating carbon dioxide (CO_2) leading to increased ocean acidification—a long-term decline in seawater pH (AMAP 2014, xi and 27). This ongoing change impacts Arctic marine ecosystems which are already affected by rising temperatures and melting sea ice. Warmer temperatures also increase the threat of midlatitude invasive species and pollutants arriving in Arctic marine ecosystems.

Sectoral Regulations for the Marine Arctic

While a specific international treaty for the Arctic has long been debated, the sovereignty of Arctic nations and the very different situation in the Arctic prevents an overarching Arctic-specific treaty that would parallel the Antarctic Treaty. The Arctic Ocean is already regulated under the international law of the sea and almost all Arctic marine coastal states are parties to UNCLOS. Like other regional seas, however, the Arctic Ocean has become the object of more detailed geographically specific norm-making in the context of the international law of the sea—but outside the Arctic Council. The two most important developments in this regard are the entry into force of the Polar Code on 1 January 2017 and the adoption of a fisheries agreement in December 2017.

The Polar Code

After substantial discussions at the International Maritime Organization (IMO), in the frameworks of the International Convention for the Safety of Life at Sea (SOLAS) and the International Convention for the Prevention of Pollution from Ships (MARPOL), the Polar Code provides legally binding standards concerning ships that operate in polar waters, that is, both in the

Arctic and Southern Oceans (Polar Code 2017). The aim of the Polar Code is to protect human life and the natural environment of polar waters. This goal is pursued through the establishment not only of binding technical standards but also of requirements regarding vessel manning, seafarer training and voyage planning.

The entry into force of the Polar Code, which followed years of debate, was timely as there is currently a boom in Arctic cruise shipping. This boom, which gained widespread public attention with the journey of the *Crystal Serenity* through the Northwest Passage, appears to be continuing unabated.[6] In light of the very limited search and rescue infrastructure available in the Arctic, a focus on disaster prevention and human safety remains essential for the foreseeable future. Since there is also minimal infrastructure for waste reception in polar regions, the Polar Code also imposes tighter restrictions for discharging food and gray water (Polar Code 2017, chap. 5). Protecting the Arctic marine areas from accidental environmental damage is also matter of increasing concern (see Overby 2014, 358). Earlier disasters, such as the oil spill caused by the *Exxon Valdez* or the loss of the *Selendang Ayu*, are reminders that the Arctic, even outside the Central Arctic Ocean, continues to provide a challenging work environment for the oil and shipping industries. Melting sea ice does not mean a complete absence of sea ice in Arctic shipping lanes. Bergy water constitutes a serious risk for vessels, for example, through damage sustained by screws or rudders. Indeed, in some areas, climate change already increases the risk posed by icebergs: an increase in Arctic temperatures leads to larger icebergs calving off glaciers in the high north. These larger icebergs take longer to melt, making it more likely that they will float further south and pose a threat to vessels in shipping lanes in the North Atlantic (a geographic area not covered by the Polar Code). Likewise, the Polar Code does not apply to fishing vessels, although incidents like that of the *Antarctic Chieftain* in 2015 are a reminder that emergencies suffered by fishing vessels in polar waters also have the potential to cause harm to human safety as well as the environment. Efforts are also underway by the International Maritime Organization (IMO) to limit the use of heavy ship fuels in Arctic waters, in addition to the sulfur content limits applicable to ship fuels globally starting 1 January 2020.

The Central Arctic Ocean Fisheries Agreement
Large parts of the Arctic Ocean fall within either the sovereign territorial seas of the coastal states Russia, Norway, Denmark (with regard to Greenland), Canada and the United States or are subject to sovereign rights, for example, in EEZs. The central part of the Arctic Ocean, which is bordered by the waters of these states, is high seas. Therefore, vessels from all flag states are permitted to engage in the classical freedoms of the High Seas there, including fishing. At

---

[6] All remaining 2018 journeys through the Northwest Passage with Polar Cruises are either full or with limited availability, see https://www.polarcruises.com/arctic/destinations/northwest-passage.

this time, very little is known about the abundance (or lack thereof) of living resources in the central Arctic marine area. In 2004, the Arctic Council's Climate Impact Assessment predicted that fish stocks would move poleward due to rising ocean temperatures. This has been the case in recent years with the northward movement of mackerel into Iceland's EEZ (see Seafish 2013). As multi-year sea ice melts also in the central parts of the Arctic Ocean, this part of the Arctic is quickly becoming more accessible, also for fishing. Currently, about 40% of the Arctic Ocean is already ice free during the summer months and it is expected that the entire ocean will be practically ice free in the summer months at some time between 2030 and 2040. Elsewhere, a lack of information about fish stocks led to delays in the adoption of measures which might have prevented overfishing (see Balton 2001). In the High Seas, the responsibility for regulating vessel behavior rests with the flag states. The aforementioned coastal states, also known as the Arctic Five (A5), joined by other actors with interests in the region (Iceland, Japan, South Korea, China and the European Union, together referred to as the A5+5), came together to establish an international agreement which prevents ships flying their flags from commercial fishing in the Central Arctic Ocean. The agreement, which follows the 2015 Declaration Concerning the Prevention of Unregulated High Seas Fishing in the Central Arctic Ocean which had been adopted by the A5, allows flag states to permit ships flying their flags to engage in fishing for exploratory purposes (Art. 3 (3) CAOFA 2017) or if there is a regional fisheries management organization which has adopted rules for the High Seas part of the Central Arctic Ocean (Art. 3 (1) (a) CAOFA 2017). This makes it clear that the agreement is not aimed at preventing fishing per se but is a temporary measure designed to prevent harm to the marine environment at a time when the region becomes accessible but vital information on fish stocks is still missing. The agreement will make fisheries management possible in a part of the seas that has never before been accessible for fishing by establishing a Joint Program of Scientific Monitoring to study the possibility of sustainable harvesting. The Central Arctic Ocean Fisheries Agreement is a rare example of international law-making for the maritime sector at a time when a problem is foreseeable but before anyone has conducted fishing in those waters (by contrast, many maritime safety rules have been established in reaction to major disasters).

How to Assess the Current Regulation for the Marine Areas and Improve it for the Future?

It seems obvious that the Arctic states and other stakeholders are reacting quickly to the vast changes that are taking place in the Arctic marine areas—the transformation that the ecosystems are undergoing and the increasing human uses of these waters. The Arctic Council, a soft-law forum, catalyzed scientific assessments, and legally binding agreements between the eight Arctic states on issues that are of crucial importance for the safety and security of seafaring and the marine environment. Search and rescue and oil spill agreements apply mostly to the marine areas of the Arctic and address issues of utmost relevance in remote maritime areas which do not have the personnel or equipment for large-scale marine emergencies.

In addition, even with no commercial fishing in the central Arctic Ocean and fairly limited vessel traffic, it was possible to push two important legally binding agreements, one through the IMO and one endorsed by the Arctic Ocean coastal states together with other invited states and the European Union. This shows that Arctic and other interested states and stakeholders are taking a proactive and precautionary approach toward regulating the Arctic marine areas.

The current regulatory measures, even if tailored to the Arctic (and Antarctic), have been based on existing global treaties. The Arctic Council-catalyzed legally binding agreements on search and rescue and oil spills draw on a range of international treaties, such as the International Convention on Maritime Search and Rescue (SAR Convention 1979), the Convention on International Civil Aviation (Chicago Convention 1944), the International Convention on Oil Pollution Preparedness, Response and Co-operation (OPRC 1990), the Copenhagen Agreement concerning Cooperation in taking Measures against Pollution of the Sea by Oil or Other Harmful Substances (Copenhagen Agreement 1993) and the International Convention Relating to Intervention on the High Seas in Cases of Oil Pollution Casualties (Intervention Convention 1969), but also on non-binding texts such as the International Aeronautical and Maritime Search and Rescue Manual (IAMSAR Manual 2007) and on recognized concepts like polluter responsibility.[7]

Even if the Arctic fisheries agreement was initiated by the Arctic Ocean coastal states, it is also correct to observe that it relies heavily on the straddling stocks convention (FSA 1995), one of the global implementing treaties of UNCLOS. The Polar Code was made mandatory by amending the existing global IMO treaties, simply because shipping is a global activity and needs to be regulated primarily via global rules. When the negotiations start between states on how to manage biodiversity beyond areas of national jurisdiction, these will be conducted under the UN auspices, with the goal as a global implementing agreement to the UNCLOS. This global treaty would also apply in the 2.8 million km^2 (Koivurova and Caddell 2018, 134) high seas area of the central Arctic Ocean (Koivurova and Caddell 2018, 137).

---

[7] This follows from the preambles of the Agreement on Cooperation on Aeronautical and Maritime Search and Rescue in the Arctic, https://oaarchive.arctic-council.org/bitstream/handle/11374/531/EDOCS-1910-v1-ACMMDK07_Nuuk_2011_Arctic_SAR_Agreement_unsigned_EN.PDF?sequence=8&isAllowed=y, and of the Agreement on Cooperation on Marine Oil Pollution Preparedness and Response in the Arctic, https://oaarchive.arctic-council.org/bitstream/handle/11374/529/EDOCS-2067-v1-ACMMSE08_KIRUNA_2013_agreement_on_oil_pollution_preparedness_and_response__in_the_arctic_formatted.PDF?sequence=5&isAllowed=y.

There are pros and cons to the current regulatory framework. It is significant that nation-states, the EU and other key stakeholders have been able to regulate activities before there are vested economic interests or major disasters in most of the Arctic marine areas. On the other hand, the downside of the current regulatory approach is that it has been advanced by states and other stakeholders via various routes (Arctic Council, Arctic 5 plus 5 and through IMO), soft- and hard-law measures, and in particular on the basis of sectoral approach to regulation rather than a holistic ecosystem approach to marine management.

How to Improve the Fragmented Landscape in Arctic Ocean Governance

What is it possible to do to improve this fragmented landscape of governance of Arctic marine areas? The way forward has already started with the Arctic Council, in particular, the PAME working group but also activities in the other working groups and task forces. PAME has identified 18 large marine ecosystems of the Arctic marine areas that serve as basis for pushing forward marine ecosystem-based governance via soft-law measures.

The current measures include PAME's adoption of an Arctic Marine Strategic Plan that guides their efforts until 2025 and encourages Arctic states and other Arctic Council actors to take concrete measures toward ecosystem-based management. Until 2013, there was a separate expert group of the Arctic Council focusing on ecosystem-based management.[8] During the United States chairmanship, one of the main themes was to draw inspiration from the regional seas agreements and other arrangements for the work in the Arctic Ocean. Currently, Finland is leading this work with the task of examining whether more integrated ocean management will be possible.

CONCLUDING THOUGHTS

In the bigger picture, it is surprising that the Arctic states and other stakeholders have been able to react to the vast transformation of the marine environment so quickly. Many of the current regulatory measures have progressed during a time when relations between Arctic states are not at their best. In addition, leaders of two of these superpowers are openly questioning the value of measures to combat climate change, which are at the core of the regulatory work in the Arctic. Despite this, various soft- and hard-regulatory measures have already occurred before extensive human economic activity has entered many of these marine areas. The Arctic Council has been able to catalyze legally

---

[8] Before this, there was a joint project between the Council working groups Sustainable Development Working Group (SDWG) and Protection of the Arctic Marine Environment (PAME) on the project Best Practices in Ecosystem-Based Ocean Management in the Arctic (BePOMAr), which was completed by 2009.

binding agreements tackling marine emergencies and has advanced via soft-law measures some marine ecosystem-based management and scientific research in the Arctic Ocean and adjacent seas.

From this perspective, quite a lot has occurred considering the current geopolitical dynamics in the Arctic and elsewhere. Yet, climate change is unfortunately moving forward and transforming the Arctic marine area at an accelerating pace. The soft-law measures toward Arctic marine ecosystem-based management introduced by the Arctic Council are a good start and will hopefully lead the international community and Arctic marine states to take more concrete steps in this direction.

References

AMAP. 2014. Arctic Ocean Acidification 2013: An Overview. Arctic Monitoring and Assessment Programme (AMAP), Oslo, Norway.

AMSP. 2015. Arctic Marine Strategic Plan 2015–2025. April 2015. https://pame.is/images/03_Projects/AMSP/AMSP_2015-2025.pdf.

AOOGG. 2009. Arctic Offshore Oil and Gas Guidelines, 29 April 2009. https://oaarchive.arctic-council.org/bitstream/handle/11374/63/Arctic-Guidelines-2009-13th-Mar2009.pdf?sequence=1&isAllowed=y.

Arctic SAR. 2011. Agreement on Cooperation on Aeronautical and Maritime Search and Rescue in the Arctic, 12 May 2011, entered into force 19 January 2013. http://www.arctic-council.org/index.php/en/our-work/agreements.

ASC. 2017. Agreement on Enhancing International Arctic Scientific Cooperation, 11 May 2017. https://oaarchive.arctic-council.org/handle/11374/1916.

Ball, Wayne S. 1996. The Old Grey Mare, National Enclosure of the Oceans. *Ocean Development & International Law* 27: 97–124.

Balton, D. 2001. The Bering Sea Donut Hole Convention: Regional Solution, Global Implications. In *Governing High Seas Fisheries: The Interplay of Global and Regional Regimes*, ed. O. Stokke. Oxford, UK: Oxford University Press.

Byers, M. 2014. *International Law and the Arctic*. Cambridge, UK: Cambridge University Press.

Canada, Arctic Waters Pollution Prevention Act. 1970. https://www.tc.gc.ca/eng/marinesafety/debs-arctic-acts-regulations-awppa-494.htm.

Central Arctic Oceans Fisheries Agreement [Agreement to prevent unregulated high seas fisheries in the Central Arctic Ocean] (CAOFA). 2017. https://www.dfo-mpo.gc.ca/international/agreement-accord-eng.htm.

Chicago Convention. 1944. Convention on International Civil Aviation, 7 December 1944, entered into force 4 April 1947. https://www.icao.int/publications/Documents/7300_cons.pdf.

Conservation of Arctic Flora and Fauna (CAFF). 2013. Arctic Biodiversity Assessment: Report for Policy Makers. CAFF, Akureyri, Iceland. https://www.caff.is/assessment-series/arctic-biodiversity-assessment/229-arctic-biodiversity-assessment-2013-report-for-policy-makers-english.

Copenhagen Agreement. 1993. Agreement between Denmark, Finland, Iceland, Norway and Sweden Concerning Cooperation in Taking Measures against Pollution of the Sea by Oil or Other Harmful Substances, Copenhagen, 29 March 1993, entered into force 16 January 1998.

Dremliuga, R. 2017. A Note on the Application of Article 234 of the Law of the Sea Convention in Light of Climate Change: Views from Russia. *Ocean Development &*

International Law 48 (2): 128–135. https://doi.org/10.1080/00908320.2 017.1290486.

Franckx, E. 1993. *Maritime Claims in the Arctic: Canadian and Russian Perspectives.* Dordrecht/Boston/London: Martinus Nijhoff.

FSA. 1995. United Nations, Agreement for the Implementation of the Provisions of the United Nations Convention on the Law of the Sea of 10 December 1982 Relating to the Conservation and Management of Straddling Fish Stocks and Highly Migratory Fish Stocks, 8 September 1995, A/Conf.164/37.

Haacke, J. 2016. Myanmar and Maritime Security. *ASAN Forum*, February 22. http://www.theasanforum.org/myanmar-and-maritime-security/.

IAMSAR Manual. 2007. *International Aeronautical and Maritime Search and Rescue Manual.* London: IMO.

Ilulissat Declaration. https://www.regjeringen.no/globalassets/upload/ud/080525_arctic_ocean_conference-_outcome.pdf.

Intervention Convention. 1969. International Convention Relating to Intervention on the High Seas in Cases of Oil Pollution Casualties, 29 November 1969, 970 UNTS 211 (entry into force 6 May 1975).

Kikkert, P., and P.W. Lackenbauer. 2014. *Legal Appraisals of Canada's Arctic Sovereignty: Key Documents, 1905–1956.* Calgary: Center for Military and Strategic Studies.

Koh, T.T.B. 1982. A Constitution for the Oceans. http://www.un.org/depts/los/convention_agreements/texts/koh_english.pdf.

Koivurova, T., and R. Caddell. 2018. Managing Biodiversity Beyond National Jurisdiction in the Changing Arctic. *AJIL Unbound* 112: 134–138.

Koivurova, T., and D. VanderZwaag. 2007. The Arctic Council at 10 Years: Retrospect and Prospects. *University of British Columbia Law Review* 40 (1): 121–194.

MOPPRA. 2013. Agreement on Cooperation on Marine Oil Pollution Preparedness and Response in the Arctic, 15 May 2013, entry into force 25 March 2016.

No author named. 2014. The Arctic: Frozen Conflict. *The Economist*, December 17. https://www.economist.com/international/2014/12/17/frozen-conflict.

Onarheim, I.H., T. Eldevik, L.H. Smedsrud, and J.C. Stroeve. 2018. Seasonal and Regional Manifestation of Arctic Sea Ice Loss. *Journal of Climate* 31: 4917–4932. https://doi.org/10.1175/JCLI-D-17-0427.1.

OPRC. 1990. International Convention on Oil Pollution Preparedness, Response and Co-operation, 30 November 1990; entry into force 13 May 1995.

Overby, L.R. 2014. Civil Liability (for Oil Pollution) in Polar Marine Environments. In *Comité Maritime International Yearbook*, 357–366, at p. 358.

PAME. 2009. Arctic Marine Shipping Assessment (AMSA) 2009 Report, 6–7. https://www.pame.is/images/03_Projects/AMSA/AMSA_2009_report/AMSA_2009_Report_2nd_print.pdf.

Polar Code. 2017. International Code for Ships Operating in Polar Waters. www.imo.org/en/MediaCentre/HotTopics/polar/pages/default.aspx.

Russian Federation revised submission to the CLCS. 2015. http://www.un.org/depts/los/clcs_new/submissions_files/submission_rus_rev1.htm.

Sands, P., J. Peel, A. Fabra, and R. MacKenzie. 2018. *Principles of International Environmental Law.* Cambridge, UK: Cambridge University Press.

SAR Convention. 1979. International Convention on Maritime Search and Rescue, 27 April 1979, entered into force 22 June 1985, 1405 UNTS 118.

Scovazzi, T. 2009. Legal Issues Relating to Navigation through Arctic Waters. *The Yearbook of Polar Law Online* 1 (1): 371–382.

Seafish. 2013. Industry Briefing Note, January 2013. http://www.seafish.org/media/750990/seafishguidancenote_mackerel_201301.pdf.

Tanaka, Y. 2015. *The International Law of the Sea*. Cambridge, UK: Cambridge University Press.

UNCLOS. United Nations Convention on the Law of the Sea, 10 December 1982, 1833 UNTS 3; 21 ILM 1261. http://www.un.org/Depts/los/convention_agreements/convention_overview_convention.htm.

UNCLOS III. 1982. Third United Nations Conference on the Law of the Sea (1973–1982), volume XVI, Summary Records of the Plenary, Eleventh Session: 163rd Meeting A/CONF.62/SR.163.

CHAPTER 26

International Indigenous Rights Law and Contextualized Decolonization of the Arctic

Dwight Newman

INTRODUCTION

Indigenous peoples constitute a very significant part of the population of the Arctic region. Indeed, across Alaska, northern Canada, Greenland, the northern part of the Nordic states, and northern Russia, Indigenous peoples are in many instances a majority of the population. Rights held by Indigenous peoples in their relationships with national state governments, then, bear significantly on policy issues in the Arctic. Since international legal frameworks increasingly protect Indigenous rights—and with the transformative adoption of the *United Nations Declaration on the Rights of Indigenous Peoples* (*UNDRIP*) by the United Nations General Assembly in 2007 (UNDRIP 2007)—it is worth examining the ways in which international Indigenous rights law affects policy issues in the Arctic.

The main claim of this chapter is that international Indigenous rights law exerts certain pressures in support of policy approaches that advance the contextualized decolonization of the Arctic. The claims on these effects are currently relatively modest—largely because of some complexities on the content and status of international Indigenous rights law—although they will most likely become even more significant over time. Involved within these influences are broader partnerships with Indigenous peoples in the Arctic but also broader implications in terms of state relationships within the Arctic region.

To understand this claim, it is useful to set out key background on the nature of international law generally and on the main sources and contents of

D. Newman (✉)
University of Saskatchewan, Saskatoon, SK, Canada
e-mail: Dwight.newman@usask.ca

K. S. Coates, C. Holroyd (eds.), *The Palgrave Handbook of Arctic Policy and Politics*, https://doi.org/10.1007/978-3-030-20557-7_26

international Indigenous rights law specifically. Doing so will illustrate that there is potential variation in the implications of international Indigenous rights law within different Arctic states because of their differing commitments in respect of particular international instruments as well as their domestic legal systems' different relationships to international law generally. This variation introduces one dimension of the contextuality of the implications of international Indigenous rights law within the Arctic.

Focusing then in turn on several key rights contained within the body of international Indigenous rights law—notably cultural rights, land rights, and self-determination rights—it will be possible to speak to the key policy implications that again vary in different states, as well as to how these might be expected to develop even further over time, with some more practical examples. These key rights all have implications that prohibit certain sorts of traditional state policies and that ultimately foreground a larger development of state action in the Arctic in partnership with Indigenous peoples.

While many of the policy implications reflect differences among the Arctic states, some of the consequences also have particular dimensions resulting from the unique vulnerabilities of the Arctic. And some have significant transboundary dimensions. Overall, international Indigenous rights law involve certain kinds of contextual decolonization of the Arctic while also highlighting some ways in which the Arctic functions distinctively as a region, not subject to simple state control from the south.

Sources of International Indigenous Rights Law and Varying Implications for Different Arctic States

International Indigenous rights law arises from several different sources. Pronouncing upon their contents and implications, however, is made complicated by some meaningful ongoing debates and developments on the very nature and sources of international law. Certain traditional sources are more secure, and it is thus worth commencing with those before then turning to important recent developments that may further extend the contents of international Indigenous rights law and support its ongoing progressive development over time. However, throughout this background section, it will actually become apparent that there is some variation in the contents of international Indigenous rights law across the Arctic, even while one might try to reach some broader characterizations of its implications.

Starting with the more secure traditional view, international law exists within a legal system that is separate from the legal systems of particular states, although that international legal system is in complex interrelationships with those domestic legal systems. The international legal system has no legislative body and arises from historic and contemporary interactions between states. In the traditional view of international law, international legal obligations arise from the voluntary commitments of states and especially from two specific legal sources. First, international legal obligations can arise from "conventions" (or international treaties), but those obligations apply only to those states that become parties by signing and ratifying a particular treaty. Second, international legal obligations can also arise from "customary international law," which consists of those norms that are found in the uniform practice of states where states engage in that practice while having an *"opinio juris."* This is a belief that the practice is legally obligatory, but with the possibility of a particular state remaining outside new norms in "persistent objector" status. While the formation of such customary international law is subject to complex discussion, the key point on the traditional voluntarist conception of international law is that customary international law also arises from the voluntary commitment of states (see generally Newman 2014: chap. 5).

One key international legal instrument setting out commitments on Indigenous rights is the 1989 *Indigenous and Tribal Peoples' Convention*, negotiated within the auspices of the International Labour Organization (ILO) and commonly known as ILO Convention 169 (ILO 1989). This treaty moved beyond earlier conceptions embodied in a 1957 ILO treaty also dealing with Indigenous peoples, but ILO Convention 169 has been ratified by only about two dozen states. Significantly for the Arctic context, however, those ratifying states include both Norway (as of June 1990) and Denmark (as of February 1996). The contents of ILO Convention 169 thus have direct legal bearing on both Norwegian legal obligations in Sápmi (the Sami region within the Nordic countries) and on Danish obligations to the Indigenous peoples in Greenland. Within the Nordic region, there have been some discussions on whether Sweden might ratify ILO Convention 169, but various internal political dynamics have continued to make it unlikely there and also something that seems to have been rejected, at least for now, in Finland (cf Semb 2012; Allard 2018). Ratification by any of the other Arctic states also appears unlikely. So, this significant international treaty legally has significant legal standing only to Norway and Denmark/Greenland amongst Arctic states.

Because international treaty obligations apply only to ratifying states, there is thus some variation in the international law landscape for Indigenous rights within the Arctic. Such variation may actually develop further, as there are some recent tendencies toward regional negotiations on Indigenous rights instruments. A prime example in that regard is the regional negotiation of the *Nordic Sami Convention*, which has been in progress for decades and reached a draft text in 2005 (Bankes and Kuivorova 2013). That treaty would involve Norway, Sweden, and Finland in making particular international treaty commitments, thus recognizing the transboundary dimensions of Sápmi. It does so without regard to the portion of Sápmi in Russia's Kola Peninsula, as the prospects of Russian involvement in the treaty were non-existent. While a final text of the Nordic Saami Convention was negotiated in early 2017, the ratification prospects remain unclear in some respects. But the Convention illustrates the possibility that international law treaties on Indigenous rights could specifically determine some different legal obligations in parts of the Arctic as compared to others.

A similar phenomenon could arise with a development of the recent Organization of American States (OAS) *American Declaration on the Rights of Indigenous Peoples* (*ADRIP*) into a convention. This seems more possible than a similar development on *UNDRIP*—although the two Arctic states of the Americas, Canada and the United States, have remained outside *ADRIP* and thus might remain outside any treaty developed from it.

The content of customary international law on Indigenous rights issues is more challenging to determine. While some scholars have attempted to identify particular norms of customary international law in the course of their work (e.g. Anaya 2004; Xanthaki 2007; ILA 2012; Ahren 2016), there is not complete uniformity on which particular norms are identified as constituting customary international law. Nonetheless, there are clearly some norms that are identifiable as meeting the requirements of relatively uniform state practice, combined with a belief by the states that they are acting on the basis of legal obligation.

That said, in both the treaty context and the customary international law context, the focus on states and on states' voluntary commitments within the traditional approach to international law obviously has highly detrimental implications for Indigenous peoples. Some accounts of international law do end up moving beyond that state focus in several different ways that are highly pertinent to international Indigenous rights law. First, one further traditional source of international law norms, those called "general principles of law," has sometimes been applied in a manner looking for principles common across different systems of law. But it has also now been applied in other ways that may permit the introduction of moral content into international law on matters like Indigenous rights, with an example being the judicial identification of consultation with Indigenous peoples as being a general principle receiving status in international law (Biddulph and Newman 2014).

Second, Indigenous peoples themselves have increasingly taken action to help to shape the international law of Indigenous rights. A key example is the role Indigenous peoples played in developing the draft version and negotiating with states on the final version of the *United Nations Declaration on the Rights of Indigenous Peoples* (*UNDRIP*) (Hohmann and Weller 2018; Barelli 2016). Parts of *UNDRIP* have sometimes been argued to reflect or shape customary international law, but its legal status is complex (Allen and Xanthaki 2011). At minimum, though, it is a significant normative statement that will have legal and political impact, largely because it resulted from negotiations involving states and Indigenous peoples—even if it is formally "soft law" (Barelli 2009).

UNDRIP is not a treaty, and many popular references to states "signing" or "ratifying" *UNDRIP* are thus category mistakes. No state is formally legally bound to the contents of *UNDRIP* simply by its passage or even by that state's support for *UNDRIP*, even though *UNDRIP* will exert significant normative force over time. Some Arctic countries voted against *UNDRIP* but have attempted to offer subsequent endorsements of it. That is true, for instance, of both Canada and the United States, which were amongst the four states that voted against *UNDRIP* (along with Australia and New Zealand) possibly largely based on a sense by these states that they already had domestic mechanisms in place to attempt to address Indigenous issues (Gover 2015). Longer-term, *UNDRIP* is likely to exert meaningful influence on the Indigenous rights discussions in most Arctic states.

That said, amongst the remaining Arctic states, Russia abstained on *UNDRIP* and may thus not be as responsive to arguments stemming from *UNDRIP* concerning Indigenous rights norms. It may look simply to maintain its distinctive legal system concerning "small-numbered peoples." Russia actually has strong formal legal protections in this context, although implementation and enforcement are mixed (cf Tomaselli and Koch 2014). That said, while those patterns have not worked out well for some Indigenous peoples, others have been able to draw on transnational networks and developing international norms and, for instance, negotiate benefit-sharing from corporations engaged in resource extraction. One example is the negotiation between the Nenets and Lukoil under which the Nenets achieved significant commitments from the company (Tysachniouk et al. 2018).

The implications of international law on Indigenous rights will vary among the Arctic states. This is partly because of their differing commitments and stances on key international Indigenous rights instruments. But it also stems from a different relationship of domestic legal systems to international law. Some states in the world are *dualist* in respect of international law, meaning that international treaties do not have effect within the state until domestic legislative steps are taken to implement them. States in this category include Canada, Norway, Sweden, and Denmark. In a *monist* state, ratified treaties become part of national law without domestic legislative steps. Russia would tend to be categorized as monist, the United States as mixed, and Finland sometimes as monist but sometimes as dualist (cf Allard 2018). This distinction concerning principles of their legal systems does not necessarily speak to implementation by states in all instances. Some states that are dualist nonetheless see very strong respect by legislators for international law, with Norway showing much national responsiveness tied to the country's commitment to *ILO Convention 169* (Allard 2018). Moreover, other distinctions cross-cut these states in respect of how their domestic legal systems regard customary international law.

Suffice it to say, though, that for a variety of reasons, the legal effects of international law instruments will vary in the different Arctic states. Developments on international Indigenous rights law occur partly in a legal system oriented to the voluntary commitments of states and thus have varying impacts based on the differing commitments of these states. At the same time, to some degree, Indigenous peoples themselves have worked to reshape the state-centric focus of international law. There are some notable tendencies toward international law operating in ways outside the narrow confines of traditional descriptions. More subtle influences from international law are sometimes discernible even in contexts where these might not have been expected.

But the patterns will vary in complex ways. We turn now to consider some more tangible examples in the context of several specific rights.

IMPLICATIONS FOR ARCTIC STATES FROM INDIGENOUS RIGHTS: CULTURAL RIGHTS, LAND RIGHTS, AND SELF-DETERMINATION RIGHTS

Although there are many rights held by Indigenous peoples, several key groups of rights reflect certain central claims and thus permit an analysis of policy implications. These concern rights to maintain and preserve cultures, rights in relation to traditionally occupied lands, and rights of self-determination that also find expression in rights to participation in decision-making on issues that affect Indigenous peoples (cf Xanthaki 2007). Closer examination can illustrate how these various rights ultimately have significant policy implications in various Arctic contexts and call for the contextualized decolonization of the Arctic.

First, while cultural rights certainly encompass many other specific rights, amongst those cultural rights that may have particular implications for state policy in the Arctic are those related to traditional harvesting and traditional relationships with particular species of animals. Such issues had actually been litigated in international forums under more general human rights instruments some decades ago even prior to more recent fuller elaboration of norms on Indigenous rights. Examples include litigation concerning reindeer herding amongst the Sami. Some other Arctic Indigenous peoples, such as the Gwich'in of northwestern Canada and eastern Alaska or even the Dene of Canada's northern Sub-Arctic, could have analogous claims arising from particular relationships with the caribou. Many coastal Inuit communities have a particular dependence on marine mammals, both for subsistence in a harsh environment and as integral parts of their cultures.

In general terms, the fact that Indigenous peoples hold cultural rights in relation to such animal species will constrain state policies that could negatively impact upon these species. For example, controversies about proposed mining developments in Sweden, such as the Kallak/Gállok iron ore mine, have centered on issues concerning potential negative impacts on reindeer herding, with an increasing influence from international norms that place meaningful weight on cultural rights related to reindeer herding as compared to what would have been the case in the past.

The Inuit relationship with marine mammals illustrates the ways in which rights held by Arctic Indigenous peoples may also have extraterritorial implications extending beyond the Arctic states. In these cases, policies adopted by states outside the Arctic could impact on rights held by Arctic Indigenous peoples. In the context of European Union regulation of seal products, for example, a limited exception for seal products coming from Inuit sustainable hunting was included, partly in response to the normative pressures of international Indigenous rights law. Similarly, international Indigenous rights law and cultural rights of Indigenous peoples gradually filtered into discussions of the Aboriginal Subsistence Whaling Working Group of the International Whaling Commission and its approval of whaling quotas for certain Arctic Indigenous peoples, particularly in Alaska and Greenland.

Cultural rights are not limited to those referenced thus far. As recognized in article 12 of *UNDRIP* (UNDRIP 2007), cultural rights may also have broader spiritual dimensions. In turn, those may relate also to such matters as sacred sites and even to sacred landscapes, which may have implications for state-approved development that would otherwise occur (cf Heinämäki and Herrmann 2017). There may be complex balances as between such rights and state claims to development, but there are nonetheless different restraints on state development that arise in light of possible claims within the international normative framework of international Indigenous rights law.

Second, international Indigenous rights law increasingly recognizes rights to lands that Indigenous peoples have traditionally occupied and used, with articles 25–28 of *UNDRIP* being illustrative of the developing position (UNDRIP 2007). There are deep complexities on what these articles mean for issues related to ownership of natural resources (Ahren 2016), but they do suggest at least surface ownership of significant areas of land as the default legal position.

Even prior to the clearer articulation of this position in international law, some land claims could be articulated in common law systems under the doctrine of Aboriginal title. In Canada, Aboriginal title also took on constitutional status after constitutional amendments in 1982. Aboriginal title claims motivated the negotiation of many modern treaties with Arctic Indigenous peoples in Canada over recent decades as the government tried to resolve land claims issues. The negotiated arrangements led to the recognition of extensive land and resource ownership, ongoing resource royalties from developments over larger areas of land, substantial compensatory trust funds, and negotiated self-government arrangements. In Canada's eastern Arctic, the establishment of a separate territorial government in Nunavut with a majority-Inuit population resulted from a land claims negotiation that was designed to address Aboriginal title.

Earlier, in Alaska in 1971, Congress warded off potential litigation on the common law doctrine of Aboriginal title so as to clear the way for oil and gas development with legislation imposing a settlement on all land claims. This legislation created various Alaskan Native corporations that received certain types of compensation and mineral rights allocations. That sort of imposed resolution might face challenges from international Indigenous rights norms if attempted today, but it paradoxically marked some significant recognition of Indigenous land claims.

The ultimate impacts of land rights on other Arctic states may not yet be entirely clear, but there are international Indigenous rights that could have implications for land policy. Indeed, land rights demarcations taking place in the Finnmark region of Norway follow on its land rights obligations under

ILO Convention 169 (Ravna 2016), thus suggesting a direct impact of international law on some contemporary land rights issues.

Some take the view that other Indigenous rights are all simply emanations of underlying rights of self-determination (cf Xanthaki 2007). Whether or not that is a helpful way of understanding Indigenous rights generally, there is no doubt that the gradual recognition in international instruments of Indigenous peoples' rights of self-determination is significant. The extension from "home rule" to self-government in Greenland under the 2009 *Greenland Self-Government Act*, as well as the real possibility of a future independent Greenland, is partly fostered and conditioned in some respects by rights of Indigenous self-determination (cf Kuokkanen 2017), although rights alone obviously do not determine all of the practicalities of full independence. However, external self-determination culminating in the establishment of a new state will likely be relatively exceptional, and the implications of internal self-determination are more widespread.

Without limiting the future implications of self-determination, one aspect of it includes various participatory rights that affect state decision-making on issues with substantial impacts on Indigenous peoples (cf Xanthaki 2007). Some mechanisms for consultation in the different states have emerged that have had sometimes unclear relations to international law provisions on consultation. In Norway, the state established the Norwegian Saami Parliament relatively contemporaneously with ratifying *ILO Convention 169*, albeit with no explicit link, although *ILO Convention 169* has surely influenced the subsequent adoption of consultation agreements between the Norwegian state and the Sami (Allard 2018). Interestingly, while Sweden and Finland are not parties to *ILO Convention 169*, they have followed somewhat similar policies to Norway. In adopting these measures, there have sometimes seemed to be connections to broader international human rights norms and possibly to Indigenous rights norms (Allard 2018).

In Canada, a proactive duty to consult doctrine emerged in the Supreme Court starting in 2004, growing out of domestic legal principles, although it may seem to go some distance in complying with international law obligations as well (Newman 2014). This latter doctrine has had enormous impacts in such contexts as resource extraction and, as project proponents have sought to avoid uncertainties related to governmental fulfillment of duty to consult obligations, it has normalized a prevalent practice of Indigenous-industry agreements (within Canada, often called Impact Benefit Agreements or IBAs) that would not have become so common without the background pressures of the duty to consult doctrine (Newman 2014). The ongoing development of international norms related to self-determination—along with cultural rights and land rights—will have meaningful implications in various Arctic contexts to support fuller economic participation of Indigenous peoples in resource extraction and other economic activity.

There would, of course, be quick agreement that certain past policies would not be permissible today, although whether that would be due to domestic law or international law would not always be definitively delineated. For example, in the post-WWII period, Canada engaged in many relocations of Inuit communities in the Eastern Arctic, purportedly to bring Inuit communities closer to service delivery, something which has had devastating effects on traditional Inuit life. Today, such relocation would be prohibited both by Canada's constitutional jurisprudence on Indigenous rights and by international law norms prohibiting relocation without free, prior, and informed consent (FPIC), as expressed in article 10 of *UNDRIP* (UNDRIP 2007).

BROADER IMPLICATIONS

As some of the examples referenced in this chapter make clear, Indigenous rights prohibit a recurrence or extension of various past state policies. In general terms, one could characterize the implications of Indigenous rights as requiring a contextualized decolonization of the Arctic. To make that concept clear, some Arctic states have had a tendency to draw resource wealth from their less populated norths for the benefit of their more populated souths, with that phenomenon sometimes taking place even within subunits of the state, such as in issues experienced in the Canada's so-called provincial north (Coates and Poelzer 2014). Such a phenomenon of what are effectively colonial relationships can be a natural result of majoritarian democracy. Indigenous rights expand the meaning of the democratic state beyond pure majoritarianism. In light of the relatively large Indigenous population in the Arctic, Indigenous rights tend to protect the Arctic region against southern political majoritarianism that would otherwise treat the Arctic as a resource extraction colony. While the degree to which this tendency has played out has varied among the Arctic states, Indigenous rights support a contextualized decolonization process that runs counter to such tendencies.

The diminishment of southern colonial power over the Arctic will take different forms in different Arctic contexts. International Indigenous rights law will continue to provide meaningful reasons in support of Greenlandic independence. In other parts of the Arctic, Indigenous self-determination will be realized in other ways, ranging from the creation of new political subunits like Nunavut or Finnmark to combinations of devolution of powers to northern subunits combined with significant rights specifically for Indigenous peoples within those northern subunits, to various other arrangements suited to the conditions of different contexts.

At the same time, international Indigenous rights law also provides certain protections for the Arctic that are responsive to some of its unique vulnerabilities. The Arctic has such vulnerabilities partly because of its extreme climate, which makes some of the life present within the Arctic more vulnerable than in other regions. It has them also because of other unique features of the Arctic. The phenomenon of 24-hour darkness during parts of the year has significant consequences for rescue operations and responsiveness to environmental disasters at those times. In general terms, best practices on consultation with

Indigenous peoples in Arctic contexts will be extended to take into account the unique vulnerabilities of the Arctic (Newman et al. 2014).

Transboundary issues within the Arctic highlight a further complex dimension of the region. Circumpolar Indigenous activism has been a powerful force. Indigenous peoples that cross state boundaries are also entitled to respect for their rights in a manner that surmounts those artificial barriers. And, as suggested earlier, non-Arctic states that intervene in ways that affect Arctic Indigenous peoples are also subject to extraterritorial implications of international Indigenous rights law. There are certain senses in which Indigenous rights of Arctic Indigenous peoples actually unite certain types of Arctic claims and present the Arctic as a region that is not to be subject to state control from the south.

International Indigenous rights law is more complex than is fully appreciated. The parts of it that apply to different Arctic states vary based on the commitments of those states. At the same time, those portions that are formally soft law—such as the *United Nations Declaration on the Rights of Indigenous Peoples* (UNDRIP 2007)—may exert longer-term normative pressures on Arctic states generally, although very possibly in more subtle ways. The main categories of rights recognized within the various Indigenous rights instruments all have implications for policy. In some instances, they would specifically prohibit certain past policies. In more general ways, they may help to shape state policies more responsive to Arctic Indigenous peoples.

References

Ahren, Mattias. 2016. *Indigenous Peoples' Status in the International Legal System.* Oxford: Oxford University Press.

Allard, Cristina. 2018. The Rationale for the Duty to Consult Indigenous Peoples: Comparative Reflections from Nordic and Canadian Legal Contexts. *Arctic Review on Law and Politics* 9: 25–43.

Allen, Stephen, and Alexandra Xanthaki, eds. 2011. *Reflections on the UN Declaration on the Rights of Indigenous Peoples.* Oxford: Hart Publishing.

Anaya, S. James. 2004. *Indigenous Peoples in International Law.* 2nd ed. Oxford: Oxford University Press.

Bankes, Nigel, and Timo Kuivorova, eds. 2013. *The Proposed Nordic Saami Convention: National and International Dimensions of Indigenous Property Rights.* Oxford: Hart Publishing.

Barelli, Mauro. 2009. The Role of Soft Law in the International Legal System: The Case of the United Nations Declaration on the Rights of Indigenous Peoples. *International and Comparative Law Quarterly* 58: 957–983.

———. 2016. *Seeking Justice in International Law: The Significance and Implications of the UN Declaration on the Rights of Indigenous Peoples.* London: Routledge.

Biddulph, Michelle, and Dwight Newman. 2014. A Contextualized Account of General Principles of International Law. *Pace International Law Review* 26: 286.

Coates, Ken, and Greg Poelzer. 2014. The Next Northern Challenge: The Reality of the Provincial North. *Macdonald-Laurier Institute Commentary*, April.

Gover, Kirsty. 2015. Settler-State Political Theory, 'CANZUS', and the UN Declaration on the Rights of Indigenous Peoples. *European Journal of International Law* 26: 345–373.

Heinämäki, Leena, and Thora Martina Herrmann, eds. 2017. *Experiencing and Protecting Sacred Natural Sites of Sámi and Other Indigenous Peoples: The Sacred Arctic*. New York: Springer.

Hohmann, Jessie, and Marc Weller, eds. 2018. *The UN Declaration on the Rights of Indigenous Peoples: A Commentary*. Oxford: Oxford University Press.

ILA (International Law Association). 2012. Report of the Committee on Indigenous Rights Accompanying Res. 5/2012, Sofia, Bulgaria.

ILO (International Labour Organization). 1989. *Convention Concerning Indigenous and Tribal Peoples in Independent Countries* (27 June 1989, entered into force 5 September 1991).

Kuokkanen, Rauna. 2017. The Pursuit of Inuit Sovereignty in Greenland. *Northern Public Affairs*, July, 46–49.

Newman, Dwight G. 2014. *Revisiting the Duty to Consult Aboriginal Peoples*. Saskatoon: Purich Publishing.

Newman, Dwight, Michelle Biddulph, and Lorelle Binnion. 2014. Arctic Energy Development and Best Practices on Consultation with Indigenous Peoples. *Boston University International Law Journal* 32: 101–160.

Ravna, Øyvind. 2016. Norway and Its Obligations Under ILO 169—Some Considerations After the Recent Stjernøy Supreme Court Case. *Arctic Review on Law and Politics* 7: 201–204.

Semb, Anne Julie. 2012. Why (Not) Commit?—Norway, Sweden and Finland and the ILO Convention 169. *Nordic Journal of Human Rights* 30: 122–147.

Tomaselli, Alexandra, and Anna Koch. 2014. Implementation of Indigenous Rights in Russia: Shortcomings and Recent Developments. *International Indigenous Policy Journal* 5: 1–23.

Tysachniouk, Maria, et al. 2018. Oil Extraction and Benefit Sharing in an Illiberal Context: The Nenets and Komi-Izhemtsi Indigenous Peoples in the Russian Arctic. *Society and Natural Resources* 31: 556–579.

UNDRIP (United Nations Declaration on the Rights of Indigenous Peoples). 2007. UN Doc. A/Res/61/295 (adopted 13 September 2007, published 2 October 2007).

Xanthaki, Alexandra. 2007. *Indigenous Rights and United Nations Standards: Self-Determination, Culture and Land*. Cambridge: Cambridge University Press.

CHAPTER 27

The Future of the Arctic Council

Matthew S. Wiseman

INTRODUCTION

Established in September 1996 as a high-level forum for promoting mutually beneficial cooperation among Arctic states and the inhabitants who call the region home, the Arctic Council (AC) quickly became the pre-eminent inter-governmental body for addressing social and environmental challenges in the Arctic. Composed of eight permanent member states, six permanent Indigenous peoples' organizations, non-permanent observer states, and a host of various other observer organizations, the Council facilitates scientific research assessments and binding agreements for the protection and promotion of Arctic peoples, lands, waters, and resources. Serving as a key venue for Arctic issues, the AC has not only advanced policy decisions but has also brought increased visibility to global issues through an Arctic lens. Although the Council has achieved considerable success framing Arctic issues as policy matters, ongoing changes in the region produce a host of issues for multilateral cooperation. Now is an opportune time to review the Council's history with an eye toward assessing the role and responsibility of intergovernmental fora in identifying and responding to Arctic issues, both regionally and globally significant.

As a space for state and non-state actors of different backgrounds, expertise, and knowledge systems, the Arctic Council represents a forward-thinking framework for multilateral policymaking and governance. In theory, the Council institutionalizes a worldwide view for meaningful coordination in the face of a drastically changing Arctic that promises to affect people living within and without the Circumpolar North. Detailed assessments of the Council consider the strengths and weaknesses of the forum's ability to facilitate communication, translation, and mediation among the many state and non-state actors

M. S. Wiseman (✉)
University of Toronto, Toronto, ON, Canada
e-mail: Matthew.wiseman@utoronto.ca

© The Author(s) 2020   439
K. S. Coates, C. Holroyd (eds.), *The Palgrave Handbook of Arctic Policy and Politics*, https://doi.org/10.1007/978-3-030-20557-7_27

tied to Arctic governance. While the current literature acknowledges the Council's growing importance, skeptics question whether the existing structures will overcome the evolving challenges of intergovernmental politics and cooperation. As the Council's momentum subsides, speculation about the ineffectiveness of the forum has led to doubt and criticism from outside observers.

This chapter examines the strengths and weaknesses of the Arctic Council as a facilitator and mediator of intergovernmental cooperation in the Arctic. As a functionalist neoliberal institution, the Council aims to address social and environmental issues in the Arctic through a multilateral mandate that responds to unique regional circumstances created by climate change and resource competition. Regional challenges in the Arctic extend beyond the Council's existing mandate and political authority, however. Moving forward, the Council needs to improve its transparency and adopt an inclusive approach that welcomes and encourages collaboration with outside bodies tied to environmental protection, sustainable development, and the promotion of human rights in the Arctic.

FORMATIVE YEARS

The Arctic Council emerged in response to the drastic militarization of the Circumpolar North that occurred during the Cold War. From the late 1940s through the 1960s, tensions between the United States and the Soviet Union led to the construction of military installations and the placement of nuclear weapons on both sides of the North Pole. Relations between the superpower rivals eased during the period of détente in the 1970s, marked by the Strategic Arms Limitations Talks (SALT) and the signing of the Anti-Ballistic Missile (ABM) Treaty. The 1980s saw the return of heightened tensions and increased military spending, leading to a US policy statement in April 1983 asserting an openness to engage international cooperation in the Arctic (Reagan 1983; Chater 2015). Two years later, the influential American political scientist Oran R. Young (1985–1986) proposed declaring the Arctic a peaceful zone to ease tensions in the region. Meanwhile, Moscow prepared its own statement on the international situation in the Arctic. The global community watched with anticipation on October 1, 1987, as Mikhail Gorbachev, the general secretary of the Communist Party of the Soviet Union, stepped to a podium in Murmansk and delivered a speech about the Arctic (English 2013, 6; Poto and Fornabaio 2017). While calling for demilitarization in the region, the Soviet leader articulated the concept for a multilateral forum. Gorbachev proposed a conference of Subarctic states to discuss the establishment of a joint "Arctic Research Council" (Gorbachev 1987). He emphasized the importance of peaceful cooperation for the successful development of northern resources and drew attention to the role of non-state actors in deliberations about environmental protection and cultural integration. Gorbachev also saw Indigenous participation as important for the success of the proposed council and stressed protecting the fundamental rights of all populations living in the Arctic.

Although officials representing the other Arctic states doubted the peaceful intentions of the Soviet Union, Gorbachev's speech provided the impetus for international action (English 2013, 6–7). Two years after the Murmansk speech, Finland organized a series of formal negotiations that led to the establishment of the Arctic Environmental Protection Strategy (AEPS), the precursor to the Arctic Council (Young 1998, 248). Formally adopted in June 1991, the AEPS marked the first multilateral accord concerned with addressing pollutants and environmental protection in the Arctic (Poto 2016). The AEPS composed four working groups: the Arctic Monitoring and Assessment Programme (AMAP), Conservation of Arctic Flora and Fauna (CAFF), Emergency Prevention, Preparedness and Response (EPPR), and Protection of the Arctic Marine Environment (PAME). Arctic states shared the responsibility for each core area with international governmental organizations and non-governmental organizations (NGOs). Senior Arctic Officials (SAOs) from each Arctic state facilitated and supervised projects carried out by the workings groups, including the organization of progress meetings and the development of reports and initiatives. Indigenous peoples' organizations did not have privileged status under the AEPS, which relegated Indigenous peoples to observers of the AEPS working groups.

Startling revelations from the AEPS spurred environmental protection groups, human rights activists, and politicians to advocate for greater intervention in the Arctic. Reports indicated dangerously high pollution levels, increasing human security challenges, and reduced life expectancy for Indigenous peoples in Russia who suffered from low economic conditions and poor health. In response, Indigenous peoples' organizations called for an increased voice and role in Arctic governance. Most notably, the Inuit Circumpolar Council (ICC), an organization representative of Inuit from Canada, Greenland, Russia, and the United States, took a leading role in lobbying the Canadian government to form a multilateral Arctic body. The ICC's president Mary Simon encouraged Canadian officials to create the Arctic Council, and contributed to the negotiation and implementation of the 1993 Nunavut Land Claims Agreement (English 2013, 183–187). In February 1995, Canadian prime minister Jean Chrétien proposed that the AEPS become a formal institution. US President Bill Clinton supported Chrétien's proposal and the Canadian government organized formal negotiations during the AEPS meeting in Ottawa that June. Three formal and complex negotiating sessions followed the preliminary talks, resulting in the signing of the Arctic Council declaration in Ottawa on September 19, 1996 (English 2013, 237–238).

LEGAL FRAMEWORK

The structures of Arctic governance differ vastly from the current system governing the opposite polar region, Antarctica. Because the 12 signatories of the 1959 Antarctic Treaty agreed not to consolidate their territorial claims over the continent, Antarctica is an international region void of sovereign states. In contrast, the Arctic composes eight peripheral nation states that, on the basis of international law, claim sovereign jurisdiction over lands, islands, and waters located within the delineated boundary of the Arctic Circle (Koivurova 2012). Initiatives for environmental protection and sustainable development in the Arctic are thus subject to complex international legalities, requiring multilateral cooperation for advancement and success. The Arctic Council functions accordingly, providing a fair and equitable platform for the negotiation and development of Arctic initiatives among local, regional, and national stakeholders.

Although the Arctic gradually emerged as a distinct region for international policy and law after the AEPS established written policies for environmental issues confronting the region and its inhabitants, the Arctic Council's founding in 1996 did not constitute a legal framework governing international cooperation in the region. A declaration is a form of soft law that binds states politically but not legally, and the authorities who established the AC did so via a declaration to avoid committing their respective state to any legal responsibility in the Arctic (Koivurova 2012). While the founding declaration laid the groundwork for a written agreement governing Arctic cooperation, the states involved in the AC did not discuss a comprehensive international treaty. The implications of that decision restricted the political latitude of the Council. Under the adopted framework, the AC does not have the authority to make binding decisions or develop policies for the pressing issues that affect Arctic inhabitants and their way of life.

In lieu of a comprehensive treaty governing international affairs in the Arctic, the AC embraced a strategy of inclusivity and provided a multilateral platform for state and non-state actors. As climate change sparked widespread advocacy for environmental protection and community-based socioeconomic development, heightened global attention fostered increasing political interest in Arctic governance. Melting ice continues to transform the region into a bastion of geopolitical competition among state and non-state actors concerned about social, environmental, commercial, and national security issues (Ebinger and Zambetakis 2009). As an intergovernmental forum, the AC facilitates discussion and action on all Arctic affairs except military security issues. The existing mandate emphasizes environmental protection, sustainable development, and Indigenous peoples' rights.

Based on broad and general language, the AC's founding declaration produced speculation and mixed expectations about the Council's ability to influence positive change in the Arctic. The mandate provides no precise definition for sustainable development, although the Council's focus in this area includes Arctic children and youth, health and social welfare, telemedicine, sanitation, and resource management, among others (Arctic Council 1998). Lacking a permanent secretariat during the formative years, the Council also received criticism as a weak institution unwilling to make strong financial commitments. This changed in June 2013 when the Standing Arctic Council Secretariat became operational. With the aid of a permanent secretariat, the Council expanded its research capacity and evolved into a policymaking forum that produced influential scientific assessments about a range of issues stemming from environmental, ecological, economical, and cultural changes in the Arctic.

In its current form, the Arctic Council consists of the eight Arctic States: Canada, the Kingdom of Denmark (including Greenland and the Faroe Islands), Finland, Iceland, Norway, the Russian Federation, Sweden, and the United States. Each member state shares the collective responsibilities of the Council, including the chairmanship duties that rotate every two years. The eight Arctic States appoint a Senior Arctic Official (SAO) to manage their interests in the Council. Representing sovereign federal governments, SAOs guide and monitor the Council's activities in accordance with the decisions and instructions of their respective foreign office. Cooperation among the eight member states is vital for the development and signing of ministerial declarations, which enable and promote multilateral action in the Arctic.

Recognizing the core value of traditional knowledge systems, the institutional structure of the Arctic Council integrates Indigenous peoples' organizations (IPOs) into the intergovernmental forum (Hossain and Maruyama 2016). Three IPOs representing Inuit (Inuit Circumpolar Council), Saami (Saami Council), and Russian Indigenous peoples (Russian Association of Indigenous Peoples of the North), held the status of Permanent Participants (PP) in the AEPS, and retained their privileged status when the AEPS evolved into the Arctic Council. The Aleut International Association received PP status two years later in 1998, followed by the Arctic Athabaskan Council and the Gwich'in Council International in 2000. As Permanent Participants in the Arctic Council, IPOs discuss ideas and policy directives with state representatives during ministerial meetings and facilitate community-based networking among governmental agencies, non-governmental organizations, and Indigenous peoples' associations and communities. Any state or non-state actor engaging in Arctic affairs must consult the six IPOs before making decisions that affect the Arctic and its inhabitants, including states and organizations seeking observer status (Koivurova and Heinämäki 2006; Koivurova 2010; Graczyk 2011). The creation of the Arctic Council also provided an internationally recognized forum for geographically disperse populations in the Arctic, advancing communication and cooperation among separate Indigenous groups.

In addition to Member States and Permanent Participants, the Arctic Council consists of Observers who represent non-Arctic states, intergovernmental and interparliamentary organizations, and both regional and global NGOs. Observer status in the AC enables state and non-state representatives to attend ministerial meetings and stay informed of policy directives. Observers participate in the Council with the consent of member state delegations, and use their affiliation and role to promote independent interests in Arctic politics (Manicom and Lackenbauer 2013; Ikeshima 2016; Hossain and Maruyama 2016). States and international institutions seek observer status in the Arctic Council to contribute to the governance of environmental issues affecting global climate change, and to gain a political stake in matters concerning the potential economic development of the Arctic (Chater 2016). Although Observers have a minimal role within the existing framework, their presence and cooperation enhances the international legitimacy of the Council.

First adopted in Iqaluit, Nunavut in 1998 and later revised at the eighth Ministerial Meeting of the Arctic Council in Kiruna, Sweden in 2013, the *Arctic Council Rules of Procedure* govern the process for admission into the Council for prospective observer states. As a condition for receiving Observer status and the permission to attend the ministerial meetings of the Arctic Council, the non-Arctic states have agreed to respect the sovereignty of Arctic countries (Riddel-Dixon 2017, 39). This condition includes respecting the rights of the five Arctic costal states as specified under the United Nations Convention on the Law of the Sea (UNCLOS), which entered into force in November 1994. Another of the seven criteria for admission as an observer state, Rule 6(d), stipulates that the state "respects the values, interests, culture and traditions of Arctic indigenous peoples and other Arctic inhabitants" (*Arctic Council Rules of Procedure* 2013, 14). Permanent Participants represent independent legal and political organizations under the collective direction of the Arctic Council, and Observers recognize the sovereign rights of IPOs. The current structure provides a platform for Indigenous peoples to express opinions and participate in decision-making processes with member states, marking a unique and important function of the Council.

Generating usable data across different knowledge systems is a primary function of the Arctic Council. As international law scholar Timo Koivurova (2012, 131) argues, any adequate examination of the Council requires "a clear understanding that the Arctic is simply an extension of existing political, economic, and environmental systems." Current international arrangements governing the Arctic cover issues specific to particular interest groups and organizations, such as the International Union for Circumpolar Health (IUCH) and the Convention for the Protection of the Marine Environment of the North-East Atlantic (OSPAR). While open to participation from states, communities, NGOs, business interests, and other stakeholders, issue-driven entities do not have the capacity to develop or support compressive policies for Arctic governance (Young 2005). Within this context, the Arctic Council emerged as a distinct and important forum for regional cooperation, an intergovernmental body ideal for promoting and facilitating governance in a drastically changing region of global significance.

PRODUCTIVE YEARS

Considering the drastic militarization of the Circumpolar North that occurred on both sides of the North Pole during the Cold War, it is not surprising that the Arctic became a focal point for international affairs in the 1990s. Nonetheless, the Arctic Council struggled for recognition out of the gate. During the Council's formative years, conflicts in Yugoslavia and Chechnya were front of mind for many world leaders who had little time to consider peaceful efforts for environmental protection and sustainable development in the Arctic (Spence 2013). The ongoing effects of climate change quickly altered the region's apparent value. No longer perceived as an inhospitable and inaccessible region of frozen ice, the Arctic attracted outside institutions as a region full of untapped potential. "Our understanding developed from a conception of the Arctic as being protected by its hostile environment to being a dynamically changing region with major economic possibilities, thus requiring stricter governance measures," reflected Timo Koivurova (2012, 134). In response, the Arctic Council produced notable assessments about the environmental, ecological, and human effects of oil and gas exploration in the Arctic, increased shipping in the region, and the consequences of climate change for biodiversity (Huntington and Arctic Monitoring and Assessment Programme 2007, 57; Arctic Council 2009; CAFF International Secretariat 2010).

Under the auspices of the Arctic Council, participants and observers cooperate on the production of scientific reports about significant pollution problems that continue to damage the changing environment and ecosystems of the Arctic. The Arctic Monitoring and Assessment Program produced the first in a series of reports in 1997, which document and explain the significance of such problems as pollutant migration and melting sea ice. The Arctic Climate Impact Assessment (ACIA), first commenced under the US chair period between 1998 and 2000, published a report in 2004 (Hassol) establishing the Arctic as a region vital for indicating and studying climate change. Warming in the Arctic, the report demonstrated, increased at twice the rate of any other region in the world.

As scientific reports revealed the serious effects of climate change for the environment, ecosystems, and human populations of the Arctic, the ACIA influenced how stakeholders perceived and approached challenges in the region. Unable to ignore the drastic and impactful transformation undergoing in the Arctic, state and non-state actors in the Arctic Council advocated for increased multilateral cooperation on Arctic issues affecting the global climate. At the invitation of Denmark, representatives of the five costal states bordering on the Arctic Ocean—Canada, Denmark, Norway, Russia, and the United States—met in Ilulissat, Greenland in late May 2008. Following discussions, state representatives adopted the Ilulissat Declaration to address the environmental, ecological, and human effects of climate change and melting ice in the Arctic Ocean for Arctic inhabitants and natural resources (Arctic Ocean Conference 2008). The declaration provided a foundation for responsible and shared management by the five coastal states and other users of the Arctic Ocean though the application of existing structures, precluding the need for a new international legal regime to govern activities in the sea.

Despite a rather inauspicious beginning, the Arctic Council emerged as a notable and influential player in intergovernmental affairs concerning the Arctic. Since its inception, the Council has assumed a leading role as a promoter of human rights and environmental protection. It has supported and produced scientific assessments of climate change and has raised awareness as an advocate for the concerns of Arctic Indigenous peoples. The May 2011 Ministerial Meeting in Nuuk, Greenland, marked a milestone achievement for the Council, inasmuch as member states adopted the first legally binding agreement on search and rescue in the Arctic. The AC has also served as a venue for the negotiation of major international initiatives, including the Agreement on Cooperation on Aeronautical and Maritime Search and Rescue in the Arctic, signed at Nuuk. Two years later, the Council facilitated the Agreement on Cooperation on Marine Oil Pollution Preparedness and Response in the Arctic, and the Agreement on Enhancing International Arctic Scientific Cooperation, both signed at the Ministerial meeting in Kiruna, Sweden, in May 2013. Most recently, the International Maritime Organization's International Code for Ships Operating in the Polar Waters (Polar Code) entered into force in January 2017. The Polar Code established international safety standards to protect ships, seafarers, and passengers operating in the harsh waters surrounding the two poles.

Limitations of the Arctic Council

Responsible for protecting and promoting environmental conservation and sustainable development in the Arctic, the Arctic Council's approach toward research and policymaking has received attention from political scientists, international relations scholars, geographers, and scientists. In 2012, Paula Kankaanpää and Oran Young published the results of a survey circulated among a selected group of 859 individuals either involved or familiar with the affairs of the Arctic Council. Hoping to "tap the knowledge of those in a position to articulate informed views regarding the activities of the council," Kankaanpää and Young created and distributed a questionnaire designed as a performance assessment tool. Although the respondents overwhelmingly suggested the Council had outperformed the initial expectations of most observers, the results of the survey indicated a need to increase its effectiveness for remaining relevant and influential in the international arena.

Focused on environmental protection and sustainable development, the circumscribed mandate of the Arctic Council excludes issues important to current and future governance in the Arctic. Policy matters concerning immigration, security, and trade—key issues important to local communities, national governments, and international organizations—are beyond the Council's scope. The existing structures of the AC also impose limitations. As an intergovernmental forum, the Council's framework does not allow for legally binding arrangements or regulations. The Council produces important research outputs about such issues as regional healthcare, pollution monitoring, and the maintenance and promotion of Indigenous cultures and languages, but converting research into action is challenging without the legal ability to enact or enforce policy. Moreover, the Council lacks an executive body. While the permanent secretariat provides administrative functions, funding is inconsistent and the rotating chairmanship creates discrepancies in the Council's primary agenda. Financial challenges create structural limitations, too. The eight Arctic states make decisions based on consensus, relying heavily on consultation with Permanent Participants. Yet PPs often do not have the financial capacity to attend ministerial meetings and engage the Council in a manner conducive to proper and effective multilateral governance. Investing in the autonomy of non-state participants will promote Indigenous participation and advance the Council's core priorities.

Creating a functional space for knowledge mobilization and policy action requires mechanisms for cultural translation and respect, as public policy scholar Jennifer Spence (2017) recently observed in an appraisal of the Arctic Council. Spence interviewed 45 people involved with the AC either directly or through an affiliated organization, interviews that revealed, in her assessment, the increasing pressures under which the Council operates. "In fact," Spence concluded, "organizational changes intended to 'strengthen' the AC are serving to weaken the systems that facilitate communication, translation, and mediation across the boundaries between Indigenous peoples, technical experts, and policy makers." The Arctic Council is a key forum for intergovernmental policymaking, but the strength of any policy system that marries domestic decision-making with international fora requires participants willing to question and challenge the status quo. As evolving issues affect the Arctic and its inhabitants, the structures underlying multilateral governance in the region will necessitate regular review and modification. Increasing the Arctic Council's transparency is important for advancing international responsibility in the Arctic, where state and non-state actors maintain an active and influential role in studying and addressing region-specific issues.

With the lure of economic gains motivating participation in the Arctic Council, questions linger about the role and intention of observer states and international institutions. Thirteen non-Arctic states are Observers to the AC: China, France, Germany, India, Italy, Japan, the Netherlands, Poland, Singapore, South Korea, Spain, Switzerland, and the United Kingdom. Because the Arctic contains approximately 90 billion barrels of oil and other untapped natural resources, international media outlets regularly claim that these countries want access to the economic trade potential of energy and resource extraction in the region (Chater 2016). Political scientist Andrew Chater (2015, 2016) recently suggested that economic opportunities in the Arctic might influence the priorities of the AC away from environmental protection. While climate change and resource competition create evolving challenges for multilateral governance in the region, the inclusive structures underlying the Council's framework provide the political backbone to resist economic pressure through intergovernmental cooperation.

The Way Forward

The changing circumstances involving climate change and resource competition shape the Arctic's international perception and determine the parameters within which the Arctic Council facilitates intergovernmental cooperation in the region. As the pre-eminent Arctic power, the Russian Federation considers sovereignty in the Circumpolar North a top domestic and foreign policy goal. The country has the longest Arctic coastline and the most populated Arctic region, supported by the largest ice-breaking fleet in the world and the biggest year-round ice-free port city in the Arctic, Murmansk (Charron et al. 2012). Yet contrary to sensationalist media reports that predict the use of military strength in the push for the Arctic's scarce resources, practical and peaceful collaboration underscores the existing international frameworks that govern independent and multilateral actions in the region. As Arctic policy expert Elizabeth Riddel-Dixon (2017, 23) argues, Arctic coastal states have demonstrated through collaboration the ability to delineate their extended continental shelves while respecting sovereign jurisdiction as established under international law. Dixon cites the Arctic Council as an exemplifier of constructive cooperation among Arctic countries, non-Arctic states, Indigenous groups, international institutions, and non-governmental organizations for addressing significant issues in the region.

Diplomatically, the Arctic Council is well positioned to avoid competitive politics and disruptive science. In May 2017, a group composed of two-dozen Canadian students and early-career scientists participated in a mock AC session held at McGill University in Montréal, Québec (Quinn 2018). The session occurred as part of *Negotiating the Arctic: A science Diplomacy Perspective*, a forum organized by the Montreal-based, non-profit group Science and Policy Exchange, which exposes students to policy issues involving science while fostering engagement and exchange among students and experts. Melody Brown Burkins of the John Sloan Dickey Center for International Understanding at Dartmouth College led the simulation, allowing participants to experience negotiation from the perspective of Member States, Permanent Participants, the six IPOs, and Observer states and organizations. "These are opportunities to get away from competitive politics and competitive science and I like the idea of moving towards more inclusiveness; not just between nations, but between things like science and traditional knowledge," Burkins said when interviewed about the mock session: "The Arctic Council is a place where all these values are clear" (Burkins quoted in Quinn 2018). The mock session simulated a Sustainable Working Group Meeting where participants made final recommendations for "Connecting UN Sustainable Development Goals to Arctic International Science Cooperation Activities," a project conceived and negotiated in advance of the next ministerial meeting. Aja Mason, a student participant from Whitehorse who represented Greenland (Denmark) in the exercise, gained a new appreciation for the policymaking process: "While I cannot speak on behalf of all northerners, or all Yukoners, I nonetheless felt compelled to draw attention to the ways in which the processes of self-governance and self-determination of Indigenous peoples in the North is so rare in comparison to down south," Mason said: "As a non-Indigenous researcher, I want to participate in and promote research that doesn't infringe on these essential capacities" (Mason quoted in Quinn 2018). Indeed, participants in the mock session concluded that fostering inclusivity and support among cooperative states and organizations creates opportunities for merging scientific epistemologies and traditional knowledge systems.

While daunting challenges face the Arctic Council moving forward, the established structures will continue to create opportunities for increased multilateral cooperation. In order to identify effective strategies for protecting and promoting the current and future interests of Arctic peoples, lands, waters, and resources, the Council should remain flexible and adopt principles that encourage active participation among state and non-state parties. This calls for a twofold approach where state- and community-driven planning precedes resource mobilization and policy action. At the same time, the Council needs to embrace outside organizations and fora for achieving positive change in the Arctic. The issues facing the region and its inhabitants are too great for a single intergovernmental body. Managing and responding to the effects of climate change and resource competition requires layers of transparency and collaboration. Adopting an inclusive outlook will strengthen the Arctic Council's capacity to champion and improve intergovernmental cooperation in the areas of environmental protection, sustainable development, and human rights.

Conclusion

During a 23-year history, the Arctic Council has demonstrated that multilateralism is an effective tool for achieving international policy directives. The Council's overall effectiveness exceeded the relatively ambiguous expectations established during its inception. As the principle forum devoted to facilitating and promoting international cooperation in the Arctic, particularly in the areas of environmental protection and sustainable development, the Council has produced valuable scientific studies that have inspired the development of legally binding agreements with regulations safeguarding the Arctic and its inhabitants. Multilateral cooperation vis-à-vis the Arctic Council has also led to large science-based initiatives for the mitigation and elimination of natural and artificial pollutants. Concurrently, the evolving complexities of climate change and resource competition in the Circumpolar North create a host of new challenges and opportunities for multilateral governance and international cooperation. The importance of the Arctic Council will increase in the coming decades, calling for a sustained investment on the part of Member States, Permanent Participants, and Observers.

As the premier forum for intergovernmental cooperation in the Arctic, the Arctic Council has received criticism from skeptics who question the effectiveness of multilateral governing structures. Despite the successful facilitation, development, and implementation of effective international policies for the Arctic, the Council's most recent years represent a period of relative stagnancy. Recent multilateral cooperation resulted in the signing of the Agreement on Enhancing International Arctic Scientific Cooperation, but neither that agreement nor the Council's Task Force on Improved Connectivity in the Arctic has resulted in tangible actions that protect Arctic peoples, lands, and resources. Furthermore, although the Council is an inclusive body comprised of governmental and non-governmental actors, domestic interests that benefit the economic elite remain key motivators for international policy choices about the Arctic. In a global political system that functions according to the needs and desires of powerful governments, the Arctic Council should continue to secure and allocate resources to protect and promote the interests of the ordinary people who live and work in the region.

Effective cooperation and knowledge exchange are prerequisites for addressing current and future challenges in the Arctic. Under the Ilulissat Declaration, the five coastal states have a stewardship role in protecting the ecosystem of the Arctic Ocean. Shipping disasters and subsequent pollution of the marine environment disturb the ecological balance of the Arctic Ocean and harm the livelihoods of local inhabitants and Indigenous communities. The increased use of Arctic waters for tourism, shipping, scientific research, and resource development will increase the risk of accidents and the need for concerted efforts to protect and promote the interests of Arctic residents. As temperatures continue to rise and ice continues to melt, new developments caused by climate change and human activity will also increase the political and economic salience of the region. In turn, the increased prominence of the Arctic will create new governance issues at the local, regional, national, and international levels. With the long-term implications of climate change and resource development still to be determined in the Arctic, the region's inhabitants require an active and effective platform for addressing current and future challenges to environmental, ecological, and cultural sustainability.

As a platform created with an inclusive framework, the Arctic Council ultimately serves as a model for future governance bodies and international scientific cooperation institutions. While the Council has achieved considerable success in identifying and responding to issues affecting the Arctic, the ongoing changes in the region necessitate a flexible framework for maintaining and improving the Council's effectiveness moving forward. This suggests there is a continuing need to assess and adjust the internal mechanisms of the Council to address the existing and future challenges of multilateral governance in the Arctic. Communication is the key. Enhancing the Council's transparency will generate feedback useful for improving the goals, efforts, and outputs of both the workings groups and task forces.

The Arctic Council is an effective intergovernmental forum, but new issues arising in the Arctic call for an increasingly diverse approach to knowledge acquisition and information exchange. Tied to formal and informal governance structures, international policymaking for the Arctic must involve considerations pertinent to the many and varied social, cultural, and environmental needs of the local inhabitants and Indigenous communities located in and near the Circumpolar North. As the pre-eminent governance body for studying, addressing, and overcoming the issues that face a changing Arctic and its inhabitants, the Arctic Council needs to assume a leading role in protecting and promoting the needs of state and non-state actors. This is a difficult task, but advancing multilateral governance will remain the first step to influencing positive change in a region of growing significance.

References

Arctic Council. 1998. *The Iqaluit Declaration*. Tromsø, Norway: Arctic Council Secretariat.
———. 2009. *Arctic Marine Shipping Assessment 2009 Report*.
Arctic Ocean Conference. 2008. *2008 Ilulissat Declaration: Adopted in Ilulissat, Greenland on 28 May 2008*. Accessed June 10, 2018. https://cil.nus.edu.sg/wp-content/uploads/formidable/18/2008-Ilulissat-Declaration.pdf.
CAFF International Secretariat. 2010. *Arctic Biodiversity Trends 2010: Selected Indicators of Change*. Akureyri, Iceland.
Charron, A., J. Plouffe, and S. Roussel. 2012. The Russian Arctic Hegemon: Foreign Policy Implications for Canada. *Canadian Foreign Policy Journal* 18 (1): 38–50.
Chater, A. 2015. Explaining the Evolution of the Arctic Council. PhD diss., Western University.
———. 2016. Explaining Non-Arctic States in the Arctic Council. *Strategic Analysis* 40 (3): 173–184.
Ebinger, C.K., and E. Zambetakis. 2009. The Geopolitics of Arctic Melt. *International Affairs* 85 (6): 1215–1232.
English, J. 2013. *Ice and Water: Politics, Peoples, and the Arctic Council*. Toronto: Allen Lane.
Gorbachev, M. 1987. The Speech in Murmansk at the ceremonial meeting on the occasion of the presentation of the Order of Lenin and the Gold Star Medal to the city of Murmansk, October 1, 1987. Moscow: Novosti Press Agency, 23–31. Accessed June 15, 2018. https://www.barentsinfo.fi/docs/Gorbachev_speech.pdf.
Graczyk, P. 2011. Observers in the Arctic Council—Evolution and Prospects. *The Yearbook of Polar Law* 3 (1): 575–633.
Hassol, S.J. 2004. *Arctic Climate Impact Assessment, Impacts of a Warming Arctic*. Cambridge, UK: Cambridge University Press.
Hossain, K., and H. Maruyama. 2016. Japan's Admission to the Arctic Council and Commitment to the Rights of Its Indigenous Ainu People. *The Polar Journal* 6 (1): 169–187.
Huntington, H.P., and Arctic Monitoring and Assessment Programme. 2007. *Arctic Oil and Gas 2007*. Oslo, Norway.
Ikeshima, T. 2016. Japan's Role as an Asian Observer State within and Outside the Arctic Council's Framework. *Polar Science* 10 (3): 458–462.
Kankaanpää, P., and O.R. Young. 2012. The Effectiveness of the Arctic Council. *Polar Research* 31 (1): 1–14.
Koivurova, T. 2010. Limits and Possibilities of the Arctic Council in a Rapidly Changing Scene of Arctic Governance. *Polar Record* 46 (2): 146–156.

———. 2012. The Arctic Council: A Testing Ground for New International Environmental Governance. *Brown Journal of World Affairs* 19 (1): 131–144.

Koivurova, T., and L. Heinämäki. 2006. The Participation of Indigenous Peoples in International Norm-Making in the Arctic. *Polar Record* 42 (2): 101–109.

Manicom, J., and P.W. Lackenbauer. 2013. *East Asian States, the Arctic Council and International Relations in the Arctic.* Waterloo, ON: Centre for International Governance Innovation.

Poto, M. 2016. Participatory Rights of Indigenous Peoples: The Virtuous Example of the Arctic Region. *Environmental Law and Management* 28 (2): 81–89.

Poto, M.P., and L. Fornabaio. 2017. Participation as the Essence of Good Governance: Some General Reflections and a Case Study on the Arctic Council. *Arctic Review on Law and Politics* 8: 139–159.

Quinn, E. 2018. Canadian Students Flex Their Diplomatic Chops in Arctic Council Simulation. Eye on the Arctic. Accessed June 16, 2018. http://www.rcinet.ca/eye-on-the-arctic/2018/05/17/canadian-students-flex-their-diplomatic-chops-in-arctic-council-simulation-science-environment/.

Reagan R. 1983. United States Arctic Policy. National Security Decision Directive, 90. Washington, DC, April 14. Accessed June 15, 2018. https://fas.org/irp/offdocs/nsdd/nsdd-090.htm.

Riddel-Dixon, E. 2017. *Breaking the Ice: Canada, Sovereignty, and the Arctic Extended Continental Shelf.* Toronto: Dundurn.

Spence, J. 2013. Strengthening the Arctic Council: Insights from the Architecture Behind Canadian Participation. *The Northern Review* 37: 37–56.

———. 2017. Is a Melting Arctic Making the Arctic Council Too Cool? Exploring the Limits to the Effectives of a Boundary Organization. *Review of Policy Research* 34 (6): 790–811.

Young, O.R. 1985–1986. The Age of the Arctic. *Foreign Policy* 61: 160–179.

———. 1998. *Creating Regimes: Arctic Accords and International Governance.* London: Cornell University Press.

———. 2005. Governing the Arctic: From Cold War Theater to Mosaic of Cooperation. *Global Governance* 11 (1): 9–15.

PART VI

Arctic Security

The Evolving North American Arctic Security Context: Can Security Be Traditional?

Heather N. Nicol

Introduction

New ways of seeing security and threat in the North American North have developed in recent years, fueled in part by developments elsewhere in the world. These have encouraged innovative exploration of the concept. But more generally, the definition of security has developed over a lengthy period, particularly with regard to how it understands the role of state and individuals (Williams 1998), the agency of sub-national and non-state actors, and the nature of sovereignty itself (Nicol 2016). It should come as little surprise, therefore, that today the rapid nature of environmental change now unfolding in parts of the world has influenced how security is understood. There is general understanding that climate change will pose many new and widespread existential threats to the state, but also the acknowledgement that the most critical new security threats will not always be military in nature. Additionally, globalization has also transformed the practice of security and made new forms of surveillance and territorial control possible in the absence of the physical presence of military or state actors, and in doing so, has implicated non-civilian actors and agencies as security operators.

Nonetheless, there is a tendency for the popular media and the public to see the Arctic in conventional ways and to assert that it should be protected by conventional security practices because of a host of imminent geopolitical and military threats—such as Russian submarines or aggressive Chinese economic agendas. The alarm has certainly been raised by prominent security

H. N. Nicol (✉)
Trent University, Peterborough, ON, Canada
e-mail: heathernicol@trentu.ca

© The Author(s) 2020
K. S. Coates, C. Holroyd (eds.), *The Palgrave Handbook of Arctic Policy and Politics*, https://doi.org/10.1007/978-3-030-20557-7_28

455

advocates that the Arctic faces imminent peril. On the other hand, many national and regional security actors argue that this is not the case. Instead, it is the broader problem of climate change, unstable weather events, and other potential environmental that constitutes the immediate problem (Alkire 2004; Nicol and Heininen 2013). There is now recognition by security actors that rapid environmental change is more likely to challenge the health and viability of northern communities and economies than any other threat. In assessing their claim, this essay traces the story in the North American North. It explores how security actors are now engaged in non-traditional and even civilian-side security work. It suggests that the practice of security in the North American North has its own history and characteristics, and the way in which security threats are now understood are specific to the region and its evolving security narratives.

RETHINKING SECURITY

For Williams (1998, 438), the origins of security are deep. It has its roots in a historical shift to "modernity": "Seen historically, the development of an 'objective' realm of personal security divorced from the question of one's personal or collective identity, and enforced through the status of citizenship within the state, is one of the central transformations of modernity." That said, this is not the most recent iteration of security. Rather than maintaining its exclusive focus on threats that challenge the existence of the state, with an emphasis on military force, security has now begun to incorporate the central concerns of what has been viewed as "soft" or non-traditional security: "The security concept is therefore being revised and broadened to include sectors such as economic, environmental and societal developments" (Williams 1998, 438). Swanström (2010) observes that since the end of the Cold War there has been growing awareness on the need to widen the concept of security and distinguish between "hard/traditional" and "soft/non-traditional" security threats. But equally important, "there has also been a failure to understand how traditional and non-traditional security threats overlap, and in many ways, reinforce each other" (ibid., 35–36). What was once thought to be an unresolvable divide between traditional objective understandings of state security, and constructed and subjective understandings of identity and individual security is no longer so: "Security is no longer viewed as referring naturally or unproblematically to the military security of the state. Rather, the goal becomes to understand how different 'referent objects'—from the economy, to the environment, to military relations—can become subject to the processes of 'securitization'" (Williams 1998, 438).

The point made in this essay is that it is not just the expansion of definition of security to include other types of security than traditional or military security that is important, but also the way in which the practice of traditional security itself has itself changed as a result. Traditional security now has new and nonconventional people, places, and situations to secure. The shift began in the late twentieth century when both the politics of the post-Cold War era and the

Bruntland Commission Report on *Our Common Future* (1987) encouraged recognition that environments and peoples were in need of protection to ensure sustainability. Subsequently, the *UNHDP Development Report* (1994), the *United Nations Framework Convention on Climate Change* (1992), and more recently 2016 *United Nations Framework Convention on Climate Change* (UN 2016b) also contributed to a sense that environments and populations were an issue for security. The broadening of the security dialogue from its traditional preoccupation with national security prompted a new way of thinking about security itself—in light of the need to incorporate environmental challenges supplement more traditional security narratives. It was not just that the existence of multiple forms of security were acknowledged in this process, but more importantly that there was a reassessment of the nature of the relationship between traditional military or national security and other forms or referents of security. Indeed, a florescence of security types, from human, to environmental, to foods security are now acknowledged. However, the point is that traditional security narratives in the North are themselves being broadened to include non-traditional goals and objectives (Hoogensen et al. 2009). Moreover, there is recognition by security actors across the North American North that the community of security operators is widening to include civil agencies as well as state actors: for example, Northern communities themselves are now seen as both security actors and entities in need of security. The interplay between strategic and environmental concerns, coupled with the geography of community settlement, governance, and services, has contributed to a reassessment of the role, meaning, and nature of traditional security in the circumpolar world. In other words, security has not just broadened to acknowledge that non-military security is one type of security among many, but that the object of military security has itself been transformed with reference to the nature of new non-traditional security threats. This approach has found its niche in the circumpolar region and, in particular, in the North American Arctic (Heininen 2004; Keskitalo 2004).

That said, it is a common complaint by North American northern security actors that their southern governments still do not understand the non-conventional nature of the current security threats in the North and consequently, do not resource accordingly. This chapter examines how current developments in the North American Arctic reflect the general development of a broader definition of traditional security within the region. It documents the fundamental shift in how traditional security is understood and practiced as reflected in security documents and implementation strategies. In doing so, it builds on Everett (2018), who argues, for example, that contemporary circumpolar governance, in general, seems less concerned with the realist politics that are dominant in much of the South, leading to a greater emphasis on socioecological issues that affect daily life rather than on the potential of hard security. In particular, Everett notes that this characteristic socioecological approach actively seeks out international cooperation rather than competition, but that circumpolar governance cannot be understood in a vacuum. There are, for example, a breadth of issues faced by individual regions within the circumpolar North and a general co-incidence of approaches because of a host of international fora and agreements. This broad trend is also true of the North American Arctic where despite claims to the continuing saliency of competitive international geopolitics, a more environmentally focused and human security dialogue has also emerged. The latter, consistent with a broader understanding of human security, directs the main focus of security activity away from the protection of state borders and assets to the protection of individuals who reside within (Hoogensen et al. 2009).

This chapter begins with a brief discussion of how Arctic traditional security goals have been redefined within national documents and strategies, and goes on to discuss how this new understanding and the subsequent incorporation of security goals are reflected in the dialogue of regional security operators and agencies in the North American Arctic. In doing so, the security dialogue and security operations within the Canadian and American North are examined.

Transforming Security Narratives in the North American North

Although melting ice and permafrost conditions are still quite variable, political actors and agencies speculate about the territorial implications of changing environments and act accordingly. Open water, increased transportation activity, and greater potential for environmental disaster are all new and looming threats for Canadian and American military agencies and operators. Consistent with broader thinking about environmental sustainability, but also as a response to regional trends, since the 1980s, if not before then, there has been recognition that the nature of security challenges in the North are increasingly broad and non-conventional in nature—for example, growing concern over climate change and long-range pollution led to dialogue about the need for environmental cooperation. Recognition of a common environmental threat unified the approach of the eight Arctic states, and in 1996 led to the development of the Arctic Council, an intergovernmental forum which shapes state policy and practice within the circumpolar region (Keskitalo 2004; Lackenbauer et al. 2017). But the Arctic Council was never a forum for discussion of traditional security. Health, pollution, conservation, and similar challenges were instead the issues under its purview. The way in which this environmental and broader human security agenda was structured under the Arctic Council, encouraged the formation of a policy and governance divide. Hard or traditional military security issues remained off-limits for discussion. This feature of the Arctic Council has had a stabilizing influence on regional governance. Rather than using a rhetoric loaded with tension and confrontation, as was common throughout the Cold War era, the new northern discourses stressed regional cooperation, human security, and sustainable development. Nonetheless, until very recently, they have discouraged thinking that large-scale humanitarian events, including oil spills and natural disasters, should also be considered as of interest to traditional security providers, or even as traditional security events.

Since then, however, the way in which security is understood by security agencies and operators has evolved. Rather than seeing traditional and environmental security as two different types of security—as did the Arctic Council at its founding—Arctic states and security agencies are now seeing environmental and military security as two sides of the same coin, and traditional and environmental security dialogues are increasingly co-constitutive. This is clear, for example, in the nature of national statements emanating from North American governments during the first decade of the twenty-first century. National strategy documents developed by both Canada and the United States between 1997 and 2015 show an increasing propensity to use traditional military and environmental security concepts in tandem when discussing the North American Arctic and offer some insights on how environmental issues found their way security policy and implementation documents at the regional level.

The Obama Administration was the first US government to introduce environmental concerns in this way. In 2009, for example, the *US Arctic Policy Directive (NSPD 66/HSPD 25)* described US national security and homeland security interests in the Arctic as "international governance; the extended continental shelf and boundary issues; promotion of international scientific cooperation; maritime transportation; economic issues, including energy; and environmental protection and conservation of natural resources" (CRS 2018). In May of the following year, the Obama Administration released a national security strategy document that reinforced the 2009 Directive, again referencing goals of human security as strategic in nature: "The United States is an Arctic Nation with broad and fundamental interests in the Arctic region, where we seek to meet our national security needs, protect the environment, responsibly manage resources, account for indigenous communities, support scientific research, and strengthen international cooperation on a wide range of issues" (ibid.). By 2013, the *US National Strategy for the Arctic* clearly referenced both communities and environments as instrumental to broader security goals. It noted that "Climate change is already affecting the entire global population, and Alaska residents are experiencing the impacts in the Arctic. To ensure a cohesive Federal approach, implementation activities must be aligned with the Executive Order on Preparing the United States for the Impacts of Climate Change while executing the Strategy" (Ibid., 10).

There was clearly pressure in Washington to develop both a national Arctic strategy and an implementation plan for this strategy. Partly this was because the United States was poised to assume control of the Arctic Council Chairmanship in 2015. The State Department believed that both a general strategy and implementation plan needed to be in place by the time that the US Chairmanship began which emphasized climate and environment as well as science-based policy making, and ocean stewardship. It is important to underscore this linked concern. Hossain and Barla (2017, 2), for example, suggest that the 2013 *US National Strategy for the Arctic* attempted to "steer the Arctic region in the right direction" to assume its intergovernmental leadership role. While this meant that the US Chairmanship agenda would ultimately be based upon these national Arctic strategic goals and priorities for the Arctic region, it also meant that there was a concerted effort to broaden the traditional strategic goals embedded in national security narratives, to include compelling environmental concerns. Key to this approach was a recognition that US security in the Arctic encompassed a broad spectrum of efforts ranging from supporting safe commercial and scientific operations to activities more directly linked to national defense. Key strategic platforms included the intention to pursue what it called "Responsible Arctic Region Stewardship" and to "continue to protect the Arctic environment and conserve its resources; establish and institutionalize an integrated Arctic management framework; chart the Arctic region; and employ scientific research and traditional knowledge to increase understanding of the Arctic" (United States 2013). There was also the intent to pursue innovative arrangements, meaning fostering partnerships with the State of Alaska, Arctic states, and other international partners (United States 2013).

These documents specifically coupled climate change with American national security interests in the Arctic in ways that highlighted the importance of environment. The *Arctic Strategy* document note, for example, that: "There may be potentially profound environmental consequences of continued ocean warming and Arctic ice melt…and the consequent increase in pollution as emissions of black carbon and other substances from fossil fuel consumption—could have unintended consequences on climate trends, fragile ecosystems and Arctic communities" (United States 2013). US national security interests, now broadened to include a clear environmental agenda clearly aligned the US vision with the work of the Arctic Council in ways that not been previously possible.

In addition to the 2013 National Strategy, was the subsequent *US National Strategy for the Arctic Implementation Plan* (United States 2014). The latter identified the process for transforming the objectives of the Arctic national strategy into concrete initiatives to be pursued by the US government within the Arctic region. Included among these action areas were initiatives for development of communications infrastructure, clean energy, maritime domain awareness and security, hazardous, material spill prevention, containment and response, ocean stewardship, black carbon reduction, scientific cooperation, and health.

The Arctic Council Chairmanship process, therefore, was instrumental in contextualizing US security interests with reference to broader security goals and practices. Weaving together the themes of security, climate change, and strategic interests, it defined US national priorities in the Arctic as including "national security, sovereign rights and responsibilities, maritime safety, environmental stewardship, scientific research, management of natural resources and preservation of indigenous culture and language" (Hoag 2016). Overall, if these documents are reflective of more general thinking, it would be safe to assume that by the end of the Obama Administration, there is clear evidence of the beginnings of what might be considered as the "securitization" of environmental concerns.

The incorporation of a broad environmental security dimension within US strategic plans for national security is similarly reflected with Canadian strategic documents between 1997 and 2015, but arguably in different ways: "In 1997, a Canadian parliamentary committee recommended that the country should focus on international Arctic cooperation through multilateral governance to address pressing "human security" and environmental challenges in the region. Committee chairman Bill Graham reported that environmentally sustainable human development was "the long-term foundation for assuring circumpolar security, with priority being given to the well-being of Arctic peoples and to safeguarding northern habitants from intrusions which have impinged aggressively on them" (Lackenbauer and Dean 2016, xxviiii). *The Northern Dimension of Canada's Foreign Policy* released in 2000, according to Lackenbauer and Dean, "revealed how environmental and social challenges now predominated as concerns."

The idea that environmental and social challenges subsequently factor heavily in Canada's Arcic security policy is open to debate. Lackenbauer and Dean (2016, xxx) go on to suggest that: "By the early 2000s, the rising tide of scientific evidence about the pace and impact of global warming in the Arctic led some Canadian academic commentators to push for a more proactive Arctic strategy. This strategy anticipated that climate change would stimulate a "sovereignty crisis," with renewed challenges to the legal status of the waters of the Northwest Passage for international transit shipping." In other words, the threat of climate change and indeed, environmental instability in general, was used as a rationale for enhancing Canada's traditional security's concern with territorial defence and sovereignty rather than seeing environment as a object to be secured.

This hierarchical relationship between human, environment and traditional security reappeared elsewhere in the same document. It called for "a framework to promote the extension of Canadian interests and values," and to "renew the government's commitment to cooperation with our own northern peoples and with our circumpolar neighbors to address shared issues and responsibilities" (Heininen and Nicol 2007, 148). Indeed, the Canadian government asserted that in promoting its Arctic foreign policy, it was continuing Canada's "long-standing foreign policy tradition" (ibid., 148). It argued that in doing so, Canada's foreign policy had become "the gateway for the incorporation of new ideas about the relevancy of human security in context of environment and civil society" (ibid., 148–149). Equally important, however, the Canadian Government asserted through its northern dimension document that the future challenges faced within the Canadian Arctic would most likely "take the shape of transboundary environmental threats—persistent organic pollutants, climate change, nuclear waste—that are having dangerously increasing impacts on the health and vitality of human beings, northern lands, waters and animal life" (ibid., 149). Here Canada was simply rearticulating the rationale that led to the establishment of the Arctic Council cooperation in 1996. Indeed, it would be fair to say that between 1997 and 2000, the Canadian

Government's primary focus on Arctic policy remained primarily upon sovereignty and hard security considerations—a "Canada first policy" developed. The emphasis was not so much environmental security as environment threat as a foil to re-enforce traditional security concerns.

Indeed, responding to this agenda, and urging a more broadly based and enlightened engagement, in 2009, Arctic expert Franklyn Griffiths argued that a Canadian Arctic strategy should exercise "due care in the exploitation and enjoyment of a shared natural environment." He suggested that its concern should be with "stewardship and sovereignty", stewardship defined as "locally informed governance that not only polices but also shows respect and care for the natural environment and living things in it." For Griffiths, stewardship enhanced national sovereignty "in the conditions of natural and human interdependence that prevail in the Arctic" (Griffiths 2009, ii). The twin pillars of stewardship and sovereignty that Griffiths identified over a decade ago continue to inform Canadian Arctic security policy today. They suggested that environment was not just a trigger for hard security responses, but a security relationship worth protecting for its own sake.

Whether directly related to Griffiths's pleas, or more broadly reflective of changing domestic and external agendas, over the next two years the Canadian Government did indeed broaden its Arctic security narrative to include community well-being, development, and environmental stewardship. Canada's 2010 *Statement on Canada's Arctic Foreign Policy*, for example, identified four pillars or themes that were to be protected and nurtured: exercising Arctic sovereignty, promoting social and economic development, protecting environmental heritage, and improving and devolving Northern governance (http://www.northernstrategy.gc.ca/index-eng.asp). Meanwhile, Canada's 2009 *Northern Strategy: Our North, Our Heritage, Our Future* announced Canada's intent to protect and manage the unique and fragile ecosystems and wildlife of the Arctic. The strategy contained language which connected and underscored the joint needs of defense and environmental security:

> Canada is taking concrete measures to protect our Arctic waters by introducing new ballast water control regulations that will reduce the risk of vessels releasing harmful aquatic species and pathogens into our waters. We also amended the Arctic Waters Pollution Prevention Act to extend the application of the Act from 100 to 200 nautical miles from our coastline, the full extent of our exclusive economic zone as recognized under the United Nations Convention on the Law of the Sea. This amendment gives us pollution prevention enforcement jurisdiction over an additional half million square kilometres of our waters. In addition, we are establishing new regulations under the Canada Shipping Act, 2001 to require all vessels entering Canadian Arctic waters to report to the Canadian Coast Guard's NOrDreG reporting system. And finally, Canada is working with Northern communities and governments to ensure that its search and rescue capacity meets the needs of an ever-changing North. (Government of Canada 2009, 11–12)

Indeed, the flurry of Arctic policy documents tabled between 2009 and 2010 suggests that the focus on environmental protection was much more prevalent than earlier in the decade. These documents described important new initiatives such as impact monitoring programs, scientific research to support regulatory decision-making, remediation of contaminated sites, and the creation of new terrestrial and marine protected areas (Lackenbauer and Dean 2016, xlii). Climate change and its impact on northern defence activities were now identified as key concerns. There was a subtle shift occurring in the way in which environment became something worth protecting and not just a rationale for defence. Reading these documents together, is reasonable to conclude that, by 2009, the seeds were planted for understanding the important relationship between defense, environment and human security more broadly. Indeed, more specifically, Everett (2018, 13) argues that Canada's 2010 foreign policy strategy was of equal important to the 2009 strategy document in that, "when read together, the two documents collectively prioritized the same four areas: social and economic development; Northern environmental heritage; and devolving northern governance". Overall, these two documents set the stage for new understandings of Arctic regional security to develop in North America during the second decade of the 21st century. In exploring beyond the era of the Harper government, for example, it is apparent that "the Trudeau-Obama [2015 Arctic Leaders] statement focused even more upon on bilateral cooperation with the US and international cooperation through the Arctic Council for environmental issues. The two leaders found that regional threats included increased maritime traffic, environmental change (loss of sea ice), and the potential for oil spills" (Everett 2018, 13). It was at this moment that the two separate environment and human security agendas, developed through independent political processes began to look much more alike.

REGIONAL SECURITY

The reworking of Arctic security concerns in North America, at the national level has, as we have seen, been an ongoing project. Today, North American governments today are in agreement that Arctic security is diverse and complex in nature. There is now recognition of the traditional economy of the region, its Indigenous peoples, and its fragile environments. While historically, traditional security narratives have generally stressed that the state is the only legitimate security actor within the region, this understanding in the North American Arctic is now challenged, at least in part, by the shifting concerns of Canadian and American strategic, defence and policy documents. Recognition of the deep connections between traditional and non-traditional security within the North is also reflected in new thinking about the delivery of security in the circumpolar world that now sees fragile communities, environments and economies as referent objects of security. It also sees the role of non-military actors and agencies differently. Climate change has raised the specter of a new type of security challenge with not just environmental, but also social, economic, and political consequences. In general, the former (environmental change) leads to a roster of new challenges for the latter (social, economic, and political systems). These, in turn, are manifested at the level of the local community.

For local communities and actors, the effects of climate change are already direct and readily apparent. Changes to both mean annual temperature and overall changes to normative weather patterns create a host of new risks ranging from loss of food species and food insecurity, to coastal slumping and loss of infrastructure, housing, and cultural assets. Similarly, climate change changes sea and freshwater ice conditions, affecting not just their importance as species' habitats, but their potential to change the accessibility of resources within the region. Similarly, on a larger scale, the effects of warming temperatures and decreasing albedo effects lead to the opening of new ice-free territories and the potential for new actors and agencies to operate within the region. Rapid environmental transformation of circumpolar environments therefore creates more general uncertainty about the future, and growing recognition, that security now means not just national security, but also includes food security, environmental security, protection of infrastructure, military security, and cultural security or human security more generally (Hoogensen et al. 2009; Nicol and Heininen 2013). To be effective, therefore, traditional security must also recognize and incorporate security in all its dimensions.

Two examples are presented to indicate the degree to which this thinking has already infused agencies and actors involved in regional security operations within the North American North in the implementation of defence activities. The first is the way in which the US government, its Department of Homeland Security, and the US Coast Guard have sought to examine and refine maritime security through a more concerted domain awareness approach. The second is the way in which Canada has used a "Whole of Government" approach to Arctic defence exercises. We begin with the discussion of the foundations for current maritime security operations in the US by referencing its origins in the 2004 *National Strategy for Maritime Security (MDA)*, one of eight plans implemented by the 2004 by the *National Security Presidential Directive-41/ Homeland Security Presidential Directive-13* (see https://www.dhs.gov/ national-plan-achieve-maritime-domain-awareness). In these directives, we see how US policy makers saw environment in relation to security in the early 21st century. According to this document, domain awareness meant creating an effective understanding of the maritime domain and any conditions or situations that might impact upon the security, safety, economy or environment. In many ways, their viewpoint was very similar to that of their Canadian counterparts. Environmental threats were a little more than factors that might affect hard security operations. While the MDA sought to define and implement domain awareness as a security strategy for US maritime regions, the role of environment was given minor mention in the document, and mainly in relation to its potential to affect broader security operations. It established national priorities for achieving maritime domain awareness and, in doing so, drew upon insights and expertise of a variety of federal agencies and departments. The MDA makes it clear that environmental threat is of real importance to domain awareness, and is, in fact, one of four pillars of threat (others including including national security, terrorism, and criminality). Moreover, all of the maritime plans which evolved from the aforementioned *Directive 13* make it clear that security threats are to be understood and in time resolved through collaboration with civilian and non-civilian partners. Indeed, Kraska (2016) reminds us that while there are significant traditional military interests within the Arctic maritime region, there are also broad civil and military interests for US security within the North American Arctic. Environmental factors will be among the most important of these to play out in the future. In particular, he notes that US Arctic policy has expressed an interest more generally in establishing risk-based capability in order to counter hazards in the Arctic environment, this including pollution prevention, and improved search and rescue capabilities. Safe navigation standards, accurate and timely environmental, and navigation information will also be crucial.

We draw upon this history of domain awareness to make the case that the US Coast Guard has, arguably, been the US defence agency most affected by the new awareness of Arctic security needs. An unprecedented emphasis on domain awareness in the Coast Guard has led to the establishment of specific programs and agencies. The Arctic Domain Awareness Center (ADAC) in Anchorage Alaska, for example, reports to Homeland Security, but its main agency of impact is the US Coast Guard. The Coast Guard's interest in domain awareness is in evolving its capacity and knowledge for new and non-conventional threats within the Arctic region—including search and rescue, humanitarian assistance, disaster response, and related security, suggesting that domain awareness is key to an evolving Coast Guard agenda: There is a "potential need for more law enforcement, search-and-rescue activities, and patrol of illicit activities in the region. The Coast Guard's Arctic fleet will need more functionality than it has had in in the past...such as improved communications systems, more aerial support from helicopters and drones, and more oil cleanup kits" (https://www.arctictoday.com/outgoing-commandant-says-arctic-become-top-priority-us-coast-guard/). Similarly, for ADAC, the security challenges are many, and diverse, if only because: "The changing physical environmental factors, including reduced ice and thawing permafrost, diminished shore-fast ice, increased storm frequency and severity coupled with increased human activity...equates to increased demands for both deliberate and emergency response by Canadian and U.S. communities of planners as well as first responders" (ADAC 2018, http://arcticdomainawarenesscenter.org/css/images/newdesign/pdf/reports/ADAC-AMESW%202018_p01_Report_v1.pdf). For these reasons, capacity to identify situational needs and to develop new research on means by which domain awareness can be enhanced and applied by regional operators. In light of such significant changes to the nature of the security threat and responding actor, ADAC suggests that:

While a relatively new term, the need to better understand and factor "Arctic environmental security" in the context of protecting national interests, advancing regional cooperation, addressing civil support to citizens, ensuring human security, providing defense, and law enforcement is timely and necessary. In fact, in terms of scale and intensity, the relatively rapid advance of a warming Arctic is seeing a marked increase in severity of weather and weather-related impacts (such as coastal storm surge and quickly accelerating erosion of soils), resulting in negative impacts to infrastructure and creating an increasingly complex physical maritime environment. Further, increases in ocean acidification of highly productive fishing regions, such as the Bering Sea, puts the harvest of much needed fish-related proteins at increasing risk. (ADAC 2018 http://arcticdomainawareness-center.org/css/images/newdesign/pdf/reports/ADAC-AMESW%202018_p01_Report_v1.pdf)

In other words, security in the North American Arctic is best maintained through high-quality domain awareness research, technologies and capacities that record and observe changes to Arctic environments and allow regional security agencies to act upon them.

A second example of how the expansion of security has influenced the implementation of traditional security and military responsibility comes from the Canadian North. The Canadian Government and the Canadian Armed Forces (CAF) have redesigned security exercises in Canada's high North to better provide comprehensive attention to a host of new security challenges and scenarios using a "Whole of Government" (WoG) approach (Lackenbauer and Nicol 2017). As we have already seen, the Government of Canada elevated the North to one of its top policy priorities under Prime Minister Harper's tenure. Canada's 2009 *Northern Strategy* for example, committed the federal government to helping the region achieve its full potential "within a strong and prosperous Canada" while "realizing this vision through a wide array of supporting objectives requires strong relationships and partnerships between federal departments and agencies, territorial governments, Aboriginal governments and organizations, Northerners, and other stakeholders." To do this, it adopted a collaborative approach which infused more general security practices and operations within the region.

Operation Nanook is an annual Canadian Armed Forces led exercise that takes place within the Canadian Arctic. Its mandate is embedded within the broader strategic mandate established by *Canada's Northern Strategy*, which, as we have seen, includes both Arctic sovereignty and environmental protection as priority areas. Manson (2017) reminds us that argues that *Nanook* is the centerpiece of several sovereignty operations conducted annually by the Canadian Armed Forces in Canada's North, and is, as well, the primary WoG operation for the Arctic region. As such, it involves both practicing response to complicated scenarios, which include search and rescue (SAR), marine disaster, and community safety, as well as more conventional security exercises designed to build capacity for defense of sovereignty and territory. *Operation Nanook 2014* was designed to meet the broad WoG parameters of Canada's Northern defense agenda and focus upon exercising to the rescue of commercial and civilian interests in Canada's High North. According to Manson (2017, 427–428), *Operation Nanook 2014* involved:

> all of the environmental services (the Royal Canadian Navy, the Canadian Army, and the Royal Canadian Air Force), all levels of the Canadian government, and sometimes international partners from the Arctic Council and/or the North Atlantic Treaty Organization (NATO). It also included foreign partners from Denmark, Finland, Iceland, Sweden, Norway, the United Kingdom, and the US. A Search and Rescue (SAR) scenario brought CAF, the Canadian Coast Guard (CCG), the US, and Denmark together in international waters beyond Canada's territorial boundaries in the Davis Strait. There, a notional missing fishing vessel had released two life rafts and the RCN Maritime Coastal Defence Vessel (MCDV) HMCS Shawinigan led the search. The overarching scenario was directed by the JRCC in Halifax, who also had training audience participants among its normal staff, including both RCN and RCAF operations personnel, CCG, and Transport Canada representatives.

Nanook 2014 was an exceptional event in that it highlighted not only the way in which Canadian government agencies use WoG to implement broad-based security scenarios and responses within defence exercise mandates, but also to degree to which there is an expectation of joint interoperability in the North American Arctic. *Nanook 2014* also highlighted the degree to which Arctic communities figure into new defense activities. Remote northern communities are, in real life, among the potential first responders to disaster and on the front lines more generally of large-scale environmental events. Their supply lines and resources can be stretched beyond repair in the event of an incident, meaning that one of the basic principles for planning operations in the North is to ensure that the local economies are not negatively impacted by the large-scale presence of either first responders or people in distress. This also creates a mindfulness concerning the role of civilian populations as responders and participants in search and rescue, as well as hosts for responders and armed forces.

The lesson learned for both planners and operators involved in the *Nanook 2014* simulation exemplifies one of the most important points about the relationship between military/traditional and human/environmental security more broadly, at least as it is understood in the Canadian North. According to participants of the exercise:

> Long gone are the days when the Canadian military operates in a theatre in seclusion, without any interaction with Other Governmental Departments (OGDs), civilian contractors, and even Non-Governmental Organizations (NGOs). This is especially true for operations in the Arctic, where the Area of Operation will be as austere (and potentially as dangerous) as any theatre outside of Canada. This austerity, the provisions in Comprehensive Land Claims Agreements, and the limited economies of northern communities combine to dictate a careful and respectful approach to planning and conducting operations, even in emergency

situations. Consultation with local and indigenous groups is essential, as are the achievement of land-use, environmental and the associated permits for water consumption, for example. More importantly for CAF Commanders and key staff is the open and respectful interaction with our civilian partners. The CAF is but one player in a large cast of diverse stakeholders that changes drastically from territory to territory, and from community to community. (Manson 2017, 242)

That said, on a more immediate level, the author's recent conversations with security agencies and actors operating throughout both the Canadian and American Arctic continue to reinforce the desire of operator agencies to better develop situational awareness and a more varied toolkit to both predict events and respond to them. The ADAC and *Nanook* examples accurately situate the new way of understanding Arctic security and the role of security operators. For operators and governmental agencies situated within the region, for example, growing security and operational risks reference several primary themes that are changing and evolving at a rapid pace. These include how to define the level of risk for a maritime incident in the Arctic, and how to respond to challenges posed by oil spill, environmental disasters, or events related to increased traffic—including criminal activity. It is also crucial to add value to military activities, by increasing scientific activity and building these types of synergies as well as a need for innovative ways of addressing problems of infrastructure and remote resourcing for security activities. Each of these insights provides new context and new challenges for security dialogue, definition, and delivery.

COMMUNITIES

Traditional security practice and definition have been transformed in the North American Arctic by national strategies and regional security actors. There is a specificity to this transformation which may or may not be transferable to other places and times. One reason is that the knowledge and the experience of local communities are key to the configuration of the security landscape. Both Canadian and American Arctic security agencies are aware that local communities have situational knowledge that is unavailable from existing technologies and surveillance tools. Local communities have been leading informants about snow and ice conditions, maritime conditions, and environmental anomalies. More specifically, Indigenous communities have apprised the larger regional and national communities more generally about the inherent changes to Northern ecosystems and lifestyles posed by climate change—and were among the first to do so. Indeed, one of the inherent differences between traditional security narratives today and those as recent as five to ten years ago, is the degree to which indigenous knowledge and the understanding of how environmental change is influencing food security and cultural survival is now acknowledged within mainstream Arctic security discussions by practitioners and agencies alike. Defense agencies openly look to communities and Indigenous residents of the North to be the "eyes and ears" of a security landscape that is increasingly subject to the potential for human and environmental tragedy.

The increasing hybridization of civil-military security activity includes, as well, the work of police and policing agencies which work closely with both communities as well as military actors. Indeed, communities are essential for law enforcement agencies, which struggle not only with remote conditions, but also with local intelligence capacities. Rather than seeing circumpolar communities solely as the targets of protection activities, however, new thinking sees communities as security actors: effective security arrangements include communities and community action. They also make reference to community knowledge as effective components of enhanced situational awareness. Communities, for their part, are interested in receiving training and equipment for these purposes, much as Alaskan natives have received from US military and coast guard units. Similarly, Canadian Inuit have become involved in patrol and search and rescue activities on a regular basis—these Canadian Rangers being an example of how Indigenous knowledge and practice have been seamlessly integrated into a distinctive mode of security practice.

Yet not all security threats occur away from the community local. Villages themselves are increasingly threatened by changing environmental conditions, and such changes are dramatically affecting the people of the region. In communities like Tuktoyuktak, there is a general unease about ability to respond to local disasters while waiting for first responders—generally American or Canadian coast guards or armed forces (ADAC 2019).

In other words, there is a distinctive brand of traditional security, which has evolved in the North American North from a unique geography and demography and very specific governance and defence processes, policies and objectives that is not replicated elsewhere. The broad reshaping of security thinking more generally has obviously contributed to the process, but overall it has evolved to meet specific North American traditional northern security threats.

CONCLUSIONS

North American states have developed strategic responses, national frameworks and foreign policies to accommodate new security imperatives and uncertainties. Strategic documents and security actors alike recognize the importance of environmental sustainability and community involvement, underscoring the fact that at all levels, the conceptualization and delivery of Arctic security is no longer "traditional". The existential threat of climate change and non-conventional environmental threats to Northern communities and environments have reshaped national security narrative applied concerning the North American North and have effectively broadened once was almost exclusively a concern with traditional military security. The process has been protracted, having its roots in the late 20th and early 21st centuries, and regional actors are still working through the implications for Arctic security. New approaches based on domain awareness, collaborative governmental initiatives, and broadly ranging security objectives are now underway. This chapter suggests that traditional security actors have retained agency, but are also engaged in more collaborative ways that engage more broadly with civil society goals and actors, environmental challenges and community well-being. Northern communities are increasingly engaged with national security actors and agencies, in efforts to better respond to and mitigate the most egregious impacts of climate change upon lifestyles and infrastructures. Security actors and agencies are openly addressing and rethinking the division between "national" and "local" security contexts, as well as the way in which security is delivered, in recognition of the need for greater precision and more nuanced situational awareness. Indeed, the threat of foreign powers plying through North American Arctic waters is not so much in the threat of occupation and military aggression, but in their potential disregard for maintaining standards and threatening sensitive Arctic environments, or their role in transporting the products of extractive industries that have had deleterious environmental impacts on northern landscapes.

Traditional Arctic military actors also understand this and have reinvented their purpose within a larger narrative. Indeed, the evidence suggests that both national strategists and Arctic security operators increasingly see security within a broader human and environmental security framework. We might be well-served to see traditional security—even when in service of the nation state—as a broad framework of options knitted together through regional needs rather than an impenetrable silo of interests to which other types of security are appended. If the examples used in this essay have done nothing else, they indicate how intricate and interwoven the connections between different types of security have become. Still, there is a common belief among security operators that policies shaping the resourcing of new security threats lag. This means that Canadian and American security actors lack the infrastructure and the hardware to respond quickly and efficiently to large-scale environmental disaster in the North. Southern policy-makers, climate change deniers notwithstanding, have indeed reoriented their understanding of security, but still lack the imagination to understand how the new security environment creates an even more compelling need to resource communities and security agencies in ways that address the diffuse and omnipresent problem of unpredictability. It is not with a sigh of relief that we can say that our security frameworks are shifting away from those with a hard and competitive geopolitical edge, unless we are willing to make the investment in cooperative and non-conventional security practices. And, as this essay has suggested, it is at the regional and community level that such practices now need to be understood.

REFERENCES

Alkire, Sabina. 2004. A Vital Core that Must Be Treated with the Same Gravitas as Traditional Security Threats. *Security Dialogue* 35 (3, Sep.): 359–360.

Arctic Domain Awareness Centre. 2018. Updated Draft: North American Arctic Maritime & Environmental Security Assessing Concern, Advancing Collaboration. A partnered workshop 18–20 September 2018, University of Alaska Anchorage, Anchorage, Alaska. http://arcticdomainawarenesscenter.org/css/images/newdesign/pdf/reports/ADAC-AMESW%202018_p01_Report_v1.pdf.

Everett, Karen G. 2018. Securitization, Borders, and the Canadian North: A Regional Approach. A dissertation submitted to the Committee of Graduate Studies in Partial Fulfillment of the Requirements for the Degree of Doctor of Philosophy in the Faculty of the Arts and Science, Trent University, Peterborough, ON.

Government of Canada. 2009. *Canada's Northern Strategy: Our North, Our Heritage, Our Future.* http://www.northernstrategy.gc.ca/cns/cns.pdf.

Government of Canada. 2010. Statement on Canada's Arctic Foreign Policy: Exercising Sovereignty and Promoting Canada's Northernstrategy Abroad. http://publications.gc.ca/collections/collection_2017/amc-gac/FR5-111-2010-eng.pdf.

Griffiths, Franklyn. 2009. Towards a Canadian Arctic Strategy. Foreign Policy for Canada's Tomorrow. CIC No. 1, Canadian International Council. http://carc.org/wp-content/uploads/2017/10/North-2030-Towards-a-Canadian-Arctic-Strategy-Franklyn-Griffiths.pdf.

Heininen, Lassi. 2004. Circumpolar International Relations and Geopolitics. In *Arctic Human Development Report (ADHR)*, ed. Niels Einarsson, Joan Nymand Larsen, Annika Nilsson, and Oran R. Young. Akureyri, Iceland: Stefansson Arctic Institute.

Heininen, Lassi, and Heather N. Nicol. 2007. The Importance of Northern Dimension of Foreign Policies in the Geopolitics of the Circumpolar North. *Geopolitics* 12: 133–135.

Hoag, Hanna. 2016. Mark Brzezinski: U.S. Arctic Policies and Priorities. *Arctic Deeply*, January 23. https://www.newsdeeply.com/arctic/community/2016/01/23/mark-brzezinski-u-s-arctic-policies-and-priorities.

Hoogensen, Gunhild, et al. 2009. Human Security in the Arctic: Yes it is Relevant! *Journal of Human Security* 5 (2): 1–10.

Hossain, Kamrul, and Harsh Barala. 2017. An Assessment of the US Chairmanship of the Arctic Council. Articles, 5 May 2017. University of Lapland. file://habitat/home/heathernicol/Desktop/kossain%20and%20Barla.pdf

Keskitalo, E.C.E. 2004. *Negotiating the Arctic: The Construction of an International Region.* New York: Routledge.

Kraska, James. 2016. Maritime Governance of the U.S. Arctic Region. In *Climate Change from a Northern Point of View*, ed. Lassi Heininen and Heather N. Nicol, 166–188. Waterloo: Centre for Foreign Policy and Federalism.

Lackenbauer, P. Whitney, and Ryan Dean. 2016. Canada's Northern Strategy under Prime Minister Stephen Harper: Key Speeches and Documents on Sovereignty, Security, and Governance, 2006–15. Calgary and Waterloo: Documents on Canadian Arctic Sovereignty and Security.

Lackenbauer, P. Whitney, and Heather N. Nicol. 2017. Arctic Sovereignty Through a Whole of Government Lens. Waterloo: Centre on Federalism and Foreign Policy.

Lackenbauer, P. Whitney, Heather N. Nicol, and Wilfrid Greaves. 2017. One Arctic: The Arctic Council and Circumpolar Governance. Waterloo; Ottawa: Centre for Federalism and Foreign Policy; Canadian Arctic resources Committee (CARC).

Manson, Deanna. 2017. Planning Operation Nanook 2014: Lessons Learned from a Joint Task Force (North) Perspective. In *Canadian Arctic Operations, 1941–2015*

Lessons Learned, Lost, and Relearned, ed. Adam Lajeunesse and P. Whitney Lackenbauer. Fredericton, NB: The Gregg Centre for the Study of War & Society.

Nicol, Heather N. 2016. From Territory to Rights. *Geopolitics* 22 (4): 1–21. https://doi.org/10.1080/14650045.2016.1264055.

Nicol, Heather N., and Lassi Heininen. 2013. Human Security, the Arctic Council and Climate Change. *Polar Record*: 1–76. https://doi.org/10.1017/S0032247412000666.

Schreiber, M. 2018. Outgoing Commandant Says Arctic has Become a Top Priority for US Coast Guard. *Arctic Today,* May 9. https://www.arctictoday.com/outgoing-commandant-says-arctic-become-top-priority-us-coast-guard/.

Swanström, Niklas. 2010. Traditional and Non-Traditional Security Threats in Central Asia: Connecting the New and the Old. *China and Eurasia Forum Quarterly* 8 (2): 35–51.

United Nations. 1992. *United Nations Framework Convention on Climate Change.*

———. 1994. *Human Development Report.* New York and Toronto: Oxford University Press.

———. 2016a. Paris Agreement. United Nations Treaty Collection, 8 July 2016.

———. 2016b. *United Nations Framework Convention on Climate Change.*

United States. 2013. National Security Presidential Directive-41/Homeland Security Presidential Directive-13. https://www.dhs.gov/national-plan-achieve-maritime-domain-awareness.

———. 2014. U.S. National Strategy for the Arctic Implementation Plan.

United States. Congressional Research Service (CRS). 2018. *Changes in the Arctic: Background and Issues for Congress.* Updated October 25, 2018. https://crsreports.congress.gov. R41153 accessed online at: https://news.usni.org/2018/11/02/report-congress-changes-arctic-2.

Williams, Michael C. 1998. Modernity, Identity and Security: A Comment on the 'Copenhagen Controversy'. *Review of International Studies* 24: 435–439.

World Commission on Environment and Development. 1987. *Our Common Future.* Oxford: Oxford University Press.

CHAPTER 29

The Arctic and Geopolitics

David A. Welch

Prelude: What Is the Arctic, and What Is "Geopolitics"?

The words "Arctic" and "geopolitics" are fixtures of the English language, although the former undoubtedly enjoys wider usage. "Geopolitics" is a term that one is most likely to encounter either in academic social science or in media commentary on global affairs. Yet neither is particularly well defined. This is no coincidence. If you look up "Arctic" on Wikipedia, you will see that it has both natural science definitions and social/political definitions. Everyone agrees that the waters of the Arctic Ocean count, but the application of the label to southward landmasses and peripheral minor seas and bays is inconsistent, and occasionally contested. The contestation is often political—which is where geopolitics comes in.

The word "geopolitics" was originally coined in 1899 by a Swedish political scientist and rapidly developed into a subfield of its own (Dodds 2007, 24). To some extent it was a case of putting old wine in new bottles; statesmen, military thinkers, scholars, and commentators had been long been aware of the impor-

An earlier version of this chapter appeared in Kimie Hara and Ken Coates, eds., *East Asia-Arctic Relations: Boundary, Security, and International Politics* (Waterloo, ON: Centre for International Governance Innovation [CIGI], 2014). I am grateful to CIGI for permission to republish an updated version, and also to Eva Buzsa, Ken Coates, Aileen Espíritu, Douglas Goold, Kimie Hara, Scott Harrison, Carin Holroyd, Thomas Homer-Dixon, Akihiro Iwashita, Whitney Lackenbauer, Keun-Gwan Lee, James Manicom, Gerald McBeath, Manjana Milkoreit, Fujio Onishi, Kai Sun, and Tamara Troyakova for their helpful comments and suggestions.

D. A. Welch (✉)
University of Waterloo, Waterloo, ON, Canada
e-mail: dawelch@uwaterloo.ca

© The Author(s) 2020
K. S. Coates, C. Holroyd (eds.), *The Palgrave Handbook of Arctic Policy and Politics*, https://doi.org/10.1007/978-3-030-20557-7_29

tance of geography in world politics, and what many regard as the seminal work of geopolitics—Alfred Thayer Mahan's *The Influence of Sea Power Upon History*—had, in fact, been published almost a decade earlier (Mahan 1890). Nevertheless, the word gave the subject a quasi-scientific cachet that helped to establish "claims to intellectual legitimacy and policy relevance" (Dodds, 26). The term was taken up with particular gusto by those whom we would identify today as "realists" preoccupied with the art and science of promoting national interest defined in terms of power by maneuvering for territorial advantage. Primarily politically conservative, early geopolitical thinkers offered justifications for hard-nosed power politics, formal and informal empire, and the high levels of armament required to pursue them (Mackinder 1904, 421–437; Dodds and Sidaway 2004, 292–297). In recent years, however, the term has been embraced by scholars in a wide variety of disciplines, including those who work in a critical or postmodern vein and whose politics are as likely to be post-colonial as early geopolitical thinkers' were pro-colonial (Dalby and Tuathail 1998; Tuathail et al. 2006; Kelly 2006, 24–53; Ciută and Klinke 2010, 323–332; Knecht and Keil 2013, 178–203).

Whatever the politics of geopolitics may be, practitioners all share a concern with relating space to politics. Politics is (or should be) about protecting things worth protecting, providing public goods, and doing today what needs to be done to enable our children to have a better tomorrow. Not surprisingly, in the field of political science, geopolitics falls squarely in the subfield of international security studies. Differences between old-style and new-style geopolitics can be understood to some extent as differences between traditional and non-traditional understandings of security. And so a good place to begin a discussion of the Arctic and geopolitics is to identify what is at stake in the region, looking through both traditional and nontraditional security lenses.

"The region," however—as I indicated at the outset—requires disambiguation. It will suffice for my purposes here simply to define the Arctic as that part of the Earth above the Arctic Circle—that is, north of 66°33′44″. This has the advantage of including territory from all eight member states of the Arctic Council. This arbitrary delineation does not mean, of course, that the issues I discuss here are not of concern to other countries or to the people or territories of the Arctic Council members, the vast majority of which lie outside the region.[1]

What *are* my purposes here? Quite simply, to show that the Arctic is completely uninteresting geopolitically from a traditional security perspective, and that while it is interesting from a nontraditional security perspective, it is truly important only in the one respect that just so happens to attract the least attention and action from policymakers. Moreover, it is the one respect that forces us to look in the other direction. Security is not at stake in any meaningful sense *in* the Arctic, but is very much at stake *because of* it.

---

[1] The one partial exception is Denmark, whose territory, if one includes Greenland, is predominantly above the Arctic Circle, but whose population is almost entirely below it.

Traditional Security

Early geopolitical thinkers were concerned almost entirely with the security of the state against military attack from another state. This understanding of security dominated International Relations as a field right up until the end of the Cold War. During the Cold War, the Arctic had geopolitical value in this traditional sense as a result of the premium the superpowers placed on the early warning of transpolar strategic bomber or ballistic missile attack, which required building, manning, supplying, and maintaining radar sites in harsh, remote northerly locations. Now that the Cold War is over, however—and in view of technological advances that have shifted the monitoring burden to space-based and unmanned sensors—the region has lost this particular "hard" security value.[2]

The Arctic never was—and for the foreseeable future never plausibly will be—a significant theater of non-nuclear war. No matter which school of military thought to which one belongs, it is impossible to imagine that significant military operations in the Arctic will ever be feasible, necessary, or desirable. Climate, terrain, sea ice, remoteness from economic and population centers, and lack of forward base infrastructure in the Arctic are all inhospitable for military operations no matter whether one favors the decisive engagement (as did Sun Tzu and Jomini), destruction of the enemy's "centre of gravity" (as did Clausewitz), or an "indirect approach" to war through flanking and maneuver (à la Liddell Hart) (Sun-Tzu 2009; Liddell Hart 1929; Clausewitz 1993; Wood 2008, 44–56; Swain 1990, 35–51; Holmes 2007, 129–151; Handel 1992). The most that can be said for the Arctic's traditional military value is that once in human history—during World War II—Arctic waters served an important logistical function. By means of convoys from Atlantic ports to Arkhangelsk and Murmansk, the Allies helped to keep the Soviet Union supplied in its fight against Hitler (Schofield 1977). It is difficult to imagine the conflict that would require Arctic transit routes in the twenty-first Century; Europe and North America are members of a "security community," (Adler and Barnett 1998)[3] and while the Northern Sea Route might be useful for shipping supplies to combatants in East or Southeast Asia, it is implausible to imagine that it would play more than a marginal role given safer Transpacific alternatives.

The Arctic is an inhospitable military environment for exactly the same reasons that it is inhospitable for large-scale human habitation. The ten largest cities in the Arctic have a combined population of fewer than 900,000—

---

[2] Arguably, even during the Cold War the Arctic had relatively little "hard" security value, owing to the fact that neither superpower harbored intentions of nuclear attack. The dangers of nuclear war were almost entirely a function of accident, inadvertence, misperception, and unintended escalation—any of which would have resulted in massive casualties south of (not north of) the Arctic Circle regardless of early warning capabilities.

[3] A security community is a group of states in whose relations the threat or use of force plays no role whatsoever.

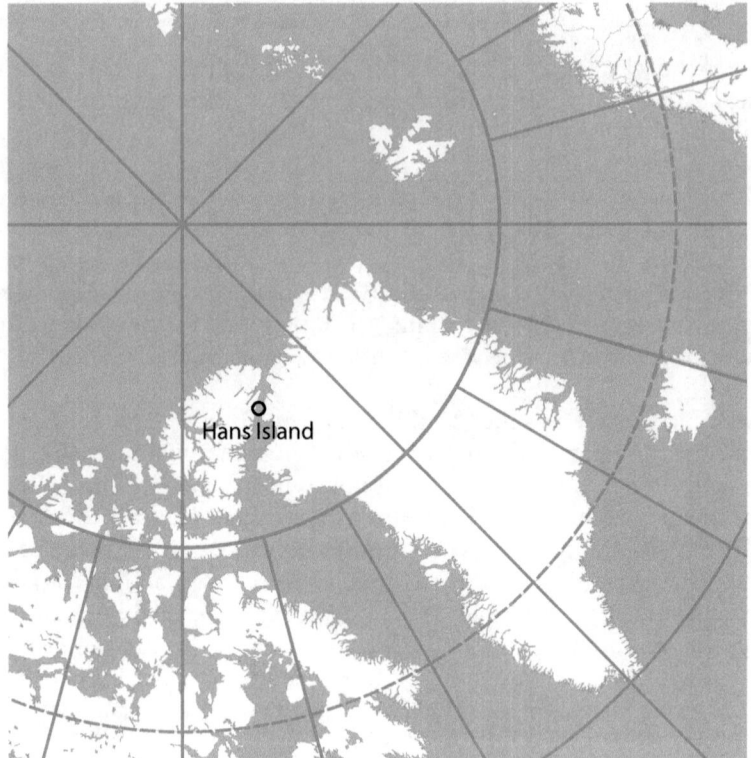

Fig. 29.1   Location of Hans Island (author's rendering)

roughly the same as that of Canada's capital city, Ottawa. Fully a third of those live in Murmansk, which enjoys the odd status of being ice-free year-round, thanks to the Gulf Stream ("10 Largest Cities" 2011). In any case, there is little in the Arctic over which to fight. There is but one territorial dispute in the entire circumpolar region—between Canada and Denmark over Hans Island, a 1.3 km² barren rock notable primarily as a source of binational mirth (see Fig. 29.1) (Byers 2010, 26–27). While there are a few maritime jurisdiction disputes, there is no indication that any of them is more than a low-grade management issue.[4]

Despite all this, one occasionally encounters alarmist accounts of traditional security threats in the Arctic. These have an air of implausibility across the board and often trade on mixing up very distinct concepts such as sovereignty and security (Huebert 2009; CBC News [2009] 2010). They can be useful in

[4] Notably, all of the countries involved in actual or potential maritime jurisdiction disputes in the Arctic are either signatories to the U.N. Convention on the Law of the Sea or are tacitly observing its provisions, which include various provisions for the peaceful settlement of disputes.

bureaucratic political games, however; the Canadian Navy, for example, used the putative US and Russian submarine threat to Canada's Arctic sovereignty to help justify the purchase of four used Upholder class submarines from Britain in 1998, never explaining to the Government, to Parliament, or to the Canadian people what exactly they intended to do with them if they encountered unwelcome foreign submarines in waters that Canada liked to think of its own, nor explaining why they were buying diesel-electric submarines that were almost entirely unsuited to Arctic operations. The purchase has proven to be a complete debacle, and yet the "submarine threat" canard refuses to die (The Economist 2012; Huebert 2011, 809–824).

NON-TRADITIONAL SECURITY

While traditional understandings of security privileged the protection of the state as the "referent object" against the threat of military attack, the field of security studies has recently embraced a variety of non-traditional conceptions with a much wider variety of threat/object pairs. We owe the useful distinction between "threat" and "referent object" to the Copenhagen School of International Relations, which also brought us the concept of "securitization"—that is, the process by which problems become elevated from run-of-the-mill political problems to "security" problems warranting extraordinary effort and resources, often justifying the suspension of normal rules (Buzan et al. 1997). Five of the more commonly discussed non-traditional security issues of potential relevance to the Arctic are *human security, cultural security, energy security, economic security,* and *environmental security.*

The concept of human security was first articulated in the United Nations' 1994 *Human Development Report*, which argued that individual human beings were the primary referent objects and that threats to their security were context-specific (United Nations Development Program 1995). Critics were quick to notice flaws in this early conceptualization, not least of which was that it was radically subjective, provided indeterminate guidance for policy, and was difficult to distinguish from both "human rights" and "development" (Paris 2004, 249–264; Howard-Hassmann 2012, 88–112; Daudelin and Hampson 1999). But it did have the effect of sensitizing both policymakers and publics to a range of issues that caused high levels of death, misery, and morbidity, but that had not attracted sustained attention and resources during the Cold War owing to the preoccupation with avoiding World War III. These issues included (*inter alia*) substate conflict, landmines, small arms and light weapons, human trafficking, food insecurity, disease, and violence against women (Davis 2009).[5]

---

[5] The three countries that took up human security most energetically in their foreign policy platforms were Canada, Japan, and Norway. Prime Minister Stephen Harper of Canada distanced his government from human security because it was so closely identified with the previous Liberal government, and in particular with former Liberal Foreign Affairs Minister Lloyd Axworthy.

In recent years, there has been a minor surge of interest in the human security of Arctic peoples (Lukovich and McBean 2009, 697–710; Heininen and Nicol 2007, 133–165; Daveluy et al. 2011; Hynek and Bosold 2010). The issues, of course, are not exactly the same as they are (say) in sub-Saharan Africa or Afghanistan, but they are real. They primarily concern the relative material quality-of-life disadvantages Arctic indigenous peoples experience vis-à-vis the non-Indigenous populations of Arctic countries. Life expectancy is shorter, and rates of infant mortality, suicide, substance abuse, spousal abuse, and sexual abuse are all typically significantly higher. To some extent these issues are a function of cultural dislocation—the loss of indigenous languages, the erosion of traditional cultural practices, and so forth—but by any measure the most important factor has been the effective colonization of the Arctic by non-Arctic peoples and the sense of disempowerment and dislocation that this brings. The experience and the pathology may not be unique to Arctic peoples—it is a sad fact that indigenous peoples everywhere suffer similar disadvantages, deprivations, and depredations—but the effects are especially noticeable in the Arctic precisely because of the delicacy of the relationship between the land and the people and the combination of small numbers and high dispersion. There are relatively few buffers to cultural conquest in the far North.

Climate change is, by all indications, an accelerant for human and cultural security challenges. With the warming of the Arctic and the retreat of the ice, traditional ways of life become harder to maintain even where there is the will to do so. One of the great unknowns is food security. The same may be said of food supplies everywhere—climate change models are notoriously sensitive to assumptions, specifications, and inputs—but in the Arctic, there is relatively little room for error. It is a particularly delicately balanced ecological zone (Wesche and Chan 2010, 361–373; Duhaime 2002; Duhaime and Bernard 2008; Waits et al. 2018, 703–713; Greaves 2016, 461–480).

Energy security is a rather different kind of problem for the Arctic than it is for the rest of the world. The Arctic's energy needs are modest in absolute terms, and climate change is unlikely to have a dramatic effect on them, either in terms of supply or demand. But climate change may well make the Arctic more important for the rest of the world's energy security and economic security. It may do so in two primary ways: first, by increasing the commercial viability of exploiting Arctic oil and natural gas deposits, fishing grounds, and ore deposits; and second, by opening up new transportation corridors. For Arctic peoples, this represents something of a mixed blessing. On the one hand, investment in resource extraction and transportation brings the promise of jobs, improvements to infrastructure, and overall wealth; on the other hand, it threatens to accelerate the erosion of traditional cultures and increase the danger of environmental catastrophe.

How likely are these? With respect to resource extraction, it is worth bearing in mind that the Arctic has always been the big payoff lurking just around the corner. Nowhere and never has it lived up to its resource extraction hype. The reasons for this are complex, but harshness, remoteness, and lack of infrastructure have always been factors (Victor et al. 2006, 128, 129, 142n, 144, 394; Moe 2012, 227–251). Climate change will certainly increase the number of days each year during which Arctic sea routes are open (Barry 2017, 69–88; Aksenov et al. 2017, 300–317; Connolly et al. 2017, 1317–1340; Theocharis et al. 2018, 112–128), which would under certain circumstances make both extraction and transport of oil and gas (particularly bulk liquid natural gas, or LNG) easier and more attractive (Schach and Madlener 2018, 438–448); but in a decarbonizing world, demand for these will shrink over time.

Climate change notwithstanding, neither the Northwest Passage nor the Northern Sea Route across the top of Russia is likely to become a major shipping artery anytime soon (Lasserre 2011, 793–808). The prospects for the latter are certainly much better than for the former, for a variety of reasons: (1) it is ice-free a larger proportion of the year; (2) it offers more of a distance savings for a dramatically higher volume of shipping; and (3) it boasts a much more highly developed shipping infrastructure in terms of available ports, ice-breaker services, and so forth (Pettersen 2013; Headland 2010, 1–13).[6] And yet even the Northern Sea Route's prospects seem modest at best. The unpredictability of sailing conditions (a) represents a deterrent to container shipping, which is highly just-in-time oriented; (b) introduces speed uncertainties, which has a strong effect on scheduling and fuel efficiency; (c) increases insurance costs; and (d) requires shippers to take on expensive Russian pilots (Ho 2011, 106–120; Khon et al. 2010, 757–768; Liu and Kronbak 2010, 434–444; Verny and Grigentin 2009, 107–117; Schøyen and Bråthen 2011, 977–983; Xu et al. 2011, 541–560; Lee 2012, 39–67). In addition to being geographically less attractive, the Northwest Passage is far less hospitable than the Northern Sea Route: climate models suggest that it is likely to be ice-free far less often and vulnerable to persistent icing at crucial choke points (Howell et al. 2008, 229–242). But relevant models are highly uncertain, a significant deterrent to planning and investment (Choi et al. 2015, 61–69; Lasserre 2014, 144–161; Tseng and Cullinane 2018, 422–438). If shipping through either route does increase dramatically, it will probably take the form of bulk rather than container shipping—which poses especially acute environmental dangers in case of accident (Lasserre, 806–807).

This last point raises the important issue of environmental security. Commentators univocally note the particular sensitivity and fragility of Arctic ecosystems both to pollution and to disruption. The Arctic and the Antarctic are the two regions of the Earth that have the lowest net energy input, owing to low solar forcing and limited interzonal energy transport mechanisms (Kageyama et al. 2010, 2931–2956; Gent 2018, 57–66).[7] This significantly

[6] The historical record thus far clearly indicates that the Northern Sea Route is more hospitable to shipping than the Northwest Passage.

[7] The major exception to this is the Atlantic Conveyor, which transfers heat from the Gulf of Mexico to Northern Europe via the Gulf Stream—the mechanism that keeps Murmansk ice-free year-round at the moment. This "heat pump" may be vulnerable to fresh-water hosing caused by glacial melting, particularly in Greenland—although relevant models are uncertain.

extends the time required for biotic adjustment. Put another way: the Arctic would take many times longer than a tropical or semi-tropical zone to recover from an oil spill or similar disaster. This unusual vulnerability points toward the importance of ensuring that development of the Arctic takes place within the context of strict environmental regulation and robust regional environmental governance.

WHAT'S MISSING FROM THE ARCTIC GEOPOLITICS DISCOURSE?

The discussion to this point would suggest that the Arctic is a region with great potential but fraught with danger. This is true enough, but the point might be misconstrued that the takeaway here is simply to ensure that the development of the Arctic takes place in a cooperative, coordinated, appropriately governed fashion. Traditional geopolitics may not be at stake in the Arctic, but non-traditional geopolitics most certainly is, and the lesson would seem to be that we must move forward gingerly to maximize the benefits and minimize the costs (Ingimundarson 2014, 183–198; Exner-Pirot 2013, 120–135; Scassola 2013, 183–204).

This is not the lesson.

It is true that human, cultural, energy, economic, and environmental security are all at stake in the Arctic, and that we must take care to avoid harm where possible. But in the grand scheme of things, these are all relatively minor problems. Owing simply to the relatively small numbers of people concerned, Arctic human security issues pale in comparison to human security challenges elsewhere. In view of the horrific levels of organized violence and exploitation that millions of people experience on a daily basis in failed and failing states around the world, the challenges Arctic populations face seem more like policy failures than acute security problems. Cultural security is, in any case, a questionable concept, as culture is inherently dynamic; it cannot be protected from change, and it is difficult to imagine the normative argument that it *ought* to be protected from change. At most one can argue that it ought to be protected from artificially rapid change, which is demonstrably psychologically disruptive. Energy security is a challenge around the world, but no more so in the Arctic, and there is little reason to think, despite the perennial hoopla, that the Arctic will be the cure for energy security challenges elsewhere. Likewise with respect to economic security. The economic security of the Arctic is more likely to depend upon the national policies of Arctic countries than on grandiose development initiatives originating elsewhere. As for the economic security of the rest of the world, the marginal contribution of Arctic resources in a warming world is not likely to have a material impact, particularly in view of the enormous increase in demand for resources that we will likely see from populous, rapidly-developing countries such as India. As far as environmental security is concerned, the Arctic is certainly a uniquely vulnerable region; but given the likelihood that it will never live up to its resource development and transportation hype, and in view of the fact that environmental disasters such as blowouts, oil spills, tailing pond leakages, and other extractive-industry accidents will only have local (if unusually persistent) effects, it is hard to make the case that the Arctic will be the site of the most egregious environmental disasters in the years to come.

What really makes the Arctic important from a geopolitical perspective is the threat it poses to *ecospheric security*. This is a conception of security that has yet to make its way into the mainstream even of non-traditional security discourse.[8] The ecosphere is that part of the Earth that does (or could) support life, and its health depends crucially upon atmospheric homeostasis and adequate biodiversity—the latter of which, according to prominent climate scientists, may be a precondition for the former (Lovelock and Margulis 1974, 1–9). At no time in known history has the planet experienced a more rapid rise in greenhouse gases or a faster increase in mean surface temperature. True, it has been hotter at times and the atmosphere has borne more carbon; but the crucial consideration is the rate of change. An elastic band will stretch much farther without breaking if pulled slowly, but we are stretching atmospheric chemistry at an unprecedented rate as a result of fossil fuel emissions. The Arctic is relevant here because vast quantities of powerful greenhouse gases—carbon dioxide and methane in particular—are locked up in permafrost. A rapidly warming Arctic has the potential to shift from a net carbon sink to a net carbon source, (Voigt et al. 2017, 3121–3138; Schaefer et al. 2011, 165–180) accelerate warming worldwide, increase the frequency and severity of wildfires (which in turn are powerful causes of warming; Mooney 2013) increase the frequency and severity of extreme weather events in other climate zones; Greene and Monger 2012, 7–9), and both alter and amplify global climate feedbacks (Sommerkorn and Hassol 2009).

Less ominous than ecospheric catastrophe, but still of concern on a scale that dwarfs any of the local security challenges facing the Arctic, is sea-level rise caused by polar warming (Rignot et al. 2011, 1–5; Hansen and Sato 2012; Kinnard et al. 2011, 509-U231; Livina and Lenton 2013, 275–286; Levermann et al. 2013, 13745–13750). Arctic sea ice is not a major issue here, except as regards (a) thermal expansion and (b) the effect of salinity dilution on heat transport mechanisms; but the Greenland and Antarctic ice caps are, as both currently sit on land and are not at the moment displacing their own weight in water. Estimates of the mean global sea-level rise that we can expect as a result of polar icecap melting are, of course, uncertain, but even a relatively modest rise will swamp island states such as Vanuatu and the Maldives wholesale, and will disproportionately affect Asia, the most populous continent and increasingly the engine of the global economy (Schleussner et al. 2018, 135–163; Woodruff et al. 2018, 48–77; Islam and Khan 2018, 297–323; Vitousek 2017).

The Arctic is not itself a site of interesting geopolitical value; but it has enormous, generally unappreciated geopolitical value to non-Arctic regions both as

---

[8] The term "environmental security" results in more than 1.2 million hits on Google; the term "ecospheric security" results in 79, many of which are my own.

a proverbial canary in a coal mine and as a potential climate change time bomb in its own right. Ironically, if we do not find a way to wean ourselves off carbon, the Arctic may itself actually become one of the few remaining places on the planet capable of sustaining human habitation (Hansen et al. 2013, 1–31; Morgan 2009, 683–693; Lovelock 2006a, b). Needless to say, there is no Arctic-governance fix to this. It is a global problem requiring an urgent, concerted global solution—a problem for which traditional geopolitical lenses and traditional geopolitical rivalries are pointless distractions, and for this very reason serious security threats in and of themselves.

REFERENCES

"10 Largest Cities Within the Arctic Circle". 2011. *The World Geography*. http://www.theworldgeography.com/2011/12/10-largest-cities-within-arctic-circle.html.

Adler, Emanuel, and Michael Barnett. 1998. *Security Communities*. Cambridge: Cambridge University Press.

Aksenov, Yevgeny, Ekaterina E. Popova, Andrew Yool, A.J. George Nurser, Timothy D. Williams, Laurent Bertino, and Jon Bergh. 2017. On the Future Navigability of Arctic Sea Routes: High-resolution Projections of the Arctic Ocean and Sea Ice. *Marine Policy* 75: 300–317.

Barry, Roger G. 2017. The Arctic Cryosphere in the Twenty-First Century. *Geographical Review* 107 (1): 69–88.

"Battle for the Arctic Heats Up". (2009) 2010. *CBC News*, August 20, 2009. http://www.cbc.ca/news/canada/story/2009/02/27/f-arctic-sovereignty.html.

Buzan, Barry, Ole Wæver, and Jaap de Wilde. 1997. *Security: A New Framework for Analysis*. Boulder, CO: Lynne Rienner.

Byers, Michael. 2010. *Who Owns the Arctic? Understanding Sovereignty Disputes in the North*. Vancouver, BC: Douglas & McIntyre.

Choi, Minjoo, Hyun Chung, Hajime Yamaguchi, and Keisuke Nagakawa. 2015. Arctic Sea Route Path Planning Based on an Uncertain Ice Prediction Model. *Cold Regions Science and Technology* 109: 61–69.

Ciută, Felix, and Ian Klinke. 2010. Lost in Conceptualization: Reading the "New Cold War" with Critical Geopolitics. *Political Geography* 29 (6): 323–332.

Clausewitz, Carl von. 1993. *On War*. Translated by Michael Howard and Peter Paret. New York: Knopf.

Connolly, Ronan, Michael Connolly, and Willie Soon. 2017. Re-calibration of Arctic Sea Ice Extent Datasets Using Arctic Surface Air Temperature Records. *Hydrological Sciences Journal* 62 (8): 1317–1340.

Dalby, Simon, and Gearóid Ó. Tuathail. 1998. *Rethinking Geopolitics*. New York: Routledge.

Daudelin, John, and Fen Osler Hampson. 1999. *Human Security and Development Policy*. Ottawa: CIDA Policy Branch Strategic Planning Division.

Daveluy, Michelle, Francis Lévesque, and Jenanne Ferguson, eds. 2011. *Humanizing Security in the Arctic*. Edmonton: CCI Press.

Davis, Jeff. 2009. Liberal-Era Diplomatic Language Killed Off. *Embassy*. Ottawa, July 1. http://www.embassynews.ca/news/2009/07/01/liberal-era-diplomatic-language-killed-off/37788.

Dodds, Klaus. 2007. *Geopolitics: A Very Short Introduction*. New York: Oxford University Press.

Dodds, Klaus, and James D. Sidaway. 2004. Halford Mackinder and the 'Geographical Pivot of History': A Centennial Retrospective. *The Geographical Journal* 170 (4, Dec.): 292–297.

Duhaime, Gérard, ed. 2002. *Sustainable Food Security in the Arctic: State of Knowledge*. Edmonton: CCI Press.

Duhaime, Gérard, and Nick Bernard, eds. 2008. *Arctic Food Security*. Edmonton: CCI Press.

Exner-Pirot, Heather. 2013. What is the Arctic a Case of? The Arctic as a Regional Environmental Security Complex and the Implications for Policy. *The Polar Journal* 3 (1): 120–135.

Gent, Peter R. 2018. A Commentary on the Atlantic Meridional Overturning Circulation Stability in Climate Models. *Ocean Modelling* 122: 57–66.

Greaves, Wilfrid. 2016. Arctic (In)security and Indigenous Peoples: Comparing Inuit in Canada and Sámi in Norway. *Security Dialogue* 47 (6): 461–480.

Greene, Charles H., and Bruce C. Monger. 2012. An Arctic Wild Card in the Weather. *Oceanography* 25 (2, June): 7–9.

Handel, Michael I. 1992. *Masters of War: Sun tzu, Clausewitz, and Jomini*. London: Frank Cass.

Hansen, James, and Makiko Sato. 2012. Update of Greenland Ice Sheet Mass Loss: Exponential?, December 26. http://www.columbia.edu/~jeh1/mailings/2012/20121226_GreenlandIceSheetUpdate.pdf.

Hansen, James, Makiko Sato, Gary Russell, and Pushker Kharecha. 2013. Climate Sensitivity, Sea Level, and Atmospheric CO_2. *Philosophical Transactions of the Royal Society A* 371 (Aug.): 1–31.

Headland, R.K. 2010. Ten Decades of Transits of the Northwest Passage. *Polar Geography* 33 (1–2): 1–13.

Heininen, Lassi, and Heather N. Nicol. 2007. The Importance of Northern Dimension Foreign Policies in the Geopolitics of the Circumpolar North. *Geopolitics* 12 (1): 133–165.

Ho, Joshua. 2011. The Opening of the Northern Sea Route. *Maritime Affairs: Journal of the National Maritime Foundation of India* 7 (1): 106–120.

Holmes, Terence M. 2007. Planning versus Chaos in Clausewitz's On War. *Journal of Strategic Studies* 30 (1): 129–151.

Howard-Hassmann, Rhoda E. 2012. Human Security: Undermining Human Rights? *Human Rights Quarterly* 34 (1): 88–112.

Howell, Stephen E.L., Adrienne Tivy, John J. Yackel, and Steve McCourt. 2008. Multi-year Sea-Ice Conditions in the Western Canadian Arctic Archipelago Region of the Northwest Passage: 1968–2006. *Atmosphere-Ocean* 46 (2): 229–242.

Huebert, Rob. 2011. Submarines, Oil Tankers, and Icebreakers: Trying to Understand Canadian Arctic Sovereignty and Security. *International Journal* 66 (4): 809–824.

Huebert, Robert N. 2009. Canadian Arctic Sovereignty and Security in a Transforming Circumpolar World. Toronto: Canadian International Council, 1 electronic text (43 p.) digital file.

Hynek, Nikola, and David Bosold. 2010. *Canada's Foreign & Security Policy: Soft and Hard Strategies of a Middle Power*. Don Mills, ON: Oxford University Press.

Ingimundarson, Valur. 2014. Managing a Contested Region: The Arctic Council and the Politics of Arctic Governance. *The Polar Journal* 4 (1): 183–198.

484   D. A. WELCH

Islam, M. Rezaul, and Niaz Ahmed Khan. 2018. Threats, Vulnerability, Resilience and Displacement among the Climate Change and Natural Disaster-Affected People in South-East Asia: An Overview. *Journal of the Asia Pacific Economy* 23 (2): 297–323.

Kageyama, Masa, André Paul, Didier M. Roche, and Cédric J. Van Meerbeeck. 2010. Modelling Glacial Climatic Millennial-Scale Variability Related to Changes in the Atlantic Meridional Overturning Circulation: A Review. *Quaternary Science Reviews* 29 (21–22): 2931–2956.

Kelly, Phil. 2006. A Critique of Critical Geopolitics. *Geopolitics* 11 (1): 24–53.

Khon, V., I. Mokhov, M. Latif, V. Semenov, and W. Park. 2010. Perspectives of Northern Sea Route and Northwest Passage in the Twenty-First Century. *Climatic Change* 100 (3–4): 757–768.

Kinnard, Christophe, Christian M. Zdanowicz, David A. Fisher, Elisabeth Isaksson, Anne de Vernal, and Lonnie G. Thompson. 2011. Reconstructed Changes in Arctic Sea Ice over the Past 1,450 Years. *Nature* 479 (7374, Nov.): 509–U231.

Knecht, Sebastian, and Kathrin Keil. 2013. Arctic Geopolitics Revisited: Spatialising Governance in the Circumpolar North. *The Polar Journal* 3 (1): 178–203.

Lasserre, Frédéric. 2011. Arctic Shipping Routes: From the Panama Myth to Reality. *International Journal* 66 (4): 793–808.

———. 2014. Case Studies of Shipping along Arctic Routes. Analysis and Profitability Perspectives for the Container Sector. *Transportation Research Part A* 66: 144–161.

Lee, Sung-Woo. 2012. Potential Arctic Shipping: Change, Benefit, Risk and Cooperation. In *The Arctic in World Affairs: A North Pacific Dialogue on Arctic Marine Issues*, ed. Oran R. Young, Jong Deog Kim, and Yoon Hyung Kim, 39–67. Seoul and Honolulu: Korea Maritime Institute and East-West Center.

Levermann, Anders, Peter U. Clark, Ben Marzeion, Glenn A. Milne, David Pollard, Valentina Radic, and Alexander Robinson. 2013. The Multimillennial Sea-Level Commitment of Global Warming. *Proceedings of the National Academy of Sciences of the United States of America* 110 (34, Aug.): 13745–13750.

Liddell Hart, B.H. 1929. *The Decisive Wars of History: A Study in Strategy.* London: G. Bell.

Liu, Miaojia, and Jacob Kronbak. 2010. The Potential Economic Viability of Using the Northern Sea Route (NSR) as an Alternative Route between Asia and Europe. *Journal of Transport Geography* 18 (3): 434–444.

Livina, V.N., and T.M. Lenton. 2013. A Recent Tipping Point in the Arctic Sea-Ice Cover: Abrupt and Persistent Increase in the Seasonal Cycle since 2007. *Cryosphere* 7 (1): 275–286.

Lovelock, James. 2006a. The Earth is about to Catch a Morbid Fever that May Last as Long as 100,000 Years. *The Independent*, January 16. http://ind.pn/12yGf1l.

———. 2006b. *The Revenge of Gaia: Earth's Climate in Crisis and the Fate of Humanity.* New York: Basic Books.

Lovelock, James E., and Lynn Margulis. 1974. Atmospheric Homeostasis By and For the Biosphere: The Gaia Hypothesis. *Tellus* 26 (2, May): 1–9.

Lukovich, Jennifer, and Gordon McBean. 2009. Addressing Human Security in the Arctic in the Context of Climate Change Through Science and Technology. *Mitigation and Adaptation Strategies for Global Change* 14 (8): 697–710.

Mackinder, H.J. 1904. The Geographical Pivot of History. *The Geographical Journal* 23 (4, Apr.): 421–437.

Mahan, A.T. 1890. *The Influence of Sea Power upon History, 1660–1783.* London: S. Low, Marston, Searle & Rivington.

Moe, Arild. 2012. Potential Arctic Oil and Gas Development: What are Realistic Expectations? In *The Arctic in World Affairs: A North Pacific Dialogue on Arctic Marine Issues*, ed. Oran R. Young, Jong Deog Kim, and Yoon Hyung Kim, 227–251. Seoul and Honolulu: Korea Maritime Institute and East-West Center.

Mooney, Chris. 2013. More Wildfires = More Warming = More Wildfires. *Mother Jones*, July 29. http://www.motherjones.com/environment/2013/07/global-warming-wildfire-permafrost-feedback.

Morgan, Dennis Ray. 2009. World on Fire: Two Scenarios of the Destruction of Human Civilization and Possible Extinction of the Human Race. *Futures* 41 (10): 683–693.

Paris, Roland. 2004. Human Security: Paradigm Shift or Hot Air? In *New Global Dangers: Changing Dimensions of International Security*, ed. Michael Brown et al., 249–264. Cambridge, MA: MIT Press.

Pettersen, Trude. 2013. Preparing for Record Season on the Northern Sea Route. *Barents Observer*. http://barentsobserver.com/en/business/2013/06/preparing-record-season-northern-sea-route-06-06.

Rignot, E., I. Velicogna, M.R. van den Broeke, A. Monaghan, and J. Lenaerts. 2011. Acceleration of the Contribution of the Greenland and Antarctic Ice Sheets to Sea Level Rise. *Geophysical Research Letters* 38 (L05503, Mar.): 1–5.

"Rock Bottom". 2012. *The Economist*, March 7.

Scassola, Andrea. 2013. All Is Well in the High North? Contemporary Sources of Tension in the Arctic. *New Global Studies* 7 (2): 183–204.

Schach, Michael, and Reinhard Madlener. 2018. Impacts of an Ice-Free Northeast Passage on LNG Markets and Geopolitics. *Energy Policy* 122: 438–448.

Schaefer, Kevin, Tingjun Zhang, Lori Bruhwiler, and Andrew P. Barrett. 2011. Amount and Timing of Permafrost Carbon Release in Response to Climate Warming. *Tellus Series B-Chemical and Physical Meteorology* 63 (2, Apr.): 165–180.

Schleussner, Carl-Friedrich, Delphine Deryng, Sarah D'Haen, William Hare, Tabea Lissner, Mouhamed Ly, Alexander Nauels, et al. 2018. 1.5°C Hotspots: Climate Hazards, Vulnerabilities, and Impacts. *Annual Review of Environment and Resources* 43: 135–163.

Schofield, B.B. 1977. *The Arctic Convoys*. London: Macdonald and Jane's.

Schøyen, Halvor, and Svein Bråthen. 2011. The Northern Sea Route versus the Suez Canal: Cases from Bulk Shipping. *Journal of Transport Geography* 19 (4, July): 977–983.

Sommerkorn, Martin, and Susan Joy Hassol, eds. 2009. *Arctic Climate Feedbacks: Global Implications*. Oslo: World Wildlife Fund International Arctic Programme.

Sun-Tzu. 2009. *The Art of War*. Translated by John Minford. New York: Penguin Books.

Swain, Richard. 1990. B. H. Liddell Hart and the Creation of a Theory of War, 1919–1933. *Armed Forces & Society* 17 (1): 35–51.

Theocharis, Dimitrios, Stephen Pettit, Vasco Sanchez Rodrigues, and Jane Haider. 2018. Arctic Shipping: A Systematic Literature Review of Comparative Studies. *Journal of Transport Geography* 69: 112–128.

Tseng, Po-Hsing, and Kevin Cullinane. 2018. Key Criteria Influencing the Choice of Arctic Shipping: A Fuzzy Analytic Hierarchy Process Model. *Maritime Policy & Management* 45 (4): 422–438.

Tuathail, Gearóid Ó., Simon Dalby, and Paul Routledge, eds. 2006. *The Geopolitics Reader*. London: Routledge.

United Nations Development Program. 1995. Human Development Report 1994. Oxford University Press for the United Nations Development Program. http://www.undp.org/hdro/hdrs/1994/english/94ch2.pdf.

Verny, Jerome, and Christophe Grigentin. 2009. Container Shipping on the Northern Sea Route. *International Journal of Production Economics* 122 (1): 107–117.

Victor, David G., Amy Jaffe, and Mark H. Hayes, eds. 2006. *Natural Gas and Geopolitics: From 1970 to 2040.* Cambridge: Cambridge University Press.

Vitousek, Sean. 2017. What is the Cost of One Meter of Sea Level Rise? *Union of Concerned Scientists*, July 19. https://blog.ucsusa.org/guest-commentary/what-is-the-cost-of-one-meter-of-sea-level-rise.

Voigt, Carolina, Richard E. Lamprecht, Maija E. Marushchak, Saara E. Lind, Alexander Novakovskiy, Mika Aurela, Pertti J. Martikainen, and Christina Biasi. 2017. Warming of Subarctic Tundra Increases Emissions of All Three Important Greenhouse Gases—Carbon Dioxide, Methane, and Nitrous Oxide. *Global Change Biology* 23 (8): 3121–3138.

Waits, Audrey, Anastasia Emelyanova, Antti Oksanen, Khaled Abass, and Arja Rautio. 2018. Human Infectious Diseases and the Changing Climate in the Arctic. *Environment International* 121 (Part 1): 703–713.

Wesche, Sonia, and Hing Chan. 2010. Adapting to the Impacts of Climate Change on Food Security among Inuit in the Western Canadian Arctic. *EcoHealth* 7 (3): 361–373.

Wood, Jason D. 2008. Clausewitz in the Caliphate: Center of Gravity in the Post–9/11 Security Environment. *Comparative Strategy* 27 (1): 44–56.

Woodruff, Sierra, Todd K. BenDor, and Aaron L. Strong. 2018. Fighting the Inevitable: Infrastructure Investment and Coastal Community Adaptation to Sea Level Rise. *System Dynamics Review* 34 (1–2): 48–77.

Xu, Hua, Zhifang Yin, Dashan Jia, Fengjun Jin, and Hua Ouyang. 2011. The Potential Seasonal Alternative of Asia–Europe Container Service via Northern Sea Route under the Arctic Sea Ice Retreat. *Maritime Policy & Management* 38 (5): 541–560.

CHAPTER 30

The Militarization of the Arctic to 1990

Peter Kikkert and P. Whitney Lackenbauer

In 1938, Austrian-born journalist (and alleged Soviet spy) Hans Peter Smolka tried to convince the readers of *Foreign Affairs* that the Arctic would soon emerge as an important military space. If Russia went to war with Germany and Japan, he argued, those two states might blockade or capture Leningrad and Vladivostok. "Russia would thus be bottled up on three sides: west, south and east," Smolka noted. "But in the North—and there only—there is an independent, continuous and all-Russian coastline, unassailable by anyone." The Russians would still be able to move men and materials along the Northern Sea Route and to collect supplies from potential allies through their Arctic waters. Accordingly, Murmansk would become the main naval base of the Russian fleet. "In other words," he concluded, "the 'backyard of Asia' is about to become the front porch of a newly oriented 'Arctic conscious' Russia" (Smolka 1938, 272; Brubaker and Østreng 1999, 301).

While Smolka emphasized the military value of the Arctic in terms of logistics and communications, explorer Vilhjalmur Stefansson depicted the region as a potential front line in any future conflict seven years later. Stefansson had already anticipated a polar Mediterranean where ships and submarines crisscrossed the Arctic carrying trade goods to all corners of the globe, and predicted that these developments would shape the "political and military future of the world" (Stefansson 1922). In January 1945, he fleshed out his prediction, explaining that "if you shoot robot bombs (as heaven preserve us from ever doing) they will cross the Arctic on their way from London to Seattle,

P. Kikkert (✉)
St. Francis Xavier University, Antigonish, NS, Canada

P. W. Lackenbauer
Trent University, Peterborough, ON, Canada
e-mail: pwhitneylackenbauer@trentu.ca

© The Author(s) 2020
K. S. Coates, C. Holroyd (eds.), *The Palgrave Handbook of Arctic Policy and Politics*, https://doi.org/10.1007/978-3-030-20557-7_30

from Pieping to New York, from San Francisco to Moscow. This is the way bombers will fly if we ever permit them to" (Stefansson 1945, 6; Lloyd 1948).

The technological and geopolitical developments of the first half of the twentieth century inspired both Smolka and Stefansson to militarize the geography of the Arctic and highlight the emerging strategic threats and opportunities posed by its geographic features. Rachel Woodward (2004, 3) notes that "military geographies are everywhere; every corner of every place in every land in every part of this world of ours is touched, shaped, viewed, and represented in some way by military forces and activities." The Arctic is a case in point. As Europeans pushed into the region, representations of the Arctic environment as a space best explored and "conquered" by the training, equipment and courage of military personnel became common—prioritizing and normalizing the role of the military in the region. Not until the early twentieth century, however, did the "transformation of assessments" and "reassessment of priorities" occur that led some strategists and polar experts—like Smolka and Stefansson— to conceptualize the Arctic as a militarized space (Farish 2013a, 248; Enloe 2004, 220; Loyd 2009, 864). Their ideas fed into the actual militarization of the Arctic—the mobilization of the region for military purposes—that began in earnest during the Second World War and intensified throughout the Cold War (Coates et al. 2011, 456). During the war and the four decades that followed, the militarization of the Arctic fits the definition offered by Chris Pearson et al. (2010, 3) of a "process that occurs through, and leaves its mark on societies, economies, culture and political structures" and that re-shapes the landscape in "both a physical and cultural sense."

The militarization of the Arctic was dictated by geography and shaped by technology. The intersection of the two transformed the Arctic from a "*military vacuum* prior to World War II, to a *military flank* in the 1950–1970 period, and a *military front* in the 1980s" (Østreng 1992, 30). During these decades, the militarization of the Arctic served as both a *response* to global geopolitical tensions and as a *source* of intensified tensions. The defense projects, military infrastructure, nuclear tests, scientific research, and technological developments that enabled and flowed out of this process irrevocably changed the region: not just militarily and politically, but also in terms of culture and environment. While militarized landscapes often look different from the civilian landscapes that surround them (Pearson et al. 2010, 4), in the Arctic defense projects often overlapped with civilian spaces or created new shared spaces. From the Second World War onwards, Northern Indigenous Peoples endured the destruction of traditional territories and hunting grounds, forced relocations, new settlements, new economies, and environmental destruction concomitant to the militarization of their homelands.

The Arctic has known conflict since humans first settled in the region. The oral histories of Indigenous peoples tell of clashes over land and resources, often against rival cultural groups, and of battles for vengeance (McGhee 1996, 223; Blondin 1997, 130–135; Abel 1993). When European explorers started to intrude into the region, small-scale conflicts often broke out between them and the Indigenous groups they encountered (Dodds and Nuttall 2016, 106). During the three expeditions privateer-turned-explorer Martin Frobisher made to Qikiqtaaluk (Baffin Island) in the 1570s, for example, violence broke out between Inuit and Englishmen, leading to the deaths and abduction of several Inuit. More sustained violence broke out between imperial Russian forces and the Indigenous peoples of Siberia during Russia's colonial expansion to the Arctic coast between the sixteenth and eighteenth centuries. In particular, the Chukchi, inhabiting the Chukchi Peninsula and shores of the Chukchi Sea and Bering Sea, faced frequent hostile expeditions in the first half of the eighteenth century from a Russian government that endeavored to destroy "aggressive Chukchi" with a "military hand" (Abryutina 2007, 329; Collins 1982, 17–44). Russian expansion into Alaska and the Pacific Northwest also led to periodic violence between the Aleut and Tlingit and the troops of the Russian American Company, supported by the Imperial Russian Navy (Black 2004, 89, 158–159). While these early examples of military activity in the region certainly reflect the violence of colonization and empire-building, they did not involve the mobilization of the region for military purposes.

While Europeans may not have understood the Arctic as a militarized space between the sixteenth and nineteenth centuries, they appreciated the geostrategic value of shorter shipping routes via Northwestern or Northeastern Passages, as well as new trade links and access to resource-rich territories—the exploitation of which occasionally demanded a military response. In the 1600s, the White Sea and the town of Arkhangel'sk took on strategic importance as the primary maritime trade route in and out of Russia in the face of Sweden's dominance of the Baltic, with the port eventually housing a small naval force to provide security. Much of Arkhangel'sk's importance was lost, however, with the capture of St. Petersburg from Sweden and the reorientation of Russian naval interests to the Baltic (Hill 2007b, 361–363). As France and Britain fought for control of the North American continent, their military forces sporadically conducted operations in and around Hudson Bay as they fought for control over the key Arctic gateway to the North American fur trade (Eyre 1987, 292). During the eighteenth and nineteenth centuries, Russian and British governments approved naval expeditions into the Arctic with geostrategic and geopolitical considerations in mind: the quest for the Passages and new territory, and the need to maintain the power balance in the region (Baev, 17–18; Lincoln 1994; Black 2004, 39–78; McCannon 1998, 15–16; Wallace 1980).

Within this geopolitical context, expeditions by the Imperial Russian Navy and the Royal Navy explored vast swathes of the High Arctic. The activities of the Royal Navy, in particular, garnered widespread publicity in the decades after the Napoleonic Wars, which inspired the cultural militarization of the Arctic environment as a place best conquered by the training, experience, equipment, and courage of military personnel (Wallace 1980). Other military forces followed into the region: the US Navy and Army personnel that ventured into the Arctic Archipelago and the waters north of Alaska; the Austro-

Hungarian expedition that discovered Franz Josef Land archipelago in 1873; and the exploratory efforts of the Danish naval officer Georf Carl Amdrup along Greenland's northeastern coast, to name just a few (Fogelson 1992; Apollonio 2008; Mills 2003). The mapping and scientific studies undertaken by these military expeditions often proved vital to the legal arguments crafted by states to support their sovereignty over parts of the Arctic. Nevertheless, while members of various state militaries played an important role in the exploration of the Arctic, no one envisioned fighting a war in the region or using it for military purposes in the nineteenth century. In their classic geopolitical studies at the beginning of the twentieth century, Alfred Thayer Mahan and Halford Mackinder ignored the Arctic's strategic potential, beyond viewing the "ice-clad Polar Sea" as a natural defensive barrier (Antrim 2010, 16–18; Zellen 2009, 32–34).

The Russo-Japanese War of 1904–1905, however, forced strategists to consider the military value of Russia's Arctic waters and the region more generally as a military space. The epic seven-month voyage of Russia's 2nd Pacific Squadron through the Baltic Sea, North Sea, and English Channel and around Africa en route to Port Arthur consumed significant time and resulted in tired crews, low morale, the massive expenditure of coal, and maintenance issues, all of which contributed to the disastrous defeat at the Battle of Tshushima. Even before the squadron departed, Russian strategists proposed sending the force to the Pacific via the Northern Sea Route (Northeast Passage), but too little was known about the passage to allow such a maneuver (Kuksin 1991, 300). The naval disaster at Tshushima emphasized the potential military value of the Northern Sea Route, which could allow Russia to deploy her naval forces to various parts of the globe relatively expediently and covertly, effectively addressing the country's limited access to the high seas (Hill 2007b, 365–366; McCannon 2007, 297; Barr 1991, 24). With these strategic implications in mind, the Tsar's government agreed to support the Arctic Ocean Hydrographic Expedition, which explored much of the Northern Sea Route and discovered new territory between 1910 and 1915 (Kuksin 1991, 300). For the first time, the Arctic was treated as a military object waiting to be exploited—even if it did not immediately lead to the deployment of the Russian Navy to northern waters.

The technological and strategic developments of the First World War amplified the potential military value of the Arctic. Wartime maritime operations inspired Robert Peary to highlight the potential of Greenland as a "naval and aeronautical base" to facilitate deployments into the North Atlantic—a way to promote his bid to Washington to explore the region and lay an official territorial claim (Peary 1917, 274). Russia's need for aid from its western allies amplified the strategic importance of Arkhangel'sk (Baev 2009, 18), which served as busy port of entry for military supplies, and led to the establishment of Murmansk on Kola Bay (connected to St. Petersburg by a newly-built railway), which provided even easier access to allied merchant ships (Hill 2007a). To safeguard their northern supply route, the Russians established the Northern Icebreaker Flotilla and then the short-lived Flotilla of the Arctic Ocean, which included a cruiser squadron and minesweeping detachment (Luzin et al. 1994, 4; Hill 2007b, 369–377). While Russia's Arctic waters provided a valuable logistical lifeline during the war, from a Soviet perspective they also served as a gateway to invasion. In the 1918–1919 Siberian intervention, British troops deployed to Murmansk; French and American troops to Arkhangelsk; and Japanese, British, and Canadian troops to Vladivostok to safeguard the stock-piled supplies in the ports. The Allies also ended up providing limited support to White Russian forces (Hill 2007b, 374; Flake 2015, 81). Understanding of the Soviet Arctic as both a strategic lifeline and an exposed flank laid the conceptual foundation for militarization in the decades that followed.

While the First World War brought heightened military activity into the Arctic, the conflict's immediate aftermath led to the partial demilitarization of at least part of the region. On 9 February 1920, 14 contracting parties signed the Svalbard Treaty, which recognized Norwegian sovereignty over the archipelago with circumscribed its rights. Article 9 of the treaty prohibited the construction of naval bases and fortifications or the use of Svalbard for "war-like purposes" (Pedersen 2011, 123). While the Soviets initially objected to the treaty, over time they came to appreciate that the "demilitarized status of the islands, located essentially at the door of Russia's Arctic sector, clearly was to Russia's advantage" (Flake 2015, 81) by removing a potential threat from its northern flank.

In the broader circumpolar north, however, the march toward militarization continued. With the rapid technological development of aircraft, the conceptualization of the Arctic as a strategic corridor providing a hemispheric short cut via the "great-circle route" slowly started to catch on, as did the role it could play in the aerial projection of power (Douhet 2009). The international tension caused by a short-lived dispute over Wrangel Island in the early 1920s placed a spotlight on these developments. Vilhjalmur Stefansson's attempts to convince the Canadian and British governments to claim the island 140 km off the Siberian coastline emphasized its potential strategic value (Stefansson 1925, 145–150). In London, an interdepartmental committee with representatives from the Admiralty and Air Ministry agreed that Wrangel could serve as an important base for an "Arctic Circle Route" which could reduce the journey from London to Tokyo by 3000 miles, and for wireless communications to support naval operations in the North Pacific. The committee emphasized that "the island is the only territory in a vast area to which Great Britain has any claim, and the Admiralty consider that it would be short-sighted policy to surrender our claims to it" (Foreign Office 1923). While neither the Admiralty or Air Ministry envisioned the Arctic as a battlefield, they understood the potential military value of northern islands.

While Britain contemplated an imperial claim to Wrangel Island, the situation awakened their American counterparts to the Arctic's military possibilities. Members of the US military suspected that Britain might renew its alliance with Japan (the Anglo-Japanese Alliance of 1902–1923). If war broke out, a British air base on Wrangel would pose a significant threat to American military assets in Alaska (Diubaldo 1967, 220–221). Ultimately, repeated American inquiries and protests about Wrangel Island cooled British interest in claiming the island, as did strident Soviet objections. To the Kremlin, a foreign capitalist claim to an island off its northern coastline, which Russia had formally claimed in 1916, had dire strategic implications. In 1924, it sent the icebreaker-turned-gunboat *Krasny Oktyabr* to the island to plant a Soviet flag and remove Stefansson's small occupying party (Diubaldo 1978, 185; Flake 2015, 82).

The development of aerial technology, the Wrangel Island affair, Stefansson's depiction of a polar Mediterranean, and polar explorer Bob Bartlett's testimony to the Navy Department in 1923 that the "flying route across the Pole is the aerial Panama Canal of the future," all sparked US Navy (USN) interest in the Arctic (Fogelson 1992, 80–89). While USN plans to deploy the airship *Shenandoah* to the Arctic were blocked by funding issues, the Navy did support the aerial expedition of Donald Macmillan and Eugene McDonald in 1925 to search for northern islands that could be used as air bases. To secure the Navy's support, the men had argued that while the United States had allowed Canada to "arbitrarily" claim islands north of the continent in the past because of the "supposed uselessness of the land," the development of aircraft had vastly increased their potential commercial and military value (Bryant and Cones 2000, 181).

While the US Navy's interest in the Arctic eventually fizzled out, American General Billy Mitchell became an outspoken proponent of the military value of the region. In 1935, he testified to Congress that Alaska was "the most central place in the world for aircraft," "the most strategic place in the world" and that "in the future whoever holds Alaska will hold the world" (Conn et al. 1964, 247; Perras, 28–53). Despite these warnings, few Canadian or American strategists took the defense of the northern approaches to the continent seriously in the interwar years. Canadian and American soldiers who went north in this era did so mainly for national development interests. The US government used service personnel and the Army Corps of Engineers for surveys, mapmaking, and construction projects to improve harbors, roads, and trails throughout Alaska, often for commercial purposes (Mighetto and Homstad 1997, 11–22). The Canadian government tasked the Royal Canadian Air Force with aerial mapping and ice reconnaissance in Davis and Hudson Straits, while the Royal Canadian Corps of Signals established a network of radio installations (Eyre 1987, 293–294; Lackenbauer 2013, 5–8). Thus, at the start of the Second World War, the North remained a distant priority for most North American defense planners. Canadian historian Charles Stacey (1940, 5) captured the sentiment well when he stated that on Canada's "northern territories those two famous servants of the Czar, Generals January and February, mount guard for the Canadian people all year round."

In sharp contrast, the Soviet Union invested a great deal of time, energy, and money into the militarization of the Russian Arctic in the interwar years. Throughout the 1920s, the military supported exploratory efforts in the High Arctic, assisting the Northern Scientific-Commercial Expedition

(Seveskpeditsiia) and the Committee of the Northern Sea Route (the Komseveroput), which controlled the core of the Soviet Union's Arctic strategy (McCannon 1998, 23–25). Stalin initiated a dramatic increase in state activity in the High Arctic in the 1930s. The Soviets wanted to achieve "physical mastery" of the entire region to pave the way for economic development and a "suitable payoff"—although much of the research and activities also had explicit military purposes (McCannon 1998, 42). Under the institutional control of the Glavsevmorput—the Main Administration of the Northern Sea Route, or the "Commissariat of Ice"—the Soviets initiated a frenzy of exploration, meteorological studies, scientific programs, aerial advancement, and permanent physical occupation in a systematic investigation of the region. In the High Arctic, their signature projects involved the improvement of Arctic aviation (including several trans-polar flights) and the development of the Northern Sea Route, which a Soviet icebreaker first transited in a single season in 1932, raising hope that it could be used as a regular shipping lane and an avenue to reinforce military forces in the Far East (McCannon 1998, 57–58, 70–80; Emmerson 2010, 31; Armstrong 2011).

The "centralization of scientific research under Soviet power … merged seamlessly with the militarization of the Soviet Arctic" (Doel et al. 2014, 63). Scientific accomplishments, especially around the study and use of the NSR, and the creation of the Baltic-White Sea canal demanded and facilitated the creation in 1933 of a Soviet Northern Flotilla, based in Murmansk and later Polyarny. Four years later, the Soviets went a step further and established a Northern Fleet to safeguard the NSR and maritime lines of communication in the Arctic, while also interdicting enemy use of the North Sea and North Atlantic (Hill 2003, 2007b, 378–379; Luzin et al. 1994, 7; McCannon 2007). The vast improvements made by the Soviets in Arctic aviation and air infrastructure also provided an additional layer of security of the Northern Sea Route (Emmerson, 106; McCannon 2010).

During the Second World War, the militarization of the Arctic began in earnest as it played an important role in the areas of supply, logistics, and meteorology. Examples include the Arctic convoys on the Murmansk Run, the Russian use of the Northern Sea Route, the establishment of Allied and German weather stations to support aerial and naval operations in the Atlantic, and the supply line between Ladd Field, Alaska, and the Soviet Union (Emmerson 2010, 106–108; Grant 2010; Doel et al. 2014; Armstrong 2011, 144; Dege 2004). The most important developments related to Allied convoys that ferried supplies to Arkhangelsk and Murmansk—the shortest and most direct route to bring aid into the Soviet Union. The Arctic route also proved incredibly dangerous, and the German Kriegsmarine and Luftwaffe based in occupied northern Norway patrolled the Kara Sea and the approaches to Novaya Zemlya in search of Allied shipping, often with deadly effect (Luzin 2007, 426; Woodman 2004). Meanwhile, the Soviet Northern Fleet provided protection to the Allied convoys and defended the Northern Sea Route, which the

Glavsevmorput operated as a supply route for lend-lease aid from the United States (Hill 2007a; Luzin 2007; McCannon 2007, 419; Suprun 2007).

The Arctic and sub-Arctic also became the scene of high-intensity conflict between 1939 and 1945, including fierce fighting in northern Norway, Finland, and on the Aleutian Islands, and much smaller operations on Spitsbergen and Greenland. The Winter War between the Soviet Union and Finland in 1939–1940 involved combat operations in Finnish Lapland, especially around the ice-free port of Petsamo (Trotter 2002, 171–174). Finnish and German troops fought together on the northern front during the invasion of the Soviet Union in a failed attempt to block access to the harbors on the Barents Sea and take Murmansk. In late 1944, the Finns made peace with the Soviets and switched sides, forcing the Germans out of their territory in the Lapland War (Emmerson 2010, 108). The Petsamo-Kirkenes Offensive of October 1944 saw the Soviet Army and Northern Fleet expel the Germans from their northern defenses in Finland and clear them from the northernmost part of Norway, Finnmark (Gebhardt 1990; Ziemke 1976; Chew 1981). Across the Arctic in the Pacific Northwest, the war brought the Japanese raid on Dutch Harbor, Alaska and the subsequent American-Canadian campaign to retake Attu and Kiska in the Aleutian Islands in 1943 (Garfield 1995; Perras 2003). Less well known are the small-scale military operations that took place on Spitsbergen, where British, Canadian, and German forces conducted raids to destroy coal mines, supplies, equipment, and weather stations, and, in the Canadian case, evacuate 2000 Soviet and Norwegian miners (Dean and Lackenbuaer 2017; Arlov 1994, 74–76).

The war also inspired the construction of military infrastructure across Arctic North America. The Battle of the Atlantic and the North Atlantic Air Route gave Iceland and Greenland great strategic importance and led to the creation of new air and sea bases. Between 1941 and 1945, the United States established extensive facilities for air and sea transportation in Greenland, as well as radio beacons, radio stations, weather stations, defenses, and search-and-rescue stations pursuant to a bilateral agreement with the Kingdom of Denmark (Grant 2010, 247–257). The American-built airbase Bluie West in Narsarsuaq in south Greenland became an essential link in the North Atlantic ferry route, taking in over 10,000 aircraft at its peak (Conn et al. 1964, 451–452). The Americans also doubled the size of their forces in Alaska and constructed the Northwest Staging Route (NWSR—a string of airfields from Edmonton to Alaska), the Alaska Highway, and a 1000-km oil pipeline from Norman Wells to Whitehorse—projects that brought nearly 40,000 American military personnel into the Canadian Northwest (Coates and Morrison 1992, 1994; Dziuban 1959; Grant 1988). American defense activities expanded into the eastern Canadian Arctic with the establishment of the Crimson Route, an alternate path for ferrying planes and material to Britain, which involved the construction of large-scale installations at various sites (Coates et al. 2008, 61; Farish and Lackenbauer 2009).

Although the Americans abandoned much of this northern military infrastructure in the latter stages of the war, this spasm of militarization had a transformative impact on some Northern Indigenous peoples. The defense projects undertaken across Alaska, the Canadian North, and Greenland represented the first intrusion of the state into the lives of many Indigenous groups and caused profound cultural and environmental changes (Gagnon and Elders 2002). Heavy fighting took place in vast swaths of Sápmi, the traditional land of the Sámi in northern Finland, Russia, and Norway. For example, during the Lapland War, the Finnish Sámi were evacuated to southern Finland and returned to a decimated homeland littered with thousands of land-mines, the result of the German scorched earth policy, which also extended to Sámi territory in Finnmark (Seitsonen and Koskinen-Koivisto 2018). After the Japanese occupation of Kiska and Attu, hundreds of Aleuts were also forced out of their traditional territory on the Aleutian and Pribilof Islands and evacuated to locations in southeast Alaska, where they endured harsh conditions (Kirtland and Coffin 1981; Kohlhoff 1995). Other Northern Indigenous Peoples volunteered or were conscripted for combat roles. Indigenous peoples in Alaska joined the Alaska Territorial Guard to watch over the territory—the first time Aleuts, Athabaskan, Inupiaq, Haida, Tlingit, Tsimshian, Yupik, and others worked together (Marston 1969). Furthermore, during the German invasion of the Soviet Union, many Arctic peoples—the Komi, Sami, Nenets, Hanti, Mansi, and Karelian—were mobilized to serve in the Red Army, often placed in special reindeer brigades serving on the northern front. They became known for their herding, transport, scouting, and marksmanship abilities (Gorter-Gronvik and Suprun 2000).

Strategic bombing, the use of the atomic bombs, and growing hostility between the US and Soviet Union conspired to intensify and accelerate militarization of the Arctic in the postwar world. Polar projection maps emphasized the United States' proximity to the Soviet Union and American and Russian strategists envisioned hostile bombers flooding over the northern approaches to wreak havoc on their national heartlands (Jockel 1987; Isemann 2009, 34–40). On 5 December 1945, General Henry H. Arnold, the retiring Commander in Chief of the US Army Air Force declared publicly that the Arctic would become "the strategic center" of a third world war (Eayrs 1972, 320). Accordingly, defense planners contemplated ambitious Arctic projects to serve their broader security interests.

In Washington, worries about the Soviets attacking across the Arctic were compounded by the knowledge that the Soviets were ahead of the Americans in every facet of northern operations: air, sea, and ground. "The United States came later to the Arctic scene, and was exceptional in seeing the Arctic primarily in military and national security terms," Doel et al. (2014, 69) recently observed. "By the early Cold War, White House and Pentagon leaders viewed knowledge about the environment of the far north as vital for waging war against the Soviet Union." US defense research focused on geophysical fields like meteorology, geology, seismology, and oceanography and the Americans initiated and funded projects across Alaska, the Canadian Arctic and Greenland (Heymann et al. 2010). As early as 1946, Washington asked for permission to perform aerial flights and build weather stations in the Canadian Arctic (Grant 1988, 2010; Coates et al. 2008). In 1947, the Office of Naval Research established the Arctic Research Laboratory at Point Barrow, Alaska, while Canada and the United States launched the Joint Experimental Station at Fort Churchill, alongside Canada's Defence Research Northern Laboratory (Doel et al. 2014; Farish 2013b; Iarocci 2008). Ground exercises were held in the Canadian North and Alaska to determine how to move and supply a force operating in the Arctic (Horn 2002; Lackenbauer and Kikkert 2016). During the Cold War, the scope of American defense research expanded to include the development of scientific stations on ice islands and Project Iceworm, a program to build a network of mobile nuclear launch sites under the Greenland ice sheet that led to the construction of Camp Century under the ice (Petersen 2008).

With increasing American knowledge about the Arctic came an expanded military footprint. The development of air bases for Strategic Air Command and interceptors, aerial surveillance, air defense systems, and early warning systems in Alaska, the Canadian Arctic, Iceland and Greenland became a pivotal part of the American nuclear deterrent. In Alaska, the military undertook a massive construction program with new installations for bombers and interceptors, forward bases, radar, and communication stations. The Americans endeavored to transform Greenland into the world's largest "stationary aircraft carrier" given that bombers based on the island could reach targets in the Soviet Union faster than from any other place in North America. The Americans constructed the air base at Thule in the middle of traditional hunting territory of the Inughuit, and the installation quickly grew to encompass 2600 acres, with 82 miles of roads, fuel tanks with a capacity of 100 million gallons, housing for 12,000, and extensive air defenses (Doel 2016, 29; Grant 2010, 311–318).

As the Cold War heated up in the 1950s and the Soviets tested their first nuclear weapons, the Americans spearheaded the effort to construct air defense systems in the northern approaches to the continent that would preserve the nuclear deterrent and protect the industrial heartland of North America. Built along the 69th parallel, approximately 200 miles above the Arctic Circle, the Distant Early Warning (DEW) Line, consisting of seven sites in Alaska and 22 in Canada, stretched over 3000 miles from Lisburne on Alaska's northwest coast to Cape Dyer on the east coast of Canada's Baffin Island. Additional stations were later added to extend radar coverage to Greenland and the eastern Aleutians by 1958. Construction of the line "required the biggest task-force of ships assembled since the invasion of Europe and the largest air operation since the Berlin airlift to take in the supplies," a Canadian official trumpeted in a 1957 magazine article, along with the transportation of a large temporary workforce to the Arctic and the development of innovative transportation systems and communications infrastructure (Lackenbauer 2013). The Soviet development of the first intercontinental ballistic missile (ICBM) in 1957 changed the strategic equation by replacing the manned bomber as the Soviets' primary nuclear delivery vehicle, however, prompting the United States to shift its focus away from the North American Arctic, although Ballistic Missile Early Warning System (BMEWS) radar stations were constructed in Alaska and Greenland (Jockel 2007).

Northern Norway also occupied an important geostrategic position throughout the Cold War because it bordered the Soviet Union's heavily militarized Kola Peninsula and strategists considered it essential to control Soviet access to the Norwegian Sea and the Atlantic Ocean. Norway enhanced its research facilities at Ny-Ålesund on Svalbard, while Sweden's Kiruna Geophysical Observatory monitored Soviet nuclear tests on Novaya Zemlya (Doel et al. 2014, 63–64). The small town of Kirkenes represented a potential front line in any future ground war with the Soviet Union, and the Norwegian military frequently patrolled the area and prepared for an invasion. In such a scenario, NATO's northern flank would be protected by the Allied Command Europe Mobile Force, and plans were made for the deployment of British, European, American, and Canadian reinforcements (Lund 1989).

On the Soviet side of the Arctic, the Cold War also intensified militarization. While Russian scientific research in the region continued to have strong economic and commercial motivations, it also served military purposes (Doel et al. 2014, 65–66). Early Soviet efforts focused on Arctic flying and building effective year-round airfields. The Kremlin also feared bomber attacks across the Arctic and invested heavily in strategic surveillance, interceptor bases, and air defense capabilities across its northern coastline. They also established a network of forward bomber bases, stretching from the Kola Peninsula in the west to Chukotka in the east, from which Soviet bombers could reach any target in the United States (Leonard 2011, 33). Some of these bases, which could accommodate fully-loaded heavy bombers, were located on the remote High Arctic islands of Franz Josef Land, Novaya Zemlya, and Severnaya Zemlya (Huitfeldt et al. 1992, 93; Åtland 2011, 270). The Kremlin also constructed an integrated, radar-based early warning system in the mid-1950s to cover the approaches to the Soviet heartland (Leonard 2011), and the Soviet ballistic warning system, known as Sistema Preduprezhdeniya o Raketnom Napadnii, was functional by the late 1960s (Åtland 2011, 270). The Soviet Navy practiced inter-theater maneuverability through the Northern Sea Route, although only a handful of ships moving between the Northern and Pacific fleets completed this difficult task (Brubaker and Østreng 1999, 301–311). In an unambiguous case of militarization, the Kremlin chose Novaya Zemlya as their primary northern nuclear test site in 1955, and they conducted 90 atmospheric tests and 41 underground tests there, including the most powerful nuclear bomb ever exploded, the Tsar Bomba (Bergman et al. 1996).

The Kola Peninsula, which was soon dotted with naval and air bases, strategic air defense sites, and army camps, became the most intensely militarized area in the Soviet Arctic. The ice-free naval bases on the peninsula afforded Russian surface vessels and submarines ready access to the North Atlantic via the Barents and Norwegian Seas. Due to the relatively short range of Soviet submarine-launched missiles, they could only hit targets in the central United States by sailing into the Atlantic. As the Cold War progressed, however, NATO's anti-submarine defenses grew more effective in the Greenland-Iceland-United Kingdom and Norway (GIUK) gap. In this gap and in the Norwegian Sea between Andøya and Bear Island, the US Navy eventually deployed its Sound Surveillance System (SOSUS) chain of underwater listening posts (Luzin et al. 1994; Åtland 2011, 269–271). These strong NATO defenses would precipitate a Soviet naval pivot to the Arctic in the 1970s and 1980s.

The construction of defense installations and the deployment of military forces to the Arctic during the Cold War had a dramatic impact on Northern Indigenous peoples. Militarization brought new economies, communities, infrastructure, and technology to the region, with modern communication and transportation networks connecting it to southern metropoles like never before. In Canada, for example, Inuit were drawn to DEW Line sites for employment and these locations became nuclei for new communities and intensified government administration (Lackenbauer and Farish 2007). Forced relocations also disrupted Indigenous populations. To allow for the construction of the Danish base at Thule, the American military relocated 27 Inughuit families 140 km north to the newly created town of Qaanaaq in 1953 (Grant 2010, 317). In Alaska, the military relocated the village of Kaktovik on Barter Island to make room for an airfield. The Soviets resettled Indigenous groups that lived along the coast of Siberia, while the "Ice Curtain" drawn by the United States and Soviet Union across the Bering Sea separated Inuit—a transnational people—for decades (Abryutina 2007, 332). Defense installations and training areas also polluted lands and waters in Indigenous homelands throughout the Circumpolar North (Heininen 1994, 2004, 219). Soviet nuclear tests also had an immense, deleterious effect on the environment and people, leading to relocations and exposure to high doses of radiation (Josephson 2005, 286; Heininen 1994, 155–157). The environmental and social legacies of militarization continue to mark the region today.

Although the introduction of ICBMs deflected primary attention away from the Arctic as a corridor route for manned bombers, its rising importance as a theater for submarine operations ensured that geostrategic interest in the region did not disappear. In 1958, the deployment of the nuclear-powered submarine USS *Nautilus* to the North Pole highlighted the prospects for naval operations in this new maritime theater. While US Navy wished to exploit the Arctic Ocean as a strategic route to attack the Soviet northern coastline or shipping on the Northern Sea Route, and as a corridor connecting the Pacific to the Atlantic, they were forced to scrap their plans after the loss of the USS *Thresher* demanded an overhaul of the American submarine fleet (Lajeunesse 2013, 509; Lyon 1963). During the period when the Americans put their Arctic submarine aspirations on hold, the Soviet Navy started to strengthen its position in the region—a move facilitated by the much-improved missile range of Soviet submarines in the 1970s, which meant that they could assume firing positions throughout the Arctic and strike at the United States, China, and Europe (Østreng 1992; Lajeunesse 2013, 516). Soviet naval doctrine's "rear deployment strategy" (Brubaker and Østreng 1999, 302) incorporated a "bastion" concept predicated on the idea that the SSBN (nuclear-powered, ballistic missile-carrying submarine) force "would be strongest when it could conduct its operations—including missile launches—from relatively secure home waters" north of the GUIK gap (Åtland 2007, 500; Bellany 1982; Breemer 1985). This Arctic-focused naval strategy prompted even more militarization of the Kola Peninsula, with new submarine bases, naval, and air defenses.

The development of the Arctic into a "submarine sanctuary"—relatively inaccessible to most naval vessels and with grinding ice that masked submarine noise—was well underway (Østreng 1982; Purver 1984, 1988; Miller 1988). In 1981, the Soviets launched the Typhoon-class SSBN, their first submarine designed specifically for under-ice operations, which coincided with the development of long-range cruise missiles with a range of 3000 km. This made areas of the Canadian Arctic perfect launching positions to strike at the North American heartland (Lajeunesse 2013, 519). By the mid-1980s, the Soviet Northern Fleet—stationed on the Kola Peninsula—boasted 203 submarines and 220 surface vessels, representing the most modern force in the Soviet "strategic submarine fleet" (Brubaker and Østreng 1999, 302).

The expansion and activities of the Northern Fleet acted as a magnet, drawing the American submarine fleet into the region. Thus, the 1980s also saw a dramatic increase in US naval attention to the Arctic (including plans to attack the Soviets in their northern bases if war ever broke out) as they developed their own "forward deployment strategy" (Brubaker and Østreng 1999, 302; Lajeunesse 2013, 519). In a wartime scenario, US submarines planned to hunt Soviet nuclear submarines under the ice and attack Russian convoys along the NSR. In response, the Soviets established underwater detection systems in the waters off the Kola Peninsula and Kamchatka, and American and Russian submarines played a dangerous game of "cat and mouse" in the frigid northern waters off the Soviet coast (Åtland 2011, 270). On the other side of the Arctic, the Canadians tested acoustic and magnetic listening systems at chokepoints in its Arctic Archipelago to detect Soviet submarines (Lajeunesse and Carruthers 2013, 4–9). Cumulatively, the American forward deployment into the Soviet submarine bastion in the Barents, alongside submarine operations across the Arctic, ratcheted up Cold War tensions in the late 1970s and 1980s (Emmerson 2010, 116) that would not abate until the fall of the Berlin Wall and the collapse of the Soviet Union signaled American hegemony and the onset of a new post-Cold War global and regional order.

In light of the threat posed by the American forward deployment strategy, President Mikhail Gorbachev gave his famous Murmansk speech in 1987 in which he decried American attempts to initiate an arms race in the Arctic. "One can feel here the freezing breath of the 'Arctic Strategy' of the Pentagon," Gorbachev announced. To avoid an arms race, he asked that the region be considered a zone of peace for international cooperation and proposed a scaling down of naval and air force activities in the Baltic, Northern, and Greenland Seas (Emmerson 2010, 117; Åtland 2008). Kristian Åtland has argued that Gorbachev's speech marked "the beginning of a new era in the Arctic. It was an important turning point in Soviet Arctic policies and it contributed in a number of ways to the desecuritization of interstate relations in the region" (Åtland 2008, 305). The Murmansk speech fostered East-West dialogue and the development of regional cooperation in the Arctic—processes facilitated by the end of the Cold War. After four decades of intensive militarization the Arctic, the thawing of tensions opened space for economic development, commercial interests, environmental protection, and political change.

Militarization would not be wiped away so easily. The post-Cold War Arctic saw a decline in defense installations and military deployments but the region's military potential did not disappear. Even Gorbachev's appeals to demilitarize the Arctic deliberately excluded the Barents Sea and Kola Peninsula from his delimitation of the region so that he could safeguard the home of Russia's Northern Fleet. Lassi Heininen (2010, 229) observed that in a "wholly militarized" region deeply embedded "threat perceptions and enemy pictures… could have multifunctional and long-lasting impacts." Basic military structures (physical and conceptual) remained in place even after the Cold War ended, and the recent resurgence of threat perceptions spurred by political jousting, academic commentary, and media coverage about Arctic boundaries, resources, and geopolitical competition invite debate about the future of regional security in an increasingly complex international order.

References

Abel, Kerry. 1993. *Drum Songs: Glimpses of Dene History*. Montreal and Kingston: McGill-Queen's University Press.

Abryutina, Larisa. 2007. Études/Inuit/Studies 31 (1–2): 325.

Antrim, Caitlyn. 2010. The Next Geographical Pivot: The Russian Arctic in the Twenty-First Century. *Naval War College Review* 63 (3, Summer): 15–37.

Apollonio, Spencer. 2008. *Lands That Hold One Spellbound: A Story of East Greenland*. Calgary: University of Calgary Press.

Arlov, Thor. 1994. *A Short History of Svalbard*. Oslo: Norwegian Polar Institute.

Armstrong, Terrence. 2011. *The Northern Sea Route: Soviet Exploitation of the North East Passage*. Cambridge: Cambridge University Press.

Åtland, Kristan. 2007. The Introduction, Adoption and Implementation of Russia's Northern Strategic Bastion Concept, 1992–1999. *Journal of Slavic Military Studies* 20 (4, Oct.): 499–528.

———. 2008. Mikhail Gorbachev, the Murmansk Initiative, and the Desecuritization of Interstate Relations in the Arctic. *Cooperation and Conflict* 43 (3): 289–311.

———. 2011. Russia's Armed Forces and the Arctic: All Quiet on the Northern Front? *Contemporary Security Policy* 32 (2): 267–285.

Baev, P.K. 2009. Troublemaking and Risk-Tasking: The North in Russian Military Activities. In *Russia and the North*, ed. E.W. Rowe. Ottawa: University of Ottawa Press.

Barr, William. 1991. The Arctic Ocean in Russian History to 1945. In *The Soviet Maritime Arctic*, ed. L.W. Brigham. Annapolis: Naval Institute Press.

Bellany, Ian. 1982. Sea Power and the Soviet Submarine Forces. *Survival* 24 (1): 2–8.

Bergman, R., A. Baklanov, and B. Segerståhl. 1996. Overview of Nuclear Risks on the Kola Peninsula. Summary Report. IIASA Radiation Safety of the Biosphere. Laxenburg: International Institute for Applied Systems Analysis.

Black, Lydia. 2004. *The Russians in Alaska, 1732–1867*. Fairbanks: University of Alaska Press.

Blondin, George. 1997. *Yamoria: The Lawmaker*. Edmonton: NeWest Press.

Breemer, Jan. 1985. The Soviet Navy's SSBN Bastions: Evidence, Inference, and Alternative Scenarios. *The RUSI Journal* 130 (1): 18–26.

Brubaker, R. Douglas, and Willy Østreng. 1999. The Northern Sea Route Regime: Exquisite Superpower Subterfuge? *Ocean Development & International Law* 30: 299–331.

Bryant, John, and Harold N. Cones. 2000. *Dangerous Crossings: The First Modern Polar Expedition, 1925*. Annapolis: Naval Institute Press.

Chew, Allen F. 1981. Fighting the Russians in Winter: Three Case Studies (PDF). *Leavenworth Papers*. Fort Leavenworth, Kansas: Combat Studies Institute, U.S. Army Command and General Staff College.

Coates, Ken, P. Whitney Lackenbauer, William R. Morrison, and Greg Poelzer. 2008. *Arctic Front: Defending Canada in the Far North*. Toronto: Thomas Allen.

Coates, Ken, and William Morrison. 1994. *Working the North: Labor and the Northwest Defense Projects 1942–1946*. Anchorage: University of Alaska Press.

Coates, Kenneth, and William Morrison. 1992. *The Alaska Highway in WWII: The U.S. Army of Occupation in Canada's Northwest*. Norman: University of Oklahoma Press.

Coates, Peter, Tim Cole, Marianna Dudley, and Chris Pearson. 2011. Defending Nation, Defending Nature? Militarized Landscapes and Military Environmentalism in Britain, France, and the United States. *Environmental History* 16 (3, July): 456–491.

Collins, David. 1982. Russia's Conquest of Siberia: Evolving Russian and Soviet Historical Interpretations. *European History Quarterly* 12: 17–44.

Conn, S., R. Engelman, and B. Fairchild. 1964. *The U.S Army in World War II: The Western Hemisphere: Guarding the United States and its Outposts*. Washington: Department of the Army.

Dean, Ryan, and P. Whitney Lackenbauer. 2017. Conceiving and Executing Operation Gauntlet: The Canadian-Led Raid on Spitzbergen, 1941. *Canadian Military History* 26 (2): Article 16.

Dege, Wilhelm. 2004. *War North of 80: The Last German Arctic Weather Station of World War II*. Calgary: University of Calgary Press.

Diubaldo, Richard. 1967. Wrangling over Wrangel Island. *Canadian Historical Review* 48 (3, Sep.): 201–226.

———. 1978. *Stefansson and the Canadian Arctic*. Montreal: McGill-Queen's University Press.

Dodds, Klaus, and Mark Nuttall. 2016. *The Scramble for the Poles*. Cambridge: Polity Press.

Doel, Ronald. 2016. Defending the North American Continent: Why the Physical Environmental Sciences Mattered in Cold War Greenland. In *Exploring Greenland*

Cold War Science and Technology on Ice, ed. Ronald E. Doel, Kristine C. Harper, and Matthias Heymann, 25–46. New York: Palgrave Macmillan.

Doel, Ronald, Robert Marc Friedman, Julia Lajus, Sverker Sörlin, and Urban Wråkberg. 2014. Strategic Arctic Science: National Interests in Building Natural Knowledge e Interwar era through the Cold War. *Journal of Historical Geography* 44: 60–80.

Douhet, Giulio. 2009. *The Command of the Air*. Tuscaloosa, AL: University of Alabama Press.

Dziuban, Stanley. 1959. *Military Relations between the United States and Canada, 1939–1945*. Washington: Office of the Chief of Military History, Department of the Army.

Eayrs, James. 1972. *In Defence of Canada Vol. 3: Peacemaking and Deterrence*. Toronto: University of Toronto Press.

Emmerson, Charles. 2010. *The Future History of the Arctic*. New York: Public Affairs.

Enloe, C. 2004. *The Curious Feminist: Searching for Women in a New Age of Empire*. Berkeley: University of California Press.

Eyre, Kenneth. 1987. Forty Years of Military Activity in the Canadian North. *Arctic* 40 (4, Dec.): 292–299.

Farish, Matthew. 2013a. Militarization. In *The Ashgate Research Companion to Critical Geopolitics*, ed. K. Dodds, M. Kuus, and J. Sharp, 247–262. Farnham, UK: Ashgate.

———. 2013b. The Lab and the Land: Overcoming the Arctic in Cold War Alaska. *Isis*, 1–29, March.

Farish, Matthew, and Whitney Lackenbauer. 2009. High Modernism in the Arctic: Planning Frobisher Bay and Inuvik. *Journal of Historical Geography* 35 (3): 517–544.

Flake, Lincoln. 2015. Forecasting Conflict in the Arctic: The Historical Context of Russia's Security Intentions. *Journal of Slavic Military Studies* 28: 72–98.

Fogelson, Nancy. 1992. *Arctic Exploration and International Relations, 1900–1932*. Fairbanks: University of Alaska Press.

Foreign Office. 1923. Memorandum on the History, Value and Ownership of Wrangel Island, 2 July 1923, NAA, A981, ARC 2, Arctic. Canadian Activity.

Gagnon, Melanie, and Inuit Elders. 2002. *Inuit Recollections on the Military Presence in Iqaluit*. Iqaluit: Nunavut Arctic College.

Garfield, Brian. 1995. *The Thousand-Mile War: World War II in Alaska and the Aleutians*. Fairbanks: University of Alaska Press.

Gebhardt, James F. 1990. *The Petsamo-Kirkenes Operation: Soviet Breakthrough and Pursuit in the Arctic, October 1944* (Leavenworth Papers, No. 17). Fort Leavenworth, Kansas: Combat Studies Institute, US Army Command and General Staff College.

Gorter-Gronvik, Waling, and Mikhail N. Suprun. 2000. Ethnic Minorities and Warfare at the Arctic Front 1939–45. *Journal of Slavic Military Studies* 13: 127–142.

Grant, Shelagh. 1988. *Sovereignty or Security? Government Policy in the Canadian North, 1936–1950*. Vancouver: University of British Columbia Press.

———. 2010. *Polar Imperative: A History of Arctic Sovereignty in North America*. Vancouver: Douglas and McIntyre.

Heininen, Lassi. 1994. The Military and the Environment: An Arctic Case. In *Green Security or Militarized Environment*, ed. Jyrki Käkönen, 153–165. Brookfield, VT: Dartmouth Publishing.

———. 2004. Circumpolar International Relations and Geopolitics. In *Arctic Human Development Report*. Akureyri: Stefansson Arctic Institute. Chapter 12.

————. 2010. Globalization and Security in the Circumpolar North. In *Globalization and the Circumpolar North*, ed. Lassi Heininen and Chris Southcott. Fairbanks: University of Alaska Press.

Heymann, Matthais, Henrik Knudsen, Maiken L. Lolck, Henry Nielsen, Kristian H. Nielsen, and Christopher J. Ries. 2010. Exploring Greenland: Science and Technology in Cold War Settings. *Scientia Canadensis* 33 (2): 11–42.

Hill, Alexander. 2003. The Birth of the Soviet Northern Fleet 1937–42. *Journal of Slavic Military Studies* 16 (2): 65–82.

————. 2007a. British Lend Lease Aid and the Soviet War Effort, June 1941–June 1942. *The Journal of Military History* 71 (3): 773–808.

————. 2007b. Russian and Soviet Naval Power and the Arctic from the XVI Century to the Beginning of the Great Patriotic War. *Journal of Slavic Military Studies* 20: 359–392.

Horn, Lt. Colonel Bernd. 2002. Gateway to Invasion or the Curse of Geography? The Canadian Arctic and the Question of Security, 1939–1999. In *Forging a Nation: Perspectives on the Canadian Military Experience*, ed. Bernd Horn, 307–332. St. Catharines: Vanwell Publishing.

Huitfeldt, Tønne, Tomas Ries, and Gunvald Øyna. 1992. *Strategic Interests in the Arctic*. Oslo: Norwegian Institute for Defence Studies.

Iarocci, Andrew. 2008. Opening the North: Technology, Training and the Fort Churchill Joint Services Experimental Test Station, 1946–64. *Canadian Army Journal* 10 (4, Winter): 74–95.

Isemann, James Louis. 2009. To Detect, to Deter, to Defend: The Distant Early Warning (DEW) Line and Early Cold War Defense Policy, 1953–1957. Ph.D. diss., Kansas State University.

Jockel, Joseph. 1987. *No Boundaries Upstairs: Canada, the United States, and the Origins of North American Air Defence, 1945–1958*. Vancouver: University of British Columbia Press.

————. 2007. *Canada in NORAD 1957–2007: A History*. Montreal: McGill-Queens University Press.

Josephson, Paul. 2005. *Red Atom: Russia's Nuclear Power Program from Stalin to Today*. Pittsburgh: University of Pittsburgh Press.

Kirtland, John, and David Coffin. 1981. *The Relocation and Internment of the Aleuts during World War II*. Anchorage: Aleutian/Pribilof Islands Association.

Kohlhoff, Dean. 1995. *When the Wind Was a River: Aleut Evacuation in World War II*. Seattle: University of Washington Press.

Kuksin, I.Ye. 1991. The Arctic Ocean Hydrographic Expedition 1910–1915. *Polar Geography* 15 (4): 299–309.

Lackenbauer, P. Whitney. 2013. The Military as Nation-Builder: The Case of the Canadian North. *Journal of Military and Strategic Studies* 15 (1): 1–34.

Lackenbauer, P. Whitney, and Matthew Farish. 2007. The Cold War on Canadian Soil: Militarizing a Northern Environment. *Environmental History* 12 (4, Oct.): 920–950.

Lackenbauer, P. Whitney, and Peter Kikkert. 2016. Lessons in Arctic Operations: The Canadian Army Experience, 1945–1956. Documents on Canadian Arctic Sovereignty and Security (DCASS) No. 7. Calgary and Waterloo: Centre for Military and Strategic Studies/Centre on Foreign Policy and Federalism.

Lajeunesse, Adam. 2013. A Very Practical Requirement: Under-Ice Operations in the Canadian Arctic, 1960–1986. *Cold War History* 13 (4): 507–524.

Lajeunesse, Adam, and Bill Carruthers. 2013. The Ice has Ears: The Development of Canadian SOSUS. *Canadian Naval Review* 9 (3, Fall): 4–9.

Leonard, Barry. 2011. *History of Strategic and Ballistic Missile Defense, Vol. II: 1956–1972*. Collingdale, PA: Diane Publishing Co.

Lincoln, Bruce. 1994. *The Conquest of a Continent: Siberia and the Russians*. New York: Random House.

Lloyd, Trevor. 1948. Aviation in Arctic North America and Greenland. *Polar Record* 1 (35): 163–176.

Loyd, J. 2009. "A Microscopic Insurgent": Militarization, Health, and Critical Geographies of Violence. *Annals of the Association of American Geographers* 99: 863–873.

Lund, John. 1989. *Don't Rock the Boat: Reinforcing Norway in Crisis and War*. Santa Monica: The RAND Publication Series.

Luzin, Dmitrii. 2007. The Northern Sea Route During World War II, 1939–1945. *Journal of Slavic Military Studies* 20: 421–432.

Luzin, Gennady, Michael Pretes, and Vladimir Vasiliev. 1994. The Kola Peninsula: Geography, History and Resources. *Arctic* 47 (1, Mar.): 1–15.

Lyon, Waldo. 1963. The Submarine and the Arctic Ocean. *Polar Record* 11 (75): 699–705.

Marston, Marvin. 1969. *Men of the Tundra: Alaska Eskimos at War*. New York: October House.

McCannon, John. 1998. *Red Arctic: Polar Exploration and the Myth of the North in the Soviet Union, 1932–39*. Oxford: Oxford University Press.

———. 2007. The Commissariat of Ice: The Main Administration of the Northern Sea Route (GUSMP) and Stalinist Exploitation of the Arctic, 1932–1939. *Journal of Slavic Military Studies* 20: 393–419.

———. 2010. Winged Prometheans: Arctic Aviation as Socialist Construction in Stalinist Russia, 1928–1939. *Scientia Canadensis: Canadian Journal of the History of Science, Technology and Medicine* 33 (2): 75–97.

McGhee, Robert. 1996. *Ancient People of the Arctic*. Vancouver: University of British Columbia Press.

Mighetto, Lisa, and Carla Homstad. 1997. *Engineering in the Far North: A History of the US Army Engineer District in Alaska, 1867–1992*. Missoula: Historical Research Associates.

Miller, Steve. 1988. The Maritime Strategy and Geopolitics in the High North. In *The Soviet Union and Northern Waters*, ed. Clive Archer, 205–238. New York: Routledge.

Mills, William James. 2003. *Exploring Polar Frontiers: A Historical Encyclopedia*. Santa Barbara: ABC-CLIO, Inc.

Østreng, Willy. 1982. Strategic Developments in the Norwegians and Polar Seas: Problems of Denuclearization. *Bulletin of Peace Proposals* 13: 108–111.

———. 1992. Political-Military Relations among the Ice States: The Conceptual Basis of State Behaviour. In *Arctic Alternatives: Civility or Militarism in the Circumpolar North*, ed. Franklyn Griffiths, 26–45. Toronto: Science for Peace/Samuel Stevens Canadian Papers in Peace Studies.

Pearson, Chris, Peter Coates, and Tim Cole. 2010. Introduction: Beneath the Camouflage: Revealing Militarized Landscapes. In *Militarized Landscapes: From Gettysburg to Salisbury Plain*, ed. Chris Pearson, Peter Coates, and Time Cole, 1–20. London: Continuum.

Peary, Robert. 1917. *The Secrets of Polar Travel*. New York: The Century Co.

Pedersen, Torbjørn. 2011. International Law and Politics in U.S. Policymaking: The United States and the Svalbard Dispute. *Ocean Development & International Law* 42: 120–135.

Perras, Galen. 2003. *Stepping Stones to Nowhere, The Aleutian Islands, Alaska, and American Military Strategy, 1867–1945*. Vancouver: University of British Columbia Press.

Petersen, Nikolaj. 2008. The Iceman That Never Came. *Scandinavian Journal of History* 33 (1): 75–98.

Purver, Ron. 1984. Security and Arms Control at the Poles. *International Journal* 39 (4): 888–910.

———. 1988. Arctic Security: The Murmansk Initiative and its Impact. *Current Research on Peace and Violence* 11 (4): 147–158.

Seitsonen, Oula, and Eerika Koskinen-Koivisto. 2018. 'Where the F… is Vuotso?': Heritage of Second World War Forced Movement and Destruction in a Sámi Reindeer Herding Community in Finnish Lapland. *International Journal of Heritage Studies* 24 (4): 421–441.

Smolka, H.P. 1938. Soviet Strategy in the Arctic. *Foreign Affairs*, 272–278, January.

Stacey, C.P. 1940. *The Military Problems of Canada: A Survey of Defence Policies and Strategic Conditions, Past and Present*. Toronto: Ryerson Press.

Stefansson, Vilhjalmur. 1922. *The Northward Course of Empire*. New York: Macmillan.

———. 1925. *The Adventure of Wrangel Island*. New York: Macmillan.

———. 1945. *The Arctic in Fact and Fable*. New York: Foreign Policy Association.

Suprun, Mikhail. 2007. Operation 'West': The Role of the Northern Fleet and Its Air Forces in the Liberation of the Russian Arctic in 1944. *Journal of Slavic Military Studies* 20: 433–447.

Trotter, William. 2002. *The Winter War: The Russo-Finnish War of 1939–1940*. London: Aurum Press.

Wallace, Hugh. 1980. *The Navy, the Company, and Richard King: British Exploration in the Canadian Arctic, 1829–1860*. Montreal/Kingston: McGill-Queen's University Press.

Woodman, Richard. 2004. *Arctic Convoys 1941–1945*. London: John Murray.

Woodward, Rachel. 2004. *Military Geographies*. Oxford: Blackwell.

Zellen, Barry. 2009. *Arctic Doom, Arctic Boom*. Santa Barbara: ABC Clio.

Ziemke, Earl. 1976. *The German Northern Theatre of Operations 1940–1945*, German Report Series. Washington: Department of the Army.

CHAPTER 31

Arctic Climate Change: Local Impacts, Global Consequences, and Policy Implications

Warwick F. Vincent

INTRODUCTION

The Arctic is warming at rates that are more than twice the global average, with pronounced effects on sea ice, landscapes, northern infrastructure, and ecosystems. This amplified warming will continue over this century and will result in perturbations that may severely disrupt Arctic food webs and the well-being of Arctic communities. The Paris Agreement target to limit warming to +1.5°C is predicted to translate into greater than 3°C warming in the Arctic, while "business-as-usual" scenarios project Arctic temperature increases in the range 8 to 12°C. The full impacts of such large-scale warming are difficult to predict; however, they are foreshadowed by changes that are already being experienced across the Arctic. These changes have begun to affect policy decisions at all levels, from local development and conservation plans, to shipping routes and safety provisions.

Arctic climate change has implications for policy makers that extend well beyond the North Polar Region. The Arctic contains large storehouses of ice, notably the Greenland Ice Sheet that if fully melted would raise global sea levels by up to seven meters. Arctic warming is likely to alter mid-latitude weather patterns and to increase the likelihood of extreme storms and droughts. The amplified warming in the Arctic and its associated impacts such as sea ice loss,

W. F. Vincent (✉)
Centre for Northern Studies (CEN), Laval University, Quebec City, QC, Canada
e-mail: warwick.vincent@bio.ulaval.ca

© The Author(s) 2020
K. S. Coates, C. Holroyd (eds.), *The Palgrave Handbook of Arctic Policy and Politics*, https://doi.org/10.1007/978-3-030-20557-7_31

ice shelf collapse, and northern coastline erosion provide striking visual evidence that the global environment is changing rapidly, and that large changes lie ahead throughout the world. Some nations and industries see these changes in the North as opportunities for improved access to markets and resources, and warmer conditions could open up possibilities, as yet uncertain, for northern agriculture, fisheries, and tourism.

Given the potential magnitude of these global as well as local impacts, many nations are now heavily investing in Arctic climate research, including European and Asian countries that lie well outside the circumpolar region. Knowledge about the northern climate and its effects on ecosystems is therefore expanding rapidly and provides opportunities for policy makers to recommend science-based actions. This essay first introduces some of the recent findings from Arctic climate change science and then examines the associated policy implications within four themes: adaptation, conservation, mitigation, and knowledge exchange.

Faster Warming in the Arctic

The more rapid warming of the Arctic relative to the rest of the world is termed "Arctic amplification" and has been highlighted in each of the Intergovernmental Panel on Climate Change (IPCC) reports. For example, comparison of the decade 2006–2015 with a pre-industrial reference period (1850–1900) shows that the global average temperatures rose by 0.87°C over this timespan, while the measured Arctic temperature rise was two to three times higher, and with large differences among different parts of the Arctic (IPCC 2018). The IPCC climate models predict that this trend will continue: a 2°C rise by 2100 at a global scale is projected to result in a 4 to 7°C rise in Arctic temperatures, while if all current national commitments for carbon reduction can be adhered to, a mean global increase of 3°C is projected, translating to 7 to 11°C in the Arctic (mean night-time temperatures; IPCC 2018). Global fossil fuel emissions rose by 1.7% in 2017 and by around 2.7% in 2018 (Le Quéré et al. 2018), indicating that the Arctic continues to be on a rapid warming trajectory toward +10°C or above by the end of this century.

Arctic amplification is the result of several feedback effects that are important in snow and ice environments (Holland and Bitz 2003). The loss of highly reflective ("high albedo") snow or ice cover on the land or sea means that less solar energy is reflected back into the atmosphere, and more goes into heating and melting, with yet more loss of albedo, thereby causing a vicious circle of continued thawing and increased warmth. Warmer air also holds more water vapor, itself a greenhouse gas, and this further amplifies the warming effect. Recent climate modeling indicates that one of the most important feedback mechanisms may simply be the transfer of heat from the increasingly open Arctic Ocean to the atmosphere (Dai et al. 2019).

ARCTIC SENSITIVITY TO WARMING

In addition to amplified warming, the Arctic is unusually sensitive to the impacts of climate change. This is because snow and ice are major features of the northern environment, and small increases in temperature across the melting point can cause massive changes. At lower latitudes, a shift of ground temperatures from say 20 to 22°C may have little perceptible effect, at least in the short term, but a shift of the same two degree magnitude from −1 to +1°C causes a transformation of solid ice to liquid water, and totally transforms the landscape and seascape. This abrupt threshold effect is dramatically illustrated each summer as the Arctic goes through its seasonal transition of snow melting and ice break-up. When thawing occurs over the summer, the region converts to a state that looks and functions in ways that differ strikingly from winter. Currently, this seasonal thaw is kept in check by the vast storehouses of ice that are contained in permanent snowbanks, permafrost (ground that remains frozen for two or more years), thick multiyear sea ice, glaciers, and the Greenland Ice Sheet. These deep-frozen stores are legacies from past cold climates and they dampen the effects of seasonal warming, but progressively warmer summers are depleting these legacies and buffers against change.

The impacts of human-induced climate warming are now apparent across all ice-containing environments in the Arctic and Subarctic. Sea ice volume and extent have decreased persistently over the last few decades, with the area of multiyear sea ice now 60% below that observed in the 1980s, and minimum summer sea ice volume now 75% reduced relative to 1979 (Overland et al. 2018). The total areal extent of sea ice in September has dropped by 45% over the last 30 years, with more than 90% loss in some areas such as Hudson Bay, the Kara Sea and the Chukchi Sea (Stroeve and Notz 2018). Full loss of summer sea ice is expected over the next few decades, accompanied by increasing extension of ice-free conditions into autumn (Lebrun et al. 2019). The thickest marine-derived ice on Arctic seas occurred in the form of ice shelves along the northern coast of Ellesmere Island, Canada that formed over a period of several thousand years. These substantially collapsed throughout the twentieth century, with loss of the largest ice shelf in 2012 (Copland et al. 2018). Only one remains intact, the Milne Ice Shelf that retains a unique lake ecosystem, but there is evidence of ongoing thinning and imminent break-up (Hamilton et al. 2017).

Glaciers are melting rapidly throughout the Arctic, with large differences among regions. An analysis of records from the Cryosat satellite showed that average rates of ice loss during the period 2011 to 2017 ranged from 2 billion tons (Gt) per year in Iceland to 59 Gt per year in Arctic Canada (Richter-Menge et al. 2018). In the Canadian Arctic Archipelago (CAA), a long-term mass balance analysis showed that the glaciers and ice caps contracted at much faster rates over the last two decades, particularly in the southern region (Baffin Island) of the archipelago where the ice caps have recently lost their protective layer of perennial snow cover (Noël et al. 2018).

Recent changes in surface features are also resulting in a more rapid melting and loss of ice from Greenland's ice caps and glaciers, which currently account for around 43% of global sea level rise. The areas that are most sensitive to warming are the peripheral glaciers and ice caps, which may lose up to 28% of their mass over the next century. Like the glaciers of the CAA, these ice features appear to have passed through a tipping point in 1997, with major loss of their surface refreezing capacity at that time (Noël et al. 2017). Rainfall events on the Greenland Ice Sheet are becoming increasingly common, and this liquid water is hastening the melting of the ice (Oltmanns et al. 2019). This process is further accelerated by pigmented microbes that grow in the surface meltwater. The microbial communities, in combination with the deposition of soot and other dust materials, darken the surface of ice and increase the extent of sunlight absorption, heating, and meltwater production (Kintisch 2017).

Arctic lakes and rivers are also showing evidence of dramatic change. Canada's most northern freshwater ecosystem, Ward Hunt Lake in Quttinirpaaq National Park (QUNP), had 4.3 m of perennial ice in the 1950s, but from 2008 onwards the ice rapidly thinned, and the lake experienced open water conditions in summer 2011 (Paquette et al. 2015), perhaps for the first time in millennia. These warm conditions in northern Canada also had an impact on Lake Hazen, a deep lake further to the south in QUNP, which showed a transition toward increased likelihood of summer ice-free conditions and evidence of concomitant biological shifts (Lehnherr et al. 2018). Arctic warming is intensifying the water cycle over northern lands, and there is an increase in river discharge to the Arctic Ocean, with potential dampening effects on marine productivity and food webs (Li et al. 2009).

Ecological Impacts of Declining Sea Ice

Loss of sea ice has a direct impact on many species that live on, in and near the ice, and that are intimately connected in ice-dependent food webs (Vincent et al. 2011). A variety of cold-adapted algae live within the saltwater channels that permeate the ice, with highest abundance at the ice-water interface. These ice algae are a food source for microscopic animals including zooplankton. Once the seasonal ice melts, the algae rapidly sink to the bottom of the sea where they are used by benthic (bottom-dwelling) animals such as clams, in turn eaten by walrus and other diving marine mammals and birds. The open waters of the ocean at the edge of the ice zone are sites of elevated algal production by phytoplankton, which are also fed on by the zooplankton at the bottom of the planktonic food web, providing food for seabirds such as auklets, and fish, including Arctic cod. The latter is fed on by seals and beluga whales, with seals as the main prey for polar bears. The zooplankton and the algae in both habitats are rich in energy and high-quality nutrients, in particular polyunsaturated fatty acids (PUFAs). The PUFAs are passed up the food web and contribute to the health of Inuit and other local and Indigenous peoples who depend on the sea for subsistence hunting and fishing. The ice is also used as a platform for calving seals and polar bears, and as a diving platform needed by walrus to reach their benthic food. In areas of major loss of sea ice, there is evidence of a shift away from walrus as the top of the food chain to an ecosystem based more on open water plankton and fish (Grebmeier et al. 2006).

Narwhals and bowheads are highly specialized for pack ice conditions and are therefore negatively affected by sea ice loss, while other whale species that have more generalist feeding habits such as belugas may be able to adjust more readily to such losses. Sea ice loss has resulted in the northern expansion of orcas that prey on the young of other whales, resulting in additional pressure on Arctic specialized species. Polar bears are especially vulnerable to declining sea ice conditions, and large reductions are predicted in their populations, including complete loss from certain areas where they are presently common. Certain terrestrial animals that use sea ice as a foraging habitat (e.g., Arctic fox, snowy owls) or as migration routes will also be adversely affected. For example, Peary caribou depend on the sea ice between islands of the CAA for migration in spring and early winter, which ensures both genetic exchange between populations and an ability to recolonize habitats; the ongoing loss of sea ice will thereby jeopardize the survival of this species (Mallory and Boyce 2019).

Sea ice decline also has broader, indirect effects, including by influencing the temperature and precipitation over land (Macias-Fauria and Post 2018). For example, open water conditions increase the likelihood of rainfall events, which can result in ice crusts on snow that prevent reindeer and other animals from feeding. Open waters around the Yamal Peninsula in northwestern Siberia and a resultant rain-on-snow event caused massive mortalities of the reindeer population, with long-term socioeconomic impacts on the Nenets herders whose livelihoods and well-being depend on these animals (Forbes et al. 2016). Although increased open water may result in increased moisture during some seasons, the warmer temperatures can also dry out the land. This drying effect may account for the measured decline in shrub growth in coastal Greenland and Svalbard (Forchhammer 2017). Sea ice loss experienced in the late twentieth century in Hudson Bay has also been implicated in the rapid warming of adjacent lands in northern Quebec (Bhiry et al. 2011) and the Hudson Bay Lowlands (Rühland et al. 2013), with associated changes in vegetation and lake ecology.

Impacts on Indigenous People and Cultures

People have lived in the perennially cold regions of the North for millennia. Many of their cultural practices require free movement across the ice on rivers, lakes, and the sea for subsistence hunting and fishing, and there is a vital sense of connectedness to the wildlife, plant life, and other natural features of the Arctic environment. These Indigenous cultures have shown enormous resilience to past and present changes, but climate warming compounded by other stressors such as rapid development and health issues is now severely testing that resilience.

Arctic climate change is affecting northern communities in multiple ways (Pearce et al. 2015). For example, the changing ice and weather conditions are causing increased travel risks, including via traditional routes over river ice and coastal sea ice. The shifting ice patterns are also affecting food security by limiting access to certain hunting and fishing resources, and decreasing the availability of important wildlife species for subsistence. These reduced ice conditions also favor rapid economic development in some locations, with associated ship traffic and possible social as well as environmental impacts. Finally, the effects of flooding and loss of permafrost stability are causing increasing challenges for the construction and maintenance of municipal infrastructure.

The combination of climate-related stresses can also elicit strong emotional reactions such as anger, sadness, frustration, anxiety, depression and despair, which Cunsolo and Ellis (2018) describe as an expression of grief for ecological loss, or "ecological grief." In a comparison of two communities affected by climate change, an Inuit community in northern Canada and family farmers in the Australian Wheatbelt, they found similar experiences of ecological grief across three categories: physical ecological losses, loss of traditional environmental knowledge, and anticipated future losses. Indigenous organizations at all levels, from municipalities to national and international bodies, recognize the need to develop and implement policies that strengthen local resilience in the face of these ever-mounting challenges. An essential starting point for these policies is recognition of the intertwined and co-evolving nature of the social, ecological, and biophysical features of the Arctic and their connections to the rest of the world (Arctic Council 2016).

Vulnerability of Northern Infrastructure

Much of the engineered infrastructure of the North was built during the twentieth century when permafrost was considered a solid concrete-like foundation for homes, roads, bridges, railways, runways, pipelines, communication towers, waste containments, and other facilities. Permafrost is warming throughout the world, with fastest rates in the Arctic (by around +0.39°C over the decade 2007–2016: Biskaborn et al. 2019), accompanied by a deepening of the seasonally thawed "active layer." As a consequence, the stability of northern permafrost lands is no longer a dependable ecosystem service, and built infrastructure is increasingly at risk (Vincent et al. 2017). Arctic coastal communities are especially vulnerable because of coastline erosion by permafrost thaw and the greater wave exposure caused by extensive open water conditions (Fritz et al. 2017).

Total precipitation on average will continue to increase over the Arctic, but with transition toward increased rainfall rather than snowfall (Bintanja and Andry 2017). This greater delivery of liquid water will speed up land erosion and snow melt, and thereby create further hazards for northern infrastructure due to flooding and permafrost subsidence. In Alaska, the financial costs of climate-related damage to public infrastructure are estimated as 4 to 5.5 billion US$ for the period 2015 to 2099, with the largest source of damage due to road flooding followed by building damage caused by thawing permafrost (Melvin et al. 2017). An analysis for one northern region of Canada (the Inuvialuit Settlement Region) has estimated that the adaptation costs for building foundations would be in excess of 100 million CAD$, and questions remain as to who would pay for such work (Pearce et al. 2015).

Arctic soils with high concentrations of ice in fine sediments are particularly susceptible to thawing and subsidence. A recent analysis has shown that one-third of infrastructure across the circumpolar North lies in such high-risk regions and will be subject to thaw instability over the next four decades. This includes 1590 km of the Eastern Siberia–Pacific Ocean (ESPO) oil pipeline, 1260 km of gas pipelines that originate in the Yamal-Nenets region and 550 km of the Trans-Alaska Pipeline System, along with more than 13,000 km of roads and more than 100 airports. The pipeline vulnerability is of special concern given the prospect of major oil spills and the impacts on energy delivery (including to Europe) and thus on economic activity and national security (Hjort et al. 2018).

Increased Northern Shipping

The diminishing sea ice is opening up new opportunities for marine transport. The most notable example is the "Polar Silk Road," a component of China's Belt and Road Initiative that involves the development of the Northern Sea route along the Siberian coast in cooperation with Russia. A subsidiary of China's largest shipping company started regular use of this route in summer 2017, and the resultant transport activity is growing rapidly: cargo shipping on the Northern Sea Route rose to 18 million tons in 2018, an increase of 80% over 2017 and 360% over 2013 (Humpert 2019). These shipping tonnages are still very small in scale relative to the rest of the world, and need to be placed in global perspective (Holroyd 2019). Furthermore, ice conditions will likely remain unpredictable and dangerous well into the future, and Arctic shipping ventures have considerable operational and commercial risks (Lasserre 2018). These factors will continue to dampen interest, and transpolar shipping across the central Arctic Ocean is unlikely in the near term. Nevertheless, the current and projected shipping activities across the region are large relative to previous transport in the Arctic, and the risk of accidents is increasing. This increased shipping and tourist activity heightens the need for improvements in Arctic marine disaster and response policies (Mileski et al. 2018).

Implications of Arctic Climate Change Outside the Region

Many countries are now paying close attention to Arctic climate impacts, and to the global influence of the changing Arctic. China's Arctic policy, for example, begins by underscoring both aspects:

Global warming in recent years has accelerated the melting of ice and snow in the Arctic region. As economic globalization and regional integration further develops and deepens, the Arctic is gaining global significance for its rising strategic, economic values and those relating to scientific research, environmental protection, sea passages, and natural resources. The Arctic situation now goes beyond its original inter-Arctic States or regional nature, having a vital bearing on the interests of States outside the region and the interests of the international community as a whole, as well as on the survival, the development, and the shared future for mankind. It is an issue with global implications and international impacts. (People's Republic of China 2018, para. 1)

As noted above, China has a special interest in the maritime transport opportunities opened up by Arctic warming and sea ice loss, with improved access to markets as well as to energy supplies from Russia. This has reconfigured international security issues (including military), providing Russia with a vast Asian market for its western Siberian gas reserves, and China with a trade route to Europe outside the influence of the United States (Liu 2018).

Similarly, the United States in its release of the funding program "Navigating the new Arctic," draws attention to the changing Arctic and its global significance:

Arctic change will fundamentally alter climate, weather and ecosystems globally in ways that we do not yet understand but that will have profound impacts on the world's economy and security. Rapid loss of Arctic sea ice and other changes will also bring new access to the Arctic's natural resources such as fossil fuels, minerals, and new fisheries, and this new access is already attracting international attention from industry and nations seeking new resources. (NSF 2018, para. 2)

The influence of the Arctic on weather patterns further to the south is currently a subject of intense research and ongoing scientific discussion (Overland et al. 2018). Noting the "expanding footprint of Arctic change" via global sea level rise, coastal erosion, permafrost carbon release, storm impacts, and ocean-atmosphere warming, Moon et al. (2019) conclude that Arctic sea ice loss may already be causing extreme weather events that are manifested in mid-latitudes across the Northern Hemisphere. Unusually cold winter weather in North America has been attributed to the increased waviness of the Polar Front, the circumpolar jet stream that separates cold Arctic air from warmer air to the south. There is evidence that this is related to warming of the North Polar Region, which weakens the north-south temperature gradients, slows the flow of the jet stream, and allows cold Arctic air to penetrate southwards. Similarly, release of summer heat from the increasingly ice-free Arctic Ocean north of Alaska may have contributed to drought conditions in California.

The conspicuous changes taking place in the Arctic provide a clear early warning that severe climate impacts are to be expected throughout the rest of the world if we continue on the current emissions trajectory, and they also raise moral issues for our global society. In her landmark volume "The Right to be Cold," Inuit leader Sheila Watt-Cloutier presents the view that the effects of climate change on northerners constitute a violation of international human rights, including the rights of Inuk hunters on the snow and ice (Watt-Cloutier 2015). The eminent philosopher Thomas De Koninck and his colleagues argue that the degradation of the Arctic associated with climate change is an ethical failure by all humankind to respect the fundamental notion of "oikos" and the dignity of our existence. They suggest that Kant's definition of dignity as "inner worth" provides a unifying principle to address the "*complex and evolving problems of the North*" and to respect the beauty of all human beings and the natural world (De Koninck and de Raymond 2019, 52).

Adaptation Policies

The Arctic is changing so rapidly that local policy decisions are urgently needed to address the present and near-future challenges posed by climate warming. For example, in their analysis of northern infrastructure on permafrost, Hjort et al. (2018) conclude that the risks will remain high up to 2050 even if there are substantial cuts to greenhouse gas emissions, and that community and regional adaptation policies to minimize and manage these contingencies must be put in place as soon as possible. A broad sweep of adaptation policies are now in development throughout the North led by local, national, or in some cases international initiatives in response to the increasing impacts of Arctic climate change.

In the Canadian North, construction engineers are placing increasing attention toward "designing for change" in which the long-term stability of the environment is no longer taken for granted (Vincent et al. 2017). This involves engineering practices and designs that may be more expensive in the short term than conventional practices, but that are economical in the longer term. Discussions with northern communities, engineers, and permafrost specialists have culminated in a set of national standards for geotechnical surveys before construction on permafrost, with additional standards for drainage systems in northern communities on thawing permafrost landscapes and for thermosiphons (permafrost cooling systems), building foundations, and snow loading in the changing Arctic climate.

Climate adaptation strategies are in rapid development within specific national regions. Integrated Regional Impact Studies across the Inuit territories of Canada have included community-specific analyses of vulnerability, defined as the susceptibility to harm relative to the capacity to adapt (Ford and Smit 2004), and the production of permafrost risk maps to define areas safe for building in certain villages (Allard and Lemay 2012). Attention is also being put toward improved surveillance methods to monitor, communicate, and respond to changes, for example, by the use of satellite remote sensing to warn of unsafe river and sea ice conditions, and multi-kilometer long, fiber optic sensors to warn of localized thaw and collapse of roads and runways. Similarly, there is a need for increased surveillance and prevention policies for aquatic ecosystems. The warming climate combined with increased transfer of invasive species may prompt harmful algal blooms in coastal regions, making shellfish dangerous to eat; some toxins in harmful algae are passed up the food chain and have direct effects on the health and reproduction of marine mammals, and inshore environments need to be monitored. For drinking water supplies, adequate surveillance and advisories are also critical to ensure water quality and safety. Protection of these essential resources requires integrated freshwater management policies, including consideration of alternate water sources as traditional supplies change in quantity or quality.

At a broader multinational level, the Arctic Monitoring and Assessment Program (AMAP) was tasked by the Arctic Council to "produce information to assist local decision makers and stakeholders in three pilot regions in developing adaptation tools and strategies to better deal with climate change" (AMAP 2017a, 4), and the resultant work has culminated in a set of reports with ongoing updates for the Barents Region (AMAP 2017a), Baffin Bay Davis Strait Region (AMAP 2018), and the Bering/Chukchi/Beaufort Region (AMAP 2017b). These reports have identified specific tools to aid local adaptation, including models, scenarios, and narratives. The detailed exploration of alternate scenarios may be especially useful given the uncertainties inherent in climate prediction, as well as in global carbon emission trajectories. For example, Walsh et al. (2018) made downscaled estimates of air temperature and precipitation for more than 4000 communities in Alaska and western Canada. They found that ongoing climate change is inevitable over the next few decades, underscoring the pressing need for adaptation strategies, and that beyond 2050 the choice of emissions trajectory made a large difference in the future climate of each community.

CONSERVATION POLICIES

Regional parks, wildlife refuges, marine protected areas, and other conservation zones play a key role in protecting northern species and ecosystems from additional stresses superimposed on the rapidly warming climate, and they are now more important than ever. Arctic ecosystems have a lower diversity of plants and animals than in the temperate zone, and loss of only one or a few species may completely disrupt their food webs. These ecosystems are underpinned by a remarkable variety of microscopic life that has unusual adaptations to the polar environment (e.g., Tsuji et al. 2019). With continued warming, many species of plants, animals, and microbes will be pushed to the upper limit of their thermal tolerances, which will increase their sensitivity to other stressors. These effects are compounded by the increasing human presence in the Arctic, the associated increase in roads, shipping, aircraft movements, and increased likelihood of arrival of invasive species and their rapid dispersal.

New and existing protection zones require ongoing policy support at all scales, from catchment conservation to safeguard local water supplies, to the creation of large wilderness areas to protect Arctic ecosystems and their migratory animals. The existing protected areas of the Arctic have been created through traditional conservation policies of protecting ecosystems, habitats, and biodiversity, before the impacts of Arctic climate change were a matter of discussion or concern. However, these lie in areas that are now experiencing increasing climate impacts, and climate-related arguments could be incorporated within their strategic conservation plans and their rationale for protection.

For example, after considerable pressure by Indigenous, research and other groups, the borders of Tursujuq Park, the largest park in the northern Quebec territory of Nunavik, were extended to incorporate and preserve a large catchment that had been previously excluded because of its great interest to the hydroelectricity industry. This extension thereby protected a unique population of freshwater seals as well as striking landscapes and ecosystems. This area lies in the discontinuous permafrost region that is now experiencing rapid thawing and landscape changes (Allard and Lemay 2012), and the park offers an important refuge against the large human presence and road-building that would accompany hydroelectric development and industrialization. Similarly, in one of the largest northern conservation zones in Canada, Quttinirpaaq National Park, studies over the last two decades have shown that the land, lake, and fjord environments are responding strongly to the current trend of accelerated warming at these extreme high latitudes (82–83°N), leading to the perturbation or even complete loss of certain ecosystem types (Copland et al. 2018). In both of these cases, the parks provide refuges from additional stressors during this period of increasing climate perturbation.

Northern parks and other protected areas are likely to come under increasing economic and political pressure as the drive to extract resources from the Arctic continues to accelerate, along with improvements in access. Ongoing vigilance is required to maintain long-term conservation policies in the face of this pressure. A disturbing example is the current precarious state of the Arctic National Wildlife Refuge (ANWR) in Alaska. This vast, undeveloped wilderness is unusually rich in species diversity, including 42 fish and over 200 bird species. It also contains many mammal species such as caribou that are culturally important to the Inupiat and Gwich'in people. This refuge was opened up for oil and gas drilling under the terms of the "Tax Cuts and Jobs Act" of 2017, which allows certain areas to be leased for oil and gas exploration, and other areas to be identified for land easements that will give oil and drilling companies the legal right to use the land.

Climate change is also resulting in larger scale regional policies on conservation. Recognizing that sea ice is diminishing rapidly and putting Arctic marine wildlife at risk, the Word Wildlife Fund initiated planning for a localized "Last Ice Area" in the far North, where the diminishing ice may still be in place in 2050 and could provide a refuge for ice-dependent marine species. This area extends from across the Canadian Arctic Archipelago to northern Greenland, and it includes several areas that are already protected such as Northeast Greenland National Park (the largest national park in the world), Quttinirpaaq National Park in northern Ellesmere Island, and a Canadian marine conservation area that is now in advanced planning, Tallurutiup Imanga, at the eastern end of the Northwest Passage (WWF 2018). In March 2016, the United States and Canada issued a joint leaders' statement in which the two nations agreed to join forces in meeting the challenges in the Arctic region, with recognition that it is on the *frontline of climate change* (The White House 2016). As part of this agreement, Canada stated its intention to launch a "new process with Northern and Indigenous partners to explore options to protect the 'last ice area' within Canadian waters, in a way that benefits communities and ecosystems" (Prime Minister of Canada 2016), including evaluation of a new conservation area in the far North called Tuvaijuittuq ("the ice never melts" in Inuktitut).

Following the US-Canada Joint Statement of Arctic Leaders, the United States in December 2016 established the Northern Bering Sea Climate Resilience Area protecting the cultural and subsistence resources of over 80 tribes and a major migratory corridor for marine animals. Russia's Arctic policy also refers explicitly to climate change as factor motivating their creation of national conservation areas:

> *In the sphere of environmental security it is necessary: to ensure preservation of the biological diversity of the Arctic flora and fauna, including by expansion of a network of especially protected natural territories and water areas, taking into account national interests of the Russian Federation, necessity of preservation of the natural environment in the conditions of expansion of economic activities and global climate changes.* (Russian Federation 2008, para. IV 8c)

The central Arctic Ocean currently lies outside territorial boundaries and is an important focus of policy discussions concerning international conservation. The prospect of this area opening up to exploitation in the future has led to a binding agreement among many nations to prevent unregulated fisheries in this 2.8 million square kilometer region. This area has never been fished commercially, but the moratorium was agreed upon as the best precautionary approach to fisheries management *"given the changing conditions of the Arctic Ocean"* (European Union 2018, para. 2). There are calls for this international conservation policy to be extended more broadly into shipping activities in general. With the increased likelihood of a transpolar sea route through the high seas of the Arctic Ocean by the end of this century and concerns about oil spillage, noise, and pollution, this area could be designated as a "Particularly Sensitive Sea Area" under international law as a precautionary shipping measure (Stevenson et al. 2019).

MITIGATION POLICIES

Climate mitigation policies that limit emissions from human activities have the potential to make a massive difference in lessening the severity of impacts on the Arctic and throughout the world. Recent analyses of records from the past show how the present trajectory may lead to a gross perturbation of our planetary environment, including the Arctic. Business-as-usual emissions would lead to a climate that has not been experienced since the early Eocene, some 50 million years ago, and would unwind the long-term cooling trend of tens of millions of years in less than two centuries. It seems unlikely that current eco-systems throughout the world could sustain this unprecedented speed of change (Burke et al. 2018). The same business-as-usual scenario predicts warming in the upper ocean in the range 35–50% of that experienced 250 million years ago. That warming is believed to have been responsible for a 96% loss of all marine species on Earth because of oxygen depletion, with the greatest effects at high latitudes (Penn et al. 2018).

The Greenland Ice Cap is known to have been unstable over much shorter time scales of warming. There is evidence that sea levels rose by six meters over 1000 years during the last interglacial period around 100,000 years ago, and that near-complete deglaciation of southern Greenland occurred in the inter-glacial around 400,000 years ago (Fischer et al. 2018). The IPCC (2018) analysis concluded that irreversible loss of the Greenland ice sheet could be triggered at around 1.5 to 2°C, indicating the urgent need to reduce emissions. Similarly, the probability of a sea ice-free Arctic Ocean during summer is substantially lower with global warming of 1.5°C compared to 2°C. The report also points out that the 1.5 rather than 2°C temperature target would make a large difference in the amount of human suffering that will be imposed by global warming, including through sea level rise, heat-related mortality, forest fires, impacts on food supplies, ecosystem services, and limits to adaptive capac-ities, and that the Arctic is especially vulnerable to the additional 0.5°C in global temperature. The engineering risk analysis for Arctic infrastructure by Hjort et al. (2018, 3) shows that while large impacts are to be expected over the next few decades irrespective of emissions control, reducing the extent of warming to the Paris Agreement's aspirational target of +1.5°C "would make a clear difference in terms of potential damage to infrastructure."

Arctic permafrost soils contain vast quantities of carbon, which if fully mobi-lized by complete thawing and decomposition could more than double atmo-spheric carbon dioxide levels. In many areas of the North, some of this soil carbon is being converted to the more powerful greenhouse gas methane via microbial processes in permafrost-derived lakes, ponds, and wetlands (Vincent et al. 2017). An analysis of moderate warming conditions, within the range of the Paris Agreement, during the last interglacial period 400,000 years ago indi-cates that a runaway mobilization of these reserves did not occur, and nor does it seem that there was a release of marine methane hydrates at that time. Greater warming, however, is a serious concern for release of this carbon and the associated feedback effects (Fischer et al. 2018). A modeling comparison of greenhouse gas emission trajectories shows that a business-as-usual scenario could shift northern permafrost lands from being a net sink to net source of carbon beyond the year 2100, indicating the importance of mitigation actions to attenuate this permafrost feedback effect on climate (McGuire et al. 2018). New factors are also coming to light that could accelerate permafrost thawing and methane production, for example, warm rainfall events in spring (Neumann et al. 2019).

The implementation of the Paris Agreement requires an urgent stepping up of national policies in three areas: energy efficiency, alternative energy sources, and climate frameworks concerning mitigation and adaptation. In an analysis of data from 18 nations that showed consistently decreasing CO_2 emissions over the period 2005–2015, Le Quéré et al. (2019) found that there was a positive correlation between the rate of decline in emissions and the number of policies passed by law in each of these categories. The urgency of such polices was underscored in the IPCC (2018) report, which concludes that overshoot of the 1.5°C target can only be avoided if global CO_2 emissions start to decline before 2030. Similarly, a recent analysis of millions of policy scenarios shows that immediate global abatement of greenhouse gas emissions is required to assure a tolerable climate for future generations (Lamontagne et al. 2019). The recovery from overshoots of the Paris Agreement target would require a geo-engineering approach such as large-scale carbon dioxide reduction (CDR) or induced changes in atmospheric reflectivity, involving technologies that are currently not feasible at a global scale and that carry huge risks for the future of humanity and the biosphere.

KNOWLEDGE POLICIES

The Arctic is changing rapidly, and the short-term and especially long-term security of its residents and ecosystems requires climate policies at all scales, from local to global. The setting and implementation of such policies can only occur if people and their governing representatives understand the nature of climate-related problems and the need for action. This requires ongoing studies to not only define the current state and functioning of Arctic, but also fundamental and applied research to address uncertainties in projections and to find new solutions toward effective adaptation, conservation and mitigation measures. It also requires policies to promote knowledge exchange at all levels, from disseminating locally relevant information (for example, explaining to northern residents, municipalities, and developers the science behind "building for change" and risk assessment maps, and linking this to Indigenous Knowledge), to effectively communicating the most recent scientific insights about Arctic change and its global implications to government policy makers and the public throughout the world.

The Arctic is now a focus of unprecedented attention by research agencies and scientists. The International Arctic Science Committee (IASC), the umbrella organization for coordinating Arctic research, was made up of eight Arctic nations at its inception in 1990, but today is composed of government nominated delegates from 23 nations, including strong representation from Asia and Europe (Rogne et al. 2015). A number of large-scale initiatives are in progress under the auspices of IASC, for example, the "Multidisciplinary drifting Observatory for the Study of Arctic Climate" (MOSAiC 2018), an over-wintering mission in the central Arctic Ocean that involves 600 science personnel supported by five ice breakers, aircraft, and satellite remote sensing to examine the causes and consequences of sea ice decline. This has given rise to the related IASC study "Terrestrial Multidisciplinary distributed Observatories for the Study of Arctic Connections" (T-MOSAiC 2018) that involves more than 100 land-based stations around the Arctic to examine the effects of Arctic sea ice and climate change on landscapes, land-based ecosystems, and people in the circumpolar North. In an analysis of Japan's Arctic Policy, Ikeshima (2016, 460) notes that an "urgent requirement is the construction of a new icebreaker or an ice-strengthened vessel" for Japan to participate more fully in Arctic climate change research given the implications for future maritime transport and the opportunities for "collaboration and cooperation between the Arctic and non-Arctic states." All of this expanding research activity and collaboration will have the most societal value if the scientific information can reach "policy makers and other people with influence" in a timely and accessible manner (Ditchley Foundation 2017).

Indigenous experience and understanding provide a knowledge stream that has enormous value for incorporation into climate-related policies. In the context of Arctic climate change, Gilligan et al. (2006) recognize three types of knowledge systems: Traditional Knowledge defined as that based on tradition and passed from generation to generation, Local Knowledge that is generated by a community based on first-hand experience, and Scientific Knowledge based on the Western or European approach toward observation and data analysis. They note that combining information from these three sources is essential, but with respectful attention to the holders of Indigenous Knowledge (Traditional and Local) and the use of such information. There are national and international calls for closer partnerships between local communities and research programs, including Indigenous-led research. The Arctic Science Ministerial (2018, 3) declared that "*Indigenous Peoples should be involved as appropriate—as they are in this Ministerial discussion—in the assessment and definition of Arctic research priorities*" and that there is "*the necessity for all States and the European Union conducting research in this region to work together, in collaboration with Arctic Indigenous Peoples and local communities.*" The increasing involvement of northern communities in research will favor the trend toward incorporating Indigenous Knowledge in Arctic policies.

There is overwhelming scientific evidence that our planetary climate is changing rapidly as a result of human activities, and the IPCC has made a clear statement that urgent action is needed to reduce greenhouse gas emissions and concentrations, and thereby prevent the ecological crisis and human suffering throughout the world that our present trajectory is leading toward (IPCC 2018). The increased frequency of wildfires, storms, and heat waves, along with declining biodiversity, decreased crop yields, rising sea levels, and coastal flooding has meant that the reality of climate change has begun to penetrate human consciousness at a global level, as witnessed by the remarkable consensus of 196 nations in the Paris Agreement. However, the urgency expressed in the IPCC (2018) report is not widely understood or accepted, with nations weakening their commitment to mitigation or (in the case of the United States) withdrawing from the Agreement. Even the Arctic Science Ministerial (2018, 6), while noting how the Arctic is "*one of the most sensitive areas to climate change on Earth*," made no reference to mitigation.

There are hopeful trends in the level of public awareness about global climate change, but much more work needs to be done in science education and outreach. A recent survey of the American public found that the majority believe it is very likely that climate change is happening (73% in December 2018, the highest since the survey began in 2008) and is mostly human-induced (62%); however, only 22% agreed that most climate scientists have concluded that human-caused climate warming is occurring (Leiserowitz et al. 2019), despite clear statements to this effect from the IPCC, national assessments, and professional scientist associations. Arctic research has a key role to play in this knowledge communication process, with its compelling visual messages that Earth's climate is changing rapidly, and that the future well-being of the Arctic, and the world, depends on urgent climate policy actions at a global scale.

Acknowledgments I thank Drs. Catherine Girard, Kanae Komaki, Connie Lovejoy and Maribeth Murray for valuable comments and suggestions on draft versions of the manuscript, and the following agencies for supporting our research on Arctic climate change and northern ecosystems: the Natural Sciences and Engineering Research Council of Canada, Fonds de Recherche du Québec—Nature et technologies, the Canada Research Chair program, the Network of Centres of Excellence ArcticNet, the Polar Continental Shelf Program, the Canada Foundation for Innovation, and the Canada First Excellence Research Fund program Sentinel North.

References

Allard, M., and M. Lemay, eds. 2012. *Nunavik and Nunatsiavut: From Science to Policy. An Integrated Regional Impact Study (IRIS) of Climate Change and Modernization*. Quebec City, QC: ArcticNet Inc.

AMAP. 2017a. *Adaptation Actions for a Changing Arctic (AACA)—Barents Area Overview Report*. Oslo: Arctic Monitoring and Assessment Programme. https://www.amap.no/documents/download/2885/inline.

———. 2017b. *Adaptation Actions for a Changing Arctic (AACA)—Bering/Chukchi/Beaufort Region Overview Report*. Oslo: Arctic Monitoring and Assessment Programme. https://www.amap.no/documents/download/2887/inline.

———. 2018. *Adaptation Actions for a Changing Arctic: Perspectives from the Baffin Bay/Davis Strait Region*. Oslo: Arctic Monitoring and Assessment Programme. https://www.amap.no/documents/download/3015/inline.

Arctic Council. 2016. *Arctic Resilience Report*. Edited by M. Carson and G. Peterson. Stockholm: Stockholm Environment Institute and Stockholm Resilience Centre. http://www.arctic-council.org/arr.

Arctic Science Ministerial. 2018. *Joint Statement of Ministers on the Occasion of the Second Arctic Science Ministerial*. Berlin: Federal Republic of Germany, October 26. https://www.arcticobserving.org/images/pdf/misc/asm-2-joint-statement.pdf.

Bhiry, N., A. Delwaide, M. Allard, Y. Bégin, L. Filion, M. Lavoie, C. Nozais, et al. 2011. Environmental Change in the Great Whale River Region, Hudson Bay: Five Decades of Multidisciplinary Research by Centre d'études nordiques (CEN). *Ecoscience* 18 (3): 182–203.

Bintanja, R., and O. Andry. 2017. Towards a Rain-Dominated Arctic. *Nature Climate Change* 7: 263–267.

Biskaborn, B.K., S.L. Smith, J. Noetzli, H. Matthes, G. Vieira, D.A. Streletskiy, P. Schoeneich, et al. 2019. Permafrost is Warming at a Global Scale. *Nature Communications* 10 (1): 264.

Burke, K.D., J.W. Williams, M.A. Chandler, A.M. Haywood, D.J. Lunt, and B.L. Otto-Bliesner. 2018. Pliocene and Eocene Provide Best Analogs for Near-Future Climates. *Proceedings of the National Academy of Sciences* 115 (52): 13288–13293.

Copland, L., A. White, A. Crawford, D.R. Mueller, W. Van Wychen, L. Thomson, and W.F. Vincent. 2018. Glaciers, Ice Shelves and Ice Islands. In *Policy in the Eastern Canadian Arctic: An Integrated Regional Impact Study (IRIS) of Climate Change and Modernization*, ed. T. Bell and T.M. Brown, 95–117. Quebec City, QC: ArcticNet Inc. http://www.cen.ulaval.ca/warwickvincent/PDFfiles/349-Copland.pdf.

Cunsolo, A., and N.R. Ellis. 2018. Ecological Grief as a Mental Health Response to Climate Change-Related Loss. *Nature Climate Change* 8 (4): 275–281.

Dai, A., D. Luo, M. Song, and J. Liu. 2019. Arctic Amplification is Caused by Sea-Ice Loss under Increasing CO_2. *Nature Communications* 10 (1): 121.

De Koninck, T., and J.-F. de Raymond. 2019. *Beauty Obliges: Ecology and Dignity*. Quebec City, QC: Presses de l'Université Laval.

Ditchley Foundation. 2017. *The Arctic at the Crossroads: Cooperation or Competition? Director's Report*, June 9–11. https://www.ditchley.com/event-programme-theme/past-events/2010-2019/2017/arctic-crossroads-cooperation-or-competition.

European Union. 2018. *Proposal for a Council Decision on the Conclusion, on Behalf of the European Union, of the Agreement to Prevent Unregulated High Seas Fisheries in the Central Arctic Ocean, Explanatory Memorandum*, June 12. https://eur-lex.europa.eu/legal-content/EN/TXT/?uri=COM:2018:453:FIN.

Fischer, H., K.J. Meissner, A.C. Mix, N.J. Abram, J. Austermann, V. Brovkin, E. Capron, et al. 2018. Palaeoclimate Constraints on the Impact of 2 °C Anthropogenic Warming and Beyond. *Nature Geoscience* 11: 474–485.

Forbes, B.C., T. Kumpula, N. Meschtyb, R. Laptander, M. Macias-Fauria, P. Zetterberg, M. Verdonen, et al. 2016. Sea Ice, Rain-on-snow and Tundra Reindeer Nomadism in Arctic Russia. *Biology Letters* 12 (11): 20160466.

Forchhammer, M. 2017. Sea-Ice Induced Growth Decline in Arctic Shrubs. *Biology Letters* 13 (8): 20170122.

Ford, J.D., and B. Smit. 2004. A Framework for Assessing the Vulnerability of Communities in the Canadian Arctic to Risks Associated with Climate Change. *Arctic* 57 (4): 389–400.

Fritz, M., J.E. Vonk, and H. Lantuit. 2017. Collapsing Arctic Coastlines. *Nature Climate Change* 7: 6–7.

Gilligan, J., J. Clifford-Peña, J. Edye-Rowntree, K. Johansson, R. Gislason, T. Green, and G. Arnold. 2006. The Value of Linking, Tradition, Local and Scientific Knowledge. In *Climate Change—Linking Traditional and Scientific Knowledge*, ed. R. Riewe and J. Oates, 3–12. Winnipeg, MB: Aboriginal Issues Press.

Grebmeier, J.M., J.E. Overland, S.E. Moore, E.V. Farley, E.C. Carmack, L.W. Cooper, K.E. Frey, et al. 2006. A Major Ecosystem Shift in the Northern Bering Sea. *Science* 311 (5766): 1461–1464.

Hamilton, A.K., B.E. Laval, D.R. Mueller, W.F. Vincent, and L. Copland. 2017. Dynamic Response of an Arctic Epishelf Lake to Seasonal and Long-term Forcing: Implications for Ice Shelf Thickness. *The Cryosphere* 11 (5): 2189–2211.

Hjort, J., O. Karjalainen, J. Aalto, S. Westermann, V.E. Romanovsky, F.E. Nelson, B. Etzelmüller, and M. Luoto. 2018. Degrading Permafrost Puts Arctic Infrastructure at Risk by Mid-Century. *Nature Communications* 9 (1): 5147.

Holroyd, C. 2019. East Asia (Japan, South Korea and China) and the Arctic. In *Palgrave Handbook of Arctic Policy*, ed. K. Coates and C. Holroyd. London, UK: Palgrave Macmillan (this volume).

Holland, M.M., and C.M. Bitz. 2003. Polar Amplification of Climate Change in Coupled Models. *Climate Dynamics* 21 (3–4): 221–232.

Humpert, M. 2019. Russia's Northern Sea Route Sees Record Cargo Volume in 2018. *High North News*, February 20. https://www.arctictoday.com/russias-northern-sea-route-sees-record-cargo-volume-in-2018/.

Ikeshima, T. 2016. Japan's Role as an Asian Observer State within and outside the Arctic Council's Framework. *Polar Science* 10 (3): 458–462.

IPCC. 2018. *Special Report on Global Warming of 1.5°C (SR15)*. Geneva, Switzerland: Intergovernmental Panel on Climate Change, World Meteorological Organization. https://www.ipcc.ch/sr15/.

Kintisch, E. 2017. Meltdown. *Science* 355 (6327): 788–791.

Lamontagne, J.R., P.M. Reed, G. Marangoni, K. Keller, and G.G. Garner. 2019. Robust Abatement Pathways to Tolerable Climate Futures Require Immediate Global Action. *Nature Climate Change* 9: 290–294.

Lasserre, F. 2018. Arctic Shipping: A Contrasted Expansion of a Largely Destinational Market. In *The Global Arctic Handbook*, ed. M. Finger and L. Heininen, 83–100. Berlin, Germany: Springer.

Le Quéré, C., R.M. Andrew, P. Friedlingstein, S. Sitch, J. Hauck, J. Pongratz, P.A. Pickers, et al. 2018. Global Carbon Budget 2018. *Earth System Science Data* 10: 2141–2194.

Le Quéré, C., J.I. Korsbakken, C. Wilson, J. Tosun, R. Andrew, R.J. Andres, J.G. Canadell, et al. 2019. Drivers of Declining CO_2 Emissions in 18 Developed Economies. *Nature Climate Change* 9: 213–217.

Lebrun, M., M. Vancoppenolle, G. Madec, and F. Massonnet. 2019. Arctic Sea-ice-free Season Projected to Extend into Autumn. *The Cryosphere* 13 (1): 79–96.

Lehnherr, I., V.L.S. Louis, M. Sharp, A.S. Gardner, J.P. Smol, S.L. Schiff, D.C.G. Muir, et al. 2018. The World's Largest High Arctic Lake Responds Rapidly to Climate Warming. *Nature Communications* 9 (1): 1290.

Leiserowitz, A., E. Maibach, S. Rosenthal, J. Kotcher, M. Ballew, M. Goldberg, and A. Gustafson. 2019. *Climate Change and the American Mind: December 2018*. Yale Program on Climate Change Communication. Report. New Haven, CT: Yale University and George Mason University. http://climatecommunication.yale.edu/publications/climate-change-in-the-american-mind-december-2018/.

Li, W.K., F.A. McLaughlin, C. Lovejoy, and E.C. Carmack. 2009. Smallest Algae Thrive as the Arctic Ocean Freshens. *Science* 326 (5952): 539–539.

Liu, N. 2018. Will China Build a Green Belt and Road in the Arctic? *Review of European, Comparative & International Environmental Law* 27 (1): 55–62.

Macias-Fauria, M., and E. Post. 2018. Effects of Sea Ice on Arctic Biota: An Emerging Crisis Discipline. *Biology Letters* 14 (3): 20170702.

Mallory, C.D., and M.S. Boyce. 2019. Prioritization of Landscape Connectivity for the Conservation of Peary Caribou. *Ecology and Evolution*. https://doi.org/10.1002/ece3.4915.

McGuire, A.D., D.M. Lawrence, C. Koven, J.S. Clein, E. Burke, G. Chen, E. Jafarov, et al. 2018. Dependence of the Evolution of Carbon Dynamics in the Northern Permafrost Region on the Trajectory of Climate Change. *Proceedings of the National Academy of Sciences* 115 (15): 3882–3887.

Melvin, A.M., P. Larsen, B. Boehlert, J.E. Neumann, P. Chinowsky, X. Espinet, J. Martinich, et al. 2017. Climate Change Damages to Alaska Public Infrastructure and the Economics of Proactive Adaptation. *Proceedings of the National Academy of Sciences* 114 (2): E122–E131.

Mileski, J., A. Gharehgozli, L. Ghoram, and R. Swaney. 2018. Cooperation in Developing a Disaster Prevention and Response Plan for Arctic Shipping. *Marine Policy* 92: 131–137.

Moon, T.A., I. Overeem, M. Druckenmiller, M. Holland, H. Huntington, G. Kling, and A.L. Lovecraft. 2019. The Expanding Footprint of Rapid Arctic Change. *Earth's Future*. https://doi.org/10.1029/2018EF001088.

MOSAiC. 2018. *Multidisciplinary drifting Observatory for the Study of Arctic Climate— Implementation Plan*. Akureyri, Iceland: IASC, April. https://www.mosaic-expedition.org.

Neumann, R.B., C.J. Moorberg, J.D. Lundquist, J.C. Turner, M.P. Waldrop, J.W. McFarland, E.S. Euskirchen, C.W. Edgar, and M.R. Turetsky. 2019. Warming Effects of Spring Rainfall Increase Methane Emissions from Thawing Permafrost. *Geophysical Research Letters* 46 (3): 1393–1401.

Noël, B., W. van de Berg, S. Lhermitte, B. Wouters, H. Machguth, I. Howat, M. Citterio, G. Moholdt, J.T.M. Lenaerts, and M.R. van den Broeke. 2017. A Tipping Point in Refreezing Accelerates Mass Loss of Greenland's Glaciers and Ice Caps. *Nature Communications* 8: 14730.

Noël, B., W.J. van de Berg, S. Lhermitte, B. Wouters, N. Schaffer, and M.R. van den Broeke. 2018. Six Decades of Glacial Mass Loss in the Canadian Arctic Archipelago. *Journal of Geophysical Research: Earth Surface* 123 (6): 1430–1449.

NSF. 2018. *Navigating the New Arctic*. Alexandria, VA: National Science Foundation, United States of America, October 24. https://www.nsf.gov/news/special_reports/big_ideas/arctic.jsp.

Oltmanns, M., F. Straneo, and M. Tedesco. 2019. Increased Greenland Melt Triggered by Large-Scale, Year-Round Precipitation Events. *The Cryosphere* 13 (3): 815–825.

Overland, J., E. Dunlea, J.E. Box, R. Corell, M. Forsius, V. Kattsov, M.S. Olsen, et al. 2018. The Urgency of Arctic Change. *Polar Science*. https://doi.org/10.1016/j.polar.2018.11.008.

Paquette, M., D. Fortier, D.R. Mueller, D. Sarrazin, and W.F. Vincent. 2015. Rapid Disappearance of Perennial Ice on Canada's Most Northern Lake. *Geophysical Research Letters* 42 (5): 1433–1440.

Pearce, T., J. Ford, F. Duerden, C. Furgal, J. Dawson, and B. Smit. 2015. *From Science to Policy in the Western and Central Canadian Arctic: An Integrated Regional Impact Study (IRIS) of Climate Change and Modernization*. Edited by G. Stern and A. Gaden, 403–427. Quebec City, QC: ArcticNet Inc.

Penn, J.L., C. Deutsch, J.L. Payne, and E.A. Sperling. 2018. Temperature-Dependent Hypoxia Explains Biogeography and Severity of End-Permian Marine Mass Extinction. *Science* 362 (6419): eaat1327.

People's Republic of China. 2018. *China's Arctic Policy.* Beijing: The State Council Information Office of the People's Republic of China, White Paper. http://english.scio.gov.cn/node_8002680.html.

Prime Minister of Canada. 2016. *Select Actions Being Taken under the United States-Canada Joint Arctic Leaders' Statement,* December 20. https://pm.gc.ca/eng/news/2016/12/20/select-actions-being-taken-under-united-states-canada-joint-arctic-leaders-statement.

Richter-Menge, J., M.O. Jeffries, and E. Osborne, eds. 2018. The Arctic [in "State of the Climate in 2017"]. *Bulletin of the American Meteorological Society* 99 (8): S143–S173.

Rogne, O., V. Rachold, L. Hacquebord, and R. Corell. 2015. *IASC after 25 Years—A Quarter of a Century of International Arctic Research Cooperation.* Potsdam, Germany: International Arctic Science Committee. https://iasc25.iasc.info/.

Rühland, K.M., A.M. Paterson, W. Keller, N. Michelutti, and J.P. Smol. 2013. Global Warming Triggers the Loss of a Key Arctic Refugium. *Proceedings of the Royal Society B: Biological Sciences* 280 (1772): 20131887.

Russian Federation. 2008. *Basics of the State Policy of the Russian Federation in the Arctic for the Period Till 2020 and for a Further Perspective.* http://www.arctis-search.com.

Stevenson, T.C., J. Davies, H.P. Huntington, and W. Sheard. 2019. An Examination of Trans-Arctic Vessel Routing in the Central Arctic Ocean. *Marine Policy* 100: 83–89.

Stroeve, J., and D. Notz. 2018. Changing State of Arctic Sea Ice across All Seasons. *Environmental Research Letters* 13 (10): 103001.

T-MOSAiC. 2018. *Terrestrial Multidisciplinary distributed Observatories for the Study of Arctic Connections—Science Plan.* Akureyri, Iceland: IASC, December. https://www.t-mosaic.com.

The White House. 2016. *U.S.-Canada Joint Statement on Climate, Energy, and Arctic Leadership.* Washington, DC: The White House, Office of the Press Secretary, March 10. https://obamawhitehouse.archives.gov/the-press-office/2016/03/10/us-canada-joint-statement-climate-energy-and-arctic-leadership.

Tsuji, M., Y. Tanabe, W.F. Vincent, and M. Uchida. 2019. *Vishniacozyma ellesmerensis* sp. nov., a New Psychrophilic Yeast Isolated from a Retreating Glacier in the Canadian High Arctic. *International Journal of Systematic and Evolutionary Microbiology* 69 (3): 696–670.

Vincent, W.F., T.V. Callaghan, D. Dahl-Jensen, M. Johansson, K.M. Kovacs, C. Michel, T. Prowse, J.D. Reist, and M. Sharp. 2011. Ecological Implications of Changes in the Arctic Cryosphere. *Ambio* 40 (1): 87–99.

Vincent, W.F., M. Lemay, and M. Allard. 2017. Arctic Permafrost Landscapes in Transition: Towards an Integrated Earth System Approach. *Arctic Science* 3 (2): 39–64.

Walsh, J.E., U.S. Bhatt, J.S. Littell, M. Leonawicz, M. Lindgren, T.A. Kurkowski, P.A. Bieniek, R. Thoman, S. Gray, and T.S. Rupp. 2018. Downscaling of Climate Model Output for Alaskan Stakeholders. *Environmental Modelling & Software* 110: 38–51.

Watt-Cloutier, S. 2015. *The Right to be Cold: One Woman's Story of Protecting Her Culture, The Arctic and the Whole Planet.* Toronto: Penguin Canada.

WWF. 2018. *Last Ice Area—Similijuaq.* Toronto: World Wildlife Fund. https://arctic-wwf.org/site/assets/files/2021/lia-brochure.pdf.

PART VII

Reflections on *Future of the Arctic*

CHAPTER 32

The Future of the Arctic: Policy Prospects for the Twenty-First Century

Ken S. Coates and Carin Holroyd

No one knows what the Arctic will look like in 2050. When even the most basic elements are unknown—the price and demand for Arctic resources, scientific innovation and societal acceptance of new technologies, the full regional impact of global climate change, and the future of Indigenous languages and cultures—it becomes near impossible to determine demographic trajectories, the fate of cities, towns, and villages, the nature of employment in the North, and the most effective political and administrative systems for the Far North. However, planning for the future need not take place in a vacuum. There are aspects policy-makers know or that they anticipate that help guide government strategic planning and policy development going forward.

The Arctic transitions of the past 50 years have been nothing short of remarkable, in terms of regional autonomy, the articulation of northern visions within and between nations, the re-empowerment of Indigenous peoples, and the development of strong and vibrant Arctic societies. The Arctic may well be affected more by impeding technological adaptation than most other parts of the world and the socioeconomic transitions promise to be dramatic. The current foundation of Arctic innovation and adaptability, however, is sustained through large-scale subsidization by national governments. There is no assurance that this support will remain in the future, particularly as national authorities cope with dramatic and comprehensive pressures on the countrywide basis. Here is the major problem. To respond adequately to the twenty-first century

K. S. Coates
Johnson-Shoyama Graduate School of Public Policy, University of Saskatchewan, Saskatoon, SK, Canada

C. Holroyd (✉)
Department of Political Studies, University of Saskatchewan, Saskatoon, SK, Canada

© The Author(s) 2020   529
K. S. Coates, C. Holroyd (eds.), *The Palgrave Handbook of Arctic Policy and Politics*, https://doi.org/10.1007/978-3-030-20557-7_32

will require more financial support, not less, and there is no reasonable expectation that enhanced funding will be available in all of the Arctic regions, with the Canadian and Russian Norths facing particular challenges (Einarsson et al. 2004; Larsen and Fondahl 2015).

The Arctic has much of the human capacity needed to imagine the future and to articulate a response to the evolving political, technological, and economic circumstances; some parts of the North, particularly in Canada, Greenland, Russia, and Alaska, face regional and local challenges with implementation. Major capacity issues remain in large parts of the Arctic and no level of well-wishing or earnest encouragement will overcome the comparative absence of trained and skilled workers and managers and multi-generational educational deficiencies. With a handful of exceptions, principally in northern Scandinavia, the North lacks the innovative and creative business cultures needed to capitalize on changing commercial conditions. In much of the North, the regional economies are dependent on resource development and government spending (including by the military), none of which are secure or totally reliable. The recent emergence of ecotourism and local Indigenous businesses is not yet sufficient to sustain or improve upon the North's quality of life over the longer term.

The Arctic, of course, is not a unified and coherent region, moving ahead with a common purpose or shared goals. It is divided between numerous national governments, various subnational authorities, many Indigenous jurisdictions, and a growing number of well-meaning international interveners. Even such innovative institutions as the Arctic Council, however, lack the ability to mobilize wide-ranging socioeconomic change across the Arctic. There is some intraregional cooperation—more on Indigenous, academic, and cultural matters than on economic or technological strategies—but there is no substantial or systematic effort being made to prepare the Arctic for the almost inevitable and dramatic transitions that lay in the offing. This is hardly unique to the Far North, of course; indeed, there is more intraregional collaboration in the Arctic than in comparable regions around the world.

The development of a powerful and effective public policy response to the challenges of the twenty-first-century Arctic requires a regional commitment to change the trajectory of the North in line with shifting global realities, the development of a realistic and comprehensive north-centric strategy for the Arctic, and regional and local plans that focus on the world of 2050 and beyond. It is important to shift, at least in part, away from the short-term focus of the present to the more creative and comprehensive approach to future making. There is nothing easy in adopting this approach. In fact, the many factors and pressures imbedded in the status quo mitigate against the likelihood of the Arctic developing a coherent, effective, and manageable transformative strategy. Current political systems are not well suited to rapid, coherent, and well-developed strategies for technological and political adaptation. The Far North's legitimate preoccupation with pressing social, economic, and cultural issues will be impeded, if not stopped, a serious contemplation of the prospects for future generations. In the end, it is most likely that the North will continue along its current trajectory, that purposeful regional strategies will not be developed, and that the North will, like much of the rest of the world, continue to respond episodically to global forces for change rather than preparing a strategy for twenty-first-century adaptation. But perhaps the contemporary Arctic has the strength, vision, and determination to create and implement a vision for the Far North that prepares the people of the Arctic for the possibilities and challenges of the twenty-first century.

INFORMED SPECULATION ABOUT 2050

Developing a policy framework for the future of the Arctic requires, at the very least, a preliminary assessment of what lies ahead (Smith 2010). These forecasts are, of course, only educated guesses, with some of the more disruptive scenarios arising more out of science fiction and technological speculation than sober anticipation of the decades ahead. Will the world be weaned from its dependence on fossil fuels and supported instead by renewable or alternate energy sources, as some believe and many hope, liberating the North from the financially crushing weight and costs of energy production and distribution? Will food factories provide fresh, nutritious vegetables and fruits in small communities across the North, addressing major food security issues? Will rampant climate change cause massive disruption in the permafrost, releasing vast quantities of methane and setting off a chain reaction of serious difficulties with Arctic buildings, roads, and pipelines? Much remains to be discovered and learned, but policy-makers contemplating the future have no choice but to make cautious and intelligent assumptions about what lies ahead.

The path forward will likely be shaped by a remarkable surge in scientific and technological research and development. The tale of humanity is, in significant measure, tied to evolving technologies and the socioeconomic impact of these technology-based changes. The twenty-first century is likely see sustained and intensified technological innovation, with the potential to bring impressive improvements in the quality of life for many of the world's citizens, unprecedented threats to personal privacy, institutional and governmental security, and formidable changes in the workplace. Drones may replace Arctic rescue systems and have already started to transform mineral exploration. Advanced sensors could transform the monitoring of Arctic wildlife and environmental change. Remote manipulation of machinery, which is already operational in parts of the mining sector, has the potential to eliminate or transform the jobs of most workers on the mines and oil fields. National advantage may well continue to transfer to technological and scientifically sophisticated countries. Areas with educational and research deficiencies, a description that applies to much of the Far North, would suffer accordingly.

The twenty-first century may see, as well, a substantial geopolitical reformulation. There is an impending tectonic shift from North America and Europe to East Asia and South Asia, with potentially significant rebounds in Africa and South East Asia. The prospects for a continued decline in Russia—the combination of a demographic plunge of historic proportion, an overdependence on the oil and gas industry, and the intersection of Russian nationalism and Arctic militarization—could see a sharp reduction in the country's international significance but a surge in the nation's Arctic adventurism. But it is India and China, the two most heavily populated countries in the world and both with surging middle classes and growing economic footprints, that have the potential to transform the global order. Both countries have Arctic interests and engagements, minimal in the case of India and increasing for China, but no direct or substantial Arctic presence. As the decades unfold, both India and China expect to be part of all global conversations, including those in the Arctic, the greatest international zone of all.

The years ahead will also see the world wrestling with a broad series of ethical and moral issues, many of them based on the technological transitions that are underway. Issues of personal privacy have surfaced in the wake of controversies about the use of data by Facebook and Google. The commercial power of the new super companies, particularly Apple, Microsoft, Google, Facebook, Huawei, Amazon, and Alibaba, has threatened national sovereignty and reshaped many sectors of the economy. The capacity to alter basic biology, through pesticides, genetic engineering, and various medical interventions, has the potential to reshape the very nature of humanity and the natural world. As the world copes with the myriad forces for change, national governments, and international agencies may be forced to choose between the interests of the sparsely populated Far North and densely settled equatorial and southern regions. These are all broad and complex issues of global importance. They will affect the Arctic.

This said, it is useful to contemplate the broad contours of global change leading to 2050, with a brief reflection on how these forces might affect the Far North. Consider some of the major issues, each of which presents complex policy challenges for northern, Indigenous, and national leaders. The budget needed to address these issues must be drawn from the funds available for all northern needs and concerns, which are numerous, already expensive and often underfunded, particularly when compared to evident need. Even when governments and administrators understand the challenges and agree on solutions, it is difficult to locate the money, policy space, and public attention needed to act. And with democratic nations well known for focusing on short-time, electoral-cycle concerns, creating the political energy for policy work on these complex global questions has proven difficult. Symbolism, as has long been the case for the Arctic, is much easier than sustained investments and real change.

Climate Change

Climate change is hitting the Arctic disproportionally. The most tangible signs are the melting of Arctic Ocean ice and the expanded opening of Arctic shipping lanes. Less well known but equally serious are such forces as the thawing of permafrost and the disruption and changes of animal, bird, fish, and plant habitats. Environmentalist hyperbole, especially around the iconic polar bears, has distorted the discussions, focusing on stories of interest to international audiences and distracting from local and regional questions of intense concern to Arctic residents. The potential consequences are substantial, ranging from the interruption of traditional harvesting to the buckling of the land under Arctic homes and buildings. Some Arctic communities, located close to the ocean front, are already being relocated to safer ground. The disruptions undercut centuries-old Indigenous understandings of homeland territories, threatening the knowledge that is foundation of Indigenous life in the Arctic. Sheila Watts-Cloutier, the well-known Canadian Inuit leader, argues eloquently about the "human right to be cold" and underscores the fundamental transitions associated with climate change in the North. The full contours of environmental transitions in the Arctic remain largely unknown although current analysis points to substantial and sustained changes.

Northerners have been quick to respond, exploring all measure of alternate energy systems, and participating in Arctic-wide monitoring and evaluation of ecological change. The people of the Arctic, while heavy users of fossil fuels, make miniscule contributions to global climate change, but stand to suffer disproportionately from the environmental disruptions that have already started across the Far North and that will likely accelerate in the years to come. There is little, from the policy or program angle, that Arctic leaders can do to offset the environmental damage, but they can be proactive in educating the world about the human impacts of climate change, seek to hold the industrial world accountable (financially and otherwise) for northern conditions, and develop responses that allow them to maintain their settlements and lifestyle in a warming and shifting environment (ACI Assessment 2004).

The Future of Work

As discussed in an earlier chapter, understanding and responding to the future of work have the attention of governments, families, and young people around the world. The scale and speed of technological displacement have brought short-term challenges for some people, forcing substantial adjustments within companies, governments, and regions. The list of transformative technologies is substantial and includes big data, artificial intelligence, robotics, augmented reality, virtual reality, and gamification. Machine-based systems are now shifting from the transformation of industrial and low skilled labor to professional work, including in law, medicine, accounting, and government services. The North has been affected by some of the changes, such as autonomous vehicles and labor-saving mining devices. The region stands to see even more changes in the future. With many jobs likely to be replaced, disrupted, or transformed by technology, it is not clear what new positions will emerge or if they will be based in northern communities or operated from a distance out of southern offices. Given the high technology/high science nature of much of the anticipated new work, the substantial limitations of non-Scandinavian Arctic education assume increased importance.

There is more at stake than the loss of jobs, even if the number is substantial. In the post-World War II era, one of the key elements in Western industrial economies was the ability to produce high salaries and secure jobs for people with limited technical abilities and largely non-specialized skills. This development rested on generally buoyant economies, strong trade unions, and a seemingly insatiable commercial demand for willing and reliable labor. Many parts of the Arctic, reliant on the natural resource economy, have provided long-term opportunities for such workers, with the major innovation being the transition from company towns in the 1950s and 1960s to a predominantly fly-in-fly-out work force thereafter.

Most of these conditions had eroded, at least in part, by the early twenty-first century. Demand for partially skilled labor seems to have declined, with the workers displaced in part by robots and various digital technologies. In the new economy, however, there is comparatively low-wage work in the service sector and high-wage work for people with advanced scientific and technological skills. Economics argue that the technological displacement of labor typically creates both short-term problems for specific regions and sectors but, for the economic as a whole, frees workers and money for other economic opportunities. For the Arctic, this means the likely loss of many currently unskilled or semi-skilled jobs through the introduction of automated vehicles, digital systems, and automated extraction processes. This change will hit men particularly hard. As male labor has long been dominant in the Far North, there could well be a series of gender and labor force changes that could cause significant transformation in the region. Put simply, it is not clear where the new work will be found if the anticipated shifts occur or how the gender balance in the northern workforce will change as a result (Brynjolfsson and McAfee 2014).

The Transformation of the Resource Economy

With the notable exception of Russia, where the emerging norms in the Western industrial world do not hold in full, the resource economy is being upended by the convergence of concerns about climate change, Indigenous rights, and environmental protection. New projects are subject to extensive, expensive, and time-consuming review and permitting processes; in North America, Indigenous concurrence has emerged as an effective requirement. Protests against resource development, driven largely by organizations and forces from outside the Far North, are now commonplace, even when Indigenous people from the North are supportive. It is getting harder to do business in the resource sector in the North, albeit for reasons that are good and important for the protection of the environment and to maximize Indigenous engagement. Add to this the expansion of resource development in many parts of the world, and the picture emerges of a resource economy in global transition. The regional impact of the resource economy is being affected, as well, by the introduction of technology-intensive mining, more autonomous systems, and a growing reliance on fly-in fly-out workers, with the workers often living in southern areas. The global emphasis on reducing industry's environmental footprint could lead to a decline in northern resource activity, particularly as work in the Far North remains difficult and expensive and as less expensive opportunities come on line in the developing world (Duhaime and Caron 2006).

Economic Inequality

The contemporary North is rife with income and economic inequality (Pikkety 2014). Most Indigenous communities, even in Scandinavia, lag behind non-Indigenous settlements in terms of income and opportunity. Part of this is due to choice, specifically the decision to stay in ancestral homelands and avoid the migration to regional centers and southern cities. In other places, particularly the Far North in North America, major educational shortcomings ensure that few Indigenous peoples make it into the Arctic professional class of doctors, nurses, lawyers, senior government officials, and resource engineers and managers. The challenge is exacerbated by the fact that a substantial number of northern workers are either only temporary residents of the North; they come to the region, in the words of former Yukon Premier Tony Penikett, "to make a killing, not to make a living." In many areas, from the Norwegian offshore oil fields to NWT diamond mines and Alaskan pipeline operations, workers fly in and fly out, typically on short-term rotations, and take much of the earned income from the Arctic to southern centers. In Canada, the fly-in-fly-out pattern increasingly now often includes northern residents who, now earning premium wages and with the income necessary to flourish in urban environments, capitalize on the opportunity to relocate to southern centers.

The result, most pronounced in the Canadian North and Alaska, is a large, well-paid non-Indigenous population, much more permanently based on the region than a generation ago, but also tied to the seasonal, work, and market cycles of the resource economy. While a substantial Indigenous middle-class has emerged across much of the Arctic, tied to the growth of Indigenous institutions, government work and opportunities in Indigenous business, health care and education, a large majority of the Indigenous peoples are either poor, in financial terms, or, as in Scandinavia, trail behind national averages. If the resource economy remains strong and if government funding for northern affairs remains high, neither of which are guaranteed, the inequality challenge will likely maintain its current contours, without significant change. But with new technologies threatening to undercut work in the resource economic and other sectors, the prospect of more pronounced inequality, between Indigenous and non-Indigenous peoples and between North and South, becomes more likely.

Governments in the more prosperous nations do not have real solutions in hand for the tech-driven rise in inequality and the continued separation of countries, regions, and peoples in terms of income and financial well-being. Some solutions are being contemplated, from taxes on robots, higher taxes on high earners and prosperous companies, and a Basic Income Guarantee to redistribute money and provide a higher level of well-being for more people. None of these have found substantial traction in any country (Finland recently abandoned a BIG experiment), although the Scandinavian commitment to the welfare state and its battery of government agencies, support programs, and community investments has offset many of the small-town and northern challenges in recent decades. The Nordic system, the substantial subsidies that support the territorial governments in Canada and the large-scale military spending in Alaska have already moderated extreme interregional economic inequality although the subsidies have not addressed the specific challenges related to the economic well-being of Indigenous peoples and remote communities.

The Global City State Economy and the Decline of Rural Areas and Small Towns

The challenges facing Arctic communities and remote villages are far from unique (Taylor and Derudder 2015). On a global scale, rural and small-town population and economic decline are substantial and are likely to accelerate in the coming decades. In countries as diverse as Portugal and Japan, India and the United States, the issue has produced acute public policy challenges. A small number of city states (cities of two million people or more and with both regional domination and extra-regional economic reach) have emerged to dominate the global economy and, in the process, to set government agendas. The Arctic states have few of these cities. Stockholm is at the lower end of the city state scale, a national economic powerhouse but less prominent internationally. Helsinki and Oslo fall well short of having such economic power. Moscow held promise in the past, but Russia's continued economic challenges have rendered it less economically important. In Canada, Ottawa, Montreal, Calgary, and Vancouver all have considerable economic clout and potential, and Toronto dominates the financing of the global mining industry. But none of these cities see themselves as being substantially Arctic in orientation. In the United States, none of the grand city states—New York City, Chicago, Houston, and Los Angeles—have significant urban interests in the Far North. The closest, and its standing as an international city state is much in question, is Copenhagen, Denmark with a strong historical and contemporary connection with Greenland that makes it perhaps the most Arctic-aware cities in the world.

With few if any city state champions emerging, the dangers to the Far North associated with the transition to a city state economy loom much larger than the potential benefits. Across the Arctic, populations are generally stagnant or declining. Small, remote communities have troubling holding young people in place. Many northerners leave for advanced education and training, often never to return permanently due to an absence of career opportunities. Regional centers—Tromsø and Bodo, Luleå and Rovaniemi, Fairbanks and Anchorage,

Whitehorse and Yellowknife—are doing quite well, although they are regional magnets drawing people and economic activity away from the smaller communities.

Put simply, governments face formidable challenges in attempting to respond to what are truly global and intense forces. Some Arctic communities will continue to thrive: those with major mineral deposits (Kiruna, Sweden), regional government and service centers (Whitehorse and Yellowknife), tourist sites (where would Rovaniemi be without Santa Claus or Svalbard without the northern lights?), and a handful of small cities noted for their high quality of life and amenities (Tromsø and Anchorage, Alaska, are excellent examples). Others will struggle. The high cost of providing even basic services in the Arctic limit the growth and stability of Iqaluit and Nuuk, Inuvik and Barrow. Small centers, lacking economic importance to the nation states and suffering with severe diseconomies of scale and distance, struggle to keep up in an increasingly technologically rich world.

Indigenous communities appear more attuned to the depopulation scenarios than most public governments, which fear being criticized for abandoning vulnerable populations. National governments pay substantially for local services in small towns while paying again, directly or indirectly, for similar services for northern people who have relocated to the cities. A tiny number of Indigenous governments in Canada have experienced substantial relocation to nearby regional centers. In many other cases, Indigenous governments maintain distinct councils in larger centers that are now home to many of their members. The Gwich'in First Nation, for example, centered in Inuvik, NWT, has councils in Yellowknife (the territorial capital), Whitehorse, Yukon, and Edmonton, Alberta. Member-centered governments, including those established through land claim settlements, are more responsive to population shifts than public governments, whose regional representations typically have to fight hard to maintain or expand local services.

SHAPING THE FUTURE: PRACTICAL STEPS FOR ARCTIC GOVERNMENTS

There is no easy or obvious path forward for Arctic politicians and policy-makers (Stokke and Hønneland 2006). There is a growing understanding of what works in terms of regional stability and community development and less awareness or discussion of which current policies are less successful. Southern control, exercised through national governments, creates substantial bureaucracies and maintains fundamental disconnections with northern residents. Military investments bring money and jobs to the Arctic, but can easily distort or undermine local economies in the process. Treating Indigenous peoples as "problems" to "solved" by national governments is a long-term policy blunder, long overdue to be replaced by sincere partnerships, Indigenous autonomy, and permanent collaboration. Arctic governments still have no solid strategies for dealing with core issues: cold and winter, isolation and distance, and the often-misguided southern perceptions of the Far North that often render the region into little more than an international curiosity.

On the constructive side, there are recent policy initiatives that do work. Indigenous empowerment through legal or political processes and the resolution of Indigenous land rights (and the allocation of land and funds to Indigenous control) are essential elements in long-term reconciliation. So, too, are substantial regional investments in education and research (the long-term positive impacts of the University of Alaska system and the University of Tromsø are little understood, just as the negative effects of the absence of such post-secondary opportunities in the Canadian North are underestimated). Training people in the North and for the North is one of the most important initiatives available to all Arctic governments and underpins the transformation of Yukon College into Yukon University and the development of a polytechnic university in the Northwest Territories.

Culturally based approaches to the law, imprisonment, and rehabilitation, including sustained efforts to support individuals and families struggling with the oppressive weight of history and external domination, are essential and have produced strong results. Understanding the current and future impacts of new technologies, particularly on the Arctic work, will be critical, as will be need to establish intense, gap-reducing educational and training processes for Arctic residents. Scandinavia has demonstrated the foundational importance of ensuring that Arctic communities have comparable services to those in southern and urban areas. Carefully planned infrastructure, and not the kinds of roads, railways, or other services hastily built to accommodate military or resource requirements, can lay a foundation for regional prosperity and stability. Of course in much of the Canadian North and Alaska, the costs of providing such infrastructure across vast Arctic and sub-Arctic lands are seen as prohibitive by national and regional governments.

Arctic public policy development is challenging and it will continue in this manner. There are many actors and forces, from Indigenous peoples and their governments, non-Indigenous peoples, commercial interests, national governments, international agencies (including the increasingly interventionist environmental groups), southern populations (who pay through their taxes for the heavily subsidized northern societies), largely southern-based academics, and national and international commentators. None of these groups are shy about stating their views about the present and future but there is often little overlap between these competing world views and policy prescription. The recent past has highlighted the importance of listening attentively to Northerners and to responding to their concerns, but with an appropriate level of fiscal responsibility and overriding practicality.

Some suggested main steps include:

- *Embrace Indigenous empowerment*: The strengthening on Indigenous political and legal rights is the most important tool available to Arctic governments and policy-makers. Where this has been done, and the Yukon is perhaps the best example, new possibilities open up and the framework for a new Arctic emerges.

- *It is absolutely vital that Arctic policy-makers be true "future thinkers"*: They have to monitor international developments, anticipate technological change, and stay abreast of global developments. What happens internationally matters in the Far North, now more than ever. Northerners need to monitor developments and anticipate the impacts and benefits of emerging technologies.

- *Arctic policy-makers have to articulate a solid vision of the future*: The plans have to be realistic, positive, honest (even hard-hitting where required) and must avoid the fatalism or unrealistic optimism that was commonplace in the past. It is vital to articulate a sense of what is at stake in terms of jobs, economic development, technological change, and cultural dissonance, but this analysis should always be combined with advice about what the communities and families can do to address the challenges and respond to the opportunities.

- *Follow developments in other jurisdictions*: Arctic governments, national and subnational, must be acute observers of other countries, regional authorities, communities, and organizations. They could look, on a grand scale, to the approaches of countries like Japan and its creative and assertive Japan 5.0 strategy that is designed to create the world's first truly digitally enabled society. They can examine the approaches to innovation in Finland and Israel. Within the Arctic, officials and planners can look at the inclusion efforts of Yukon, alternate energy systems in Alaska, the role and impact of Indigenous economic development corporations across northern Canada, the impressive institution building in Troms County, ecotourism development in Rovaniemi, "new economy" regional planning in Luleå and Skellefteå (both in Sweden), governance innovation in Greenland, co-management of natural resources in the Northwest Territories, and cultural and linguistic initiatives in Nunavut. Many times, the best solutions to northern challenges can be found in other Arctic jurisdictions.

- *Encourage local institutions*: The future adaptability of any society depends, in substantial measure, on the quality and innovative capacity of institutions, specifically schools, colleges, polytechnics, universities, research units, hospitals, government agencies, and local economic development organizations. The success of Troms County, Norway, is due in substantial measure to the high quality and creativity of the University of Tromsø, Norway's Arctic University. Arguably, Yukon's greatest strength lies in the depth and professionalism of the territorial civil service. Similarly, the resurgence in the quality of life in Alaska owes a considerable amount to the adaptability and resilience of the Alaskan native corporations. Where institutions are strong, societies have the opportunity to respond to evolving technological and economic reality. In the absence of such institutional strengths, regional societies face serious difficulties in their attempts to respond to twenty-first-century realities.

- *Create local opportunities to discuss the future*: It is vital that northern governments simply not accept the inevitability of rapid and negative change. Transitions are coming, and it is vital that Arctic authorities create many opportunities for citizens to contemplate what lies ahead. There are many ways that this can be done. Governments can bring provocative and controversial thinkers to northern towns, not for strategic planning in the first instance, but rather initially to provoke debate about the future. Informal planning groups, the wide circulation of important reports and studies on future and prospective developments, and frequent references to future challenges and opportunities facing the region and the Arctic as whole in political speeches are but a few of the steps that can be taken to spark public debate about the years ahead. An informed citizenry, engaged in open dialogue about the future of the Arctic, is essential to the development of creative and constructive policy.
- *Identify vulnerabilities*: It is vital that governments make a systematic effort to identify communities, economic sectors, companies, and labor groups at risk. Further, governments have to speak, bluntly, and constructively about the challenges that lay ahead. They need to challenge organizations and groups, along with associated government agencies, to develop plans for adaptation to rapid change.
- *Recognize the role of local governments as trend setters*: While national innovation strategies continue to attract a great deal of political and administrative attention, the reality is that local and regional authorities and business communities can be more responsive to imminent and future threats. Regional authorities, combining the strengths and responsibilities of government and business, have excellent opportunities to develop realistic and sustainable solutions to contemporary challenges.
- *Develop longer term plans for organizations, communities and regions*: Governments in democratic systems suffer from the inevitable constraints of short-term electoral cycles, which tie the hands of politicians and administrators as they develop complex and experimental economic, technological, and political responses to immediate challenges while keeping a careful eye on the winnability of the next election. Given the speed and comprehensiveness of global changes, it is vital that national and regional leaders look to the middle of the century and beyond as they develop administrative and government plans and strategies. While the imperatives of the present are real and impending challenges have to be taken seriously, all societies require effective and carefully developed plans for the 2030s and 2050s. This is a serious problem for politicians and administrators, for the pressure to act quickly in response to current issues, but having a realistic approach to long-term developments has become an essential element in public policy development for the twenty-first-century Arctic.

CAN THE ARCTIC BE A FUTURE MAKER?

Arctic nations and regions have some real strengths, including strong national governments, highly levels of state subsidization, effective Indigenous organizations, and solid regional and local authorities. There are a handful of robust, largely administrative centers, from Nuuk to Bodo and from Yellowknife to Fairbanks. The Arctic hosts a small number of excellent universities and colleges, including the University of Alaska, Yukon College/Yukon University, University of Nuuk, University of Lapland, Luleå University, Umeå University, and Bodo University among others. There are some successful businesses, particularly in the resource sector operating off Norway's coast, the diamond mines of the Northwest Territories, and throughout Alaska. Perhaps more importantly, all of the Arctic regions save for Russia have robust civil institutions, a commitment to the role of law, respect for Indigenous rights, and deep democratic systems, adapted in most instances to the special circumstances of Arctic regions.

The Far North, however, also has serious shortcomings, all of which have could shape and affect the region's ability to respond quickly and creativity to the realities of the twenty-first century. The Arctic in general is facing demographic decline, in part due to outmigration (including among Indigenous peoples), the technological and economic turmoil in the resource sector, and the differential family conditions of Indigenous and non-Indigenous peoples. Many northern and Indigenous communities are small and remote; these are among the most vulnerable to the socioeconomic and technological pressures of the coming decades. The majority of these settlements will continue to decline, some to the point of total collapse, with fly-in communities most vulnerable. The northern regions are all, to a greater or lesser extent, colonies of national governments, with development shaped by national priorities and with preponderant economic and political power resting in the Southern and urban centers. Furthermore, the collision between Indigenous rights, growing environmental oversight and intervention, and the North's dependence on resource development has been working against the interests of Arctic peoples and regions.

The future is uncertain, for all of humanity and not just for the Far North. But the challenges facing the Arctic are among the most acute in the world, with regional authorities generally lacking the resources or ability to tackle profoundly global forces. Arctic peoples have long been noted for their resilience. The achievements of the past half century make it clear that the Far North cannot be counted out. Indeed, if the positive and constructive changes in recent years continue and if the policy innovations of the past provide a foundation for further, proactive change, the region's many challenges and opportunities seem more manageable. There is no reason to believe that the path to 2050 will be smooth and many reasons to think that the pace of externally imposed change will accelerate rather than decline. Effective policy-making and thoughtful political action will be required for the peoples and jurisdictions of the Far North to achieve their aspirations and to overcome the region's socioeconomic, climatic, and other challenges.

References

ACI Assessment. 2004. Impacts of a Warming Arctic-Arctic Climate Impact Assessment. In *Impacts of a Warming Arctic-Arctic Climate Impact Assessment*, by Arctic Climate Impact Assessment, 144. Cambridge, UK: Cambridge University Press, December. ISBN 0521617782 2004.

Brynjolfsson, Erik, and Andrew McAfee. 2014. *The Second Machine Age: Work, Progress, and Prosperity in a Time of Brilliant Technologies*. WW Norton & Company.

Duhaime, Gérard, and Andrée Caron. 2006. The Economy of the Circumpolar Arctic. In *The Economy of the North*, 17. Oslo: Statistics Norway.

Einarsson, Niels, Joan Nymand Larsen, Annika Nilsson, and Oran R. Young. 2004. *Arctic Human Development Report*. Copenhagen: Nordisk Ministerråd.

Larsen, Joan Nymand, and Gail Fondahl. 2015. *Arctic Human Development Report*. Vol. II. Copenhagen: Nordisk Ministerråd.

Pikkety, T. 2014. *Capital in the Twenty-First Century*. Translated by A. Goldhammer. Cambridge: Belknap Press of Harvard University Press.

Smith, Laurence C. 2010. *The World in 2050: Four Forces Shaping Civilization's Northern Future*. Penguin.

Stokke, Olav Schram, and Geir Hønneland, eds. 2006. *International Cooperation and Arctic Governance: Regime Effectiveness and Northern Region Building*. Routledge.

Taylor, Peter J., and Ben Derudder. 2015. *World City Network: A Global Urban Analysis*. Routledge.

CHAPTER 33

Dotting the Ice

Tony Penikett

CROSSING LINES

A visitor from outer space might conjecture that the Arctic Council is a club for urban elites from the eight states of the Circumpolar North. For its members, Club Arctic is a world of tailored suits from the air-conditioned cubicles of centrally heated capital cities. From homelands beset by shifting game populations and melting permafrost, Arctic governors, hunters and miners would recognize Club Arctic only as a place where they do not belong.

For centuries, nation-states and their statesmen have drawn lines on maps, marking boundaries and staking claims to territory. In 1763, France ceded lands east of the Mississippi to Britain, and George III's cabinet[1] drew a line down the "spine" of the Appalachians[2] dividing the continent into British North America and Indian Territory.

US President George Washington and his successors moved that line steadily westward with hundreds of Indian treaties. The United States and Canada further subdivided Indian "hunting grounds" into Indian reserves and reservations, as often as not trapping First Citizens on poor lands and in pockets of permanent poverty. The United States also created legal color lines between

---

[1] The *Royal Proclamation* is a document that set out guidelines for European settlement of Aboriginal territories in what is now North America. The *Royal Proclamation* was initially issued by King George III in *1763* to officially claim British territory in North America after Britain won the Seven Years War. https://indigenousfoundations.arts.ubc.ca/royal_proclamation_1763/.

[2] [Colin G. Calloway, *The Indian World of George Washington: The First President, the First Americans, and the Birth of the Nation*, Oxford University Press, 2018].

T. Penikett (✉)
Simon Fraser University, Morris J. Wosk Centre for Dialogue, Vancouver, BC, Canada

K. S. Coates, C. Holroyd (eds.), *The Palgrave Handbook of Arctic Policy and Politics*, https://doi.org/10.1007/978-3-030-20557-7_33

blacks and whites. In 1867, Russia moved its boundary line with the United States from Alaska to the Bering Sea. Today, Greenland is creating a national boundary, a new line between itself and Denmark.

THE *CLUB ARCTIC* FANTASY

When the Arctic states formed the Arctic Council in 1996, the group had to decide on certain lines. Would their Arctic's southern boundary be the tree line, the Arctic circle or the 60th parallel? Climate change was moving the tree line steadily northward, and that might justify the exclusion by national politicians of the Dené (Athabascans) and the Aleut as Arctic peoples. Choosing the Arctic Circle would embrace the Inuit, certain Russian Indigenous peoples and the Sámi, but probably not many Dené villages.

Eventually, for the Arctic's boundary in North America,[3] the eight Arctic states chose the 60th parallel. The Arctic Eight also invited in six international Indigenous organizations as Permanent Participants with *voice* but no *vote*. The Arctic Council decision encompassed the geography of Canada's three northern territories but not all of the Arctic's political realities, excluding particularly regional governments such as those of Alaska, Finnmark, Lapland, Northwest Territories (NWT), Nunavut, and the Yukon Territory, which do most of the Arctic's governing (airports, colleges, hospitals, roads and schools, etc.).

In 1998, the Northern Forum, the organization of Arctic regional governments, applied for one seat on terms similar to that of the Indigenous Permanent Participants, but the Arctic Council rejected their application. In so doing, the Arctic Council left northern settlers, who count as 85% of the Arctic's population, with neither a vote nor a voice. At this magical moment, the Arctic Council also became an organization representing a fantasy Arctic, one composed of nation-states and Indigenous villages plus covetous Observers, with no representation from the regions or 85% of Arctic residents. With this action, the "club" really crossed the line.

As it happens, Permanent Participants won seats at the Arctic Council's ministers' table in large part because of their successful rights struggles over the previous generation. Land claims and self-government agreements permanently changed the region. The scale of the land settlements in these modern treaties dwarfed the nineteenth-century treaties and made Indigenous peoples the largest private land owners in Alaska, Yukon, NWT, and Nunavut. The scope of the self-government agreements in Canada astonishes southerners, who cannot believe that villages enjoy quasi-provincial powers. Greenland has,

---

[3] The Arctic Monitoring Assessment Program (AMAP), which predates the Arctic Council, created its "AMAP area" as the territory where it would carry out environmental monitoring under the Environmental Protection Strategy. AMAP has defined a regional extent based on a compromise among various definitions. The "AMAP area" essentially includes the terrestrial and marine areas north of the Arctic Circle (66°32′N), and north of 62°N in Asia and 60°N in North America, modified to include the marine areas north of the Aleutian chain, Hudson Bay, and parts of the North Atlantic Ocean including the Labrador Sea, excluding the Baltic Sea.

in the same period, gone from colony to Home Rule to Self-Government and now, perhaps, nation-state status.

Currently, there is much talk in Canada about reconciliation between Indigenous and settler populations. But the only region where reconciliation has taken root with land treaties and tribal self-government agreements is the Far North. Most of these accords took decades to negotiate and have now operated for two or three decades. Generally, however, southerners still know little of these community- and nation-building achievements.

Without the modern treaties and the Arctic awareness their negotiation raised, neither the Arctic Council nor the Permanent Participants might have existed. Yet the capitals of the federal Arctic States exist thousands of miles to the south of the Arctic, and national governments have short memories. Among other things, the national governments completely forget or conveniently ignore the role of Arctic settlers in making and implementing modern land and political settlements.

Fox Fables

Our rate of growth [in the North] is very slow, and there is a reason for this. We live in the midst of the world's smartest animals—the polar bear and the Arctic fox. To live well, we exercise precaution, we plan our activities carefully.[4]
—former Greenland Prime Minister Aleqa Hammond

Once upon a time, the Arctic was a fearsome, frozen, and forbidding place—at least that was its image in the global South. Nowadays, for TV viewers in the Far North, the Great Cities of the south may appear far more frightening with their yawning income gaps between the billionaires atop the tall towers and copper-toned[5] immigrant workers chasing down McJobs and affordable squats on the garbage-choked streets below, all the while dodging threats of deportation or worse from trigger-happy cops.

In such moments, Arctic homelands might seem like a saner, safer place. Northerners love a southern holiday in the United States. They enjoy the faux-royalty of Hollywood and the White House, the quaint tradition of buying gas by the gallon, and the titillating possibility of a drive-by shooting. When the holiday is done, they head home to the Arctic and to the Arctic fox (*Vulpes Lagopus*).

The only place one can find an Arctic fox in the world-class city is in a zoo or a TV documentary about the melting polar ice cap. Still, the Arctic species

---

[4] Aleqa Hammond in Penikett (2017), *Hunting the Northern Character*, UBC Press, 2017, 10–11 [The adventures of a FOX NEWs "genius" at Helsinki, Singapore and Windsor Castle brought to mind Vilhjalmur Stefansson's comment about "adventure was a sign of incompetence."].

[5] Wikipedia: "The original Coppertone logo was the profile of an Indian chief... Sometime around 1965, Jodie Foster made her acting debut as the Coppertone girl in a television commercial, when she was three years old."

has a cousin, the red or "quick brown" fox (*vulpes vulpes*) famous for jumping over lazy dogs on prairie grasslands or in wooded areas. Its by-blow namesake Red-State Fox (*vulpes vulgaris*) seems most at home in the Deep South.

Very much a dominant predator species, Red-State Fox is so angry, so aggressive that it creates its own ecology. As with *Drake* or *Oprah*, North American television viewers know this beast by its first name—encoded in three capitalized letters—as *FOX*! Arctic elders may remember the old RCA Victor trademark of the terrier listening to the speaker of a disc gramophone beside the cutline "His Master's Voice." In the twenty-first century, FOX does the talking, not the listening, but the vocalizations are still very much issued in its master's voice.

Every North American has heard the totemic FOX bark about Mexican walls, Muslim bans, and fake news. FOX expertly vocalizes the normal views of the billionaire oligarchs who now oversee American government, just as their overseas counterparts do in the kleptocracies of Asia, Eastern Europe, and the Middle East. Since the end of World War II, with only a tiny percentage of the world's population, wealthy America has fought to keep the lion's share of the world's riches. Through the animal spirits[6] of charging bull markets and bearish stumbles, billionaires always want more than their share. That has not changed.

The Arctic fox is highly territorial and generally solitary, but it has an inter-dependent relationship with other species in its hunting grounds. It thrives in cool, noiseless, uncrowded spaces and sometimes communicates with subtle body gestures. In air far from the noxious cloud created by FOX, the little Arctic cousin has long subsisted. Yet now FOX's inexhaustible appetites threaten the Arctic fox's habitat through the melting of the polar icecap, the bit-by-bit drilling of lands and seas, and the pipelines that snake across tundra grasses to suck black gold from the dinosaur bones buried deep below the melting permafrost.

SURVIVAL

Greenland's Aleqa Hammond credits her people's survival to lessons learned from the world's intelligent animals, the polar bear and the Arctic fox. Far from FOX's front lines, on Arctic snow and ice, human northerners share the land-scape with ravens, crows, and wolves, plus unique northern species like cari-bou, muskox, and walrus.

Every Inuit community has a raven creation myth. Each Athabaskan child is born into either the Wolf or the Crow clan. Because ancient wisdom dictates that a Crow may only marry a Wolf, each knows they need each other for their community to survive. For thousands of years, emblematic crows announced the arrival of migrating caribou. In time, some of the iconic wolves that tracked

---

[6] "Animal spirits" a term from John Maynard Keynes 1936 book *The General Theory of Employment, Interest and Money* for the instincts and passions motivating human behavior, investor enthusiasm and consumer confidence, supposedly inspired by "animal spirits."

the caribou herds started feeding on food scraps at the edge of human camp fires and eventually evolved into "man's best friend,"[7] a distant relative of the wolf.

As descendants of ancient communities that endured in the world's harshest climate, northerners are survivors. For ages, the Arctic has been an arena of human struggles for food, housing, and meaningful work, as well as education, health, and precious cultural experience. Having survived the state-sponsored travails of colonialism, the Cold War, land rights struggles, outside exploitation of their resource wealth and despoliation of their environments, northerners have been finding new ways of living and working together.

For several decades now, Northerners have been breaking trail toward peaceful accommodations between old Indigenous villages and newer settler cities, adapting and growing community in a changing environment of scattering game and diminishing snow and ice. Indigenous and settler locations have starkly different histories, but increasingly, they understand they are facing a future together.

Facts on the Ground

As experiential learners, Arctic peoples know they must live with the facts. Indeed, they trust the facts they find on the ground and learn the lessons derived from the long view and broad horizons. Students of Indigenous law— laws in effect long before the creation of Canada, Norway, Russia, or the United States—learn that if animals, birds, and fish are mistreated, they will abandon their human neighbors, so there must be clear rules about allocation, harvest limits, and food preparation (Borrows 2006).

This body of law reflects the real-world Arctic, where people endure through the hard work of hand and brain, not the manipulations of the real estate sharks and dot.com predators that dwell in FOX's world. Rather than fake news or tweeted falsehoods, Arctic communities deal with hard facts on the ground: melting permafrost; shifting game populations; the historic cultural traumas of hunger, homelessness, and suicide.[8] FOX has little to say on such subjects.

---

[7] Tim Flannery, "Raised by Wolves," *New York Review of books*, April 5, 2018 Issue

One day around 26,000 years ago, an eight-to-ten-year-old child and a canine walked together into the rear of Chauvet Cave, in what is now France. Judging from their tracks, which can be traced for around 150 feet across the cave floor, their route took them past the magnificent art for which the cave is famous … Whatever happened, the pair's adventure certainly became famous in 2016, when a large radiocarbon dating program that included the smear of charcoal discarded by the child confirmed that the tracks constitute the oldest unequivocal evidence of a relationship between humans and canines.

[8] Israel Rosenfield and Edward Ziff, "Epigenetics: The Evolution Revolution," *New York Review of Books*, June 7, 2018, p. 38 "Epigenetics has also made clear that the stress caused by war, prejudice, poverty, and other forms of childhood adversity may have consequences both for the persons affected and their future—unborn—children, not only for social and economic reasons but also for biological ones."

In the face of North Korea's gulags, human rights nightmares and threats of nuclear war, FOX offers only speculation about future ocean-front condos. The best this voracious beast can do is amuse (Postman 1985),[9] anger, and distract. FOX craves arguments and polarity; it likes the hierarchical pyramid and square-table face-offs between contending fictions. Northerners are more comfortable around the council fire, in the community meeting and the dialogue circle.

FOX pushes for megaprojects to cross Indigenous lands. In Norway's Finnmark County, Sámi and municipal politicians co-manage land and resource decisions. Finland, Sweden, Norway, and the Sámi parliaments of those three countries have discussed a Sámi Convention, potentially the first international treaty to be signed by Indigenous people. This new Arctic "constitutional" order is the antithesis of FOX's ambitions.

While FOX defends the "roughing-up"[10] of suspects and the longest-possible prison sentences, Arctic visionaries imagine regions without jails and "de-carceration," or land-based alternatives. A generation ago, the Yukon Territory borrowed the Indigenous idea of "circle sentencing," in which community elders advise a court's presiding judge. Finland jails only two kids per million of its population[11] and proudly operates "open prisons."[12] FOX blames the victims for military or police killings of unarmed black and brown people. In the Arctic, where rifles are tools, not weapons, legislators fret about gun safety and suicide prevention.

For Madeleine Albright, the anti-democratic FOX agenda is a harbinger of fascism[13] but, perhaps, we should view FOX distractions as "fake fascism." Such diversions might constitute engaging propaganda, but they are not politically serious ideas. FOX styles the climate crisis as a "scam"[14] and thus considers "the environment" merely a place for jaded elites to shoot endangered species: elephants, polar bears, and tigers—not for food but for fun. Climate change

---

[9] Neil Postman, *Amusing Ourselves to Death: Public Discourse in the Age of Show Business*, Viking Penguin, 1985.

[10] David Cole, "Trump's Inquisitor," *New York Review of Books*, April 19, 2018.

[11] Eric Allison, The Guardian international edition, Tuesday 10 February 2009, "Prison is no place for children; In the UK, we lock up 23 children per 100,000 of the population, compared to six in France, two in Spain and 0.2 in Finland."

[12] Rae Ellen Bichell, *"In Finland's 'open prisons,' inmates have the keys,"* April 15, 2015, 1:45 PM EDT *"Everyone at the Kerava open prison applied to be here. They earn about $8 an hour, have cell phones, do their grocery shopping in town and get three days of vacation every couple of months. They pay rent to the prison; they choose to study for a university degree in town instead of working, they get a subsidy for it…".*

[13] Madeline Albright, *Fascism: A Warning*, HarperCollins, 2018.

[14] Scott Pruitt, FOX News June 1, 2018: "One year ago, on June 1, 2017, President Trump boldly and courageously announced that the United States would withdraw from the Paris Climate Agreement. This was a historic moment that upheld the president's campaign promise and demonstrated to the world that he puts the American people first. The president's decision, together with his decisive actions through regulatory reforms and tax relief, is unleashing the American economy."

has hit northerners twice as hard as southerners, and Arctic peoples have already learned some painful lessons that might help them to survive a FOX-sponsored planetary poisoning. In quiet but determined opposition to the FOX dystopia, Arctic Fox, Raven, Crow, and Wolf and their human clans have no option on their homelands but to resist.

Migrants

While FOX gratuitously insults First Nations[15] and wants to wall off the United States from families of Aztec, Mayan, and Olmec ancestry, Indigenous senators and state representatives hold the balance of power in the Alaska legislature today. In the Yukon Territory, self-governing First Nations are achieving levels of peace and prosperity unimagined by most Indian reserves in southern Canada. A Northwest Territories innovation has seen the formation of a hybrid municipal-tribal form of government. In Nunavut, non-Indigenous legislators have debated in Inuktitut. As noted, Greenland is on its way to becoming North America's first Indigenous nation-state.

FOX calls for immigrants from Norway rather than "shithole" African countries, but Northerners realize that people have been moving around the planet for millennia. Immigration is a very old and complicated story. Long before an immigrant American adventurer erected an Arctic Restaurant and Hotel at Whitehorse[16] during the Klondike Gold Rush, far more interesting people were moving into the area.

In 2013, in the Tanana River Valley, an archeology team led by Ben Potter from the University of Alaska Fairbanks uncovered the 11,500-year-old remains of an infant girl, who they named Sunrise Girl-Child ("Xach'itee'aanenh T'eede Gaay" in the Middle Tanana Athabaskan dialect). The remains offer genetic evidence of a 20,000-year-old migration and the possibility that *all* Indigenous Americans may be traced back to this migration (Fig. 33.1).[17]

---

[15] Eli Rosenberg, "Native Americans Called Andrew Jackson 'Indian killer.' Trump Honoured Native War Heroes in Front of His Portrait," *The Washington Post*, Tuesday, November 28, 2017, *"A slave owner, Jackson spoke about Native Americans as if they were an inferior group of people. 'Established in the midst of a superior race,' he said of the Cherokee, 'they must disappear'."* https://www.thestar.com/news/world/2017/11/28/native-americans-called-andrew-jackson-indian-killer-trump-honoured-native-war-heroes-in-front-of-his-portrait.html.

[16] *Trump's* grandfather changed the family name from "Drumpf" (Blair 2000).

[17] Michelle Z. Donahue, *National Geographic*, January 3, 2018, "Her genome is the oldest-yet complete genetic profile of a New World human. But if that isn't enough, her genes also reveal the existence of a previously unknown population of people who are related to—but older and genetically distinct from—modern Native Americans. This new information helps sketch in more details about how, when, and where the ancestors of all Native Americans became a distinct group, and how they may have dispersed into and throughout the New World. Ben Potter, the University of Alaska Fairbanks archaeologist, who unearthed the remains at the Upward River Sun site in 2013, named this new group 'Ancient Beringians'."

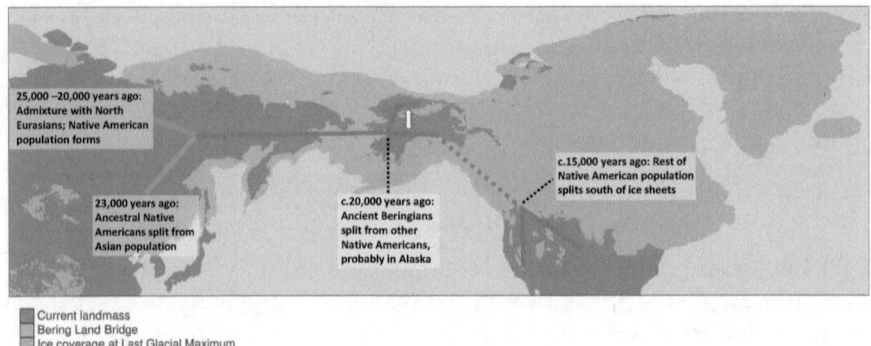

Current landmass
Bering Land Bridge
Ice coverage at Last Glacial Maximum

Fig. 33.1   Indigenous American population map

Many northerners are highly curious about their DNA relationship with Sunrise Girl-Child.[18] That's not all. In March 2018, media reported the discovery of 13,000-year-old human footprints on the shoreline of Calvert Island, British Columbia, the earliest of their kind in North America. Animal bone fragments from a 19,000-year-old specimen at the Bluefish Caves site near Old Crow, Yukon, also show signs of human tool use. University of Montreal researchers Lauriane Bourgeon and Ariane Burke think that humans may have lived in Bluefish Caves in northern Yukon 10,000 years earlier than previously thought. Another theory suggests that humans from Central Asia were stranded for a long time in Beringia 24,000 years ago, during the last ice age (Fig. 33.2).

New news of olden days: these are exciting finds. The discoveries keep Athabaskans and Inuit, Alaskans, and northern Canadians thinking about what was and what could be. Rather than it existing in the mind of the global south as the end of the trail on a road to nowhere, can we now imagine the Arctic as the portal to a very old world and perhaps a cradle where a new one is being born?

SYNCRETISM

Anthropologist Ronald Wright writes that the surviving North American Indigenous Nations have followed a path of "syncretism,"[19] a process that allows a minority community to adopt or borrow useful features of the dominant society (cars, highways, hospitals, high-tech surgery) in order to guarantee the survival of the community's core cultural values, that is, land, language, and law. See syncretism, then, as a survival strategy.

---

[18] *Not least this author, the father of three Tanana children.*

[19] "…the dominant form of Mexican resistance to the invaders would turn out to be syncretic. Syncretism—the growing together of new beliefs and old—is a way of encoding the values of a conquered culture within a dominant culture. It would allow the Franciscans to think they had succeeded—and allow the Aztecs to think they had survived" (Wright 2009).

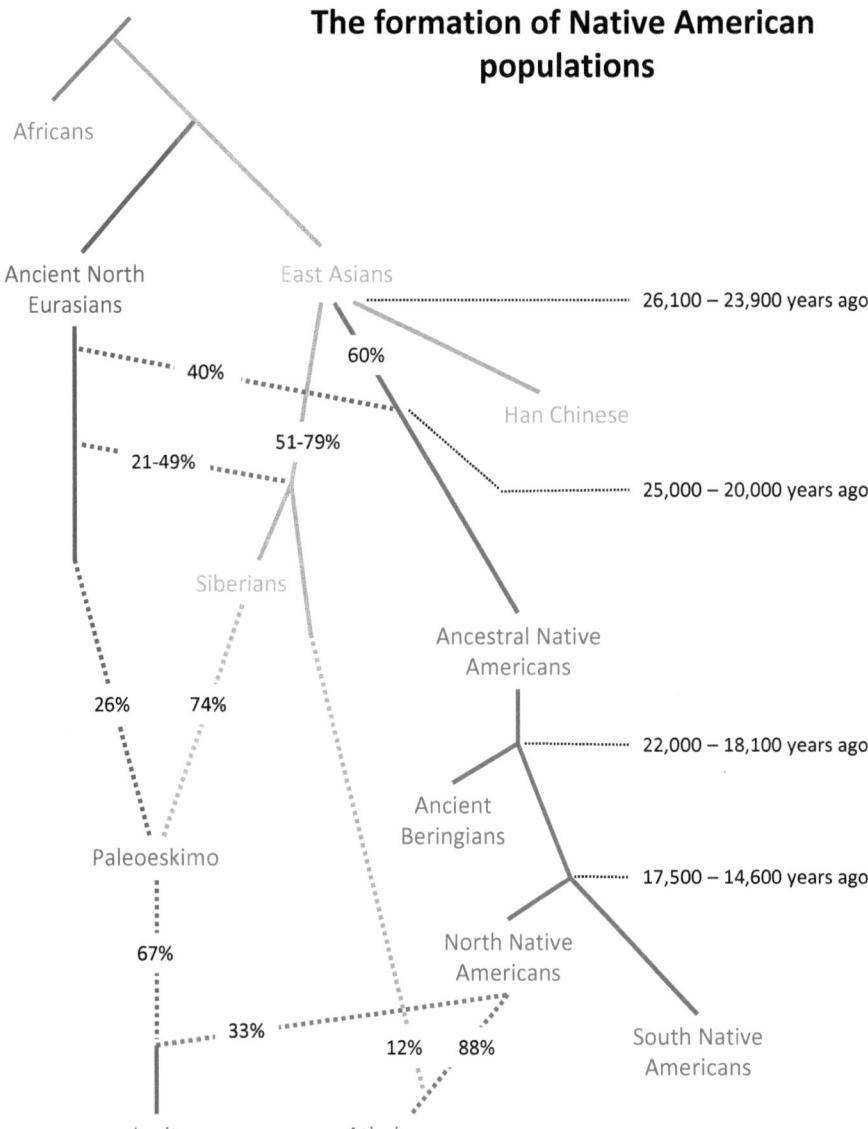

Fig. 33.2 The formation of Native American populations

Out of sight and mind of the national and international media, Indigenous and settler leaders have negotiated an astonishing array of such syncretic compromises over the last generation in the Arctic, to, for example:

- forge the great treaty compromise that Indigenous peoples from Alaska and Arctic Canada get not reserves or reservations but titled land, under the English legal tradition but collectively held, according to Indigenous practice;

- reform education to incorporate Indigenous languages and learning, with parent councils at schools having powers over the appointment of principals and the choosing of cultural curricula;
- create treaty-based co-management boards in fish, game, and land management, boards that reverse old priorities of commercial, sport, and subsistence harvests to create a new hierarchy: conservation, subsistence, sport and, last, commercial uses of fish and game;
- marry modern science and traditional Indigenous knowledge to improve fish, game, and land management;
- bring Indigenous representatives into regional government; and
- build new hybrid institutions in Alaska, Canada, Greenland, and Finnmark that respect both Indigenous and settler traditions.

Devolution, land claims, and self-government agreements permanently changed the politics and governance of the Arctic region. The making of Indigenous land settlements and self-government agreements involved strategic compromises between regional governments, which do most of the governing in the Arctic, and Indigenous villages, which are the major private landowners. Might we not see this process as syncretic, though, a "growing-together" toward a distinct Arctic community and culture, quite unlike the FOX fostered divisions between rich and poor, black and white, old and young.

Tame reindeer and wild caribou—domesticated stock and migrating herds—live in separate realities but they are not that different. If a migrating caribou herd passes close to a reindeer farm, domesticated animals may run off to join the herd. With the loss of their sea-ice hunting platforms, polar bears are learning to hunt from land and, even, to mate with grizzly bears (Smith 2011). FOX makes lots of noise, but the Arctic Fox is a quiet, clever, and monogamous creature. All Arctic creatures have to be very smart to survive, because their habitat is constantly threatened by the louder beast.

Fortunately, the Arctic is a mainly billionaire-free zone. The North still offers places of quiet beauty, seasonal certainties for the fisher and hunter, and time to think for the young worker and the older teacher. How will people survive the Arctic winter in times not only of poverty but now of climate crisis? On the deepest level, they will do what they have always done. On treaty lands and in lively democracies, community, ancient and emerging, thrives.

Above the political lines created by the Arctic Club, the circumpolar world has become a different place: not just different from the South but quite at odds with the Old North. Arctic peoples have discarded the old colonial dividing lines and are moving to create new orders. For a generation now, they have been on the front line of sustainable economics, inter-societal conflict resolution, and climate change adaption. The Arctic communities dotting the ice have crossed the old lines, and they are not going back.

Imagine now the Arctic not as the periphery, nor as Superman's Fortress of Solitude—an icy retreat from urban excitements—but rather as the home of an evolving society. Of course, at the center of the circumpolar north is not land but an ocean. Our visitor from outer space might view the Arctic Ocean, as Mediterranean-like, but we cannot yet imagine that.

Most of the time, FOX forgets the Arctic fox is out there. The Arctic fox may want to keep it that way. That too is a survival strategy. Yet with its environment eroding, how can the Arctic fox thrive?

Everybody knows that longitudinal lines converge at the North Pole, but northerners are pointing to a newer "true north" direction. To southerners, Arctic cities and villages may look like tiny black dots on vast expanses of snow and ice. But northerners can foresee the end of that ice, and they are busy breathing life into the dots. The Arctic map's black dots are not yet the nuclei of a new civilization. Connecting the black dots across the snow and ice will take time. It will be an ongoing process of linking up, of creating vital new east-west lines of communication to displace the old "south talks-north listens" rule of Arctic history. This will be the Arctic's remedy for southern "snow blindness."

By combining the "small is beautiful" (Schumacher 2010) virtues of village life, empowered by devolution agreements, modern treaties, and self-government agreements to solve their own problems using dialogue, debate, and the technologies of southern cities, the Arctic region has become an arena for sustainability, inter-societal conflict resolution, climate adaption, fair dealing, and lively democracy. In time, the planet will need the Arctic community to show that there is an alternative to FOXy disruptions.

References

Blair, Gwenda. 2000. *The Trumps: Three Generations That Built an Empire*. Simon & Schuster.

Borrows, John. 2006. *Indigenous Legal Traditions in Canada: Report for the Law Commission of Canada*. Ottawa, ON: Law Commission of Canada, January.

Calloway, Colin. 2018. *The Indian World of George Washington: The First President, the First Americans, and the Birth of the Nation*. Oxford University Press.

Penikett, Tony. 2017. *Hunting the Northern Character*. UBC Press.

Postman, Neil. 1985. *Amusing Ourselves to Death: Public Discourse in the Age of Show Business*. Penguin.

Schumacher, E.F. 2010. *Small Is Beautiful: Economics as If People Mattered*. HarperCollins, Paperback.

Smith, Laurence. 2011. *The World in 2050: Four Forces Shaping Civilization's Northern, Future*, New York: Penguin Group.

Wright, Ronald. 2009. *Stolen Continents: Conquest and Resistance in the Americas*. Penguin.

Index[1]

---

[1] Note: Page numbers followed by 'n' refer to notes.

© The Author(s) 2020   555
K. S. Coates, C. Holroyd (eds.), *The Palgrave Handbook of Arctic Policy*
and Politics, https://doi.org/10.1007/978-3-030-20557-7